W9-COJ-798

THE SOLAR HEATING DESIGN PROCESS

ACTIVE AND PASSIVE SYSTEMS

ABOUT THE AUTHOR

Jan F. Kreider, Ph.D., P.E., is a consulting engineer specializing in the design and economic analysis of solar energy and energy-conservation systems. He has been the solar consultant for the largest solar heating system in the world, and has assisted in the design of many other active and passive solar systems. He received his M.S. and Ph.D. degrees in engineering from the University of Colorado and his B.S. from Case Institute. He is a registered professional engineer and a lecturer. The author of five books and several dozen technical papers on solar technologies, he is well known in the field.

THE SOLAR HEATING DESIGN PROCESS

ACTIVE AND PASSIVE SYSTEMS

Jan F. Kreider, Ph.D., P.E.

Jan F. Kreider and Associates, Inc.
Boulder, Colorado

Jan F. Kreider
December, 1984

McGraw-Hill Book Company

New York St. Louis San Francisco Auckland Bogotá Hamburg Johannesburg London Madrid Mexico Montreal New Delhi Panama Paris São Paulo Singapore Sydney Tokyo Toronto

Library of Congress Cataloging in Publication Data

Kreider, Jan F., date.
The solar heating design process.

Bibliography: p.
Includes index.
1. Solar heating. I. Title.
TH7413.K74 697′.78 81-2338
AACR2

1 2 3 4 5 6 7 8 9 0 KPKP 8 9 8 7 6 5 4 3 2 1

ISBN 0-07-035478-2

The editors for this book were Jeremy Robinson and James T. Halston, the designer was Elliot Epstein, and the production supervisor was Sally Fliess. It was set in Electra by University Graphics.

Printed and bound by The Kingsport Press.

CONTENTS

PREFACE

In the past decade, approximately 140,000 buildings have been constructed in the United States with solar space- or water-heating systems. A small percentage of these systems were paid for in part by the federal government with various residential and commercial demonstration funds. During the course of construction and operation of this modern generation of solar buildings, the design procedures which worked well for both active and passive systems were identified. It is the purpose of this book to describe these procedures in detail for all common space- and water-heating systems.

The method chosen to present the design and construction information is to relate it to the five steps of the building design process. The five steps with which this book is concerned are programming, schematic design, design development, construction documents, and contract administration. The inputs required at each step of the process from the solar viewpoint are delineated in detail. Field experience in the design of solar systems has been predominantly concerned with active systems with much less emphasis on passive systems until the late 1970s. However, in the past few years sufficient design and performance information for passive systems has become available to permit their confident design for residential buildings. Design of other classes of passive systems is only in its infancy. The system type/design phase table inside the front cover directs the reader to text pages treating a specific system class at a specific design phase.

The book begins with an introductory chapter describing air- and liquid-based solar heating systems for buildings. This book is not a textbook and therefore does not describe the theoretical basis of many of the statements and suggestions made in subsequent chapters.

The remaining chapters of the book deal with the steps of the design process. As building designs progress, the amount of required detail increases. This is reflected by the size of later chapters and the amount of information contained in each. Necessarily, each system type must be treated in each chapter. It may, therefore, appear to a person paging through the book that the information is repetitive. However, the difference between one chapter and the next is the level of detail presented and not the subject matter.

The design of solar heating systems uses well-established heating, ventilating, and air-conditioning (HVAC) principles applied to a different energy resource. Many of the mistakes made in the past in the design of solar heating systems have resulted from the improper application of these well-established principles. Most of the lessons learned in the federal demonstration program and by private builders verify that these HVAC principles apply to solar design. Many mistakes could have been avoided if the HVAC backgrounds of the designers had been more extensive. In this book, the design of solar systems is approached primarily from the mechanical engineering and architectural points of view, with less emphasis on electrical and structural engineering. Most of the problems in the design and installation of systems in the past decade have been of a mechanical nature; hence the emphasis on the mechanical and architectural design.

Since the building industry is unfamiliar with solar design and installation, it is emphasized throughout this book that extra care will be required in the near future in the design of solar heating systems. Much detail will be required in specifications and drawings, and extra clarity will be essential in discussions with contractors and equipment fabricators. Further, special care should be exercised in the next decade since one consequence of widespread failure of solar heating systems may be the demise of solar heating.

In order that every detail of the design of active and passive systems may be considered by the designer, each chapter treating a design phase ends with a detailed checklist of items to be considered during that phase. The checklist is based on material covered in the text and can be used routinely after the reader has understood the reason for the inclusion of each item.

English engineering units have been used throughout the book since they are the standard set of units used in the HVAC and solar heating design industry in the United States. Conversion factors to the SI set of units are contained in the appendix and are included at important points in the text.

The author wishes to acknowledge the assistance of several people in the preparation of this book. Darleen McGovern typed and assembled the manuscript. Sherwood Peters and Kennon Stewart reviewed the manuscript and provided many useful comments, adding to the completeness and lucidity of the end product. William Shurcliff carefully read the manuscript and generously made hundreds of useful suggestions that improved the technical and stylistic quality of the book. Phillip Tabb generously shared his expertise in the architectural design process and also gave many illustrations of his work over the past decade. Dottie Lang supported and encouraged the author throughout the lengthy process of writing, editing, and assembling the manuscript.

Jan F. Kreider

1 INTRODUCTION

Eyes, though not ours, shall see
Sky high a signal flame
The sun returned to power above
A world but not the same

C. DAY LEWIS

Active and passive solar space heating are two of a few mature solar technologies for which detailed design guidance can be presented. The purpose of this book is to delineate accepted methods of solar architectural and engineering design and to specify all information required. Specific information is required at each step of the design process as owner, designer, and consultant inputs are synthesized into the final design of a building; therefore, in this book the chapters dealing with each step of the design process present progressively more refined information.

The solar heating systems treated in this book are commercially available systems using commercially available components. The book, as a result, is not an exhaustive survey of the many methods of solar heating which have been devised by inventors over the course of many years. Rather, we deal with systems composing the mainstream of solar heating design in the United States and systems using standard configurations. For example, we will deal with liquid- and air-based collectors using sensible heat storage and pumps or fans to circulate fluids. Such systems have been studied intensively over the past 8 to 10 years. Several tens of thousands of these systems have been built. Field experience has become available on what designs work and what designs do not work. *The designs presented in this book are designs which are reliable and which work.* The specific systems introduced in the next chapter are used throughout the book as the basis of the design process.

Most users of this book are expected to be members of the architectural and engineering design community involved with building design.

The topics dealt with are only one component of the overall design process, however. The details of the broader subject of the full design process are not dealt with in the book. It is assumed that the user is familiar with the phased architectural design process and with the design of all standard components used in conventional heating, ventilating, and air-conditioning (HVAC) systems in all types of buildings. The nonsolar items with which the mechanical engineer should be familiar are treated in the quadrennial series of handbooks prepared by the American Society of Heating, Refrigerating and Air Conditioning Engineers (ASHRAE). In addition, daylighting is described in the *Illumination Engineering Society Handbook* (IES), structural engineering is described in the *Structural Engineering Handbook* (McGraw-Hill), and architecture is described in many design guides available from most large publishing houses. There are also a number of solar energy textbooks which present the theoretical basis of solar energy systems. See, for example, Ref. 1. These textbooks for the most part give a survey of active systems for space and water heating with less detail devoted to the design of passive systems. The use of theory will be minimized, in this book, and only the results of textbook analyses which are needed for specific design steps will be used.

A solar system delivering approximately three-quarters of the annual heating demand in the northern part of the United States will cost about 10 percent of a building's construction budget. For example, one prominent solar consultant in the West has designed approximately $4 milion worth of solar systems on approximately $42 million worth of construction since 1974. This construction mix varies from residential buildings up to the largest solar heating systems in the world. Ten percent of a construction budget is a significant amount of money. Hence, space-heating systems must be designed carefully.

Solar energy systems utilize a diffuse energy source. Therefore, systems using solar heat require more care in their design and specification than do nonsolar heating systems which use relatively intense or relatively high-density energy sources. For example, an electric baseboard heater for a residence is rarely designed as carefully as a solar heating system because the amount of money involved in a 20 percent oversizing is of little consequence when viewed over the lifetime of the system. However, under- or oversizing a solar system by 20 percent can result in significant penalties in a life-cycle economic context. Erroneous sizing may make the use of solar heat infeasible.

The design of solar energy systems is not as simple as some authors would have you believe. It involves much more than finding the cost-optimal solar faction. Most of the subtleties of a solar system design result from the diffuse nature of the solar resource and from the lack of high-temperature energy available to correct for designer inaccuracies or errors. *It has been found that the rule-of-thumb approach to the design of solar energy systems is entirely inadequate.* Rules of thumb are useful, at best, only in the programming phase and in early schematic design. However, the design of solar heating systems, although more complex than it might first appear, does not require the introduction of much new material beyond that used in nonsolar HVAC design. However, careless, nonsolar HVAC calculations (upon which solar sizing is based) may penalize solar life-cycle benefits.

The components used in solar systems are for the most part the same as those used in other kinds of space-heating systems. For example, storage tanks, pumps, control valves, blowers, ducts, pipes, and so forth are all used in both solar and nonsolar systems. Accordingly, the design procedures presented in this book apply these familiar components in a new way.

The only two components of solar systems which are new and unfamiliar to the architecture/engineering (A/E) professionals are solar collectors and solar control systems. Solar collectors are devices for efficiently converting solar radiation into heat and are dealt with in detail in Chap. 2. Solar control systems are the key to the efficient performance of any solar-based thermal system. The lack of understanding of and attention to solar controls has caused some otherwise well-designed systems to operate inefficiently or not at all. The proper design of control systems is emphasized throughout this book.

We also concentrate on the design of solar systems, not of components alone. It is the integration of well-designed components into a well-designed system which assures the best performance. The best collector connected to an improperly sized pump or an undersized heat exchanger cannot deliver the amount of energy for which it was designed. The approach taken in this book, then, is to describe the components which compose the solar thermal system and then to give a system description.

Since the oil embargo of 1973, perhaps as many as 140,000 solar systems have been installed on buildings in the United States (exclusive of solar heating systems for swimming pools). In the 7 or 8 yr since the installation of modern solar systems began, many operational results have been published. However, much of this information has been disseminated to only a small group of solar designers and federal contractors. One of the purposes of this book is to interpret the data gathered over the last decade in a design context and to make the results available to designers. Much experience in the 1970s was gained at the expense of errors in solar designs. Many of these errors could have been avoided if the designer had been familiar with good HVAC design practices and had made use of them. In this book, we will emphasize the application of long-standing HVAC design methods to the solar energy resource.

THE SOLAR HEATING DESIGN PROCESS

The solar *design process* can be summarized as a *matching process:*

Matching heat source to the load (magnitude, rate, location, phase)
Matching costs to benefits (dollar saving, comfort)
Matching complexity to context (residential—smaller and simpler; commercial—larger and more involved)

The process, as defined in this book, consists of the five principal phases of the standard A/E process used in the preparation of design documents for the construction of any building.

Over the past 20 yr the architectural profession has developed a building design procedure consisting of several phases. During each phase, a design at a specific level of detail is developed and an estimate of project costs based on that design is prepared. The owner of the building reviews the output of each design phase, accepts the results, and authorizes the next phase with a formal, written approval of the current phase. The design of building solar systems *must be well integrated* into the established design process.

The engineering design team includes several disciplines—architecture and electrical, civil, mechanical, and structural engineering. For example, active solar systems are designed by HVAC engineers; structures for collector supports and for passive wall designs are the responsibility of structural engineers; daylighting is within the purview of electrical engineers (or architectural engineers); and so on. Each group must be aware of its new responsibilities associated with solar energy use. Fees charged will reflect the new design responsibilities.

At each step in the solar design process, the level of detail will increase as more and more characteristics of the building become known. For example, early in the design process it will not be clear whether an active system or a passive system is most appropriate. However, as the design proceeds it may become obvious that a passive system is unsuitable because of the lack of vertical south-facing wall, for example. Then the decision would be to use an active system with collectors mounted on the roof. At a subsequent phase in the design process, the decision whether that system should be air-based or liquid-based would be made. During the next step in the process, a specific type of system, having been selected on the basis of many criteria, would then be optimized. Finally the optimized system becomes the basis of the preparation of construction documents. *Each step in the solar design process is an iteration on and refinement of the preceding step* and must be carried out in phase with the work of all design disciplines involved.

Programming

Programming is the first phase of the solar heating design process. At this level, the type of building to be built is quantified with respect to its volume, usable floor space, operating schedule, interior temperatures, ventilation rates and many other criteria. Building orientation is roughly determined and a site survey is used to assess the solar energy resource. At this level, it is not possible to carry out a solar design in any significant detail, but well-established rules can be used to estimate active or passive collector size. Experience regarding costs can then be used to prepare a rough cost estimate.

Schematic Design

Schematic design (SD) is the next step of the design process. In schematic design, the sizes of major components of active and passive systems are identified. For example, the thickness of a passive storage wall, the size of the thermal circulation vents, and the amount of the insulation for control of nighttime heat loss are calculated. For an active collector required to supply a specific portion of the building heating load (usually 60 to 80 percent), the associated heat exchangers, pumps, and storage and terminal devices can all be roughly sized using well-established rules or relatively elementary calculations. A detailed heat load calculation, used in later phases as the basis for the tentative solar design, is normally not done during SD. However, a load *estimate* is essential.

During schematic design the first attempt at optimizing the solar system is made; that is, the most cost-effective mix of solar heat with conventional heat is established. This determination is based on an estimate of performance of the solar heating system and an estimate of the cost of backup energy and its expected rate of inflation. The SD economic analysis is not detailed but includes only the five or six principal parameters.

Design Development

Design development (DD) is the third step of the design process. At the end of SD, one type of solar heating system will have been selected. During this phase all technical questions regarding all components of the solar system are resolved. *Design development is the heart of the solar heating design process.* Studies of the size of storage relative to collector area or flow rate relative to annual energy delivery, for example, are carried out during DD. Detailed economic analyses during this phase are based on quotes collected from the major solar component suppliers. A key difference between solar and nonsolar DD is that solar system prices will need to be known for a *range of system sizes*. These costs are used in a life-cycle cost analysis to find the optimum system size. No such optimization is made for conventional heating systems since the system is purchased to carry 100 percent of the heating load throughout the year. As explained later, 100 percent solar systems are rarely economical.

Construction Documents

Making *construction documents* (CD) is the fourth phase of the solar design process. Written specifications and working drawings are outputs of this phase. Solar installation drawings, piping diagrams, control procedures, and equipment specifications will be prepared and integrated into the project construction documents. The solar part of this package consists of solar system drawings and details. Specifications for solar collectors, storage tanks, heat exchangers, and so forth will be made a part of the CDs. In this text, we use the Construction Specification Institute (CSI) format for the preparation of specifications.

Construction documents are the basis for the bidding phase. The bidding phase is not treated separately in this book but is considered to be part of CD work. Bid reviewing guidelines, types of information required, and the importance of a prebid meeting for contractors (who may underbid because of unfamiliarity with solar work) are contained in Chap. 5.

Contract Administration

Contract administration (CA) is the final phase of the process dealt with in this book. During CA, the successful bidder is selected and the building is constructed. This is not a design phase specifically but is as important as the preceding phases. During contract administration, material lists and shop drawings are reviewed and the solar system is installed on the building. Since many contractors installing solar systems are unfamiliar with their function, extra guidance will be required. Therefore, installation suggestions are given in Chap. 5 on contract administration.

As CA progresses and the solar system installation is completed, start-up takes place and a formal acceptance test is run. Suggested formats for these tests are given, along with the contents of an owner's or user's manual. A user's manual is standard for any type of heating system in a large building, and the solar features should be well integrated. Even in the case of residences, the owner should have an operator's manual containing some basic troubleshooting information, flow diagrams, equipment specifications, and service contacts.

Each chapter dealing with a phase of the design process contains a lengthy checklist. The first time this book is used, the designer must study the details and the bases for recommendations. However, once the bases are understood, it will not be necessary to use the book page by page in later designs. Instead the checklists based on chapter contents can be used. In many cases, these checklists will be more lengthy than required for a specific job, since they must be comprehensive and deal with all possible solar system configurations.

The contents of this book are not theoretical but are based on well-established physical principles or design rules used in the HVAC industry. The theoretical basis for design is not reiterated in detail. A few sections of theory have been included in the book, however, and these sections will be preceded by a circle O. The user need not be concerned routinely with these theoretical details; but where a nonstandard design or a nonstandard size of a component is required, the supplements may be helpful.

Retrofits

When the design of a retrofit solar system for an existing building is undertaken, the early design steps can be omitted. The existing building will dictate the appropriate type of solar system. The availability of space for collectors will dictate collector type and orientation. The design of retrofit systems is not treated in this book as a separate topic. The retrofit designer can use the book by referring to specific sections at the design development level to size components of the selected system. The five-step design process deals primarily with new buildings.

Passive vs. Active

The relative amount of solar design effort expended by phase is quite different for active and passive systems. Since passive features are much more a part of the building itself than are active ones, passive design is "front-loaded." That is, more effort is expended in programming and SD on passive systems than on active ones. Only during the formative stages can passive features (and energy-conservation features) be fully incorporated. The decision to add a massive storage wall and skylights to a building during DD would destroy the entire DD-level design. If inserted at the proper earlier stage, these two features could be smoothly incorporated into the building.

SOLAR ECONOMICS

Solar heat is expensive heat, and further, the investment in a solar system is made at the beginning of the life of the building. Therefore, system cost becomes part of the initial cost of the building. The economics of solar and nonsolar sources of heat are fundamentally different. For example, neither solar heat nor petroleum has an inherent cost when in its natural state. That is, sunshine shining on the roof of a building is no more and no less costly than oil in a geological formation deep in the earth. It is the extraction and conversion of energy sources to useful forms which accounts for their cost. With natural gas heating, the cost is associated with resource retrieval, distribution, and providing a profit to the energy company.

In solar applications, the owners of the solar system are their own energy company, so to speak. Therefore, the cost of extraction, conversion, and collection of solar heat is borne by the owners of the system. The cost of solar energy then can be viewed as the prorated or amortized cost of the energy collection and conversion system over the expected lifetime of the system. The capital investment in a nonsolar heating system is smaller than the investment in a solar system. However, the cumulative cost of gas or oil throughout the lifetime of the building is much larger for a nonsolar building than for a solar-heated building. In the two systems the distribution of costs occurs at different times.

The concept of life-cycle costing is used in this book to provide a meaningful comparison between the high initial cost of a solar system and the high year-by-year fuel cost for the nonsolar system. Using the standard methods of engineering economics, we can compare these fundamentally different types of costs. The comparison determines whether the solar system is economically attractive relative to nonsolar systems.

Energy Costs

Energy costs in this book are expressed in terms of dollars per million British thermal units (Btu). In many cases the cost of energy in this particular set of units is not known initially and conversions will need to be made. For example, fuel oil in the winter of 1980–1981 sold for about $1 per gallon. A gallon of fuel oil provides 140,000 Btu. If this fuel oil is burned in a furnace with 70 percent efficiency, it delivers approximately 100,000 Btu of useful heat. Thus, fuel oil heat costs $10 per million British thermal units. This type of calculation is used for all energy sources. Figure 1.1 shows how to find the net cost of fuel if the gross cost per unit is known.

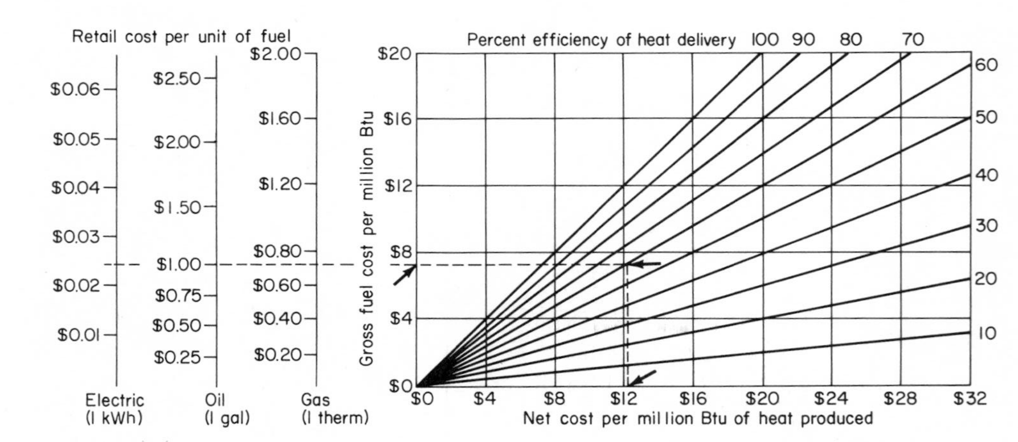

FIG. 1.1 Net costs of fuel accounting for furnace or boiler efficiency. The dashed line shows that fuel oil costing $1 per gallon provides heat at a cost of about $12 per million Btu if burned at 60 percent efficiency.

Maximizing Solar Savings

If the cost of solar heat is less than the cost of a competing energy source and energy-conservation methods, there is an economic incentive to use solar heat. The purpose of economic optimization is to maximize the difference between the life-cycle cost of solar heat and the life-cycle cost of fossil fuels. The design goal is to maximize savings to the owner of a solar system subject to any constraints present.

Savings from investments in solar systems are treated somewhat differently, from a tax viewpoint, than are savings from other sources. For example, if an investment in a solar water heater saves several hundred dollars per year in fuel costs, the return on this investment is not taxed as long-term capital gains; thus, if capital gains are taxed at 40 percent, the value of the solar saving is the same as an equivalent return on an investment in stocks or real estate of 1.7 times the size. The economics of corporate, private, and public projects differ and are described later.

Wide-scale use of solar energy to replace other energy forms will occur only if there is significant economic incentive. Careful calculations of the costs and benefits of owning a solar system are therefore required. The economic analysis methods given in the book will show how to predict the size of the most cost-effective solar system.

The use of an arbitrarily sized solar system based on an arbitrary solar percentage of annual heat load or an arbitrary relationship of collector area to heated floor area, for example, will not give the system owner the best return on investment. Although appealing because of their simplicity, arbitrary methods of selecting solar system size should be avoided.

PROSPECTUS

The remaining chapters of this book describe the several steps of the design process. Before treating the phased-design approach, Chap. 2 describes the major components and functions of active and passive solar systems. But specific sizing criteria and costs are not presented in this chapter.

Chapter 3 deals with programming and schematic design. These are the first two steps of the design process during which rough estimates of solar system size and cost are made and selection criteria are developed. In addition, a simple method of determining economic feasibility is described. If solar heating is economically infeasible, this must be established very early in the design process, because the use of solar energy has significant impacts on building mass, systems, orientation, and configuration.

Chapter 4 deals with design development. In design development, the details of every component and every solar subsystem used on the building are delineated. As a result, this chapter describes all factors which affect each component of active or passive systems. This chapter contains the most detail and the most quantitative support information.

Chapter 5 deals with construction documents and contract administration. The preparation of construction documents is not dealt with in great detail, since preparation of working drawings and specifications is a standard function of architects and engineers. However, the chapter describes the solar aspects of construction documents. Examples of specifications and working drawings from typical projects are presented.

The key task in the CD phase is establishing details of the interface between the solar system and other building systems—structural, electrical, and mechanical. (Of course, the awareness of interface needs must exist as early as schematic design.) In the CA portion of Chap. 5, many recommendations are given for system installation, system start-up, control debugging, and acceptance test procedures.

The appendix contains all reference material required to execute solar heating design. Included are tables of R factors and heat-loss factors for various types of construction, tables of solar radiation and degree-day quantities for more than 200 locations in the United States, and tables of economic functions. In addition, hourly solar radiation tables prepared by ASHRAE are included.

One list of references is used for the entire book. References are denoted in the text by parenthetical numbers.

2 SOLAR HEATING SYSTEMS

There is no energy crisis, but only a crisis of ignorance, selfishness, and fear.

R. BUCKMINSTER FULLER

Here we review the major components and the operation of active and passive systems for space and water heating. The first part of the chapter deals with air- and liquid-based solar collectors, thermal storage, heat-transport subsystems, pumps, fans, heat exchangers, controls, and other subsidiary components. The second part of the chapter deals with active systems—domestic hot-water (DHW) heating, liquid-based space heating, and air-based space heating. Finally, passive and hybrid solar heating systems of three types are described. Energy conservation, prerequisite to the cost-effective use of solar energy, is described briefly. Readers familiar with the function of solar systems can skip this chapter and proceed to Chap. 3.

THE GENERAL SOLAR HEATING SYSTEM

Figure 2.1 shows the components of any active or passive solar heating system. Five specific components are included—solar collector, thermal storage, nonsolar auxiliary system, distribution system, and control system. The manner in which these five components are interconnected and the specific configuration of each determine the nature of the solar heat production process. For example, an active liquid-based system uses a solar collector from which heat is removed by a liquid. That heated liquid is transferred to thermal storage. Some of that heat is distributed to satisfy building energy demands.

The controller determines system operations including solar collection and distribution. Also, the nonsolar auxiliary is activated if solar storage is depleted. The control system, although physically small, is a very important component.

The same five components present in active systems are present in passive systems. For example, in a passive system of the thermal-storage-wall type the solar

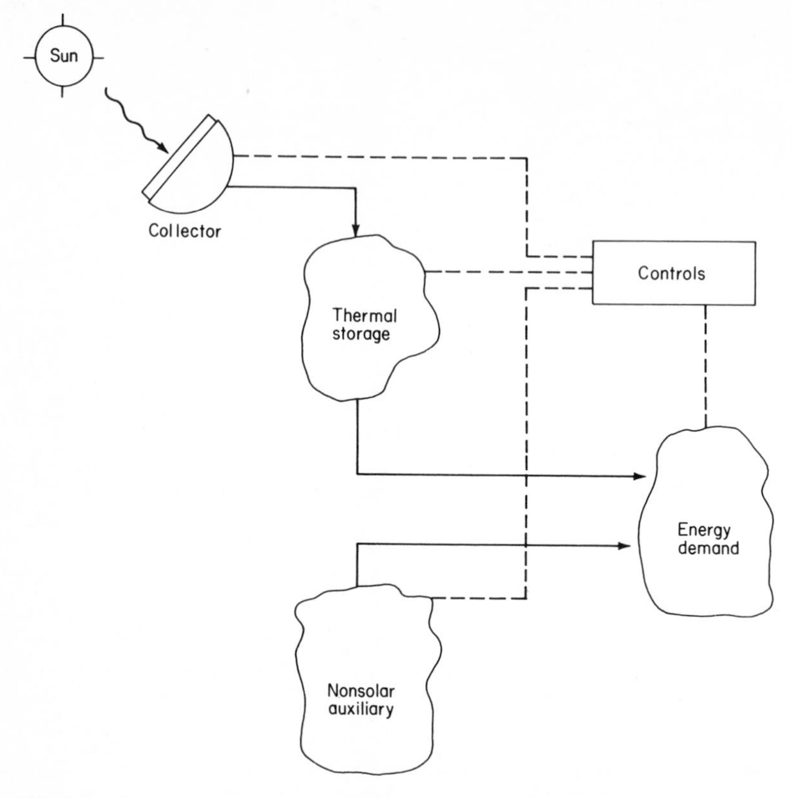

FIG. 2.1 General solar thermal system diagram showing the five components present in any solar system.

collector consists of the glazing and wall surface. Storage is integral with the absorber surface and can be either a concrete or water wall immediately behind and in contact with the dark-colored collecting surface. The demand is the heating demand of the building. The nonsolar auxiliary is used to satisfy the demand if passive storage has been depleted by a long period of cloudy or very cold weather. The control elements of a passive system do not use electronic signals and actuators but are integrated into physical components of the passive system. Backdraft dampers are one of the control features of a storage wall system. They prohibit a reverse thermocirculation through the wall air slots used during the day as a heat supply.

Solar systems of all types respond to the environment and to certain characteristics of system components. For example, the location of a solar system, the imposed heat load, ambient solar intensity levels, and ambient temperature all determine the response of the system to its load. Identical solar systems will function entirely differently in different locations. Major characteristics of solar systems are

Collector size
Collector tilt angle
Collector orientation

Collector flow rate
Storage size
Heat-exchanger size
Required delivery temperature
Interconnection and control of these various components.

This chapter first describes major components and then describes their functioning in standard systems.

COMPONENTS OF SOLAR HEATING SYSTEMS

In this section, we will describe common solar collectors of various types—both flat-plate and tubular, thermal-storage, heat-transport, and control systems.

Solar Collectors

The majority of systems treated in this book use the flat-plate collector—either air or liquid type. These collectors provide an adequate outlet temperature to drive most space and domestic hot-water heating systems and provide an excellent thermodynamic match between the load and the collection process. Tubular collectors also will be described briefly, but the major emphasis will be on flat-plate devices. In the future, when mass-produced, focusing collectors may deliver more energy per unit cost than flat-plate types. Already some concentrators are competitive with common flat-plate collectors.

The flat-plate collector was first developed in the eighteenth century as a research apparatus used by physicists studying the conversion of solar radiation to heat. Solar collectors of the modern era were first fabricated just before World War I. Between the wars they were widely used for water heating in California and Florida.

Many types of flat-plate collectors have been built and Fig. 2.2 shows cross sections of those commercially available today.

Covers. The uppermost component of a flat-plate collector is the transparent cover of glass or plastic. One or two sheets are used in commercial units. The purpose of the cover(s) is to trap a layer of air ½ to 1 in thick, providing thermal resistance to heat loss from the absorber plate. Glazing must have a high transmission coefficient for solar radiation. From a practical point of view, the material must also be durable in its severe environment; it must have dimensional stability; and it must be capable of withstanding environmental stresses such as hail, rain, and snow.

The decision whether to use a plastic or a glass cover depends on cost and durability. A subsidiary consideration is weight. Although plastic collector covers show considerable promise and low cost, they have been subject in many installations to damage by wind and thermal deformation. Solar collector temperature may range from −20°F to well above 300°F in the course of a year. Expansion places stress on plastic film covers and can stretch the material. As the collector cools, the

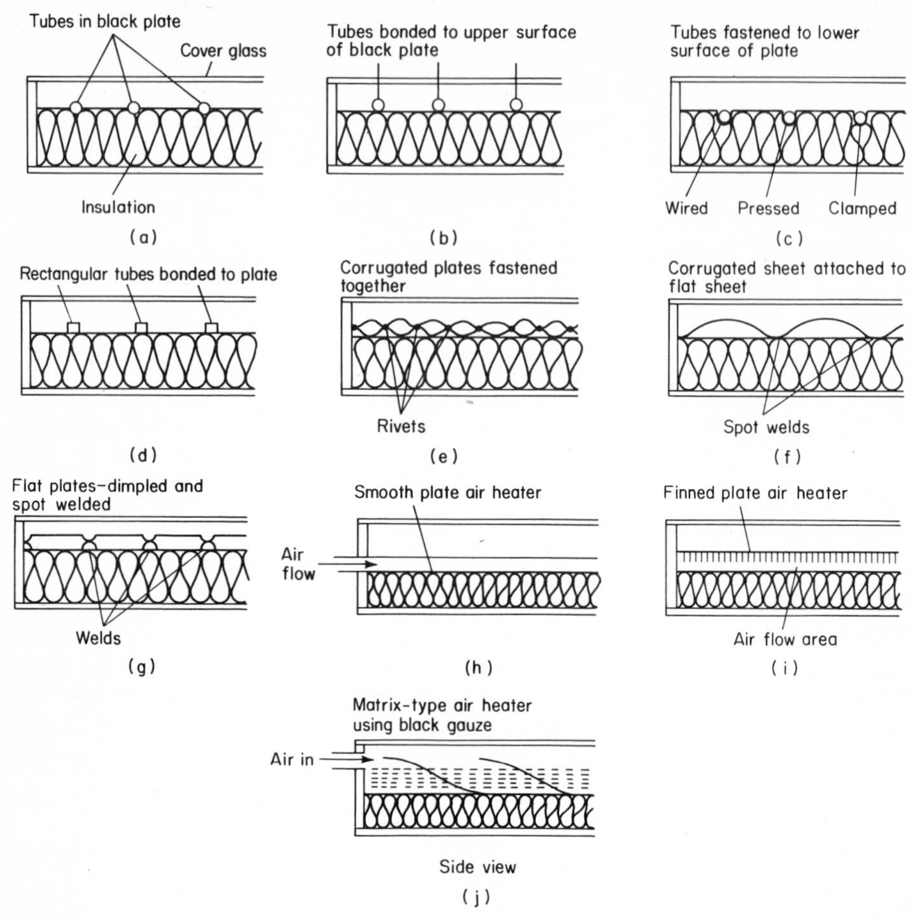

FIG. 2.2 Cross sections of various liquid- and air-based flat-plate collectors in common use: (*a*)–(*g*) are liquid types; (*h*)–(*j*) use air.

outer cover may sag and flutter when the wind blows. This fluttering will lead to fatigue, and certain plastic materials will ultimately fracture. Even if fracture does not occur, the material may sag so close to the absorber plate that it will melt.

Glass has many advantages. It does not discolor or stretch, and it has good resistance, if tempered, to large temperature changes. To reduce thermal stress effects, the edges of the glass should be polished before tempering. Glass with very low quantities of ferric oxide has high solar transmittance equivalent to that for thin plastic films. Glass is highly resistant to snow, wind, and hail and has consistently good performance history under exposure to sunlight. Certain plastics tend to deteriorate after several years with reduced transmittance, whereas glass more than 100 yr old has the same transmittance as it did when new. (However, one solar installation used photochromic glass which darkened with increasing sunlight—an unusual and hard-to-discover error.)

The glass cover attachment to the collector housing is one of the most important features of the collector. The glass seal must be able to withstand very high temperature, very low temperature, the differential expansion and contraction of both the cover and the housing. Few materials satisfy these criteria. For example, typical rubber caulking compounds, although flexible and capable of withstanding low temperature, fail at very high temperature. Highest temperatures are encountered in flat-plate collectors during periods when fluid flow ceases. Under these conditions temperatures above 400°F have been measured. A promising seal material is called EPDM (ethylene-propylene diene monomer). Certain silicone materials also work well.

Seals must permit easy removal of the covers. In case a cover is damaged, it is essential that the glazing be replacable without removal of the collector from the building and return to the factory. Therefore, seals and their retainers must be designed accordingly.

Absorbers. Absorber plates in nearly all high-quality flat-plate collectors are metal. Three metals are commonly used—aluminum, copper, and steel. Aluminum is light and has high thermal conductivity. However, aluminum is very difficult to repair in the field; if a leak develops, the collector must be removed and delivered to the factory or service shop for repair.

Copper has very high thermal conductivity and corrosion resistance. It is easy to repair in the field. Copper is the choice of many manufacturers. (Repairs to selective surfaces on any metal absorber plate are presently not possible.)

The third solar collector material is carbon steel. Carbon steel has a thermal conductivity about one-tenth that of copper, but is much less expensive than copper. Corrosion control in steel is well developed; corrosion inhibitors in steel conduits carrying water are well known. Since the thermal conductivity of steel is relatively low, the amount of steel required in a tube-in-sheet absorber plate is higher than in either aluminum or copper absorbers. This increases the weight and the time constant of a solar collector but probably has little cost impact because of the lower price of steel relative to copper and aluminum.

Collectors used in air systems are of steel. The most common air collector is shown in Fig. 2.2*h*. It consists of a shallow steel duct, painted black on one side. Rectangular ducts are easy to make from steel. The lower conductivity of steel is not as important in such an air collector as in a liquid collector. The governing resistance to heat flow in air collectors is in the air boundary layer inside the duct; hence, the convection-heat-transfer coefficient is relatively low. Therefore, the thermal properties of the absorber metal do not matter much. Since steel is cheapest, it is universally used in air collectors.

Absorber Surfaces. As shown in Fig. 2.2, there are many functional configurations of flat-plate collectors. Each type of collector shown in this figure is made by at least one company in the United States on a commercial basis. The cost differences among collectors of a specific type seem to be determined in part by the surface treatment of the absorber. The surface of an effective solar absorber must convert at least 90 percent of the incident solar flux to heat. The surface must be durable over a very wide (−50 to 400°F) range of temperatures in an environment which may be humid as a result of trapped moisture in the collector housing. The most common absorber surface used in collectors is high-temperature flat black paint. It must not outgas at temperatures up to 400°F. Outgassing is a process by which solvents and other components of paints and materials in solar collectors vaporize or pyrolyze at high temperature. The gases condense on the inner glazing surface. Many of these condensates are not entirely transparent. Continued accumulation can reduce the cover transmittance. Some type of curing process is normally used for paints. Typically, the entire absorber plate is baked at a temperature of 400°F for at least 30 minutes to drive off all solvents and other materials which could later vaporize.

A second type of surface used on flat-plate collectors is chemically deposited. For example, a steel collector might first be covered with a layer of copper. The copper is used as a base for chromium plating. The chromium, in turn, is converted to black chrome by a specific sequence of chemical baths with current passed between the bath and the absorber plate. The final step in the process is washing and drying of the absorber plates.

The two most common types of plated surfaces are black chrome and black nickel. Black chrome has excellent durability up to 450°F, well above any temperature encountered in a solar space-heating system. Black nickel also has acceptable high-temperature stability but is easily deteriorated by humidity unless very carefully treated with a surface "passivation" layer. If the passivation layer eventually deteriorates in the presence of humidity, white zinc oxide is formed, greatly reducing the surface absorptance.

The plated surfaces described above are used to increase the efficiency of a flat-plate collector by reducing thermal radiation losses. Plated surfaces have relatively low infrared-radiation emittance. As a result, selectively plated surfaces lose less heat. The low emission factor for plated metals is most important at high temperatures. In most space-heating systems operating at 140°F or below, the extra cost of selective surfaces must be justified by extra heat collection.

The construction of absorber plates must accommodate the highest expected temperature to which the surface may be exposed. For example, if copper tubing is soldered to the collector headers, high-temperature silver or 95/5 solder is required. If the tubes are not soldered or welded to the absorber plate but merely clamped, problems may develop over the lifetime of the collector because of differential thermal expansion between the plate and the tubing. For example, if an absorber plate is aluminum but the waterways are copper, differential expansion may eventually deteriorate the thermal contact between the waterway and the absorber plate. Any collector using a tube which is not very tightly clamped to, welded to, or an integral part of the absorber should be avoided.

Insulation. Losses through the insulation depend on its conductivity and thickness. An R value of 5 to 10 is normally specified as the minimum for "back" insulation on flat-plate collectors. If insulation is increased much above R-10, the collector is too thick. Too much insulation is not cost-effective since the majority of the heat loss occurs via the "front" (glazed side) of the collector.

Collector insulation should be designed to operate at 400°F continuously. That is, it must not deteriorate, outgas, expand, or contract excessively at temperatures between −30 and 400°F. The insulation must also have structural integrity. That is, it must not compact and settle when the collector tilts. Therefore, loose fiber-fill insulation is not satisfactory since it will settle to the bottom of a sloped collector.

Insulation should not attract moisture. During collector temperature cycles, it is possible for some moisture to accumulate within the insulation. It is important that that moisture not be permanently absorbed by the insulation nor decrease the R value of the insulation. Several types of insulation are commonly used. Mineral and ceramic fiber insulations are designed for higher temperatures than required in flat-plate collectors and are more expensive than necessary. Polyurethane foam and polystyrene foam have very poor performance at high temperature. Some early solar collectors using polyurethane foam were destroyed when the foam expanded under high-temperature conditions and deformed the glazings and absorber plates.

Fiber glass is the most common insulation used in solar collectors. However, ordinary construction-type fiber glass (yellow) *must not be used* since it contains binders and other materials which can outgas and reduce the transmittance of the inner covers. Fiber glass without binders (white) should be used.

Housings. The collector housing is used to protect the insulation and the absorber plate from the environment. It also provides collector mounting. The amount of labor used to install the solar collector is very much determined by the type of mounting and piping and how well both interface with the building.

Various metals (aluminum, galvanized steel), fiber glass, or high-temperature thermal plastics are commonly used. Wood has also been used in a few housings. When metal is used, corrosion can be a problem. For example, if an aluminum collector housing is attached to a steel support structure, the difference in the electrochemical potential between the two will cause the aluminum to corrode and leak within a few years. If the housing is steel, it must be protected by galvanizing and coatings of an etching primer and paint.

Fiber glass may deteriorate at high temperature but is light and easy to attach to any structure. Wood is not a recommended housing material. Wood will outgas under high-temperature conditions, depositing tars and other components on the inner surface of the glazing. Wood warps, loses strength at elevated temperature, and has poor dimensional stability. Wood also presents a fire hazard and its use is restricted by building codes (48). The nominal ignition temperature of wood is 392°F, but it can self-ignite below 250°F after long exposure to heat, according to the International Association of Arson Investigators.*

*In May 1980 a solar collector made from wood self-ignited after a long period of stagnation. Foam insulation behind the absorber cracked and exposed the wood housing to high temperatures, producing pyrophoric carbon, which self-ignites. See also Ref. 68 for UL wood collector specification.

Seals. At least two pipes or ducts pass through openings in the housing. It is essential that these openings be sealed to prevent dust from entering the collector. Any dust in the collector will partially deposit on the absorber surface and reduce its absorptance. The seals must also block rain or snow, both of which can deteriorate the insulation, the absorber plate, and even the collector housing.

The material used for sealing must be resistant to ultraviolet light, ozone, and high temperatures. Seals around pipes and ducts must be flexible to accommodate expansion and contraction of the absorber plate and the repeated motion of pipe or duct connections. EPDM or silicone seals are recommended.

The solar system designer normally is not involved in specifying the material of solar collector housings and seals, but the material used by proposed manufacturers should be reviewed.

Efficiency of Flat-Plate Collectors. One of the most common indexes of performance of a solar collector is its efficiency. The efficiency is defined as the amount of useful energy delivered to the working fluid divided by the solar radiation striking the collector. The efficiency can be measured by a formal test or calculated. It is preferable to use test data, since the calculation of efficiency requires the use of mathematical modeling procedures with certain assumptions which may not be valid for some commercial collectors. Therefore, in this book the use of test data is emphasized in preference to data which have merely been calculated.

Efficiency Equation. The efficiency of a collector can be predicted from a steady-state energy balance on the absorber plate. The useful energy carried away by air or liquid flowing through the collector is the difference between the energy absorbed on the absorber plate and the heat lost from it through the front covers, the edges, and the "back" (unglazed) surface of the collector. If the heat loss is proportional to the temperature difference between the absorber plate and the ambient air, the instantaneous, steady-state energy balance in equation form is:

$$q_u = (\tau\alpha)I_c - U_c(\overline{T}_c - T_a) \tag{2.1}$$

Equation (2.1) states that the useful energy q_u [Btu/(h · ft^2) or W/m^2] delivered by a collector is the percent of radiation transmitted through the covers, denoted by τ, multiplied by the absorptance of the absorber plate, denoted by α, times the incident solar radiation at the collector, I_c, minus the heat-loss term. Since we have assumed that the heat loss is proportional to the temperature difference, it is given by the product of the U value and the collector plate–ambient-temperature difference ($\overline{T}_c - T_a$). The U value of the collector is called the loss coefficient or the thermal-loss coefficient and is analogous to the U value of the wall of a building. It includes the various thermal resistances between the absorber plate and the environment—conduction through the back insulation, and convection and radiation through the glazing. The efficiency of a solar collector η_c can then be expressed as the ratio of q_u to I_c as shown in Eq. (2.2).

$$\eta_c = q_u/I_c \tag{2.2}$$

If Eq. (2.1) is divided by I_c, the standard efficiency equation for a flat-plate collector follows:

$$\eta_c = \tau\alpha - U_c\left(\frac{\overline{T}_c - T_a}{I_c}\right) \tag{2.3}$$

The equation consists of two terms. The first term $\tau\alpha$ represents the optical efficiency of the flat-plate collector—the amount of radiation absorbed relative to that striking the surface of the collector. The second term of the efficiency equation represents the heat losses from the collector.

Two types of quantities are present in the efficiency equation. The first—$\tau\alpha$ and U_c—are properties of any given collector. They are determined by the design and materials of the collector and are fixed for a particular collector manufacturer. The second set of quantities—the collector temperature $\overline{T}_c$, the ambient temperature T_a, and the ambient solar radiation I_c—are characteristic of the conditions under which the collector is used.

Efficiency Factors. Equation (2.3), although embodying all the parameters required to calculate instantaneous energy delivery of a collector, is not a practically useful equation. The difficulty is in finding the average collector plate temperature $\overline{T}_c$. The collector plate temperature is impossible to measure, since it varies both in the direction of flow as fluid is heated and between the fluid conduits as heat is conducted from the absorber plate to the tubes. Therefore, $\overline{T}_c$ is not a good temperature reference for collector efficiency.

Two alternative expressions for efficiency have been developed in Ref. 1. If the reference temperature for collector efficiency is the average fluid temperature $\overline{T}_f$ instead of the average collector temperature, Eq. (2.4) results.

$$\eta_c = F'\left[\tau\alpha - U_c\left(\frac{\overline{T}_f - T_a}{I_c}\right)\right] \tag{2.4}$$

This equation is identical in appearance to Eq. (2.3) except that the average fluid temperature $\overline{T}_f$ has replaced the average collector plate temperature $\overline{T}_c$. Since the fluid temperature is always less than the collector temperature (as required for heat to flow from the collector plate to the fluid), the heat-loss term based on the fluid temperature is smaller than it should be. Therefore, the right-hand side of Eq. (2.4) is multiplied by F', the plate efficiency factor. F' is not an empirical correction factor simply inserted in the equation to make the efficiency value come out right but is a measure of the heat-transfer properties of the absorber. F' depends upon the thermal conductivity of the absorber plate, the distance between tubes, the heat-transfer coefficient inside the tubes, and the relative size of the tube and the absorber plate. In addition, the heat-transfer coefficient within the tube depends on the fluid and its flow rate.

Another way of viewing F' is as the ratio of the actual heat delivered to that which could be delivered if the entire absorber plate were operating at the average fluid temperature. Table 2.1 shows the value of F' for various tube spacings, sheet

TABLE 2.1 Plate efficiency factors F' for various tube spacings, sheet materials, and sheet thicknesses*

	Tube spacing, in					
Plate thickness, in	*3*	*4*	*5*	*5.6*	*6*	*7*
			Copper sheet			
0.010	0.945	0.920	0.890	0.870	0.855	0.805
0.014	0.950	0.925	0.900	0.880	0.870	0.825
0.018	0.955	0.930	0.910	0.890	0.880	0.850
0.022	0.960	0.935	0.915	0.900	0.890	0.865
0.028	0.960	0.935	0.920	0.910	0.900	0.875
			Aluminum sheet			
0.020	0.945	0.920	0.885	0.870	0.860	0.825
0.030	0.955	0.930	0.905	0.890	0.800	0.850
0.040	0.955	0.935	0.915	0.900	0.890	0.870
			Steel sheet			
0.020	0.890	0.825	0.750	0.715	0.680	0.625
0.030	0.920	0.870	0.805	0.770	0.740	0.685
0.040	0.930	0.885	0.835	0.810	0.775	0.725
0.064	0.950	0.890	0.840	0.810	0.795	0.745

*From Ref. 2 with permission.

thicknesses, and sheet materials for liquid collectors. The data shown are for copper, aluminum, and steel plates with tube spacings between 3 and 7 in. The value of F' is seen to decrease as tube spacing increases since heat must travel through a greater length of absorber plate before it reaches the tubing. F' increases with metal thickness. For example, if the steel sheet thickness is increased from 0.020 to 0.064 in with 5-in tube spacing, the plate efficiency increases by about 12 percent, from 0.75 to 0.84. Finally, the plate efficiency increases with thermal conductivity always being greater for a given thickness of copper sheet than steel sheet.

Equation (2.4) presents the collector efficiency in terms of properties which can easily be measured—the ambient temperature, the solar radiation level, and the average fluid temperature. For liquid collectors, the average fluid temperature can be taken as the arithmetic average of inlet and outlet temperatures. Although Eq. (2.4) is entirely adequate for measuring collector performance in the field, a different version of this equation is in common use and is presently the standard equation used by all testing codes in the United States. This equation *relates collector efficiency to the collector fluid inlet temperature only:*

$$\eta_c = F_R \left[\tau\alpha - U_c \left(\frac{T_{f,i} - T_a}{I_c} \right) \right] \tag{2.5}$$

This form of the efficiency equation is very useful in calculating the long-term performance of solar systems (described in later chapters).

The factor F_R which precedes the right-hand side of Eq. (2.5) is called the heat-removal factor. The heat-removal factor depends upon the plate efficiency F' from the preceding equation as well as the mass-flow rate through the collector and the heat-loss coefficient U_c. F_R is encountered in the analysis of all types of heat exchangers. It is a measure of the effectiveness of the solar collector when it is viewed as a heat exchanger between solar flux and the working fluid. Figure 2.3 shows F_R/F' as a function of the dimensionless group $(\dot{m}c_p)_c/U_cA_cF'$ shown along the x axis of the curve. The dimensionless group is the ratio of the mass-flow-rate–specific-heat product often called the fluid capacitance rate, $(\dot{m}c_p)_c$ per unit area A_c divided by the collector loss coefficient U_c and the plate-efficiency factor F'. If the ratio is greater than 10, very little gain is achieved in heat-removal effectiveness. Typical values of the dimensionless ratio are between 8 and 10 for most liquid collectors, yielding an F_R value above 0.95. For air collectors, the value on the x axis is somewhat lower (3 to 4) and values of F_R in the vicinity of 0.6 to 0.7 are common.

Figure 2.3 is an example of the physical laws which determine the performance of many components of a solar system including the collector. As the collector flow rate $\dot{m}$ increases beyond a certain level, the effect on heat-removal factor becomes vanishing small but parasitic pumping losses continue to increase. This phenomenon is called the *law of diminishing returns*. It also applies to heat exchangers, pumps, storage, and all other major components of a solar thermal system. The important consequence of this law is that it is not worthwhile, in the case of the collector, to pump fluid at a rate which gives values of the dimensionless ratio much above 10. Larger pumps are more costly, use more energy, and only slightly increase collector efficiency. The precise flow rate depends on an economic analy-

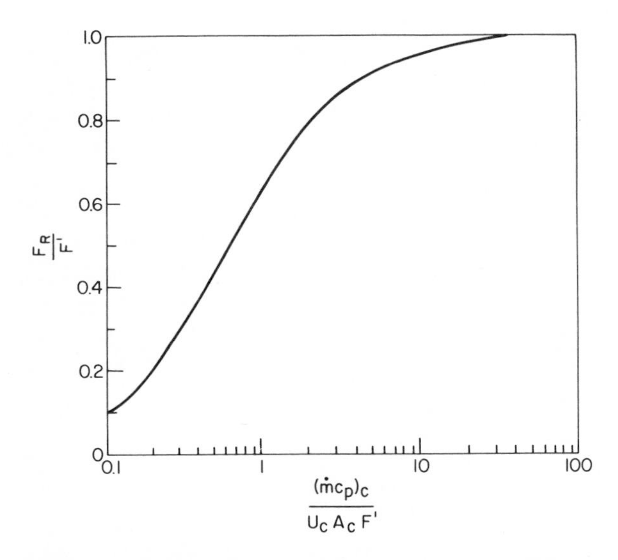

FIG. 2.3 Effect of collector fluid flow rate $\dot{m}$ on collector heat-removal factor F_R; $F_R = [(\dot{m}c_p)_c/U_cA_c][1 - \exp(-F'U_cA_c/\dot{m}c_p)]$.

sis. The trade-off is the cost of extra pumping energy vs. the dollar value of extra solar heat collected. An example in Chap. 4 illustrates the trade-offs.

Qualitative Parameter Trends. The quantity $\tau\alpha$ in all three efficiency equations depends upon the value of glazing transmittance and surface absorptance. Reference 1 shows that these two quantities depend on the angle between the sun's rays and the collector surface. See page 29 for details.

Figure 2.4 shows in qualitative form the effects of several parameters on efficiency. The higher the inlet temperature, the higher the collector overall temperature and the greater the heat loss. Therefore, as fluid inlet temperature increases, the efficiency drops off. Contrariwise, the effect of increasing ambient temperature is to increase collector efficiency.

At given operating and environmental temperatures, efficiency increases with solar radiation level as shown in the third part of Fig. 2.4. Heat losses are fixed at

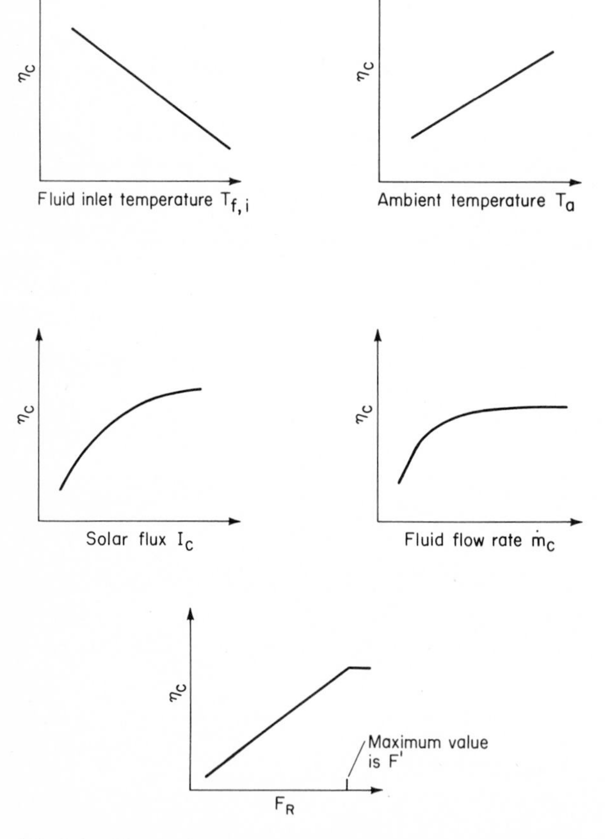

FIG. 2.4 Effect of major operating conditions of a solar collector upon collector efficiency η_c shown qualitatively.

a given temperature difference; therefore, the amount of extra energy available for conversion increases as the solar radiation level increases. Another factor that affects the performance of a collector is the fluid flow rate. However, it has a very small effect above a certain level and is considered a second-order variable as far as liquid collector efficiency calculations are concerned. The effect is of greater importance in air collectors. Collector flow rate is contained in the parameter F_R.

Collector efficiency is one measure of thermal performance. However, it is too restrictive an index to be used when selecting a solar collector for a specific task.

In later chapters, we will show that the solar collector properties $F_R\tau\alpha$ and F_RU_c can in fact be used in calculation procedures to determine the annual (not instantaneous) heat production of a given solar system (not collector) in a given location. The annual *useful* energy delivery, when coupled with the annual cost of a solar system, will be used to calculate the cost of solar energy in dollars per million British thermal units. This is the key cost index used in determining solar energy viability. The efficiency is then a starting point for system analysis, but must be incorporated with other information before the selection of the most cost-effective collector can be made.

Solar Collector Orientation. One of the major decisions which must be made regarding the installation of a solar collector is its orientation relative to the sun. Figure 2.5 shows the position of the sun at three times of year for a latitude of 40° north (e.g., Columbus, Ohio; Boulder, Colorado). At the summer solstice the sun is at its highest elevation, having an angle of 73° 30′ above the horizon at noon. In winter, by contrast, the sun is only 26°30′ above the horizon at noon. Therefore, the orientation of the collector must be related to the position of the sun at the time of year during which solar collection is to be maximized. For example, if space heating is to be done, the collector must be oriented at a fairly steep angle with respect to the horizontal. However, collector orientation for DHW heating should be such that the amount of radiation intercepted by the collector is a maximum on a year-round basis, not during the winter only. DHW collectors should roughly face the sun at the equinoxes (September 22 and March 21). They should be tilted up from the horizontal at an angle equal to the latitude.

Collectors used for active solar heating should tilt at an angle equal to the latitude plus 15°. These orientations will give the best performance over the course of a year. The angles are not critical. They can vary by ±5° and the annual performance will show very little effect. In some cases horizontal or vertical collectors will have smaller structural costs. If so, the rules must be adjusted. In fact, Fuller (64) has shown that even east- or west-facing tilted collectors can work relatively well.

The diagrams of the sun's position on the right of Fig. 2.5 show that the sun is south of an east-west line most of the year. Obviously, therefore, the collector must face roughly south. A precise, due south orientation is not required, however. Variations of ±20° from due south have very little effect on the annual performance of most systems. If off-building-axis collector structure costs are high for a non-south-axis building, azimuth angles beyond the 20° range may be more economical

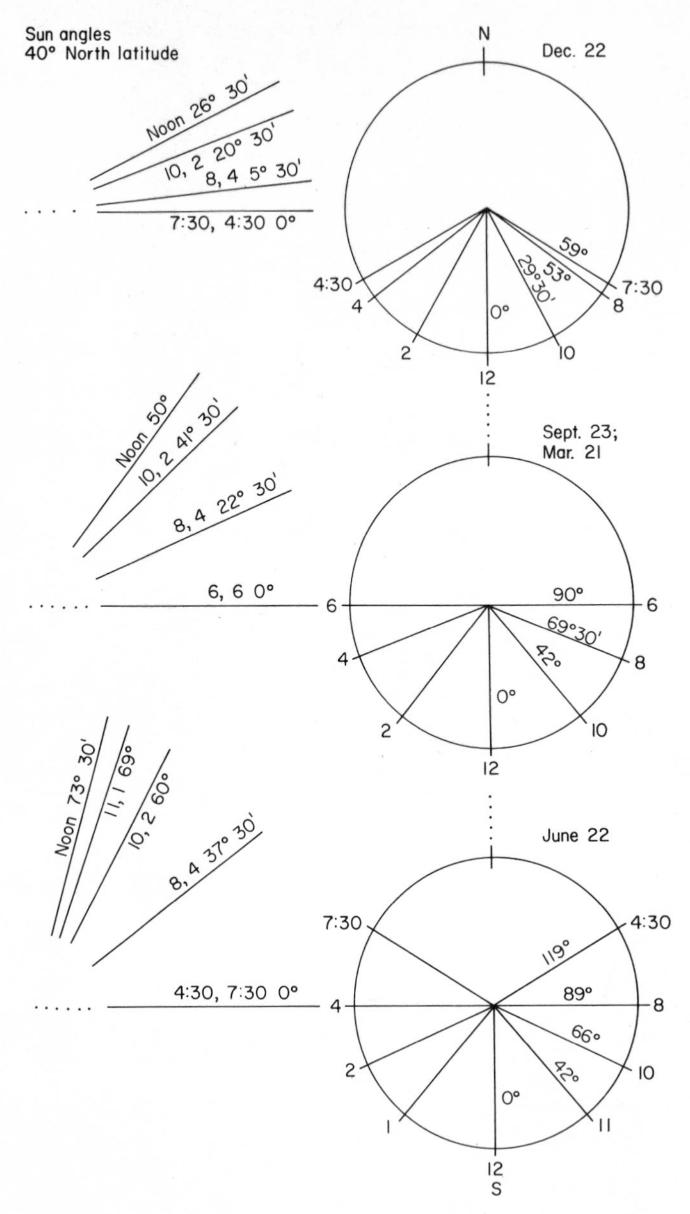

FIG. 2.5 Solar elevation and azimuth angle for three times of the year at 40° north latitude.

from a life-cycle cost viewpoint. Off-south orientations should use somewhat reduced tilt angles.

The tilt and orientation rules given are typical of many solar system design rules. There are specific best values of certain parameters, but small variations from the nominal values will have very little effect on annual system performance. Far from these nominal values, however, a very drastic falloff in performance can be expected which must be carefully studied from a cost-benefit viewpoint.

Solar Collector Testing. At the outset of this section, it was recommended that test data be used rather than calculated data. The test procedure used must reflect actual operating conditions which will be experienced in the field. The National Bureau of Standards has collaborated with ASHRAE in formulating the standard flat-plate-collector test procedure for the United States. This procedure, very briefly summarized, requires the formal measurement of collector efficiency during a sunny day. Collector heat production is measured over a 15-min period and divided by the solar radiation intercepted during the same 15-min period. Similar measurements are made at several values of fluid inlet temperature so that a collector efficiency curve (Fig. 2.6) may be plotted. Collector output is calculated from the product of fluid flow rate, fluid temperature rise, and fluid specific heat. Heat produced is divided by the collector area and the solar flux measured by a pyranometer.

After the formal test procedure, the collector is subjected to a stagnation test in which fluid is not permitted to flow through the collector. During each sunny day the absorber plate temperature rises such that the heat losses from it are equal to the absorbed solar flux. This temperature may be as great as 400°F. After the stagnation test, the performance test described earlier is repeated. If there is a difference between the two efficiency tests, it is noted.

The intercept of the test efficiency curve is the value of the optical efficiency $F_R\tau\alpha$, and the magnitude of the slope of the curve F_RU_c. The slope and intercept of this test curve with stated statistical uncertainties embody all the data collected by the formal test.

Another feature of the standard ASHRAE test is the measurement of incidence-angle effects which are quantified by the *incidence-angle modifier* $K_{\alpha\tau}$. The incidence-angle modifier is the ratio of the optical efficiency $F_R\tau\alpha$ at any incidence angle to $F_R(\tau\alpha)_n$, the normal-incidence value. Figure 2.7 shows the qualitative

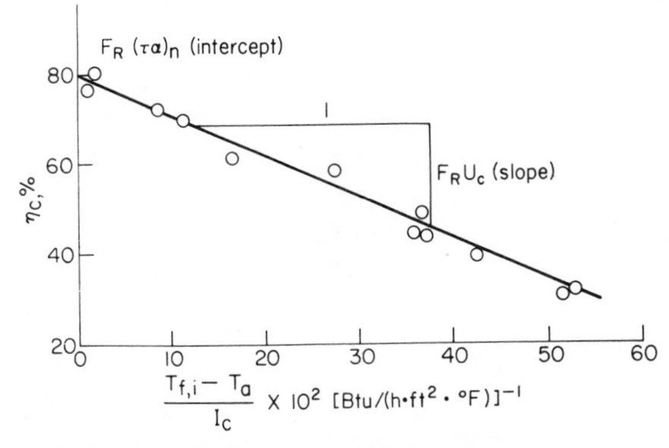

FIG. 2.6 Typical flat-plate collector efficiency curve showing method of evaluating $F_R(\tau\alpha)_n$ ("normal incidence" value) and F_RU_c.

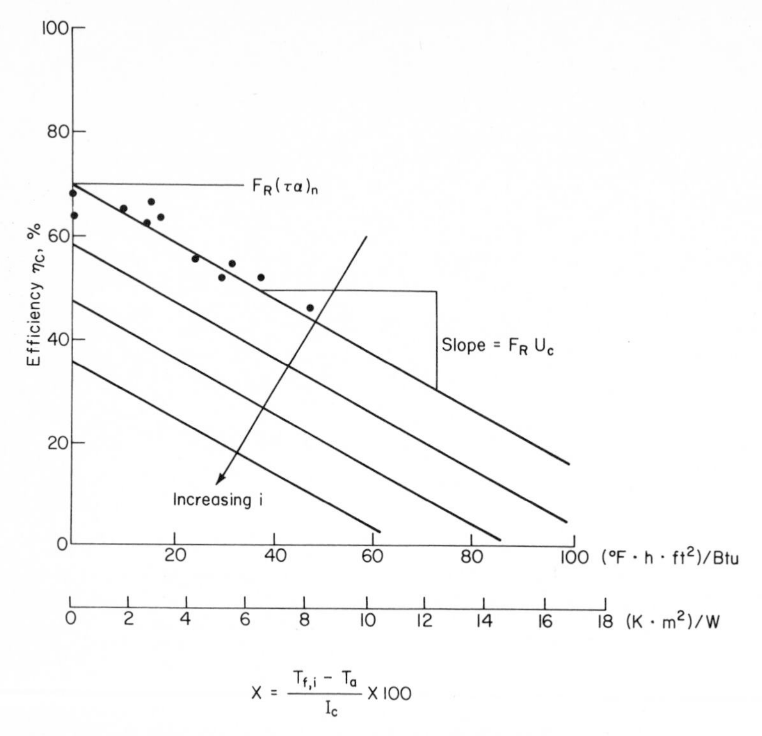

FIG. 2.7 Several efficiency curves showing the effect of increasing incidence angle *i* on flat-plate collector efficiency.

effect of incidence angle *i* on the collector efficiency curve. With increasing incidence angle, the intercept of the curve is closer to the origin, but the slope of the curve, which is a measure of heat losses only, is unchanged.

Slight modifications to the standard ASHRAE procedure have been made by some trade organizations or testing facilities. Many of the changes made are for testing convenience and have very little effect on the important results.

Evacuated Tubular Solar Collectors. One method of improving the efficiency of flat-plate collectors is to reduce the heat loss from the absorber plate to the environment. The heat loss occurs primarily through the glazed surface of the collector. If the air gaps could be completely evacuated, the heat loss by convection and conduction could be completely eliminated. This heat loss in a flat-plate collector accounts for between one-third and two-thirds of the heat loss from the front surface of the absorber plate. Flat-plate evacuation is not practical. Atmospheric pressure—approximately 225 lb/ft^2—would exist on the glass covers. Glasses will break under this load. In addition, special seals would be required to maintain the vacuum within the flat-plate collector over a long period of time.

Contrary to a rectangular glass prism, a glass cylinder has a very high compressive strength when evacuated. Several solar collector manufacturers have used cylindrical glass enclosures, as shown in Fig. 2.8. In order to completely suppress convection and conduction losses in these collectors, an internal vacuum of the order of 10^{-3} to 10^{-4} mm of mercury is required. This level of evacuation is within the abilities of commercial fluorescent tube manufacturers. Note that vacuum levels below this level permit conduction losses to continue unaffected.

The collector shown in Fig. 2.8*a* is a flat absorber surface enclosed in a glass cylinder. The absorber plate is treated with a selective surface to reduce radiation losses. As a result, conduction, convection, and radiation are small. However, there is a slight optical penalty incurred with this collector. Sunlight strikes the outlying portions of the cylindrical tube at off-normal incidence, and reflection losses are high (see page 29). A second optical penalty arises from the portion of the collector plane that is consumed nonproductively by the glass tube. The net absorber area of a tubular collector assembly is less than its gross area by twice the number of tubes multiplied by the tube thickness. Typically the tubes used in this type of collector are 2 to 4 in in diameter.

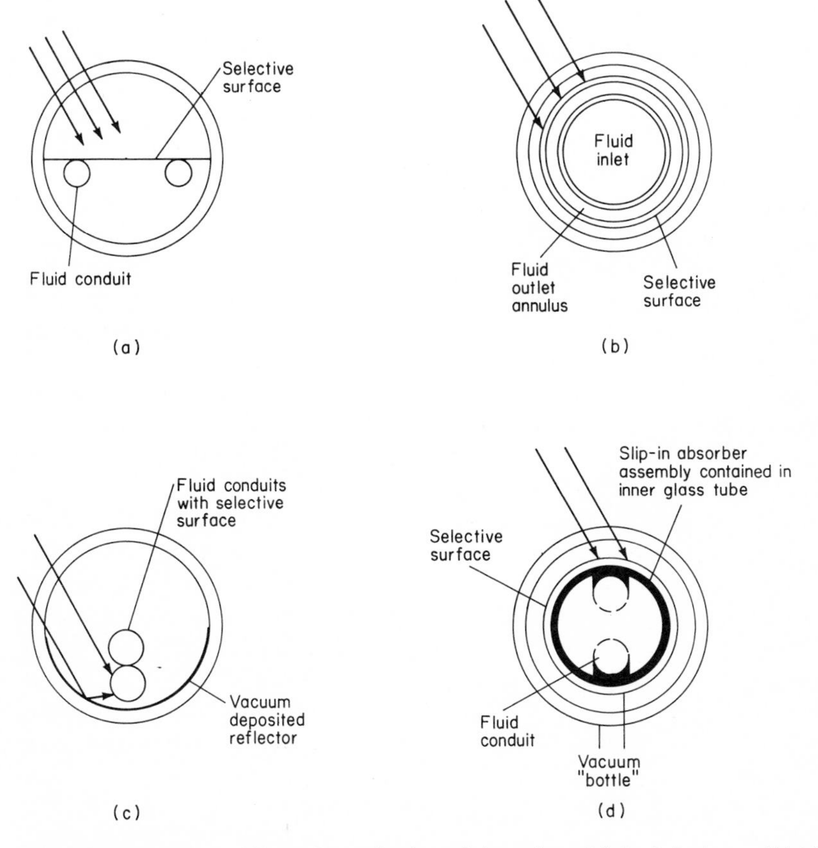

FIG. 2.8 Evacuated-tube solar collectors: (*a*) enclosed flat-plate type, (*b*) triple-concentric-tube type, (*c*) concentrating type, (*d*) dual-concentric-tube with slip-in absorber assembly.

The concentric tubular collector shown in Fig. 2.8*b* is manufactured by the Owens-Illinois Co. It consists of three concentric tubes. The innermost tube is the fluid inlet, connected to a manifold in the collector assembly. The annulus between the inner tube and the center tube is the fluid outlet, also connected to a manifold. The outermost glass tube is the vacuum jacket. A selective surface is used on the absorber tube. The glass used in this collector is Pyrex™; it has very high resistance to thermal stress at elevated temperature. One of the difficulties with this type of collector is the large amount of fluid contained in it. The fluid must be heated to a useful temperature each morning before *useful* energy delivery can take place. Warm-up lasts approximately 5 times as long as it would in a flat-plate collector. Energy used to warm fluid in a collector is not useful heat.

The tubular collector shown in Fig. 2.8*c* consists of a pair of fluid conduits onto which sunlight is reflected. The reflector is a thin silver surface deposited on the lower half of the glass tube during manufacture. Silver, when confined in a vacuum, can have a reflection coefficient in excess of 90 percent. Selective surfaces are used on the absorber tubes to control heat loss.

A fourth type of tubular collector is shown in Fig. 2.8*d*. It consists of only two concentric glass cylinders forming an evacuated annulus similar to a wide-mouthed Thermos™ bottle (of course, the silver coating is not present). A copper heat exchanger is placed in contact with the inner surface of the inner tube. This absorber consists of a sheet of copper rolled to form a press fit against the inner surface. The fluid conduit is attached to the copper sheet. The slip-in absorber assembly is attractive since the system need not be shut down if a glass cover should break. A replacement evacuated jacket can be slipped over the metal absorber without interfering with fluid flow. This cannot be done with the concentric tubular collector shown in Fig. 2.8*b*; breakage of the glass tubes requires shutdown of the affected collector array for repair.

Evacuated collectors have not been widely used for space heating and water heating. These collectors have their greatest efficiency advantage over flat-plate collectors at temperatures above 200°F, normally not required by solar space-heating systems.

Of the installations which have been made using tubular collectors, most have operated satisfactorily. One of the difficulties with the use of tubular collectors in very cold climates is the buildup of snow and ice between tubes and on top of tubes. Snow buildup on flat-plate collectors is usually not a problem since it slides easily from collector after the sun has warmed the glass beneath by a few degrees. Tubular collectors have very low heat losses and the glass tube is, therefore, close to the environmental temperature. As a result, snow normally does not slide from these tubular collectors as quickly as from flat-plate collectors. Ice on tubular collectors does not completely stop energy collection but does reduce the amount of solar radiation. When snow or ice does ultimately melt from a tubular collector, there is a further problem in channeling the melted snow and ice away from the collector array surface. In some cases drains may be required.

○ **Analytical Equation for Collector-Heat-Loss Coefficient U_c.** Previously it was recommended that U_c be determined by a test. If performance data on a collector are not available, it is possible to calculate the thermal loss. In the context of this book, this should be done only if a manufacturer has not measured performance by the standard method.

The front surface heat loss consists primarily of radiation and convection. The loss through the back of the collector, q_{back}, is primarily a result of conduction. It can be calculated by standard methods according to the thermal resistance idea. For example, a 3-in layer of fiber glass would have an R value of about 9 and a U value of about 0.11.

Calculation of heat loss from the front of a flat-plate collector requires an empirical approach. Klein (4) has developed this closed-form equation for the heat loss from a flat-plate collector [in metric (SI) units only]:

$$q_{loss} = \frac{(\overline{T}_c - T_a)\,A_c}{\dfrac{N}{(C/\overline{T}_c)\,[(\overline{T}_c - T_a)/(N + f)]^{0.33}} + \dfrac{1}{h_{c,\infty}}} + \frac{\sigma\,(\overline{T}_c^4 - T_a^4)A_c}{1/[\epsilon_{p,i} + 0.05N(1 - \epsilon_{p,i})] + (2N + f - 1)/\epsilon_{g,i} - N} + q_{back} = U_c A_c(\overline{T}_c - T_a) \qquad (2.6)$$

where $f = (1 - 0.04h_{c,\infty} + 0.0005h_{c,\infty}^2)\,(1 + 0.091N)$
$C = 365.9(1 - 0.00883\beta + 0.00013\beta^2)$
N = number of covers
$h_{c,\infty} = 5.7 + 3.8V \qquad V = \text{m/s}$
$\epsilon_{g,i}$ = infrared emittance of the covers ≈ 0.88

The equation includes all parameters which determine the thermal properties of a flat-plate collector—the emittance of the collector plate, the emittance of the collector covers, the number of covers, the tilt angle, the velocity of wind over the surface of the collector, and so forth. U_c can be calculated from Eq. (2.6) by dividing q_{loss} by the collector area A_c and by the temperature difference $(\overline{T}_c - T_a)$. This equation is accurate over a wide range of collector cover and ambient temperatures as well as over a broad range of wind speeds, collector emittance values, numbers of collector covers, and tilt angles. Approximately 1000 different computer simulations were used in determining the numerical constants in Eq. (2.6).

Effect of Solar Incidence Angle on Collector Transmittance and Absorptance. The optical efficiency of a flat-plate collector, is determined by standard tests at sun angles near normal incidence. However, solar collectors rarely operate at normal incidence. If measurements of the incidence-angle modifier are not available, they can be calculated from properties of glass and the absorber surface.

The transmittance of any transparent material is very strongly dependent upon the angle at which light strikes the surface beyond about 50° incidence. This characteristic of transparent media can be calculated from the Snell and Fresnel equations given in Ref. 1. Figure 2.9 shows the effect of incidence angle on the transmission of a single glass plate; it is approximately 92 percent for incidence angles less than 40°. Beyond 50° the transmittance begins to fall off very rapidly. Figure 2.9 is based on a negligible absorption of solar radiation within the glass cover. This is approximate if glass of low iron oxide (Fe_2O_3) content is used. The iron content of glass can be determined qualitatively by looking at its edge. If the edge has a greenish tint, it contains excessive iron oxide. If the edge appears white, it is low-iron glass and will not absorb sunlight.

The optical efficiency of a flat-plate collector is the product of the transmittance τ shown in Fig. 2.9 and the absorber plate absorption coefficient α. The absorptance of several surfaces is given in Fig. 2.10. There are two fundamentally different types of surfaces. The circles in Fig. 2.10 represent a flat-black surface whose absorptance is relatively independent of incidence angle. The $\tau\alpha$ product for a flat-black collector can be calculated by multiplying the nominal absorptance by the transmission function given in Fig. 2.9.

Selective surfaces have a fundamentally different angular characteristic. Figure 2.10 shows that many types of selective surfaces fall on a universal curve which is similar in shape to that for glass transmittance. It has a significant roll-off beyond 40 or 45°. The transmission-absorptance product for a selective surface can be calculated by multiplying values from Figs. 2.9 and 2.10 together.

Although measurement of the incidence-angle modifier by formal testing is preferred, the preceding method can be used to calculate $K_{\alpha\tau}$, the incidence-angle modifier $K_{\alpha\tau}$. Data can be correlated by a simple equation:

$$K_{\alpha\tau} = 1 + b_0 \left(\frac{1}{\cos i} - 1 \right) \tag{2.7}$$

One empirical constant b_0 is used. It can be measured experimentally or calculated from Figs. 2.9 and 2.10. The value of b_0 for all flat-plate collectors is negative. In

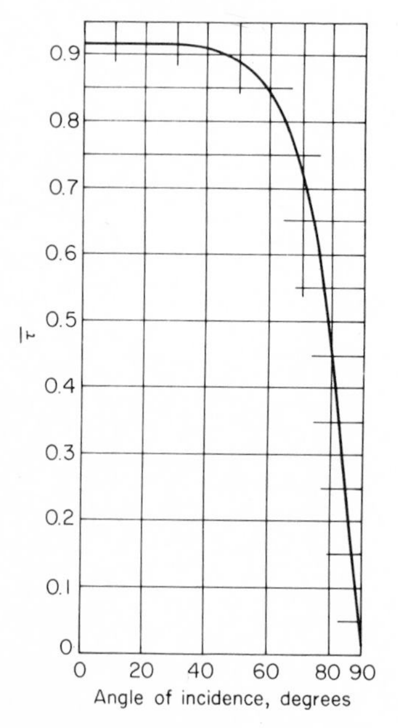

FIG. 2.9 Transmission coefficient τ for low-iron glass as a function of solar angle of incidence.

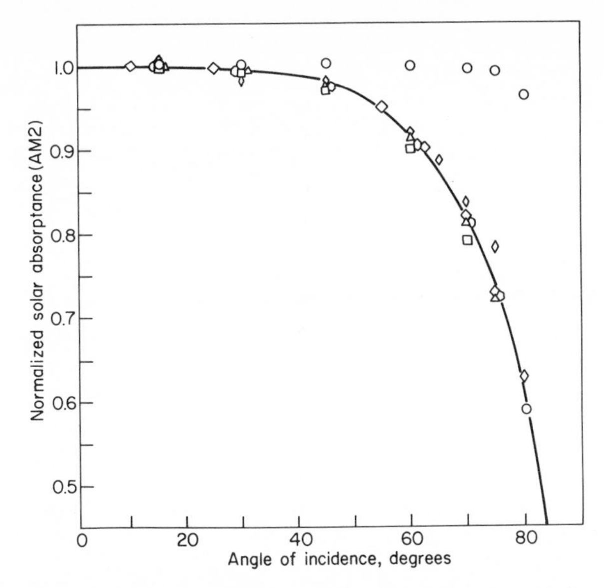

FIG. 2.10 Normalized universal absorber plate absorptance vs. incidence angle function for selective and nonselective surfaces. Flat black surface data—circles; other surfaces—other symbols. *(From Ref. 5 with permission.)*

other words, the incidence-angle modifier is a number less than 1 and a number which decreases with increasing incidence angle. This behavior is to be expected in view of the preceding discussion on the angular decrease of both absorptance and transmittance in flat-plate collectors. The effects of increasing the number of collector covers from one to two and of using multiple reflections between the covers are described in detail in Chap. 3 of Ref. 1.

Thermal Storage for Solar Heating Systems

Nearly all practical solar heating systems contain storage sufficient to heat the building over a 1- or 2-day period. Storage is required in any solar thermal system because of the variability of solar flux. In winter when heating loads are significant, solar energy can be collected for only 6 or 8 h. This heat will be used to supply a relatively uniform load over a 24-h period. During poor weather, solar heat can be used only if stored from a prior sunny period.

Sensible Storage. Two types of thermal storage are widely used for solar space-heating systems—water for liquid systems and rock for air systems. A third type of storage involving the freezing and melting of special chemical salts has also been used on a limited basis.

The amount of sensible heat stored Q_{STR} in a given volume is given by Eq. (2.8):

$$Q_{STR} = V(\rho c)\Delta T \tag{2.8}$$

Q_{STR} is the product of the storage volume V, the volumetric heat capacity (the product of the density ρ times the specific heat c), and the useful temperature change in storage ΔT. The quantity ΔT is determined by the range of usable temperatures. In DHW heating systems, this temperature could range from 80 to 140°F, for example (even higher if a mixing valve is used; see Chap. 4). The temperature range is approximately 100 to 200°F in heating applications. Equation (2.8) shows that the maximum amount of storage per unit volume is achieved by those materials which have the highest density–specific-heat product. Water has one of the highest density–specific-heat products. The density is 62.4 lb/ft^3, and the specific heat is approximately 1 Btu/(lb · °F). Therefore, the density–specific-heat product is 62.4 Btu/(ft^3 · °F).

The density–specific-heat product of rock is somewhat lower. The density of rock, with void spaces between the rock accounted for, is approximately 90 to 95 lb/ft^3. The specific heat of river gravel, commonly used for storage in air-based heating systems, is approximately 0.18 to 0.22 Btu/(lb · °F). The density–specific-heat product for rock is approximately 18 Btu/(ft^3 · °F). Therefore, to store the same amount of heat in rock as in water at the same ΔT, 3 times the volume of rock is required. Although the heat capacitance of gravel is only one-third that of water, it is nearly universally used in air-based space-heating systems because of its large surface area, low cost, and excellent performance (because of its good stratification ability, noted later).

Latent Storage. Phase-change materials rely on either the heat of fusion or the heat of solution evolved in a change from solid to liquid. The most familiar phase change involves the freezing and melting of water. To melt 1 lb of ice, 144 Btu is required. Of course, the freeze-melt process occurs at 32°F, a useless space-heating temperature. Other materials have phase changes at usable temperatures—80 to 130°F.

A few buildings have been built using phase-change storage. A number of difficulties have precluded the broad use of phase-change storage, however. Many of the materials are salts that are corrosive or sharp and tend to cut seals or soft plastics. They must, therefore be held in containers made from plastic or expensive inert metals. The thermal transport properties of the salts are poor and a relatively large surface-to-volume ratio is required in order to transfer heat sufficiently rapidly.

The repeatability of the freeze-melt cycle is poor for some materials. Some chemicals used require a progressively increasing amount of subcooling and superheating with age in order to initiate phase change. This difficulty has been resolved for the most part by using special additives to the salt materials to eliminate solid-phase settling and to initiate the phase change at a repeatable temperature. Since phase-change materials are not widely used and experience is very limited, they are not described further in this book.

Storage Size. The amount of storage used is determined by economic and not by technical criteria. Technically if sufficiently large storage were available, 100 percent of load could be met by a solar heating system. Solar heat could be collected during the long, sunny days of summer and stored for use in the short, cold days of winter. It will be shown later that the amount of storage needed by this approach is too large to be economical except in special cases, in special climatic areas, or in very large district-heating systems. Seasonal storage is uneconomical for three reasons: the relatively high losses from storage because of large area, the high cost of very large storage tanks, and the catastrophic effects of major leaks. Offsetting large storage cost is a possible collector array saving, since seasonal systems are sized according to annual-average solar conditions, not lower, winter-average conditions.

Water Storage. Water is the least expensive and highest-heat-capacitance medium used for thermal storage. Although water is inexpensive, the tank required to confine it for storage is not. Two types of storage systems can be used—open and pressurized. A relatively simple tank construction is used in open systems since only hydrostatic water pressure is involved. Open systems can have high evaporation rates. Evaporation is undesirable for several reasons. First, it represents a loss of 1000 Btu per pound of water evaporated, or approximately 8300 Btu per gallon of water evaporated. This is a significant heat leak and is the major difficulty with open storage. As water evaporates from storage, the salinity or hardness originally contained in that water remains in storage. Therefore, the concentration of water hardness increases. In order to maintain clean heat-transfer surfaces and avoid buildup of scale, it is necessary occasionally to chemically treat the water or drain part of it and replace it with fresh water. The water which was drained, of course, contains heat and represents another loss from storage.

If open tanks are inside the building, they cause a significant sensible and latent cooling load during the summer. Finally, open tanks admit dissolved oxygen, with its well-known corrosion problems, to the stored fluid and hence to all piping components in the storage system. Increased water treatment costs result.

Closed pressurized storage tanks, on the other hand, are more expensive. They cost \$1 to \$3 per gallon depending on size and material. The tanks can be metal (steel or aluminum), concrete, or fiber glass. Metal tanks are most common because of their reliability and availability. Fiber glass tanks are widely used to store liquid fuels but are not usable in solar systems above 160°F. Above this temperature fiber glass deforms under pressurization.

Concrete tanks are easily poured in place during construction but special care is required in their insulation and sealing. The entire surface of the tank must be insulated. If the tank is buried, the insulation must be waterproof and capable of supporting the entire weight of the concrete tank and its contents along with earth loads. This weight for a residential application usually is several tons. Concrete tanks are exposed to a wide range of temperatures (possibly between 60 and 210°F), causing expansion and contraction and leading to cracks. The entire internal surface of the concrete tank must be sealed with a flexible seal material to prevent leaks. Concrete storage tanks must use elastomeric seals. Pressurization of concrete tanks is difficult; most have been open.

Metal tanks are usually cylindrical and can operate reliably for 20 or 30 yr with proper corrosion control. The applicable building code usually does not make it necessary to have an American Society of Mechanical Engineers (ASME) stamp on the tank, but the tank will probably require an ASME rating. The position of connections to a metal tank is very important. It is essential that the entire tank volume be used. Any tendencies for fluid short circuits to occur between inlet and outlet fittings must be eliminated. These matters are discussed in detail in Chap.4.

Heat Loss. One of the errors in many early solar storage installations was the lack of attention paid to insulation. Although heating storage is not high-temperature storage, the amount of heat loss can be significant because (1) tanks have a large area and (2) storage is above its environmental temperature for every hour of the heating season. A common guideline for the amount of insulation required states that no more than 2 percent of the stored heat shall be lost during one day. In the next technical section the temperature drop in storage is calculated.

Storage tank insulation is done either in the field or at the factory. It is simpler to order a steel tank insulated with foam from a factory, whereas concrete tanks must be insulated in the field at the same time they are poured. Such insulation tends to be less reliable since insulation joints may not be sealed as carefully as if factory-installed.

In addition to insulation on the tank itself, *thermal breaks must be provided at each contact between the tank and its environment.* For example, the supports for a tank located in a building must include a thermal break between the tank and the supporting concrete slab. The break can be rigid glass insulation or other load-bearing insulation. Pipes entering and leaving storage are also to be insulated.

Buried Storage. Buried outdoor versus indoor storage trade-offs are dealt with in detail in Chap. 4. The two criteria which determine the location of storage are (1) the availability and cost of space and (2) the controllability of heat loss from the buried tank. In the past excessive heat loss to underground aquifers or to moist surrounding soil occurred because of insufficient or improper insulation. In one building 50 percent of the solar heat collected by that system during a year was lost. The pump used with buried storage must be located below the level of the tank outlet in a dry well adjacent to the tank. Servicing of such pumps is difficult and expensive.

Expansion. The volume of any liquid varies with temperature. Fluid expansion and contraction in hydronic systems is compensated for by an external expansion tank. The volume change in the storage fluid over the expected range of temperatures is the basis of tank sizing.

Bladder type expansion tanks consist of a rubber bladder which separates an inert gas charge above the bladder from the storage liquid. As the storage fluid is heated and expands, additional fluid is forced into the expansion tank, and the gas above the diaphragm is compressed.

As storage fluid cools, the gas charge forces fluid from the expansion tank back into the storage tank, keeping the storage volume completely filled. Over the range of temperatures involved in most water-based solar systems, the volume change will be 10 to 11 percent. Contraction and expansion characteristics for the liquids are different. For example, a glycol water solution is typically used in the solar collection loop. Some glycol solutions expand more than water. Therefore, the expansion tank in a glycol system must be larger than that in a pure water system.

Rock-Bed Storage for Air-Based Systems. The heat-removal factor F_R for air collectors is substantially lower than that for liquid collectors. Therefore, the air collector efficiency is lower than liquid collector efficiency at the same inlet temperature. An air system must operate at a much lower inlet temperature than a liquid system. Collector inlet temperature is dictated by the temperature of storage. Therefore, the temperature of the zone of storage from which air passes to the collector must be low. Usually the lowest temperature is roughly the same as room temperature (unless storage is fully charged), as seen later when the operation of air-based systems is explained. A material which allows significant *stratification* must be used if useful heat is to be stored at 130°F in one part of storage while another remains at 70°F. A rock bed works well since heat is transferred slowly from one rock to the next via a very small contact point between the two. It is, therefore, possible to maintain one zone of a rock bed at high temperature and the other zone at low temperature. It is not possible to use liquid storage with an air-based space-heating system and still achieve acceptable performance.

Materials. In a typical rock bed, clean river gravel approximately 1½ in in diameter is used. Size is not as important as uniformity. The rock is usually confined in a cubical container in residences, and the volume of storage is typically ½ to 1 ft^3 of rock per square foot of solar collector area. The sizing of rock-bed storage is considered in detail in Chap. 4.

The container must be designed for long-term service. The storage unit can be constructed from wood or concrete. All the walls of a concrete storage unit can be poured at the same time as the basement. Wood is easy to work with, but it will warp at elevated temperatures if the wood is not completely dry (causing leaks) and it will also have reduced strength. As wood dries, its ignition temperature drops, creating a possible fire hazard and code compliance problems.

The major practical problem with rock storage is air leakage. Concrete seems to be less leak-prone than wood.

Rock-bed seals and airflow distribution design are important. Uniform flow through a storage bed is essential. This is accomplished if the pressure drop through the rock is much larger than the pressure drop in the inlet or outlet plenums. If the bed pressure drop is 2 to 4 times that in the plenum, uniform flow will be assured. If bed pressure drop is not much higher than the plenum pressure drop, flow will be channeled through the least-resistance paths of the rock bed. Then only a small portion of the rock will be functional.

Duration of Storage. A common but erroneous index of storage size is the duration of storage, measured in days or weeks. *The duration of storage is not a useful index of storage size,* since the amount of heat removed from storage depends upon the heating load. On a mild day suppose 250,000 Btu might heat a building for 20 h. On a very cold day, the same amount of stored heat would be consumed in 6 or 7 h. Therefore, the use of storage duration as a measure of solar storage size is inappropriate. The proper method for characterizing the size of storage is by its volume or mass and thermal capacity.

Storage for Passive Systems. The physical relationship of storage in a passive heating system to the collector—south-facing apertures—for solar gain is much more important in passive systems than active systems. Heat produced in active collectors can be transported to any location for storage. Passive storage is often an integral part of the collector absorber. Therefore the size and types of material have an important impact on the design and temperature swing of the building. Passive systems may use either water or masonry for storage.

Directly illuminated masonry storage consists of floors or vertical interior walls placed adjacent to and inside south-facing apertures. These walls or floors are dark on the sun-facing surface. The thickness is matched to the required storage amount and the required timed release of stored heat at night. Masonry storage must be solid and continuous so that heat is readily transported from the illuminated surface to the interior. For example, a concrete block wall with large gravel-filled voids is not acceptable storage since heat migrates very slowly from pebble to pebble.

Direct-gain passive systems do not solely rely on directly illuminated storage mass of either water or masonry. Instead, sunlight is converted to heat inside the building. Heated air thus produced circulates by gravity and natural circulation through the building and contacts interior building surfaces. Part of the heat contained in the warmed air will be transferred and stored in these elements. The transport of heat into nonilluminated storage is slower than for illuminated storage.

Therefore, to store the same amount of heat for the same mass of storage a much larger surface—3 to 4 times as great—must be provided. Large storage masses used in passive systems require larger foundations. This is particularly true in areas where the compressive strength of soils is low; a spread-type foundation may be required to support the weight of passive storage elements. Some hybrid systems use rock-bed storage along with passive mass storage.

O **Heat-Loss Equation for Thermal Storage.** The temperature history of storage in active systems can be calculated from elementary heat-transfer principles. If the heat loss from a storage tank is proportional to the difference between the temperature of storage and the temperature of the environment T_∞, the temperature of storage $T_s(t)$ will undergo exponential decay given by Eq. (2.9):

$$\frac{T_s(t) - T_\infty}{T_0 - T_\infty} = e^{-(U_s A_s t / m_s c_s)} \tag{2.9}$$

The time-temperature history depends on the heat-loss coefficient U_s, the tank area A_s, and the mass and specific heat product of storage (including the tank) $m_s c_s$. It is possible to evaluate the efficacy of storage insulation experimentally by measuring storage temperature over a day or two. If $T_s(t) - T_\infty$ is plotted on semilog paper, the magnitude of the slope of the curve will be the reciprocal of the time constant τ of the exponential decay. The value of U_s can be calculated from τ. A_s is easily determined from dimensions of storage; the mass of storage and its specific heat are likewise easy to calculate. The 24-hr heat loss from storage, as stated earlier, should be less than 2 percent of the energy stored. The loss is the product of the mass, specific heat, and 24-hr temperature drop in storage. This heat loss is compared to the total heat stored, given by Eq. (2.8), and the relative size of the loss is determined by simple ratio.

The approximation that a single temperature $\overline{T}_s$ accurately represents the effective temperature of rock storage is probably not accurate, therefore, Eq. (2.9) should be viewed only as an approximation.

The 2 percent rule is based on generalities regarding the value of stored solar heat relative to the cost of insulation to contain that heat. This rule will be modified in subsequent chapters on the basis of the expected cost of insulation and the value of retained solar heat. In most cases, the economic optimum amount of insulation will result in less than 2 percent heat loss.

Heat-Transport Subsystems

Heat-transport subsystems are present in all solar systems. Heat transport is required, for example, in an active solar heating system between the collectors and storage and between the storage and the end use. Pumps, fans, heat exchangers, and control valves operated by the control system compose active heat-transport systems. In this section, we will examine the fundamental characteristics of pumps and fans, types of piping used in solar heating systems, the nature of working fluids, the performance of heat exchangers, and controls. In addition, the backup system is briefly described.

Pumps and Fans. Several pumps and fans are typically present in any active solar heating system. These devices move fluids from place to place in pipes or ducts and are sized according to conventional design rules of the HVAC industry. Only the basic characteristics of these devices are described in this section.

Figure 2.11 is a sample characteristic curve for a pump and the system in which the fluid flows. The same types of curve apply for fans and ducts. The curve labeled "system curve" represents the resistance to fluid flow at various flow rates. The pressure drop in a system is roughly proportional to the square of the flow rate. That is, if the flow rate is doubled, the resistance to fluid flow is quadrupled. The horsepower of a pump or fan is given by the product of the flow rate and the pressure drop; therefore, if the flow rate is to double, the horsepower must be increased by a factor of 8.

The pump capacity curve shown in Fig. 2.11 indicates that the highest pressure available occurs at the lowest flow rate. As flow rate increases, the available pressure decreases. At the intersection of the system curve and the pump curve, the available flow and pressure drop match the system requirement at point P, called the operating point. Various pumps are capable of providing a given flow rate. The selection of the best pump is based on efficiency. It is desirable to have the operating point correspond to the maximum efficiency point. Therefore, of the many

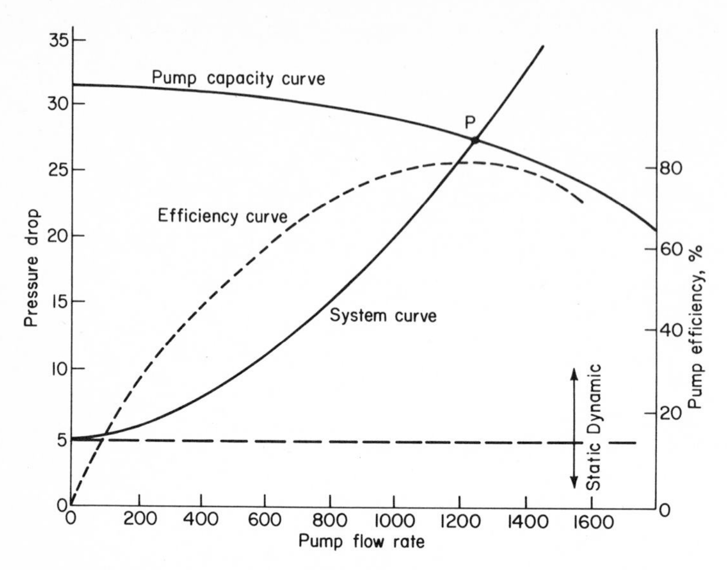

FIG. 2.11 Typical pump-capacity curve and system pressure drop characteristic. The operating point is *P*.

pumps rated at flow rate *P*, only one has the maximum efficiency. This is the pump to specify.

In some systems, one pump may operate under different load conditions. For example, in the collector loop of a liquid system, solar collector-to-storage and collector-to-heat-rejector modes are both used. Two operating points *P* may result. If so, the shape of the pump curve (determined by impeller design) becomes important. If one mode is of lower head loss than another, increased flow and pump power can result. The designer accounts for this condition by selecting a motor rated below full power (with a resulting poor power factor, however).

Fan Laws. If the efficiency of a pump or fan is roughly constant over a range of operating conditions, several relationships between flow rate, speed, pressure drop, and horsepower can be developed. These relationships are called the pump or fan laws and are described in standard textbooks on HVAC design. Briefly, the three fan laws are as follows. For constant efficiency,

1. Flow rate is proportional to revolutions per minute.
2. Pressure rise is proportional to speed squared.
3. Horsepower is proportional to the cube of the speed.

The pressure against which a pump operates consists of the static and frictional pressure drops shown in Fig. 2.11. For example, a pump may be required to lift collector fluid from a mechanical space in the basement of a building to the top of the solar-collector array. This is called the static load since it exists whether there is fluid flow or not. Static pressure is shown in Fig. 2.11 as the pressure level at no flow. Static pressure is to be added to the dynamic pressure arising from fluid friction, in order to determine the total pressure requirement.

Another static load exists when fluid must be pumped from one reservoir into another at higher pressure. Static heads are unimportant in air systems.

A fourth type of pressure drop is occasionally important in pump or fan design. This is called the velocity head and is given by the fluid kinetic energy. For example, when fluid is pumped from a pipe into a large storage tank where the fluid velocity is approximately zero, the kinetic energy is dissipated. In most solar designs, the velocity head is small and can be ignored.

The three pertinent types of pressure head described above are present in every solar fluid loop. Therefore, the total head against which the pump operates arises from the static and dynamic head across each component. In a liquid collector loop, the components include pipes, collector, pipe fittings, heat exchanger, air scoop or air separator, check valves, balancing valves, filters, and flow-control valves. The calculation of required pump capacity is easy to do in tabular form. Various nomographs and design guides are available for this purpose (see Ref. 6).

Materials. In addition to specifying the flow rate and pressure rise requirements of a pump, it is also necessary to specify other parameters. For example, in a solar domestic hot-water system, a pump may be exposed to city water pressure in excess of 125 lb/in^2; therefore the pump casing must be able to withstand this pressure. Temperatures in excess of 200°F may occur in solar heating systems in summer, and the seal material, impeller material, and motors in a pump or fan designed to operate with fluid at this temperature must be specified.

The material from which a pump impeller is made is also determined by the type of fluid pumped. If city or well water is to be pumped, a bronze or stainless steel impeller must be used to avoid corrosion. However, if an inert working fluid is used, a cast-iron impeller is satisfactory, since no chemical reaction should occur. Since corrosion is not a problem in air heating systems, steel fans are used.

Operating Costs. A proper economic analysis includes the cost of operating pumps and fans. The power consumed by the pump and the number of operating hours are multiplied to find the total power used per year. Then this figure is multiplied by the cost of electricity to find the annual operating cost. For example, a collector pump on a residential system may be rated at ¼ hp; 400 W will be required to operate the pump if its efficiency is 50 percent. In a liquid space-heating system, the collector pump operates about 2000 h per year, thereby consuming 800 kWh of electricity per year. If power costs 4 cents per kilowatt hour, the annual operating cost will be $32 per year. A ¼-hp collector pump would be required in a 1000 and 1500 ft^2 system. The annual pump operating cost per square foot is 2 to 3 cents.

Most of the inefficiencies in pumps result from frictional fluid losses and electrical losses. These losses produce heat that helps to offset heating loss from the building if the pump is located within the building. Further, the pump input causing fluid to flow is dissipated in fluid friction in the piping. Friction produces heat, and if the piping is within the building, this heat, too, is helpful.

Pipes, Ducts, and Fluids. The major design parameters which must be specified for fluid conduits are material and size. Ducts of fiber glass or galvanized steel are commonly used. Piping may be copper, black iron, galvanized iron, and infrequently, aluminum. Steel requires proper water treatment and dielectric protection from copper components. Materials are subject to corrosion, as described in Chap. 4.

Pipe and duct sizes are determined by an economic trade-off. For large ducts, the pressure drop is relatively smaller. As a result, the fan is smaller and the annual cost of operating the fan is relatively smaller. However, the cost of large ducts, and space consumed by them is relatively higher. At the other extreme, if small, relatively inexpensive ducts are used, higher pressure drops are encountered, and a relatively larger fan with associated larger annual operating cost is required. The economic optimum lies at some point between the two extremes, where the total cost of the duct, the fan, and its operation over the equipment life cycle is a minimum. The HVAC industry has developed duct and pipe sizing guidelines which give sizes close to the economic optimum.

Other parameters enter into the sizing of pipes and ducts. Very high velocity liquids or gases produce objectionably high levels of noise. Also, in liquid conduits, high velocity may cause erosion. A velocity less than 6 ft/s in copper pipes is recommended for this reason. Air velocities below 1000 ft/min are used. Another practical determinant of conduit size is flow balancing. If ducts or pipes are very large, the pressure drop in them is so small that they are difficult to balance. Other criteria for pipe and duct sizing are given in Chap. 4.

Expansion. Flow networks are subject to expansion and contraction as working-fluid temperature changes. For example, during the middle of winter the piping in a liquid collector loop at night may be below 0°F; in midsummer on a sunny day the pipe temperature will exceed 220°F. Resulting expansion and contraction must be accommodated by the design of fluid headers. If expansion and contraction is not provided for, large forces will be developed which can buckle and destroy pipes, ducts, structural members, and other components of solar systems.

Table 2.2 lists the expansion coefficients $\bar{\alpha}$ for a number of materials present in solar energy systems. The values given are averages over temperature ranges encountered in solar systems. The coefficient of linear expansion is defined as the fractional change in length $(\Delta l/l_o)$ per unit temperature change ΔT, or in equation form,

$$\bar{\alpha} = \frac{\Delta l}{l_o} \times \frac{1}{\Delta T} \tag{2.10}$$

where Δl is the length change, l_o is the original length, and ΔT is the temperature change.

The values in Table 2.2 indicate that common materials expand at different

TABLE 2.2 Coefficient of linear expansion $\bar{\alpha}$ for materials in solar systems

Material	$\bar{\alpha}$ (°C^{-1}) × 10^6	$\bar{\alpha}$ (°F^{-1}) × 10^6	*Elastic modulus, lb/in^2 × 10^{-6}*
Aluminum	23	13	10
Brass	19	11	15
Copper	17	9	19
Glass	9	5	10
Steel (carbon)	15	8	29
Concrete	14	8	

rates. Therefore, not only must expansion of solar system components be accounted for, but also *differential expansion* must be considered. For instance, if a glass collector cover is attached to an aluminum housing, the housing expanding at a rate roughly triple that of glass, could cause openings to occur between the cover and the aluminum housing. Or if an aluminum housing were attached to a steel structure, the expansion of aluminum, being twice that of steel, could cause deformation of the collector supports and mounting brackets. In practice, an aluminum collector would not be mounted to a steel frame without some sort of isolation.

Heat Loss. Since pipes and ducts carrying fluids above environmental temperatures present a significant surface area through which heat loss may occur, all piping and ducting in solar systems must be insulated. This is equally important whether the fluid conduits are outside or inside. Although heat losses from piping within a building help offset heat losses from the building, the heat transfer is uncontrolled and not related to the demand. It is, therefore, preferable to insulate all solar piping and ductwork so that solar heat is conveyed to storage. Then, when it is required, heat may be transferred from storage to the zones of the building where it is needed.

Insulation on solar piping exposed to the environment must be jacketed. The jacket protects the insulation from deterioration by rain and snow. It may be plastic (polyvinyl chloride, for example) or aluminum. Whatever the jacket material, it must be unaffected by ultraviolet radiation and must provide a durable seal against weather (including hail) for many years. The components of a solar system which require insulation include piping to and from collectors, collector headers, piping within walls and unheated garages, piping in mechanical spaces, plus other items such as heat exchangers. In addition, heat-delivery piping and ducting to load should be insulated, if located in unheated spaces beneath or within the building. Since solar heat is expensive, it is important that it be used where required and nowhere else.

Hoses. The majority of liquid-based solar systems use rigid piping. However, hoses are convenient in some cases for connecting one collector to another relatively quickly. These hoses are exposed to a very severe environment including ultraviolet radiation, very high temperature, very low temperature, expansion, contraction, and low surface tension fluids which may slowly dissolve the hose material. Although hoses appear to save labor in the original installation, it is doubtful if this initial saving is worth the risk of future material degradation and leaks. According to *The Wall Street Journal* 800 hoses required replacement on the Cary Arboretum system in the first year.

Fluids. The most common fluid used in liquid-based collectors is water. If antifreeze protection is required, water is usually mixed with a glycol antifreeze. Nearly any location with a significant space-heating load will require antifreeze protection, since the existence of a heating load implies some subfreezing weather. Other types of nonfreezing liquids may be used, including organic heat-transfer oils and silicone fluids. The positive and negative attributes of each of these fluids are discussed in detail in Chap. 4 along with economics of fluid selection.

The important properties of a liquid to be used in a solar system include high specific heat, low viscosity (particularly at low temperature), low vapor pressure/high boiling point, relatively high surface tension (to avoid excessive leaks), high density, and high thermal conductivity. If an aqueous fluid is used, the pH should be slightly basic to reduce corrosion of dissimilar metals within the fluid stream.

Water has the best combination of heat-transport properties for any liquid-based solar application. The only reason for not using water is its high freezing point and corrosive nature. The transport properties of antifreezes should approach those of water as closely as possible. For example, ethylene glycol solutions, although excellent from the freeze-protection viewpoint, are not optimal from the heat-transfer viewpoint, since the specific heat and thermal conductivity are lower than those of water and viscosity is higher. The surface tension is less than that of water, indicating that glycol solutions will tend to leak through small pores in piping systems more quickly than water would.

Organic heat-transfer oils and silicone fluids have relatively high viscosities, specific heats about half that of water, and lower densities. The effect of reduction in specific heat and density is that a greater fluid flow rate must be used to transfer the same amount of heat. Properties of heat-transfer fluids are given in Table A.11 in the appendix.

Liquid loops, which are not subject to freezing, will normally use water. These fluid loops include the heat-storage loop and the heat-delivery loop as well as the backup loop. In these water circuits deionized water is used with a corrosion inhibitor. If the local water is free from "hardness," it may be used in place of deionized water. Boiler water is treated in the usual way.

Air used in active and passive systems is less troublesome since it does not freeze or boil and does not corrode. It is the ideal fluid from a reliability standpoint. However, air does have very low density, relatively low specific heat, and low thermal conductivity. As a result, its heat-transfer capability is below that of water. By proper design this difficulty can be overcome, and air heating systems can operate as effectively as liquid systems.

○ **Heat-Loss Equations for Circular and Rectangular Ducts and Pipes.** The heat-loss rate from rectangular and cylindrical ducts and pipes is controlled by the thermal conductivity of the insulation material. A small thermal resistance does occur at the boundary layer between the flowing fluid and the pipe wall as well as at the boundary layer between the outer surface of the insulation and environment. Any heat-transfer textbook (1) gives equations for calculating these small resistances. The heat flow through the principal resistance—the insulation—is given by Eqs. (2.11) and (2.12).

$$q_{\text{rect}} = kL\left(\frac{2w + 2h}{t} + 2.16\right)\Delta T \qquad (2.11)$$

$$q_{\text{cyl}} = \frac{2\pi kL}{\ln r_o/r_i}\Delta T \qquad (2.12)$$

Equation (2.11) applies to a rectangular cross section of width w and height h with insulation of thickness t. In Eq. (2.11), k is the thermal conductivity and L is the length of the duct. The temperature difference ΔT is that between the inner and outer surfaces of the insulation.

The heat loss through cylindrical insulation is given by Eq. (2.12). The insulation is of length L, of outer radius r_o and of inner radius r_i. Values of thermal conductivity for piping and building insulation are given in appendix Table A.5. The most common insulations for ducts are bat, rigid fiber glass, and foam. The most common size used for solar piping is formed insulation between 1½ and ½ in thick. The optimum thickness of insulation must be determined by an economic analysis of the type described in Chap. 4. A typical rule is R-4 for 1-in pipe or smaller and R-6 for larger than 1 in. (Refer to ASHRAE Standard 90 for tables of sizes.)

Heat Exchangers. Heat exchangers are devices which ideally execute a no-loss heat transfer between two fluid streams which must not be mixed. Heat exchangers are standard HVAC and industrial items and are not of special design for solar energy systems. In solar systems heat exchangers are required, for example, to isolate glycol solutions used in collectors from water used in storage. A heat exchanger is also required in air-based systems to heat domestic water.

Heat exchangers require a temperature difference between the two streams in order that heat transfer may take place. The temperature difference will cause the outlet stream of the heat exchanger to be at a lower temperature than the inlet stream. This temperature decrement is the chief operating penalty incurred by a heat exchanger. If a heat exchanger is properly insulated, almost no heat loss will occur.

The example below (Fig. 2.12) indicates the effect of the temperature penalty across a heat exchanger on the performance of a solar collector used to heat an outlet stream to 145°F. As the solar collector is forced to operate at progressively higher temperatures associated with smaller heat exchangers, its efficiency is reduced. Therefore, the effect of the heat exchanger is to decrease the amount of solar energy which can be collected.

The larger the heat exchanger, the smaller the temperature difference needed to cause the specified heat transfer. One measure of the required temperature difference is called the approach temperature difference, labeled in Fig. 2.12. The approach is the temperature difference between the return from load and the supply to the collector. If no heat exchanger were present, the situation would be the same as that for zero approach. If a 20° rise were required in the load stream, that 20°F rise would be supplied across the collector. The average fluid temperature in

Case	Approach ΔT (°F)
1	0
2	5
3	10
4	20
5	40

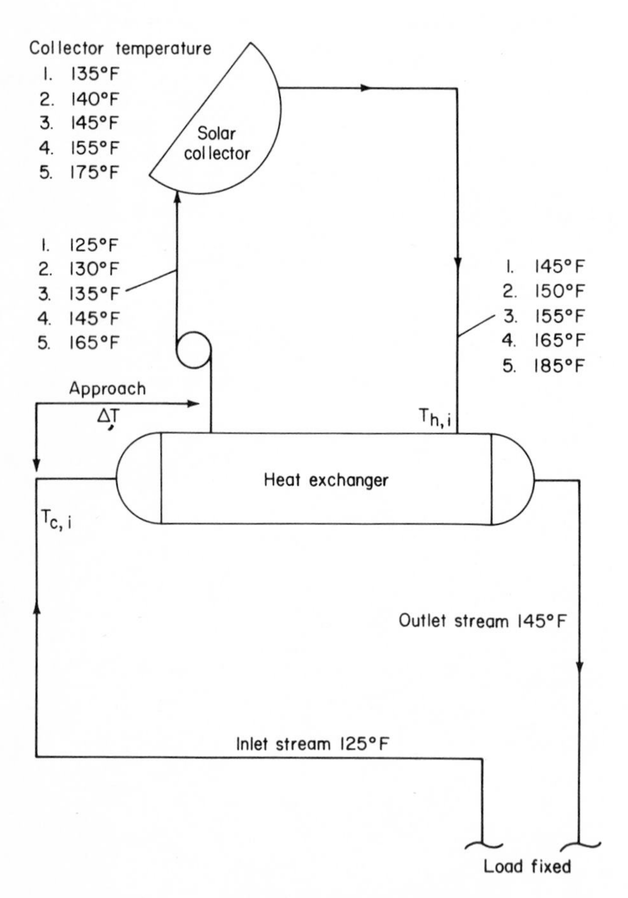

FIG. 2.12 Flow arrangement and operating temperatures for heat-exchanger example. Case 1 is the no-heat-exchanger case.

the collector would then be 135°F and the efficiency of the collector would be evaluated at this temperature from Eq. (2.4).

If the heat exchanger is relatively large, the approach can be small. In case 2 in the example, the approach temperature difference is only 5°F, the fluid inlet to the collector being 130°F. The 20°F rise across the collector results in a 150°F outlet temperature and an average collector temperature of 140°F. The efficiency penalty from a 5°F increase in collector temperature is relatively small. The designer may question whether a very large heat exchanger capable of providing a 5°F approach is economical, since the solar collection penalty is not sizable. It may be better to use a somewhat smaller and less costly heat exchanger, requiring a larger approach. In case 4, for example, a 20°F approach is used. The average solar collector temperature is 155°F and the efficiency penalty is large.

The question of precisely what heat-exchanger size should be used is again a question of economics. The sum of the cost of the heat exchanger and the value of the foregone solar heat is to be minimized. A very large heat exchanger is expensive, but the value of noncollectable solar heat will be small. A small heat exchanger is less expensive but the amount of useful solar heat foregone will be large. The method of selecting the proper solar collector heat exchanger will be described in detail in Chap. 4.

Effectiveness. In most cases, the size of solar heat exchangers, particularly between collector and storage, will be larger than might at first be expected. This is a result of the relatively high economic value of solar heat when compared with the cost of a heat exchanger. A heat exchanger with effectiveness above 80 percent will often be specified. In solar system design, the measure of heat-exchanger size is called the heat-exchanger effectiveness, defined as the amount of heat transferred divided by the amount which could be transferred if the heat exchanger were infinitely large. In equation form, the effectiveness can be written as

$$E_{hx} = \frac{Q_{hx}}{(\dot{m}c)_{\min}(T_{h,i} - T_{c,i})} \tag{2.13}$$

where Q_{hx} is the amount of heat transferred expressed in British thermal units per hour or watts, $(\dot{m}c)_{\min}$ is the minimum capacitance rate in the heat exchanger which controls the maximum amount of heat which may be transferred, $T_{h,i}$ is the hot-stream inlet temperature, and $T_{c,i}$ is the cold-stream inlet temperature. The difference between these two temperatures is the maximum temperature difference across the heat exchanger. The calculation of heat-exchanger effectiveness is covered in the next section.

Heat-exchanger manufacturers will not specify heat exchangers by heat-exchanger effectiveness but rather by the heat rate, the flow rates in the two fluid streams, and their inlet temperatures. The following example indicates how the heat-exchanger effectiveness can be calculated from the type of information provided by a heat-exchanger manufacturer.

Example. Calculate the effectiveness of a heat exchanger which has a heat rate Q_{hx} of 125,000 Btu/h. The shell-side flow rate is 25 gal/min of water and the tube-side water flow is 40 gal/min. The shell-side fluid inlet is at 125°F and the tube-side inlet is at 108°F.

Solution. To find the effectiveness, the minimum capacitance rate ($\dot{m}c$) must be found. For this example, it is on the shell side and equal to

$$(\dot{m}c)_{min} = (\dot{m}c)_{shell} = 25 \times 500 \text{ Btu/[h}\cdot{}^\circ\text{F (gal/min)]} = 12{,}500 \text{ Btu/h}\cdot{}^\circ\text{F}$$

From Eq. (2.13), the effectiveness is

$$E_{hx} = \frac{125{,}000}{12{,}500(120 - 108)} = 0.83$$

The example shows how to calculate heat-exchanger effectiveness from a manufacturer's information. Of course, the reverse operation would be used to specify the heat exchanger for a solar energy system. Based on economic criteria, heat-exchanger effectiveness is specified. Then a heat rate would be calculated on the basis of a typical set of heat-exchanger inlet temperatures. In contrast to industrial heat exchangers, solar heat exchangers operate over a range of temperatures. However, the effectiveness of heat exchangers is relatively insensitive to operating temperature. Therefore, specifying a heat rate and terminal temperatures at a *typical* operating condition is sufficient information for selection of a heat exchanger.

○ **Heat-Exchanger Penalty Factor and Effectiveness Calculation.** The effectiveness of all common heat exchangers can be calculated from fundamental heat-transfer principles using data contained in Ref. 7. Heat-exchanger effectiveness depends upon the geometry of the heat exchanger and a dimensionless measure of the heat-transfer coefficient called the number of transfer units (NTU) defined below:

$$\text{NTU} = (UA)_{hx}/(\dot{m}c)_{min} \tag{2.14}$$

The NTU is the ratio of the heat-transfer coefficient U multiplied by the effective heat-transfer area A divided by the minimum capacitance rate. Figure 2.13a shows the effectiveness for several different types of heat exchangers for the special situation in which the capacitance rates on both sides of the heat exchanger are equal. It is seen that the counterflow heat exchanger has the highest effectiveness at a given NTU value. In most liquid-based solar energy systems, a counterflow heat exchanger is specified for this reason, since high effectiveness levels are required. For the special case in which the capacitance rates on both sides of a counterflow heat exchanger are equal, Eq. (2.15) may be used to calculate the effectiveness:

$$E_{hx} = \frac{\text{NTU}}{\text{NTU} + 1} \tag{2.15}$$

If the capacitance rates are not equal, the values of effectiveness can be read from Table 2.3. In Table 2.3, the largest capacitance rate in the heat exchanger is denoted as $(\dot{m}c)_{max}$. The calculation of heat-exchanger U values can be done either by methods given in Ref. 1 or more easily by using the heat-exchanger manufacturer's computer programs.

The previous example, which related collector operating temperature to heat-exchanger size, can be generalized using an analytical result from deWinter (8), who showed that the heat-exchanger penalty factor F_{hx} can be related to heat-

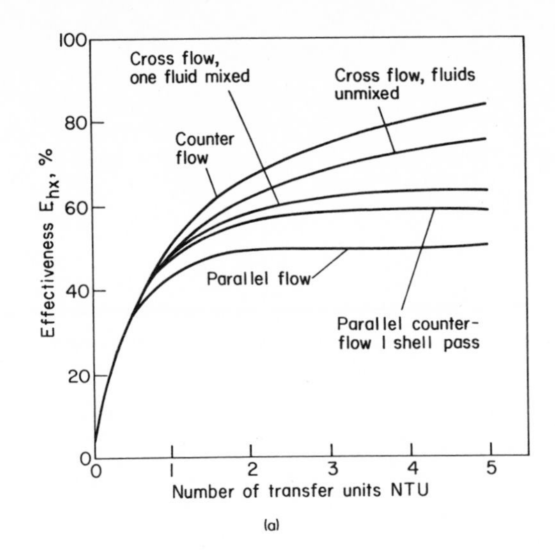

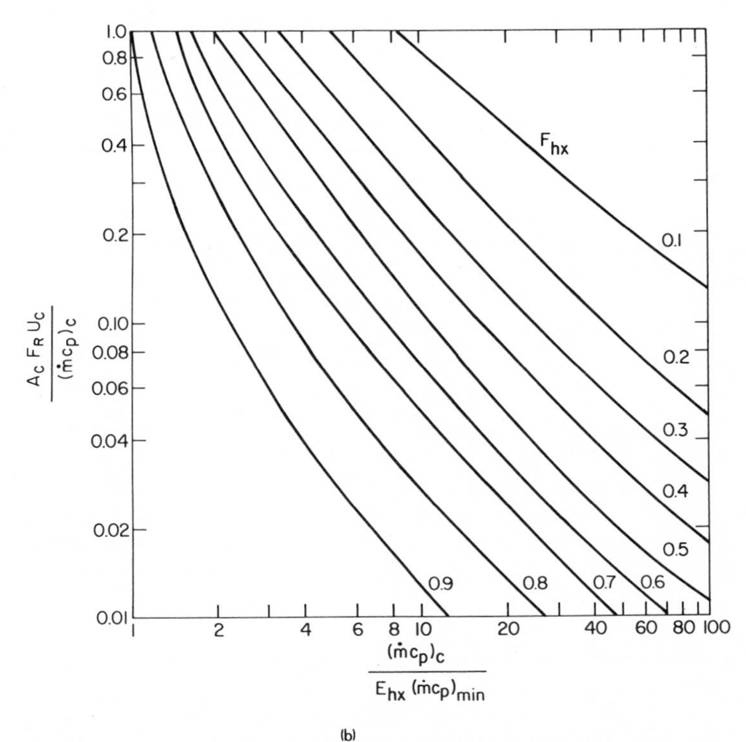

FIG. 2.13 (*a*) Heat-exchanger effectiveness for equal shell and tube capacitance rates as a function of number of transfer units NTU for five generic types of heat exchanger. *(From Compact Heat Exchangers by W. M. Kays and A. L. London. Copyright 1964 McGraw-Hill Book Co. Used with the permission of McGraw-Hill Book Co.)* (*b*) Heat-exchanger penalty factor F_{hx}. When no exchanger is present $F_{hx} = 1$.

exchanger and solar collector properties. The effect of a heat exchanger is more pronounced for solar collectors that have a large heat-loss coefficient U_c and for those heat exchangers which have a small effectiveness. Equation (2.16) is the heat-exchanger penalty factor expression to be used for systems with flat-plate collectors.

$$F_{hx} = \frac{1}{1 + [F_R U_c A_c/(\dot{m}c_p)_c][(\dot{m}c_p)_c/(\dot{m}c_p)_{min} E_{hx} - 1]} \tag{2.16}$$

Parameters in Eq. (2.16) have been defined earlier in the chapter. Figure 2.13*b* is a plot of Eq. (2.16) and shows the heat-exchanger penalty factor as a function of the two dimensionless groups in the equation. In most liquid-based systems, the penalty factor is above 0.95, indicating a 5 percent loss because of the use of the heat exchanger.

TABLE 2.3*a* Counterflow heat-exchanger performance*

	E_{hx} for indicated capacity-rate ratios $(\dot{m}c)_{min}/(\dot{m}c)_{max}$							
NTU	*0*	*0.25*	*0.50*	*0.70*	*0.75*	*0.80*	*0.90*	*1.00*
0	0	0	0	0	0	0	0	0
0.25	0.221	0.216	0.210	0.206	0.205	0.204	0.202	0.200
0.50	0.393	0.378	0.362	0.350	0.348	0.345	0.339	0.333
0.75	0.528	0.502	0.477	0.457	0.452	0.447	0.438	0.429
1.00	0.632	0.598	0.565	0.538	0.532	0.525	0.513	0.500
1.25	0.713	0.675	0.635	0.603	0.595	0.587	0.571	0.556
1.50	0.777	0.735	0.691	0.655	0.645	0.636	0.618	0.600
1.75	0.826	0.784	0.737	0.697	0.687	0.677	0.657	0.636
2.00	0.865	0.823	0.775	0.733	0.722	0.711	0.689	0.667
2.50	0.918	0.880	0.833	0.788	0.777	0.764	0.740	0.714
3.00	0.950	0.919	0.875	0.829	0.817	0.804	0.778	0.750
3.50	0.970	0.945	0.905	0.861	0.848	0.835	0.807	0.778
4.00	0.982	0.962	0.928	0.886	0.873	0.860	0.831	0.800
4.50	0.989	0.974	0.944	0.905	0.893	0.880	0.850	0.818
5.00	0.993	0.982	0.957	0.921	0.090	0.896	0.866	0.833
5.50	0.996	0.998	0.968	0.933	0.922	0.909	0.880	0.846
6.00			0.975	0.944		0.921	0.892	0.857
6.50			0.980	0.953		0.930	0.902	0.867
7.00			0.985	0.960		0.939	0.910	0.875
7.50			0.988	0.966		0.946	0.918	0.882
8.00			0.991	0.971		0.952	0.925	0.889
8.50			0.993	0.975		0.957	0.931	0.895
9.00			0.994	0.979		0.962	0.936	0.900
9.50			0.996	0.982		0.966	0.941	0.905
10.00			0.997	0.985		0.970	0.945	0.909
∞	1.000	1.000	1.000	1.000	1.000	1.000	1.000	1.000

*From *Compact Heat Exchangers* by W. M. Kays and A. L. London. Copyright 1964 by McGraw-Hill Book Co. Used with permission of McGraw-Hill Book Co.

TABLE 2.3*b* Cross-flow (one fluid "mixed," the other fluid "unmixed") exchanger effectiveness as a function of capacity. rate ratio ($c_{mixed}/c_{unmixed}$) and nummber of heat-transfer units NTU*

	E_{hx} for indicated capacity-rate ratios, $c_{mixed}/c_{unmixed}$							
NTU	*0*	*0.25*	*4.00*	*0.50*	*2.00*	*0.75*	*1.333*	*1.000*
0	0	0	0	0	0	0	0	0
0.25	0.221	0.215	0.213	0.209	0.209	0.204	0.204	0.198
0.50	0.393	0.375	0.375	0.358	0.357	0.341	0.341	0.325
0.75	0.528	0.495	0.494	0.465	0.463	0.463	0.435	0.410
1.00	0.632	0.587	0.585	0.545	0.542	0.505	0.503	0.469
1.25	0.713	0.658	0.654	0.605	0.600	0.556	0.552	0.510
1.50	0.777	0.714	0.706	0.652	0.644	0.594	0.589	0.540
1.75	0.826	0.758	0.747	0.689	0.677	0.623	0.616	0.562
2.00	0.865	0.793	0.778	0.715	0.702	0.645	0.636	0.579
2.50	0.918	0.844	0.820	0.760	0.736	0.677	0.663	0.601
3.00	0.950	0.879	0.846	0.789	0.756	0.697	0.679	0.613
3.50	0.970	0.903	0.861	0.808	0.768	0.710	0.689	0.621
4.00	0.982	0.920	0.870	0.823	0.776	0.718	0.695	0.625
4.50	0.989	0.933	0.876	0.834	0.780	0.724	0.698	0.628
5.00	0.993	0.942	0.880	0.841	0.783	0.728	0.700	0.630
∞	1.000	0.982	0.885	0.865	0.787	0.736	0.703	0.632

*From *Compact Heat Exchangers* by W. M. Kays and A. L. London. Copyright 1964 by McGraw-Hill Book Co. Used with permission of McGraw-Hill Book Co.

The heat-exchanger penalty factor is defined as the actual energy collected divided by the heat which could be collected if the fluid inlet temperature of the solar collector were the same as the heat-exchanger fluid inlet temperature from storage or load. Equation (2.17) shows how the heat-exchanger penalty factor is used in the linear collector model developed earlier in the chapter.

$$\eta_c = F_{hx}F_R\left[\tau\alpha - U_c\left(\frac{T_{hx,i} - T_a}{I_c}\right)\right] \tag{2.17}$$

All parameters in Eq. (2.17) have been defined previously, with the exception of $T_{hx,i}$, which is the heat-exchanger inlet temperature. In the example above, $T_{hx,i}$ is 125°F.

Overview of Controls for Solar Heating Systems. Control systems consist of three subsystems: sensors, controllers, and actuators. Sensors are used to determine the state of control-point variables such as collector temperature, storage temperature, sunlight level, building room temperature, time, and other inputs required to cause the solar system to operate in the prescribed manner. The controller subsystem determines what operations are required on the basis of information from the sensors. Actuators such as mode selector valves or pumps carry out decisions made by the controller. Controls are used to operate solar-collector pumps, storage pumps, backup systems, mode selector valves, energy-delivery pumps and fans, dampers, heat rejecters, freeze-protection units, and other components.

Most controllers are solid-state devices using operational amplifier chips and thermistor or resistance-temperature-detector (RTD) sensors to measure temperature. In addition, some solar control systems use solar intensity level measured by a solar cell or a pyranometer.

Both temperatures and temperature differences are often required. For example, the temperature difference between collector and storage, when above a certain level, will cause the collector and storage pumps or fans to operate. If the collector temperature itself rises above a maximum design point, irrespective of storage temperature, a heat-rejection device may be actuated in liquid systems. Various functions of this type are programmed into most commercially available controls. The designer (not the control manufacturer) must specify the desired modes of operation, the criteria for selecting a given mode, and other characteristics of the control system. In Chap. 4, controls for specific types of solar applications will be covered.

A simple solar control system is shown in Fig. 2.14. A Wheatstone bridge is used to determine the difference between two temperatures $T1$ and $T2$ measured by thermistors. Temperature $T1$ could be the collector-panel temperature and $T2$ the storage temperature. When there is sufficient differential between $T1$ and $T2$, the associated resistance difference causes Wheatstone bridge imbalance sufficient to cause the relay to engage. Contacts in the relay operate the collector and storage pumps or other devices. In this example system, $T1$ and $T2$ are the sensors; the Wheatstone bridge and operational amplifier are the controller; the actuator is the relay plus whatever electromechanical components are controlled by the relay.

The required difference between $T1$ and $T2$ for engagement of the relay is determined by the variable resistance $R1$ and the gain resistor $R2$. $R1$ will be selected based on physical characteristics of the system such as pump flow rates,

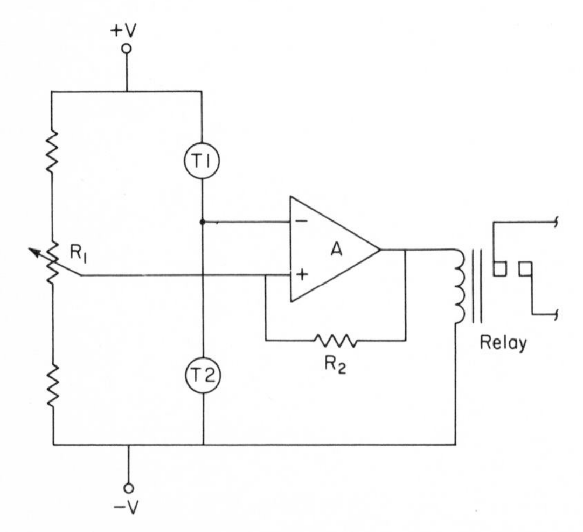

FIG. 2.14 Schematic diagram of simple differential-temperature controller for liquid- or air-based space-heating systems.

collector area, and so forth. $R2$ is selected to give the proper gain to pull in the relay at the desired temperature difference.

During system operation $T1$ and $T2$ may approach each other in value. This may occur, for example, if solar intensity is reduced by a cloud passing before the sun. In this case, the controller will determine that it is no longer worthwhile to operate and the relay will open. In the simple controller shown in Fig. 2.14, the relay is engaged for all temperature differences greater than a fixed turnon temperature difference and the relay is disengaged for all temperature differences below it. In actual practice, it will be seen that different turnon and turnoff temperature differences are required. This hysteresis effect is programmed into the controller and will be seen to be essential to proper system function.

The design of control systems for solar heating systems is one of the most critical parts of the design process. Many solar systems built in the past have operated far below capacity as a result of improper function or inappropriate application of the control system. Therefore, a large section in Chap. 4 is devoted to the proper design of controls for both active and passive systems. In Chap. 5, installation of control systems is discussed.

Backup Heating Systems for Solar Heating Systems. In subsequent chapters, it will be shown that a 100 percent capacity solar heating system is almost never cost-effective. This is because a 100-percent system must be capable of carrying the maximum expected heat load ever to occur during the lifetime of a system. By definition this maximum heat load will occur only once and during the balance of the system life the system will be too large. If a solar heating system is too large, it collects more energy than can be used and this excess energy must be discarded. Since solar heat is expensive, cost-effective solar systems cannot exist when substantial amounts of heat are dissipated.

In most parts of the United States, it will be found that a solar system delivering between 60 and 80 percent of the annual heating requirement will provide economically optimum performance. The remaining 20 to 40 percent of the annual heating load will be supplied by a nonsolar auxiliary system. This system does not differ in any substantial manner from that which would be used in the same building if no solar heating were provided.

The backup system may range from a gas furnace to an electric boiler to a heat pump to a wood-burning stove. The design of the backup unit is usually taken to be unaffected by the existence of a solar heating system in the same building. It may be possible in some circumstances to slightly undersize the nonsolar heating system assuming that interior temperatures below the design temperature can be tolerated several hours per year. The concept of not meeting the load for a few hours per year is not widely used, but in some cases a cost-benefit study will show this strategy to be viable. Some passive systems require smaller backup sizes than indicated by a heat-load calculation for a nonsolar design. This is an important result since solar systems cannot make up backup system stack losses which increase with backup unit size. The only known way to identify this effect is by detailed building computer modeling, however.

Conventional heat-loss calculations are summarized in Ref. 6. The selection of the type of backup should be based on expected availability and cost of fuel over the lifetime of the building. For example, in a particular location natural gas may be considered as the best backup energy source because of its relatively low current cost. However, natural gas is expected to increase in price in certain parts of the country and, later, to become available only intermittently. If this situation were to apply, it might be more economical in the long run to use off-peak electricity with a storage tank as the backup heating system. Off-peak electricity is currently available in some parts of the country at a price below that of gas, and no cutoffs of this energy resource are expected. This is a region-specific question which must be determined on a case-by-case basis. The criteria are long-term availability and low cost.

SOLAR DHW SYSTEMS

Solar DHW heating is the most common form of active solar heating in the world. Systems for DHW heating are relatively simple and inexpensive. Water heating is required for the entire year in most applications; therefore, the annual load factor on the solar system is high. This is distinct from space-heating systems which are used in only 6 to 8 mo of the year although they must be paid for during the entire 12 mo. In this section, the method of calculating solar water demands is outlined and various types of systems are described. Details of one system are given as an example of the functional modes of all systems. Specific design details and sizing rules are given in Chaps. 4 and 5.

DHW Heating Loads

Calculation of energy demands can be carried out very accurately if the amount of hot water required is known. The energy required to heat a given volume of water is the product of the water volume, the density, the specific heat, and the required temperature rise. In equation form, the amount of heat, Q_{hw}, is given by

$$Q_{hw} = V(\rho c)(T_{set} - T_{source}) \tag{2.18}$$

where V is the volume of water required per day, ρ is the density, c is the specific heat equal to 1.0 Btu/(lb · °F), T_{set} is the thermostat set point and delivery temperature, and T_{source} is the inlet water temperature. The source temperature will be equal to the city-water or well-water temperature (see appendix Table A.10).

A second thermal demand exists. It is the heat required to maintain the water tank and recirculation piping, if used, at the set-point temperature. This load may represent 15 to 35 percent of the annual heat used. The standby loss is given by the product of the thermal conductance of the insulation U_{hw} multiplied by its surface area A_{hw} multiplied by the difference of temperature between the water and the surroundings. In equation form, the amount of standby heat is equal to

$$Q_{standby} = U_{hw}A_{hw}(T_{set} - T_a) \tag{2.19}$$

where T_a is the air temperature in the vicinity of the water heater and delivery piping. The water-heating demand plus the standby demand constitute the total demand.

It is always cost-effective to reduce standby losses from a water heater by decreasing the value of U_{hw} in Eq. (2.19). This may be done by adding insulation to an existing water-heating system or by purchasing a water heater which has an increased insulation level. If an insulation blanket is added to an existing water-heating system, it is important that this insulation not block air entry to the burner or insulate the control module to such an extent that it overheats and malfunctions. Improper insulation of water-heating systems can cost more than adding no insulation at all.

Table 2.4 gives the average gallon-per-day requirement for service hot-water installations in various types of buildings including apartment houses, schools,

TABLE 2.4 Hot-water demands in use for various types of building*

Type of building	*Maximum hour*	*Maximum day*	*Average day*
Men's dormitories	3.8 gal/ student	22.0 gal/ student	13.1 gal/ student
Women's dormitories	5.0 gal/ student	26.5 gal/ student	12.3 gal/ student
Motels: number of units†			
20 or less	6.0 gal/unit	35.0 gal/unit	20.0 gal/unit
60	5.0 gal/unit	25.0 gal/unit	14.0 gal/unit
100 or more	4.0 gal/unit	15.0 gal/unit	10.0 gal/unit
Nursing homes	4.5 gal/bed	30.0 gal/bed	18.4 gal/bed
Office buildings	0.4 gal/ person	2.0 gal/ person	1.0 gal/ person
Food service establishments:			
Type A—full meal restaurants and cafeterias	1.5 gal/max. meals/h	11.0 gal/max. meals/h	2.4 gal/avg. meals/day‡
Type B—drive-in, grilles, luncheonettes, sandwich and snack shops	0.7 gal/max. meals/h	6.0 gal/max. meals/h	0.7 gal/avg. meals/day‡
Apartment houses: number of apartments			
20 or less	12.0 gal/apt.	80.0 gal/apt.	42.0 gal/apt.
50	10.0 gal/apt.	73.0 gal/apt.	40.0 gal/apt.
75	8.5 gal/apt.	66.0 gal/apt.	38.0 gal/apt.
100	7.0 gal/apt.	60.0 gal/apt.	37.0 gal/apt.
200 or more	5.0 gal/apt.	50.0 gal/apt.	35.0 gal/apt.
Elementary schools	0.6 gal/ student	1.5 gal/ student	0.6 gal/ student‡
Junior and senior high schools	1.0 gal/ student	3.6 gal/ student	1.8 gal/ student‡

*Reprinted with permission from the 1976 Systems Volume, *ASHRAE Handbook & Product Directory*.

† Interpolate for intermediate values.

‡ Per day of operation.

office buildings, and motels. The numbers are approximate averages for many facilities. If a solar water-heating retrofit is anticipated, the hot-water consumption should be measured over a period of typical usage. This value is used to size the solar system. The set-point temperature for water-heating systems is specified in some locations by local energy codes and is not at the option of the building owner. Reduction of T_{set} will reduce standby losses but will not reduce water demands, since the volume of water V used per day will increase as a result of the lowered delivery temperature. If only a few high-temperature use points exist, it is better to boost temperature at these points than to mix hot and cold water for all other lower temperature uses. A clock control for the recirculation loop pump can be used to reduce losses from this pipe loop. The return line should be insulated. The values of water usage presented in Table 2.4 are based on a delivery temperature of 140°F, which is above that permitted by some energy-conservation codes in the United States.

Types of Solar Water-Heating Systems

Solar DHW heating systems are classified in two generic types, direct and indirect. Direct systems are those in which the potable water is heated directly in the solar collector. Indirect systems use a separate heat-transfer fluid in the solar-collector loop. Direct systems theoretically have a higher thermal performance than the comparable indirect system, although the annual difference may be only a few percent. These direct systems require a solar collector capable of withstanding city-water pressure. If the water contains hardness, the collectors may scale and require occasional cleaning. Local swimming-pool manufacturers or city-water authorities can advise on local water hardness.

Indirect systems require an increased initial investment because of the heat exchanger, extra pump, and special collector fluid. However, indirect systems have increased reliability in freezing climates since the working fluid has a freezing point below the lowest ambient temperature expected. The heat-exchanger penalty factor F_{hx} [Eq. (2.16)] can be used to accurately determine the heat-exchanger thermal performance penalty. The remainder of this section will explain the function of six reliable types of solar water-heating systems.

Single Tank, Direct. The simplest type of direct water-heating system is shown in Fig. 2.15a. Here, potable water is circulated directly through the collectors and stored in a single tank. This tank also contains the backup heater required to maintain the outlet water (indicated by the symbol HW) at the set-point temperature. The set point may frequently be higher than the temperature of fluid delivered by the solar collector; therefore, it is important that the backup heating system not heat water which is to be introduced into the solar-collector inlet header. The simplest way of accomplishing this in a single-tank system is to place the backup heating element above the return pipe in the water-heating tank. This location of the

FIG. 2.15 Active water-heating systems: (*a*) single-tank direct, (*b*) double-tank direct, (*c*) double-tank drain-down, (*d*) single-tank indirect, (*e*) double-tank indirect, (*f*) heat-pipe backup.

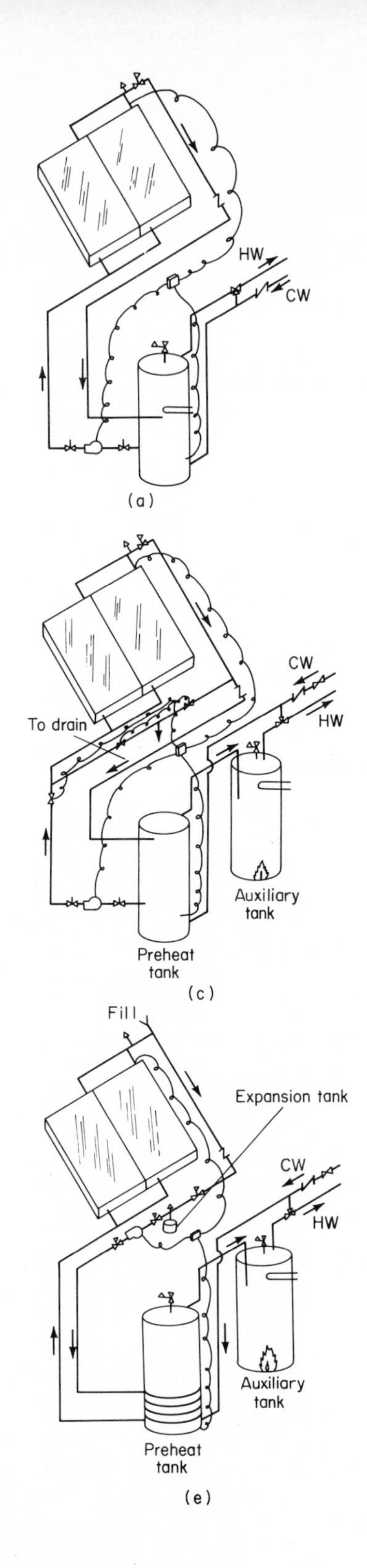
HW
CW
(a)
CW
HW
To drain
Auxiliary
tank
Preheat
tank
(c)
Fill
Expansion tank
CW
HW
Auxiliary
tank
Preheat
tank
(e)

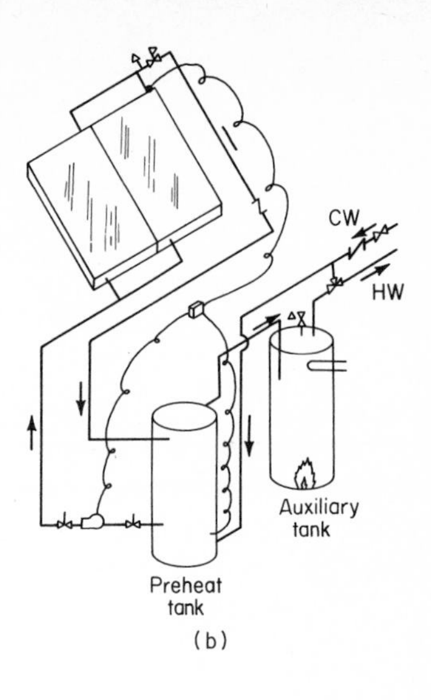
CW
HW
Auxiliary
tank
Preheat
tank
(b)

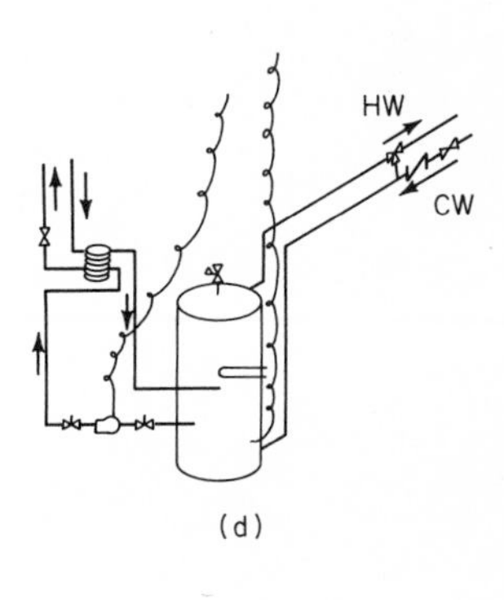
HW
CW
(d)

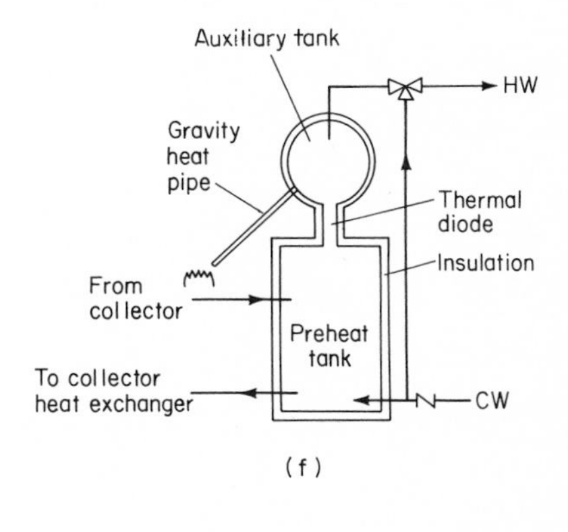
Auxiliary tank
HW
Gravity
heat
pipe
Thermal
diode
Insulation
From
collector
Preheat
tank
To collector
heat exchanger
CW
(f)

backup heating element restricts the backup energy source to electricity in most commercial systems. One system will be described later which can use natural gas or fuel oil as the backup and still maintain the required stratification in the storage tank.

The principal difficulty with the single-tank direct system in freezing climates is the difficulty in preventing collector freezing. In relatively mild climates, the collector pump can be operated at night circulating warm water through the collectors to prevent freezing. However, cold water introduced into the top of the tank may turn on the auxiliary heater and consume additional energy. A second method of freeze protection, to be described later, involves draining the collectors at night. When properly designed and built, drain-down systems will protect against freezing; but in many hundreds of systems installed in the 1970s the drain-down feature did not operate properly and many collectors were destroyed. The positive and negative attributes of drain-down systems are described in detail later (page 69).

The use of potable water in collectors restricts the collector material to copper. If galvanized piping or aluminum collectors are used, severe corrosion is inevitable. In addition, the pump impeller must be bronze or stainless steel if untreated tap water is used in the solar collector.

Double Tank, Direct. A direct solar water-heating system using two tanks is shown in Figure 2.15*b*. The advantage of the two-tank system is that any type of auxiliary heat may be used—either electricity or natural gas as shown. As a result, a lower backup fuel bill may result for those locations where gas is less expensive than electricity. In addition, extra storage capacity is provided by the preheat tank and a higher recovery rate in the auxiliary tank is possible. The auxiliary tank is simply a standard water heater, the only difference being that the fluid inlet to that tank is preheated solar water instead of cold city water.

Figure 2.15*c* is a two-tank system with drain-down. The increase in controller complexity is apparent. Freeze-up is possible if power fails. Both two-tank systems have higher standby losses than single-tank systems because of the increase in tank surface area.

Single Tank, Indirect. Figure 2.15*d* shows a single-tank indirect system using a heat exchanger between collector and storage loops. As in all single-tank systems, the backup heating element must be placed in the top of the tank.

The design of heat exchangers in indirect systems is the key to high thermal efficiency. Jacketed-tank heat exchangers or internal-coil heat exchangers can be used instead of the external exchanger in Fig. 2.15*d* but are less effective. However, the separate heat exchanger requires an additional pump. A slight increase in operating cost over the year will result. The difficulty with jacketed tanks is that heat removal from the fluid within the coil or jacket must rely on the relatively inefficient mechanism of free convection inside the tank. In the external heat-exchanger design, forced convection is used on both sides of the exchanger; therefore, a physically smaller unit may be used. Comparable effectiveness is possible by increasing the area of coil or jacket heat exchangers. Most commercial units use coil sizes and lengths which result in heat-exchanger effectiveness below 0.4. As a result, the heat-exchanger penalty factor is greater than for the external heat-exchanger approach.

Air may be used as the collector working fluid in indirect systems. Air is attractive because it presents no freezing or corrosion problems. However, air is a relatively ineffective heat-transfer medium with heat-transfer coefficients 20 to 40 times less than those for liquids under the same operating conditions. Therefore, the performance of air systems relative to liquid systems is less. This expected performance loss was verified by National Bureau of Standards tests in 1979. The only method of improving the performance of an air-based water-heating system is to increase the collector area relative to that for a liquid system serving the same load.

Double Tank, Indirect. Figure 2.15*e* shows the two-tank design for indirect solar water heating. This system uses a flow configuration similar to that for the direct two-tank system except that a heat exchanger is present. A jacketed-tank exchanger is shown in the figure. The increased surface area of two tanks relative to one tank increases standby losses.

Heat Pipe. Figure 2.15*f* shows a new concept in solar water heating developed by deWinter. A single-tank indirect system is used with auxiliary heat applied to the upper portion of the tank. The auxiliary zone of the tank and the preheated solar zone of the tank are separated by a thermal diode which completely eliminates mixing of solar-heated water and water heated by the auxiliary. The backup energy is supplied to the auxiliary zone at the top of the tank by means of a gravity heat pipe. The heat pipe is designed so that its condenser surrounds the auxiliary tank, resulting in a very efficient heat transfer. The gravity heat pipe permits any backup energy source to be used in a single-tank system. Solar performance on an annual basis is thereby improved, since the heat loss from a single-tank system is always less than the heat loss from a double-tank system.

A particularly interesting feature of the deWinter design is the ability of solar heat to provide some of the standby losses for this system. Solar-heated water rises through the heat diode and can maintain the auxiliary tank at its delivery temperature if solar capacity is sufficient. In addition, if no hot water is being used for a period of weeks, the auxiliary will never operate and the consumption of fossil fuels will be eliminated.

Other System Configurations

Many early solar water-heating systems used the thermosiphon design, which requires that the tank be placed above the top header of the solar collector. Circulation of fluid from collector to storage in this direct-heating design results from a density difference between the relatively lighter fluid in the collector and that in storage. No pumps or controls are required. When properly sized, these systems may perform as well as pumped systems. Somewhat larger pipe sizes are required in both the collector and the connecting piping, which must be well insulated to avoid convection-loop blockage. To reduce piping friction, pipe bends rather than short elbows are recommended (see page 83).

The principal difficulty with thermosiphon systems is the requirement that the tank be placed above the solar collector to prevent reverse flow at night. If solar collectors are located on the roof of a building, the tank could become the highest component of the building with negative architectural and aesthetic impacts. In addition, the structure required to support a large solar storage tank could be prohibitively expensive in large installations. Circulation pumps used in pumped systems are small, with an operating cost per year of only a few cents per square foot of collector. Therefore, most commercial systems used have been of the pumped-circulation design.

It is possible to place the storage tank in a thermosiphon system at the same level as the collector instead of above it if a check valve is used. Of course, a conventional check valve requires too much pressure to hold it open, but at least one company manufactures a thermostatic check valve that senses collector temperature. This valve acts like an automotive thermostat and opens above a specific temperature but closes when no useful solar heat is available and the collector cools.

Another idea used to produce hot water from solar energy involves the use of a heat pump. The collector is the evaporator, and the heat pump can heat water in the preheat tank. The heat pump can operate at temperatures below freezing, since the working fluid is a refrigerant with a very low freezing point. This system can even extract heat from the ambient air when no sunlight is present. The heat pump must be designed for very low temperature operation to eliminate the possibility of liquid refrigerant entering the compressor. In conventional heat-pump compressor designs, this will destroy the compressor, but special compressors capable of handling small amounts of liquid refrigerant exist. However, no commercial solar system of this type is presently available.

A third water-heating idea requires only a single water pass through the collector per day. In the more common designs described above, the cycle time is 45 to 60 min, not 1 day. With the single-cycle approach proposed by H. Tabor, pump power is greatly reduced. Since the water source temperature is usually below ambient temperature most of the year, collector efficiency is high because the collector inlet temperature is low for the entire day. For inlet temperatures roughly equal to ambient temperatures, collector efficiency $\eta_c = F_R\tau\alpha$, according to Eq. (2.5). For low flow rates, F_R is reduced vis-à-vis other designs but the absence of the second (heat-loss) term in Eq. (2.5) compensates for this effect. The single-pass design has not been widely used, but certainly merits serious consideration.

All systems shown in this chapter use closed storage tanks, preferred since parasitic heat losses and vapor release are eliminated. In most domestic heating systems, the price of closed water tanks is so low that the use of an open tank is not considered. The buildup of water hardness, the loss of heat, the introduction of air into the system, and the requirement for oversized pumps has nearly eliminated any interest in open storage systems for commercial installations as well.

Operational Modes of Solar Water-Heating Systems

As an example of the operation of solar water heaters, the function of the indirect two-tank system shown in Figure 2.15e will be described. The controller shown in the figure senses two temperatures: the temperature of fluid in the collector panel and the temperature at the bottom of the preheat tank. When the temperature of the collector panel rises 15 or 20°F above the storage-tank temperature, the pump in the collector loop begins to operate and heat is transferred from the collector fluid through the heat exchanger into the preheat tank. If an external heat exchanger is used, the storage pump is on whenever the collector pump is on. Flow continues until the temperature in the preheat tank approaches that in the collector. At this point, fluid circulation ceases. It resumes when a sufficiently great temperature difference exists between collector and storage.

City water is introduced into the bottom of the preheat tank. This assures that the coldest temperature fluid present anywhere in the system is introduced into the collector inlet header. Low inlet temperatures result in high collector efficiencies.

As hot water is drawn from the top of the auxiliary tank, it is replaced by preheated water introduced into the bottom of the tank. Preheated water normally will be at a temperature of 110 to 120°F in winter. Therefore the amount of nonsolar heat will be much smaller than if cold city water at 50°F were supplied to the inlet of the water heater.

During periods of no hot-water use, the preheat tank temperature may rise to 190 or 200°F. Then a preset pressure/temperature (P/T) control valve releases hot water to the drain. This safety device is required by most building codes, but is not the best form of heat rejection. Most P/T valves are not designed for repeated cycling but are to be used only infrequently. Therefore, a separate heat-rejecter may be required. The most reliable type of heat-rejection system is a drain valve operated by the controller. Above the set point the controller opens the drain valve located in the *top* of the preheat tank. The drain valve dumps some hot water which is replaced by cold city water introduced into the bottom of the preheat tank. The cold water flows into the heat exchanger, cooling the inlet fluid to the collectors. Cold water causes the collector temperature to drop, and the overheat control feature of the controller is deactivated. It is recommended that water not be boiled in the collector as a method of rejecting heat.

Auxiliary. Auxiliary tank design requires no special consideration. The auxiliary tank operates much as it would if no solar heating whatsoever were used. The backup energy source, whether it be steam, gas, fuel oil, or electricity, is used if the temperature in the auxiliary tank falls below the set point. The auxiliary heating unit is sized to carry the full water-heating and standby loads. The peak heating load for domestic water heating can be estimated from the information contained in Table 2.4. These peak rates, when multiplied by the expected temperature rise in winter when source water temperatures are the coldest, will determine the size and recovery rate of the backup heater.

All solar DHW-heating systems using a liquid working fluid require a check valve in the collector loop. This blocks reverse circulation in the loop and the removal of heat from storage at night. During the night, the collectors are the coldest portion of the collector fluid loop. The fluid in the collectors, as a result, is the most dense, tending to sink to the bottom of the loop. It is replaced by less-dense fluid rising to the loop high point—the collector array. The reverse flow at night causes backflow through the heat exchanger and consequent removal of heat from the storage tank. If the weather is particularly severe, the collector fluid may be below freezing at night and the water on the storage side of the collector heat exchanger could freeze destroying the heat exchanger. Therefore, the inclusion of a collector circuit check valve in any solar system is essential.

Detailed sizing rules for solar water-heating systems are given in Chap. 4. In general, 1 gal of hot water per day will be delivered per square foot of collector (averaged over the year) by most commercial solar heating systems. This rule is too crude for system sizing but can be used in the programming phase to roughly estimate area requirements. Heat exchangers in domestic water-heating systems will normally have an effectiveness greater than 0.8 if a shell-and-tube exchanger is used. Jacketed tank exchangers have lower effectiveness, but the low cost of these heat exchangers may justify their selection. Collector pumps flow approximately 1 $gal/(h \cdot ft^2)$. At these very low flow rates, the pressure drop through the piping is relatively small and the amount of pumping energy consumed is small.

Double-Wall Heat Exchanger. Many building codes require a double separating wall between a toxic fluid and a potable water source. Ethylene glycol and other chemicals used in collector loops as freeze protection are toxic.

Double-wall heat exchangers are commercially available but are more expensive than standard shell-and-tube exchangers. The designer should consider the use of a nontoxic circulating fluid in the collector loop provided that its accidental replacement by a toxic fluid sometime in the future is certain not to occur. The safest approach in many cases is to accept the extra cost and performance penalty of a double-wall heat exchanger to ensure a safe water supply.

Leakage of glycol into potable water supplies is highly unlikely, since the pressure in preheat tanks and auxiliary tanks is much higher than the pressure in the glycol loop—10 to 30 lb/in^2. If a leak were to develop in the heat exchanger between these two fluid circuits, the flow would occur from the preheat tank into the collector loop and not vice versa. However, under special conditions, for example, if a fire truck were to pump water from a local water main, the pressure in that main could drop below that of the collector loop and a small flow of glycol into the water system could occur. Therefore, the code requirement of a double-wall heat exchanger can be justified under special conditions.

Standby Recirculation Loop Losses. Two-tank systems have significant standby losses because of the requirement that the auxiliary tank be maintained at the set-point temperature for 8760 h/yr. Therefore, the standby losses given by Eq. (2.19) can be significant. Some of the standby losses can be provided by solar heat if a circulator between the preheat tank and the auxiliary tank is used. When the preheat tank temperature rises above the set point of the auxiliary tank, the circulator is activated, causing fluid circulation between the preheat tank and the auxiliary tank. This circulation rate can be very small, and the circulator used to provide it can be inexpensive. A check valve downstream of the circulator is needed.

An alternative method of providing part of the standby losses by solar heat is to locate the auxiliary tank above the preheat tank as in the deWinter design. Then an automatic gravity flow of warm water from the preheat tank to the auxiliary tank will occur any time preheat-tank temperature exceeds auxiliary-tank temperature.

Many commercial systems use a circulated loop to maintain the temperature of all hot-water taps in a building. It is recognized that circulating loops are net heat losers in water-heating systems, and some energy-conservation codes do not permit them. By proper design, it is possible for solar energy to carry a good part of these standby losses. This can be done by sensing the temperature of the preheat tank and operating the circulation pump if that temperature is sufficiently high to offset heat losses from the circulating loop. Alternatively, subject to economic viability, a heat exchanger between the return loop and the solar preheat tank can heat return loop water before its reentry into the water heater. Whether or not solar energy is used, the delivery-loop pump should be sized and then controlled, based on maintaining temperature at the most remote point of the water-circulation loop. When this temperature is present at the terminal point of the loop, the circulation pump ceases operation until a temperature below the set point reoccurs.

LIQUID-BASED SOLAR SPACE-HEATING SYSTEMS

The first liquid-based solar space-heating systems were simply enlarged water-heating systems connected to a load heat exchanger which delivered heat to a building. Solar space-heating design has progressed since the first experiments in the 1930s, and today is a mature solar energy technology. In this section, the calculation of space-heating loads is briefly summarized and the operation of liquid-based systems for space heating is described in detail. The next section will describe the operation of air-based space-heating systems.

Space-Heating Loads

The standard method of calculating building heating loads is described in detail in Ref. 6. The details are not presented here, and the reader may consult this reference or other heat-load calculation manuals prepared by professional or trade societies. Building heating loads arise from several sources. The first type of heat load is that associated with heat loss through walls, windows, and roofs. This heat loss is usually proportional to the temperature difference between the interior of the building and the environment, the proportionality constant being the *U* value multiplied by the area.

The second source of building heat loss is the replacement of warm air in the building with cool air from the environment. This exchange may be intentional or unintentional. When intentional it is called ventilation and when unintentional, infiltration. The calculation of ventilation heat loads is very simple. The volume of air replaced is multiplied by the specific heat of air and the temperature rise required to heat outdoor air to indoor air temperature. The density of air depends upon altitude.

Infiltration rates are much more difficult to estimate. They depend upon the quality of construction of a building—upon the number of door openings, the wind speed, and the amount of air displaced by the chimney effect. The chimney effect is the result of stratification of spaces, which causes a higher temperature to exist near the top of an enclosure than at the bottom. The density difference will cause an exchange of air with the environment if openings to the environment exist either through window or door cracks.

The third type of heat loss is typically independent of environmental temperature and occurs through floor slabs. The heat loss depends upon the difference between ground temperature which is normally considered to be uniform during a heating season and the interior temperature of the building which is determined by the set point of the heating system.

The heat load may be expressed in several different ways, depending upon the calculation method. The purpose of calculating heat loads in the past has been to size heating systems for the worst set of climatic conditions to be expected in winter in a given location. This form of the heat load is called the *peak heat load* and is expressed in kilowatts or British thermal units per hour. *The peak heat load is not of interest in designing solar heating systems* since solar systems are rarely designed to carry the peak heat load.

The *unit heat load* value pertinent to solar system design is more useful. It is expressed in terms of British thermal units per hour per degree Fahrenheit or kilowatts per degree Celsius. This form of the heat load can be calculated very easily from the peak load value by dividing it by the design temperature difference. The design temperature difference is the temperature difference between the building interior and exterior on a design day, defined as that day on which the outdoor temperature is so low that it exists only 1 percent or 2½ percent of the year. If the 1 percent criterion is used, the 99 percent design temperature from handbooks is selected. If the 2½ percent condition is chosen, the 97½ percent design temperature is used (6). In large buildings, the peak heat loss may not correspond to the peak heat demand because of internal heat source and mass effects.

The Degree-Day Approximation. The unit heat load expressed in British thermal units per degree Fahrenheit is a characteristic of the building and is independent of the location of the building except for the effect of wind. Location dependence is introduced by the design outdoor temperature. The unit heat load can be converted to the form used in solar designs by multiplying its value by 24 if no night setback is used. The result will be a heat load expressed in British thermal units per degree-day. The heating degree-day is the standard measure of residential heating demand used throughout the United States and is tabulated for many hundreds of locations. Most degree-day values in the United States have been based on 65°F. Before 1975 this design temperature base for new buildings was satisfactory. However, with increasing insulation levels and reduced thermostat settings, the 65°F base must be modified. The method of modifying the degree-day base described in detail in Ref. 1 also accounts for internal heat sources. Applying an empirical factor to degree-days to the 65°F base is not a satisfactory method of calculating heat losses for other bases. If internal heat sources are not considered in finding the *net* load to be met by solar heat, grossly overdesigned systems will result.

A simple method of correcting annual degree-day totals for bases other than 65°F was developed by Harris (Ref. 52). The annual heating degree-days to any base DD_b are given by

$$DD_b = [1 - k_d(65 - T_i + q_i/UA)]DD_{65} \tag{2.20}$$

where T_i = building interior temperature, °F
q_i = *algebraic* total of internal heat sources (>0) and sinks (<0), Btu/h
UA = unit building-loss coefficient including air change, Btu/(h·°F)
DD_{65} = annual total of standard heating degree-days (base 65°F)
$k_d = 6.398 DD_{65}^{-0.577}$

Equation (2.20) should be used only for annual degree-day totals. Reference 1 describes the method for finding monthly degree-days to any base.

Many designers *incorrectly express heat loads of existing buildings as the ratio of space heat used in a year by the concurrent degree-day total* (65°F base). This is incorrect since internal gains are not included. The proper method uses degree-days to the correct base as the denominator of the ratio.

The annual heat loss of a building can be calculated by multiplying the heat load expressed in British thermal units per degree-day times the number of heating degree-days per year. The product of these two quantities will give the number of British thermal units required to heat the building per year in an average year. The average year may rarely be encountered, however, and variations from predictions based on the degree-day method of ±20 percent should be expected. Large buildings with many zones are not well represented by degree-day calculations, and more detailed load procedures must be used (Ref. 6) to account for internal gains and building mass effects (see Chap. 4). The bin-hour method may give somewhat better results than the degree-day method (Ref. 61). See page 90 for more information on local calculations.

The cost of heating a building with conventional fuels can be calculated once the annual heat loss is known from the calculation described above. The annual load is multiplied by the cost of energy to find the annual cost of providing space heating. If this cost is greater than heating the same building to the same temperature with a solar system, then there is an economic incentive to consider the use of solar energy for space heating. The details of these economic comparisons are described in Chap. 4.

Space-Heating Systems

Figure 2.16 shows a typical design for solar space heating. The diagram shows the relationship of the several components and the fluid loops used for heat collection, storage, and transport to load.

The description of liquid-based solar heating systems can best be done by examining each fluid loop. The system shown in Fig. 2.16 is seen to consist of three space-heating fluid loops. The first loop is the collector fluid loop and is identical in function to that of indirect solar water-heating systems. The pump P1 causes fluid flow through the solar collectors; the heated fluid transfers its heat to the storage loop through the collector heat exchanger. An expansion tank is present to accommodate changes in fluid volume. The check valve is used to prevent reverse flow through the collector loop and a flow control valve is used in some systems for modulating flow to control the collector-fluid outlet temperature. In many systems this valve is not used, since it requires a larger pump P1 and its benefits are questionable. The heat-rejection unit (HRU) dissipates excess collected energy during low-load periods.

The storage loop consists of pump P2 and the tube side of the collector heat exchanger. P2 operates whenever collector pump P1 operates except during heat rejection. Since the storage tank is part of a closed fluid loop it also requires an expansion tank. A relief valve is present in any closed fluid loop to accommodate overpressure or overtemperature conditions.

The third loop present in liquid-based solar heating systems is the energy-delivery loop. Pump P3 circulates storage fluid through the load heat exchanger, which may be a baseboard unit, a forced-air heat exchanger, or a radiant panel. The fluid is returned to the bottom of the storage tank. If there is insufficient stored solar heat, the control system activates the auxiliary boiler and pump P3 causes fluid flow through the boiler to carry the load. The auxiliary boiler size is based on the peak heating load. Other heat-delivery systems including heat pumps may be used between the solar storage tank and the load.

Load Devices. The load heat-exchanger effectiveness determines the rate at which heat is delivered from the solar storage tank. Figure 2.17 shows the heat capacity of several terminal devices as a function of ambient temperature. The vertical axis is the fluid temperature required by the load at ambient temperatures shown along the bottom axis. The design temperature to the left is the coldest temperature expected. A standard baseboard heater requires much higher storage fluid temperatures than a forced-air system. As ambient temperature increases, the required storage fluid temperature decreases since the building heat loss is less. At the no-load temperature, where losses just balance internal heat gains, no heat is required.

Figure 2.17 shows modified baseboard and modified forced-air units specifically designed for solar systems. A modified baseboard system, for example, might consist of an increased length of baseboard or a dual-pipe baseboard system replacing a standard single-pipe unit designed for operation at a 180°F. The modified base-

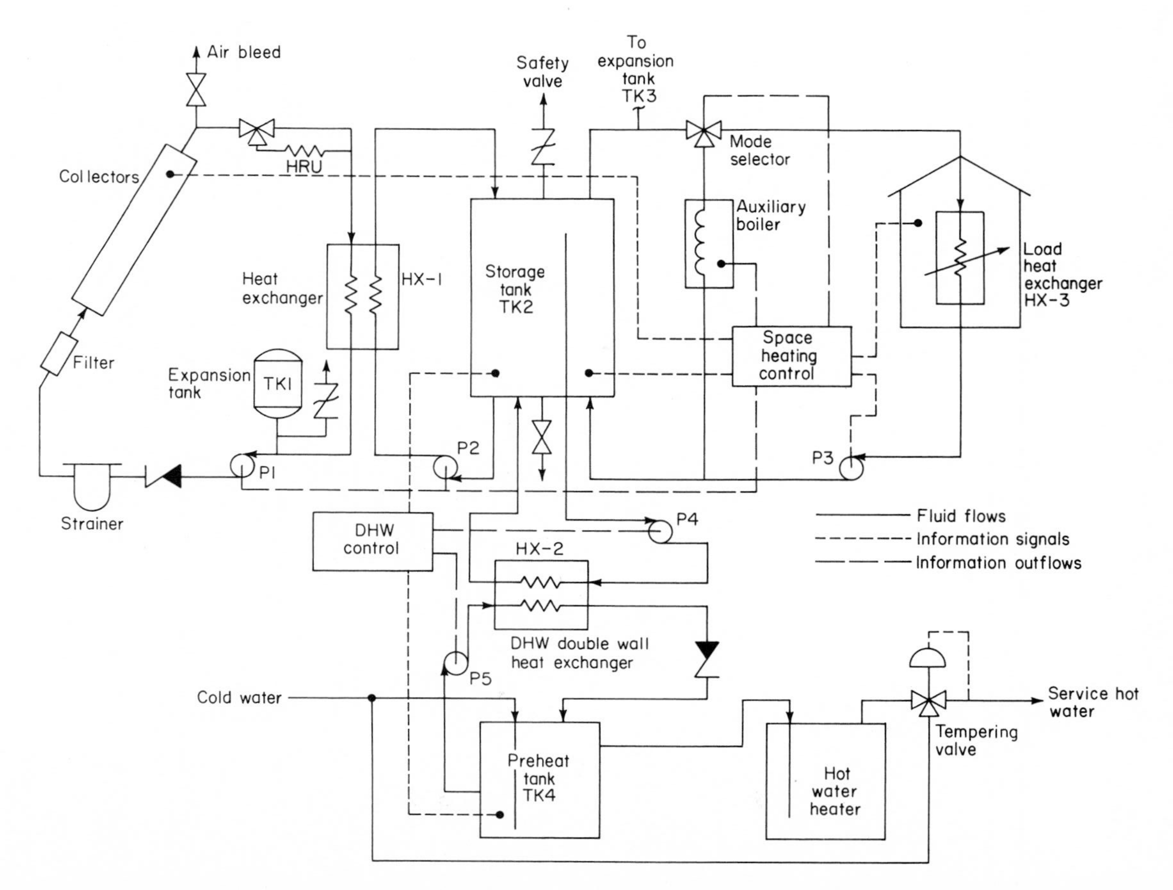

FIG. 2.16 Schematic diagram of a typical liquid-based space-heating system with domestic water preheat.

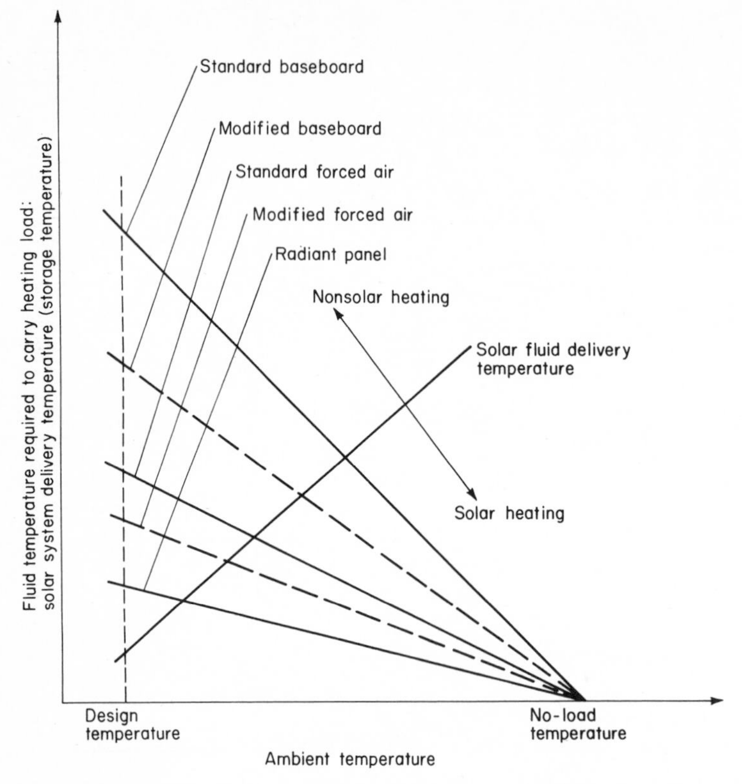

FIG. 2.17 Heat-delivery characteristics for baseboard, forced-air, and radiant space-heating systems.

board system can operate at reduced temperatures and is, therefore, able to use solar heat a greater fraction of the time. Likewise, a modified forced-air unit would be a fan coil or fin-tube heat-exchanger unit oversized to accommodate lower solar fluid temperatures. The solid line sloping up to the right in Fig. 2.17 indicates a typical solar fluid delivery-temperature characteristic. As ambient temperature increases, the output temperature of a solar collector will increase, since heat losses from the collector are reduced. The slope of this line depends upon the size of storage and the specific level of solar energy collection, and therefore, this line should be viewed as only one example of many such characteristic lines which occur under different conditions. At the intersection of the solar-heated-fluid line and the downward-sloping load line, the solar system is able to carry the heat load. For ambient temperatures above the point of intersection, 100 percent solar heating is possible. For ambient temperatures below the intersection point, solar heat will not be able to carry the entire heating load, and some backup energy will be required. The sloping load lines are the basis of ambient-reset boiler controls on larger heating systems.

Auxiliary Heat. The auxiliary boiler shown in Fig. 2.16 represents any type of backup system, including a steam coil, a gas furnace, an electric heat pump, a wood stove, or a fuel-oil-fired boiler. The auxiliary source is always connected in parallel with the storage tank. A series connection between storage and the load should not be used. If such a connection were used, it would be possible for heat from the auxiliary boiler to heat the storage tank. If the storage is heated by auxiliary heat, it cannot be heated by solar heat; further, if storage-tank temperature were elevated by fuel heat, the operating temperature of the collector would be increased and the efficiency of the collector would be decreased. Therefore, complete isolation of auxiliary heat and solar heat, via a mode-selector valve (either a three-way valve or two two-way valves), is required.

An alternative backup-system connection when heat pumps are used is to provide solar storage heat to the evaporator of the heat pump to increase its coefficient of performance (COP) over that which would occur if outside air were used as the evaporator heat source. Several systems of this "series-heat-pump" design have been built. Economic studies have indicated that the series-heat-pump idea is uneconomical under all conditions, even when coupled to very inexpensive solar collectors (Ref. 51). Hence, this approach to solar heating is not treated in this book.

Other problems with heat pumps exist, including: (1) zoning is difficult (restriction of airflow for zone control degrades performance and increases head pressure); (2) night setback may be a net energy user since morning reset will activate strip heaters (see Chap. 4); (3) exterior noise is an environmental problem, particularly in the residential urban environment.

Controls. A solar system controller will select several operational modes for the system. The first mode to be considered is solar heat collection. If the temperature of the solar collector is above that of the storage tank by a given amount, pumps P1 and P2 operate. Collection continues until solar collector and storage tank differ by only a few degrees, at which point collection ceases. The solar controller determines when P3 delivers heat. A thermostat placed inside the building determines whether heat is required to maintain the space at its design temperature. If the temperature in the heated space falls below the design point, a contact on the thermostat closes. The contact closure causes P3 to operate and an attempt is made to heat the building with solar heat. If there is insufficient capacity in the solar storage tank to provide space heating, the building interior temperature continues to fall and a second contact on the thermostat closes. The closure on the second thermostat causes a mode switch to take place. The fuel valve and igniter on the auxiliary boiler (or compressor and fan on a heat pump) are activated. The backup system continues to operate until both contacts on the thermostat open and pump P3 ceases to operate.

The system shown in Fig. 2.16 also contains a domestic water-heating system. A two-tank preheat system is shown, although a single-tank system could be used. The domestic hot-water system controller will cause pumps P4 and P5 to operate whenever the preheat tank temperature is below the main storage tank temperature by a few degrees. Preheat-tank temperature is sensed near the tank bottom but not in the immediate vicinity of the cold-water inlet, since a false signal at this point could cause the pumps to operate unnecessarily. The solar-storage tank does not contain potable water; therefore, a double-wall heat exchanger is used.

The two-tank system is identical to that described in the previous section. Note that pumps P4 and P5 can be eliminated if tank TK4 and exchanger HX2 can be positioned so that gravity flow causes fluid circulation. Solar DHW heating is often used with solar space-heating systems so that the solar collectors can be used during summer months when no space-heating demand exists. However, questions persist about the economic viability of operating the large space-heating collection system in summer just to serve the DHW load.

Heat Rejection. If more solar heat is produced in summer than required for DHW heating, heat rejection is required. Heat rejection may be done in several ways. One method of rejecting heat from the solar-collection loop is to use an air-to-liquid heat exchanger which is activated by the controller if storage temperature rises above a maximum set point, 200°F, for example. Pump P2 ceases to operate and the blower starts in the heat-rejection unit. This continues until the storage temperature has dropped below the maximum set-point temperature.

Another method of rejecting heat is to draw off hot water from the top of the preheat tank. This is the heat-rejection method described for solar water-heating systems in the preceding section. Although this method will work in space-heating systems, the time lag between the drawoff of hot water and the consequent cooling effect on the solar collectors is relatively long. The cooling effect must be transmitted through the hot-water heat exchanger, the solar-storage tank, and the collector heat exchanger. This lengthy process may result in a complete drawoff of hot water from the preheat tank before the collector temperature drops sufficiently for the heat-rejection mode to end. The loss of water is also undesirable.

Other heat-rejection methods include boiling fluid in the collectors or running the collector and storage pumps P1 and P2 at night. Neither of these methods is recommended. Boiling of heat-transfer fluids in solar collectors can result in chemical degradation of the fluids unless expensive silicone fluids are used. Operation of pumps P1 and P2 to reject heat at night is not recommended since collectors are designed to be efficient heat traps and not efficient heat rejecters. The rejection problem is an infrequent problem in most solar-heating systems but must be included for those periods when no space-heating load and no water-heating load exist. A method of sizing the solar heat rejecter is given later.

Heat rejection during a power outage may be necessary. One method of achieving this is to use two energized-closed solenoid valves—one to dump the antifreeze solution into its holding tank upon power outage and the other to circulate city water (via a pressure reducer) through the collectors and to a drain to reject heat. This method is useful only if the antifreeze is in aqueous solution, if the collector can withstand city-water pressure, and if water pressure does not depend on electric power. Interlocks are required between the solenoid valves to prohibit dilution of the drained antifreeze solution (use a float valve in the receiver tank), to avoid loss of glycol to the drain (size holding-tank properly), and to divert warmed city water to the drain (use a third energized-closed solenoid valve).

Drain-Down Systems. The solar heating system shown in Fig. 2.16 uses a nonfreezing collector fluid for freeze protection. One method of improving the performance of a solar system by a few percent is to eliminate the collector heat exchanger and use water directly in the solar collectors. Freeze protection in this system can be accomplished either by operating the collector pump during freeze spells or by draining water from the collectors. Circulation of warm fluid through the collectors is not a fail-safe freeze-protection procedure in the event of a power failure and should not be used.

The drain-down method of freeze protection can be made fail-safe in the event of a power failure but has other difficulties. Theoretically, the drain-down system has several features which would indicate that it might be practical for use in liquid-based systems. At night, the amount of heat lost when a collector cools is significantly lower if the collector is empty than if both water and collector drop in temperature. (Water left in a collector must be rewarmed the next day.) Other positive features of drain-down systems include the elimination of extra capital cost and maintenance cost for heat exchangers, extra pumps, and collector-loop fluids. The operating cost of pump P2 is eliminated and the somewhat lower collector efficiency which results inevitably from the use of the heat exchanger is avoided. Higher collector efficiency will result in a somewhat longer operating period for the collector during the day. Since most antifreeze liquids have higher viscosity and lower specific heat than water (see appendix Table A.11), higher pumping costs and larger pump sizes are required than if water were used as the circulation fluid.

These theoretical advantages of drain-down systems are offset by practical difficulties. Failure of a drain-down system can have catastrophic effects. Collectors and collector piping may freeze, requiring replacement in the dead of winter or deactivation of the solar heating system until weather moderates. In the winter of 1978–1979, 250 drain-down systems in Phoenix, Arizona, froze during an unexpected period of cold weather when the drain-down controller malfunctioned.

In order for a liquid collector system to drain, every collector and every liquid header and every collector pipe must be sloped at a sufficient angle to guarantee drainage. Any plugging of a collector or entrapment of water in a small orifice, for example at a balancing valve, can result in incomplete drainage and collector freeze-up. Heat exchangers of the size used in most residential solar systems cost the same as one or two solar collectors. Therefore, if two collectors were to freeze in one freezing episode, the cost penalty would be the same as if the heat exchanger, extra pump, and antifreeze fluid had been installed to begin with.

When water is drained from a collector, air from the environment or nitrogen from a tank replaces the water. If air is introduced into a solar collector each time the system is drained, corrosion will occur within the collector if the collector is steel. Air also causes the storage tank to corrode. Potential power outages must be considered and the drain-down system must be designed to drain if a power failure occurs.

One of the methods of introducing air into a draining collector bank is the installation of vacuum breakers at the topmost header of the collector array, a location similar to that of the air bleed in Fig. 2.16. Vacuum breakers installed on collector arrays are exposed to the environment and can freeze. If a vacuum breaker freezes, one of two things will happen: (1) the drainage process will cease and the collectors will remain partially or completely filled with water; (2) if the vacuum breaker remains open during collector filling, collector fluid will flow through the vacuum breaker and out onto the roof, where it will freeze.

A subtle malfunction in drain-down systems may occur if a small pinhole leak exists in the solar collector. This failure mode was identified by engineers at Bell and Gossett. If a pinhole occurs in a solar collector, warm vapor from the storage tank will rise into the solar collector at night. The vapor will freeze in the collector and can cause the collector to plug. If this fluid conduit plugging is complete, the pressure in the storage tank will cause fluid to flow from the storage tank up into the collector at night. The portion of the collector connected to the pinhole is, of course, at atmospheric pressure. The storage water pumped into the collector by this pressure differential will freeze and the collectors will be destroyed.

The preceding paragraphs have indicated that *there is a significant risk involved in the use of drain-down systems.* The risk can be minimized by very careful design and very careful installation. The ultimate decision about whether a drain-down system is used should be based on a risk-benefit analysis. That is, what are the odds of collector freeze-up and the cost of such a freeze-up relative to the cost of preventing freeze-up completely by use of an indirect heating system with nonfreezing working fluids and a heat exchanger? The decision of the majority of solar system designers in cold regions of the United States has been to not use the drain-down system, since its reliability cannot be assured and the potential cost of system failure can be large. Therefore, the drain-down idea will not be described in detail in subsequent steps in the design process.

Illustrative Solar Systems

Figure 2.18 shows a liquid solar heating system on the University of Colorado Credit Union building located in Boulder, Colorado. An area of 2250 ft^2 of liquid-based solar collector is used in arrays mounted both on the roof and along the south side of the building. The system is designed to carry approximately two-thirds of the heating load of this 15,000-ft^2 building in the 6200 degree-day climate of central Colorado. High levels of wall and roof insulation are used to reduce the load.

Figure 2.18*b* is a schematic flow diagram of the system. The solar heating system has functioned as designed. During a recent heating season, the heating bill for this building was less than the heating bill for the home of the credit union manager located in the same city.

AIR-BASED SOLAR SPACE-HEATING SYSTEMS

Air as a working fluid in a solar space-heating system naturally suggests itself, since most of the space-heating systems in the United States are based on the use of forced air. With air, the heat exchanger between the collector system and delivery system is avoided. In addition, air is noncorrosive, nonfreezing, and does not boil or require a heat rejecter during periods of excess solar collection. One strike against air is its poor heat-transfer characteristic. However, by proper design this difficulty can be overcome and air-based solar space-heating systems can deliver roughly the same amount of energy per year as liquid-based systems. In this section, the function of air-based heating systems is described and the operational modes are delineated. The functions of certain components of the system different from those used in liquid systems are described, and special design considerations for air-based systems are given.

Air-Based System Configuration

Figure 2.19 is a simplified schematic diagram of an air-based solar heating system. The system does not require the use of either collector or load heat exchangers, both of which are present in liquid systems. Otherwise, the function of the system is similar to that described earlier. Variable-air-volume systems require additional fans, bypass ducts, and controls. See Fig. 2.19*b*. Many other system arrangements have been used for multizone applications.

Solar heat may be used directly from the collector to heat the building without the necessity of passing through storage. This mode is shown in Fig. 2.19*a* where return air passes upward through the collector, to the right through damper D1,

(a)

FIG. 2.18 (*a*) University of Colorado Credit Union building showing 2250-ft² liquid-based space-heating system capable of providing approximately two-thirds of the annual heating requirement of the building. (*b*) Schematic flow diagram for system. *(Courtesy Jan F. Kreider and Associates, solar consultant and C. Lee, AIA, architect.)*

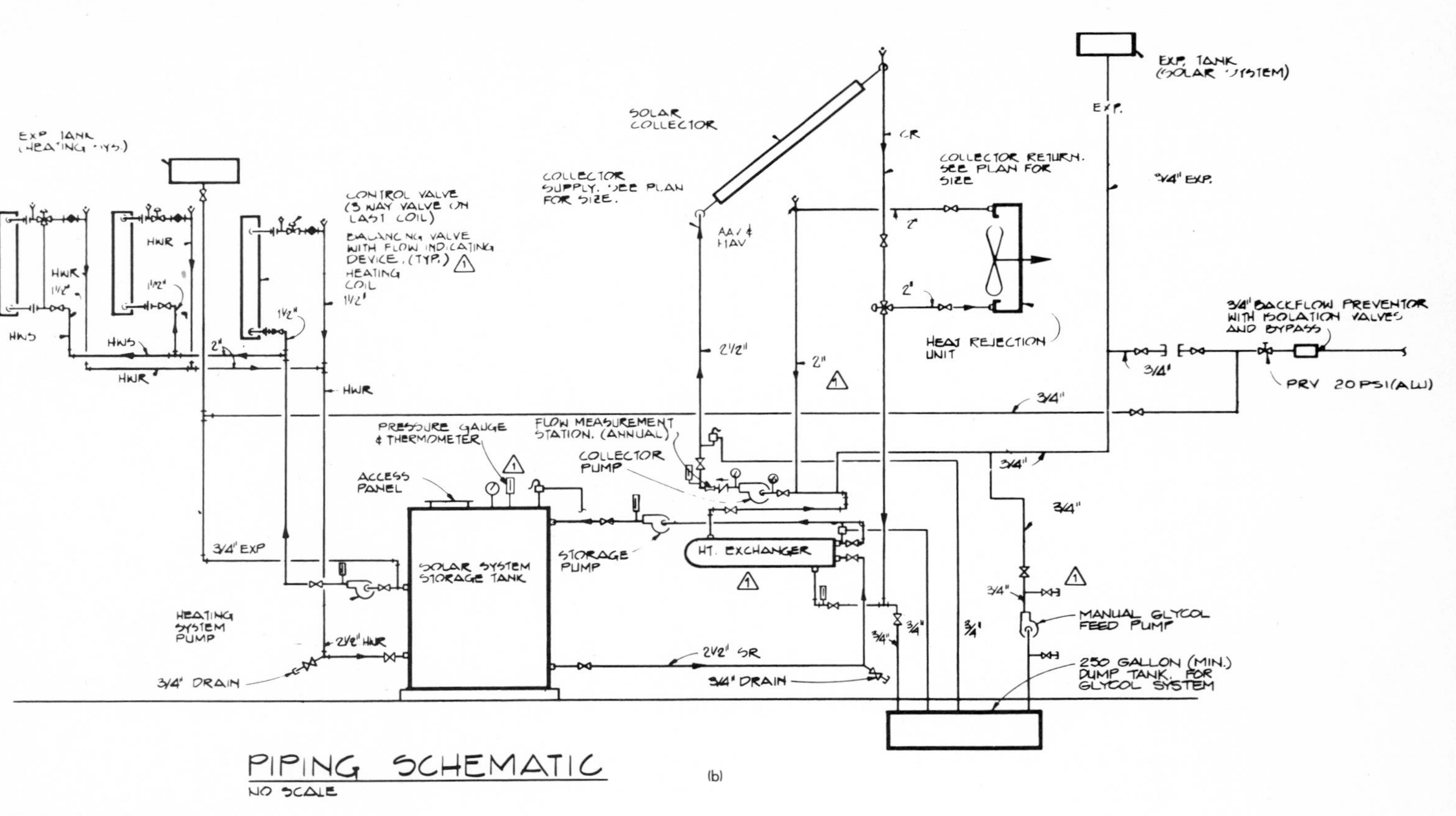

(b)

FIG. 2.18 (*Continued*)

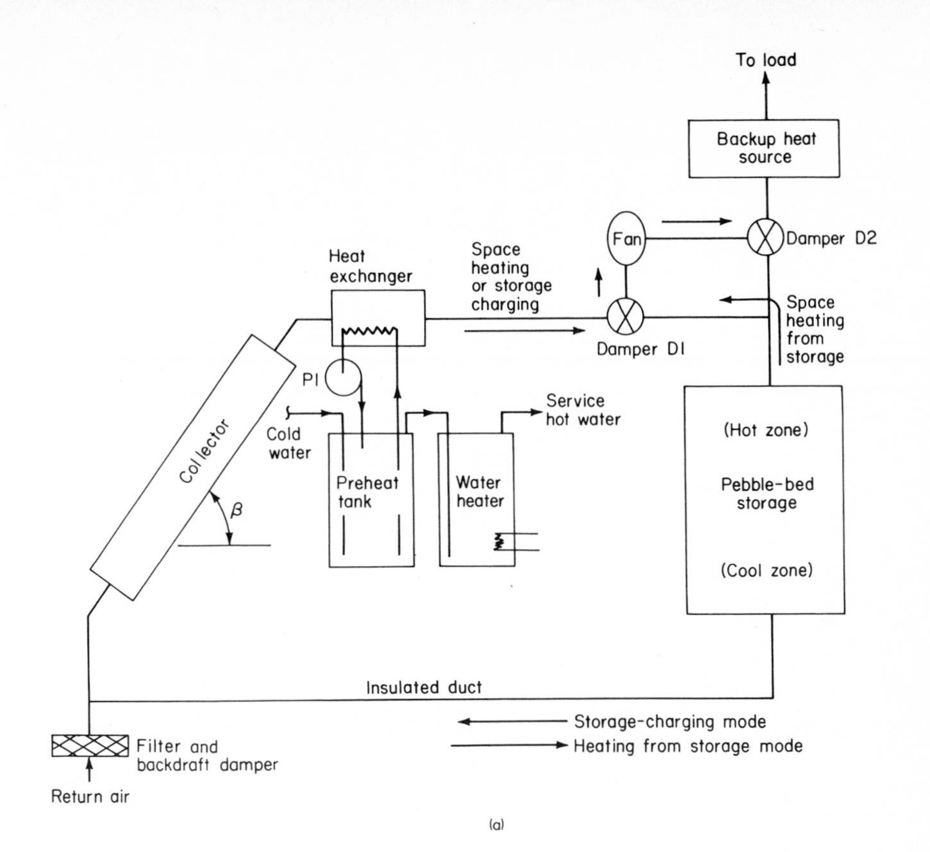

(a)

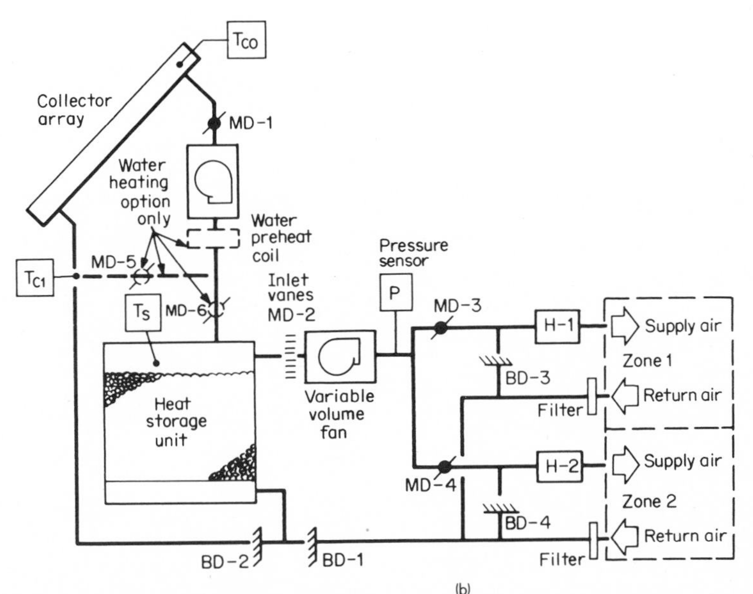

(b)

FIG. 2.19 (*a*) Air-based solar space-heating schematic diagram showing direction of fluid flow in various modes for simple-zone systems; (*b*) for multizone variable-air-volume system. Series backup shown but parallel backup can also be used (*Courtesy of Solaron Corp.*)

through the fan, and to the load through damper D2. If the space-heating load is satisfied, solar heat is stored in the rock bed. This is accomplished by airflow through the collector, through damper D1, through the fan, and then to the right through damper D2 and into storage. Air leaving the cool zone of storage on its way to the collector is approximately at room temperature. This low temperature ensures that the solar collector operates as efficiently as possible. Liquid storage for air-based systems is unsatisfactory since the temperature of returning air to the collector would by only 10 or 15°F below that of the collector exit temperature.

The third mode in which the system in Fig. 2.19 operates is heating of the building from storage. In this mode, return air passes through the filter and to the right through the insulated duct and upward through storage. Storage-heated air then passes through dampers D1 and D2 to the load. During this mode, it is critical that damper D1 seal completely. If any leakage were to occur through damper D1, some return air would pass through the solar collector and through damper D1 and would then mix with the storage-heated air entering damper D1 from the right. Mixing of very cold air passing through a nonoperating collector will reduce the temperature of air delivered to the load and may cause the backup system to start. In addition, the flow of subfreezing air through a collector at night may freeze the domestic hot-water heat exchanger and cause it to burst. This problem has occurred in many systems of this type which have used a poor-quality damper for D1.

Domestic hot-water preheating is normally done in air-based systems by an air-to-liquid heat exchanger in the collector exit duct. The heat exchanger can be located elsewhere to avoid the above-noted freezing problem, but by use of an appropriate design for damper D1, freezing can be eliminated. Pump P1 can be eliminated if the preheat tank is placed above the heat exchanger so that gravity flow will occur between the heat exchanger and the preheat tank any time the water in the heat exchanger is warmer than that in the preheat tank. The water flow through the heat exchanger is somewhat slower and the heat-transfer coefficient is below that which would occur if P1 were present. Most solar heating designs use pumped circulation for the preheat tank.

The system design shown in Fig. 2.19 does not permit space heating simultaneously with storage heating. In the liquid system described earlier, solar heat could be sent to storage at the same time heat was being removed from storage to satisfy a space- or water-heating load. The practical implications of this limitation are that the match between heat load and collector delivery temperature may not be appropriate. For example, in spring and fall, when only moderate heat loads exist, the solar collector may deliver air at an excessively high temperature. This may cause overheating of the space before a thermostat can respond and turn off the blower or divert flow to storage.

In commercial air-based systems, an additional mode may be used. Where large ventilation-air requirements exist, it may be advantageous to heat this air directly from the collectors by bypassing storage and introducing filtered outside air into the collector inlet stream. During solar outage, the makeup air could pass through storage for heating. The inlet region of storage could thereby be cooled to well below room temperature. If storage is discharged this way, the cold inlet to the collector during later solar-heat collection will increase collector efficiency. It is important that return air from the heated space not be introduced at a later time (in a space-heating mode rather than in a ventilation mode) into this "subcooled" storage, since it would, in effect, heat the cold zone of storage with warm room air instead of with solar heat. This problem can be avoided by requiring the proper temperature relationship to exist between room air and the storage zone, as described.

Other System Configurations

Although the system shown in Fig. 2.19 will operate satisfactorily, certain constraints imposed by building characteristics may require other system features. For example, the fan used to circulate collector air is also used to heat the building from storage. If the pressure drop through the collector-to-fan duct is different than that through storage and its connecting ducts, the airflow to the load will be different in the collector-to-load mode and in the storage-to-load mode. In many systems, this requires the use of a two-speed fan or two separate fans.

In summer, it is sometimes desirable not to have the entire storage bed heated (to reduce air-conditioning levels). To avoid heating the entire storage bed and still use solar heat for water preheating in summer, a bypass duct is installed around storage. The duct dampers are moved manually in spring and fall, permitting water heating without pebble-bed storage heating. Of course, the seasonal changeover dampers must be moved in advance of the deep heating season. A separate smaller blower can also be used in the summer mode to produce hotter collector outlet air for more efficient heat transfer at the DHW coil with reduced operating cost.

Since many air systems use room air as the collector fluid, a problem arises in the use of a single large system for multifamily residences or apartments. Building codes prohibit use of mixed supply air from several apartments. This problem occurs in the collector-to-load or storage-to-load modes of air systems. Therefore, a single air system cannot be used to heat more than one residential unit. This problem does not arise with liquid-based systems, which can use the same hot water in any terminal unit.

The position of the backup heat source in Fig. 2.19 in series with the collector or storage may not be optimum with fossil-fuel backup systems. Two specific difficulties have been observed in systems using this approach. The first is the likelihood of exposing the second fan motor (if one is present in the gas furnace) to excessively high temperatures. Blowers in residential gas- or oil-fired furnaces are designed to be exposed to ambient air not in excess of 100°F. Under solar operating conditions the motors may overheat, and in some systems they have ceased to operate because the thermal circuit breaker within the motor has opened.

The second difficulty with the series location for fossil-fuel backup is that stack losses through the furnace can be significant, since all solar-heated air is delivered by way of the firebox of the backup furnace. There is no simple way to modify a conventional furnace to avoid this problem, and it appears that the best design is either to place the backup heat source in parallel with the solar heat source instead of in series with it, or to use an electrical backup, which is inherently without stack losses. The parallel backup would be accomplished in Fig. 2.19 by connecting the inlet to the furnace to the storage inlet fitting instead of to the outlet fitting. Two additional backdraft dampers are required to avoid backflow through the storage.

Typical Component Sizes

The details of air system design are given in the next three chapters of this book. General characteristics and component sizes are described herein. Typical flow rates through air collectors are 2 to 4 standard ft^3/min per square foot of collector or 10 to 20 $L/(s \cdot m^2)$. The flow rate through collector ducting and fittings should cause a pressure drop less than 0.3 in of water in residences. Since large volumes of air are moved in air systems, it is important to control the pressure drop through both the collectors and the ducts. Likewise, the pressure drop through pebble-bed storage, balancing dampers, and the return-air filter must be minimized. Typical pressure drops are approximately twice those in a standard residential heating system. If filters and storage are not designed properly, the increased total pressure drop could result in a significant extra operating expense during the heating season.

Storage normally present in air-based heating systems is approximately one-half to three-quarters of a cubic foot of pebbles or rock per square foot of collector. Pebbles must be uniformly sized and very clean in order to ensure proper heat transfer without choking off airflow through the storage box.

The only heat exchanger present in air systems is used for domestic water preheating. It is usually sized to have an effectiveness of 0.5 to 0.6. The preheat system pump P1 is designed to operate when the collector air temperature is sufficiently above the preheat tank temperature to cause heat flow to occur. The turnon temperature difference can be calculated from Eq. (2.13).

Dampers are used at various points in air systems as mode controls. Dampers serve the same purpose as valves in liquid systems and require special attention if an air system is to operate satisfactorily. The damper D1 shown in Fig. 2.19 is particularly critical, as described above.

Dampers must be able to handle hot air in summer and cold air in winter. Dampers must be properly adjusted and installed, and seals must be attached to dampers with an adhesive capable of handling the broad temperature excursions to be expected. Another source of inadequate performance in air systems is leakage in ducts and collectors. As shown later, air leakage near the inlet of solar collectors is not a major problem. However, leakage in outlet ducts is a significant cause of poor performance. Outlet ducting in solar heating systems must be carefully sealed and installed, with turning vanes and smooth matching between one duct section and the next.

Illustrative Air Solar System

Figure 2.20 is a photograph of the largest air-based space heating system in the world, located on a large bus maintenance and storage facility in Denver, Colorado.

FIG. 2.20 Regional Transportation District of Denver air-based solar space-heating system. This 9-acre building is heated by an open-loop system using approximately 50,000 ft^2 of collector (mechanical engineer, K. Stewart, P. E., solar consultant, J. F. Kreider, P. E., and architect, RNL, Inc.). *(With permission by the Regional Transportation District.)*

Approximately 50,000 ft^2 of collector are used in this installation to provide about half the annual heat load of the building.

The principal heat load on this building is ventilation air; therefore, the solar collectors are used in an air-preheating mode without recirculating building air through the collector. The open-loop collector mode consists of introducing ambient air (instead of return air from the building) into the inlet of the collector (see Fig. 2.19). Then this air is delivered to the building. If excess solar capacity is available, some of the heated air is passed through pebble-bed storage, where heat is removed; the air is finally exhausted to the environment. Storage for this project has been sized larger than the nominal rule given above to make certain that nearly all useful heat is removed from the solar airstream before storage outlet air is exhausted to the environment.

Open-loop air-preheating systems are expected to have broad application in buildings with large ventilation rates as well as in agricultural and process environments where combustion air or crop-drying air is used only once. In some open-loop air-preheating designs, it is possible to recover the residual heat content of exhausted air and use it to preheat incoming air. For example, in the system shown in Fig. 2.20, air handlers contain air-to-liquid heat exchangers that remove heat from the exhaust air. This heat is transferred via a liquid-circulating loop to the inlet airstream. Therefore, the heat is repeatedly recycled although the air is continuously being renewed.

PASSIVE SYSTEMS FOR SPACE HEATING

Passive solar energy systems are those which do not use a significant input of mechanical energy to perform their heating function. In passive systems, for example, no collector pump or fan is required and no active control system is used. Passive solar energy systems have been built in the United States at a significantly smaller rate and in smaller sizes than have active systems; therefore, the amount of field experience with these systems is proportionately less. In addition, they are more difficult to monitor and to characterize quantitatively. However, because of their somewhat simpler design and lack of mechanical apparatus, these systems are capable of operating as reliably as active systems.

One of the distinctive features of passive systems is their unavoidable integration with the building to be heated. Since most passive elements are integral parts of the building, the building itself becomes the solar heating system. Therefore, building dynamics, orientation, and design are somewhat more important than in active solar heating systems. Of course, energy-conserving design is a prerequisite for passive systems as well as for active systems. There are many types of passive solar heating systems which will be described below, but a listing of all designs which qualify as strictly passive is not possible. Various combinations of passive features can be used to provide space heating. In addition, passive systems can be used with active systems forming a hybrid system with the most attractive features of both.

Generic Types of Passive Heating Systems

Direct Gain. The simplest type of passive solar heating system is called direct gain. Spaces are directly heated by sunlight passing through relatively large south-facing (i.e., SE- to SW-facing) glazings. Solar radiation is converted to heat by absorption on components within the building such as carpets, furniture, walls, or other building elements. Direct-gain systems can be very effective in producing high temperature levels during daytime hours, levels which may be excessive on sunny days in spring and fall.

A necessary adjunct to direct-gain systems is thermal storage. If thermal storage is used properly, daytime overheating can be avoided and solar-produced heat can be stored for use later. Figure 2.21 shows two realizations of the direct-gain idea. In Fig. 2.21*a*, the basic direct-gain concept is applied to a space immediately to the north of a large south glazing. Thermal storage in this application is provided by a relatively massive floor slab which has insulation beneath it.

Figure 2.21*b* shows a bilevel direct-gain application in which sunlight is used to heat both lower and upper levels. Heat produced in the upper portions of buildings is not useful in heating floor-level areas. The blower shown is used to avoid the excessive stratification which can occur in multistory spaces.

Although direct-gain passive systems are very effective in producing heat during daylight hours, they are also very effective in losing heat from the building at night. In winter, nighttime is roughly twice as long as daytime, and the potential for significant heat loss through large south-facing glazings is great. The accepted method of eliminating such heat loss is to install some type of insulation either within or behind the south-facing aperture. This insulation can be an attractively finished movable array of rigid insulation boards or a shade type of insulation which rolls up near the roof during daytime. The thermal resistance (R value) of movable insulation should be in excess of R-5. The R value of a double-glazed window itself is only 2; therefore, the addition of movable insulation at the R-5 level will reduce the heat losses through direct-gain glazing by a factor of 3.5. Shurcliff's book (Ref. 63) is the most complete book on thermal shutters and shades.

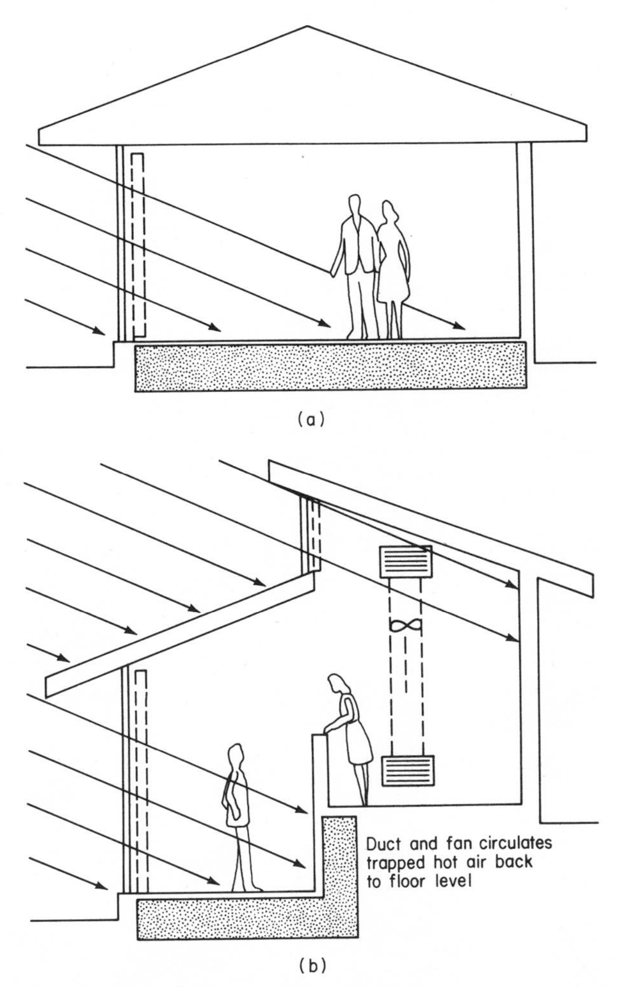
(a)
Duct and fan circulates trapped hot air back to floor level
(b)

FIG. 2.21 Direct-gain passive heating systems: (*a*) adjacent space heating, (*b*) clerestory for north zone heating.

Indirect Gain. The second class of passive solar heating systems is called indirect gain. This system converts solar flux to heat at an intermediate location and then releases the heat to the space to be heated by means of radiation, conduction, and convection processes. Figure 2.22 shows two types of indirect-gain heating systems. Figure 2.22*a* is known in the architectural literature as the Trombe design although it was originally developed by E. S. Morse in the nineteenth century. The idea of indirectly heating a space by means of intermediate conversion of sunlight to heat at a collector surface has also been practiced for centuries in the Middle East.

The system sketched in Fig. 2.22*a* also includes an air-circulation loop. In this type of system, heat is not only transferred slowly through the thermal-storage wall (TSW) but also into the heated space by means of gravity-induced air circulation. Typically 50 percent of the solar heat produced at the south-facing wall is trans-

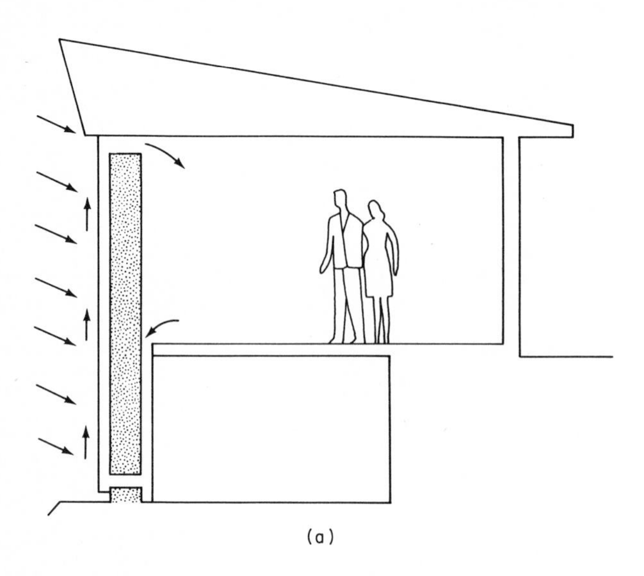

(a)

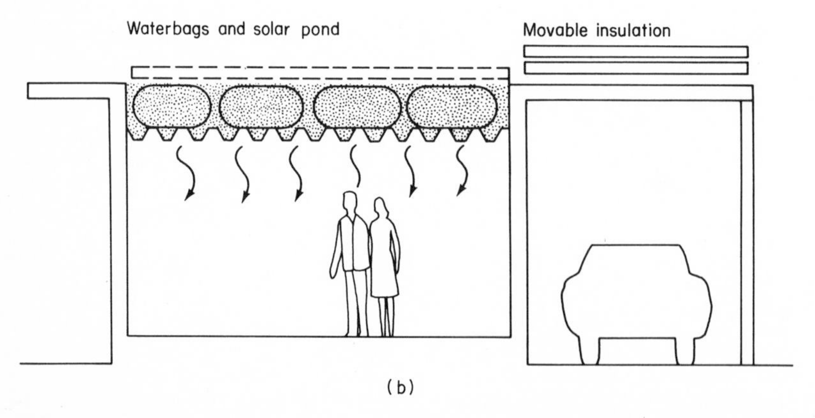

(b)

FIG. 2.22 Indirect-gain passive systems: (*a*) TSW system, (*b*) roof-pond system.

ferred into the air-circulation loop. The remaining 50 percent slowly diffuses through the thermal wall. In many climates the addition of air-circulation vents is useful in balancing the delivery of solar energy between daytime and nighttime periods. Heat produced and delivered by the airstream is available immediately, whereas heat diffusing through the thermal-storage wall is available between 8 and 12 h after being produced at the south-facing surface. Appropriate physical properties of the wall thickness and vent size are all important if a proper time lag is to be achieved.

The indirect-gain system developed by Harold Hay and shown in Fig. 2.22*b* is based on an idea long used in the American Southwest for building conditioning. In this roof-pond system, a layer of water 8 to 12 in deep is in intimate contact with a metal roof painted a dark color. As sunlight passes through the water, some heat is absorbed within the water and the balance is converted to heat at the black roof surface or at the black lower surface of the water container. This heat can be stored partially in the water pond and can be partially conducted and radiated to the space below. In order to eliminate heat loss from the roof ponds during winter nights, insulation is placed over them. Since the sun is quite low in the sky in winter, the horizontal surface of the roof pond shown in Fig. 2.22*b* is not at an optimum orientation for solar collection in northern latitudes. In addition, in freezing climates water is not a suitable storage medium and some type of freeze protection would be required. Since the volume of water used in roof-pond heating systems is rather large, the amount of antifreeze and its cost could be substantial. Increased structure to support the large storage mass is also costly. The approach has been successfully used in moderate heating climates, however.

Water also can be used in passive systems in a vertical orientation instead of in a horizontal one. Vertical tubes, metal tanks, or metal drums have been used. Water has the advantage of very high energy density but requires corrosion control if used for extended periods in open metal tanks.

Greenhouses. A third type of indirect-gain system uses a combination of direct-gain features and a thermal-storage wall. This concept is the attached greenhouse or attached sun space and is shown in Fig. 2.23. The greenhouse itself is heated directly by sunlight and functions as a direct-gain space. However, heat is transferred into the building by means of conduction through the thermal-storage wall. Although not shown in the figure, air circulation from the greenhouse uses vent slots in the top and bottom of the wall, much as in the system in Fig. 2.22*a*. The common wall serves the purpose of storage as well as heat transfer and is commonly constructed of masonry or of water containers. Movable insulation is required to prevent excessive heat loss from the greenhouse at night and during overcast periods. In addition, if plants are grown in the greenhouse, insulation is required in deep winter months to avoid freezing of the flora.

Although not widely used, many other passive heating systems have been invented and installed in a few buildings. These are not treated here, however.

Daylighting. Although not a subject of this book, lighting of interior spaces can be accomplished very effectively by sunlight (36). Natural daylighting is a design

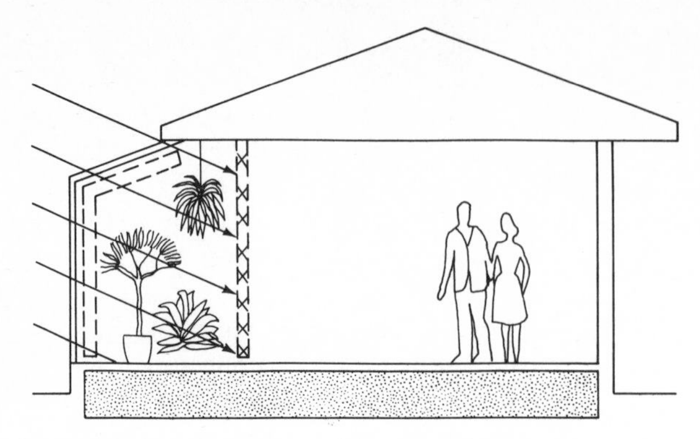

FIG. 2.23 Greenhouse or attached sun-space passive heating system using a combination of direct gain into the greenhouse and indirect gain through the thermal-storage wall, shown by cross-hatching, between the greenhouse and the living space.

discipline currently being rediscovered by illumination engineers. The calculation of natural lighting has been done for many years by architects using design tools collectively known as the lumen method. This method was popularized by the Libbey-Owens-Ford Glass Co. which has prepared a collection of nomographs and charts useful in predicting average daylighting levels. However, average daylighting levels usually do not provide sufficient information for designing natural daylighting systems.

Recent research in daylighting has resulted in computer codes which use hourly sunlight records for a specific location to calculate illumination patterns and intensity levels within an arbitrary space for each hour of daylight of the year. This type of design procedure is not suitable for residential purposes, of course, but is useful in the design of daylighting systems for large commercial buildings and office spaces. Significant reductions in electric power use are achievable by the proper use of natural daylighting.

A secondary effect of reduced electric lighting is the reduction of light-produced air-conditioning loads. In some natural daylighting projects described in Ref. 12, electrical consumption levels for lighting have been reduced from a typical value of 3 W/ft^2 to between 1 and 1½ W/ft^2. Daylighting design must be integrated with the design process because of the intricate relation between fenestration size and shape and net heat load. Daylighting design is not treated in this book since it is not a solar heating application.

Hybrid Heating Systems and Passive Systems for DHW Heating

Hybrid heating systems use a combination of active and passive features. For example, in the system shown in Fig. 2.21*b*, a fan is used to reduce stratification in a solar-heated space. The thermal-storage-wall system in Fig. 2.22*a* can be used to heat spaces other than the space immediately to the north. A fan is used to transfer solar-heated air from the glazing-to-wall cavity to north rooms or to basement rooms. In many cases, the addition of an air-circulation fan, whether it be only a roof fan or a more sophisticated circulation blower, can improve the performance and comfort level of a passive heating system.

Another hybrid heating concept is the use of a fully active solar heating system in combination with a passive heating system. In this type of hybrid system, the passive system is the baseload system and carries perhaps 30 or 40 percent of the heating load. Enlarged air-circulation vents concentrate most of this heating during daylight hours. Heat produced by the active system is stored for use at night. The load on the active system is reduced by the amount of heat provided by the passive system; therefore, the active system is small. Few such systems have been built, and the dynamics of the interaction between active and passive systems have not been clarified.

Passive DHW heating uses the thermosiphon system described in an earlier section on water heating. Such a system is shown in Fig. 2.24. Although this system has been used extensively for water heating, it has not been applied to space heating except in a few cases. The removal of heat from storage requires an active distribution system, and in that case the system could not be classified as totally passive. The design of large thermosiphon systems is an art requiring very careful sizing of manifolds and collector piping to ensure that uniform thermal-circulation flow occurs in each collector. The design of pumped systems for space heating is much more straightforward.

The thermosiphon concept shown in Fig. 2.24 can be applied to space heating,

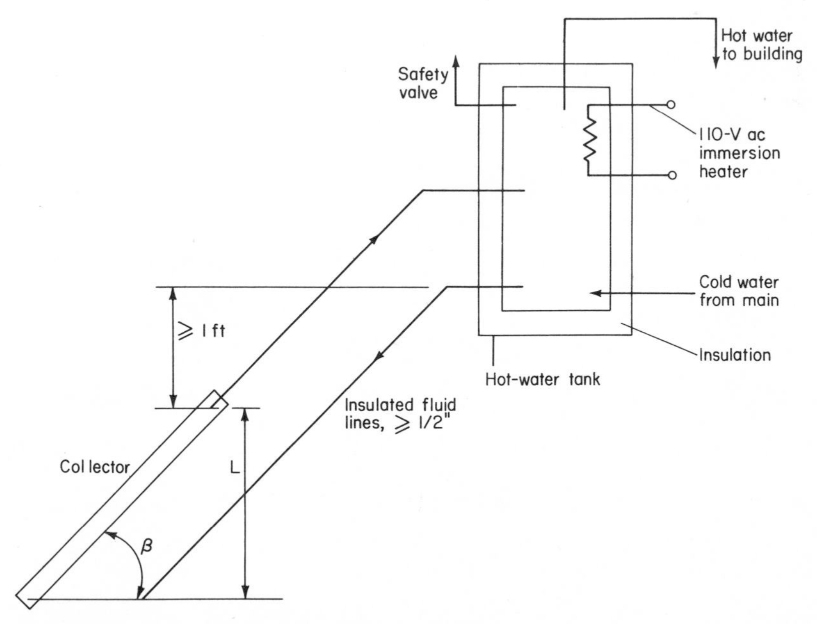

FIG. 2.24 Thermosiphon system for water heating. The collector is placed obliquely below the tank to prevent backflow.

however, if air is substituted for water as the working fluid. The collector is located obliquely below the space to be heated. Heat produced in the collector causes air to rise into the space to be heated; cool air from the return duct system enters the bottom of the collector to complete the loop. Very small systems of this type have been designed for retrofit. In such systems, a single collector is inserted in an existing window opening in the building. Solar flux striking the collector causes a circulation of air, and some space heating can be accomplished. However, since the window area in most buildings is too small to provide collector area for more than 20 or 30 percent of heating, this type of system cannot be used as a baseload solar heating system.

Large gravity-flow systems using several hundred square feet of collector area have been built for heating of residences. One difficulty is the interface with storage. If the solar system is to carry more than one-third of the heating load, storage is required. Rock-bed storage presents significant pressure drop to be overcome only by natural circulation, although the use of larger-than-normal (6- to 7-in diameter) rocks can reduce the storage-bed pressure drop.

Operating Modes for Passive Heating Systems

Although most passive systems have no active control system which selects the mode of operation, there are several distinct operating modes. Thermal circulation in indirect systems causes space heating to occur when solar intensity levels are sufficiently high. At night, a reverse flow can occur if the cavity between the south-facing glazing and the wall becomes cool. A number of ways of controlling reverse circulation have been used. The most common method of backflow control is a backdraft damper at the outlet of the upper wall air slot. This damper is made of a very light material (e.g., 1-mil polyethylene) and is pushed open easily by the forward circulation of warm air. When air tends to back-circulate, the backdraft damper closes.

The control of movable insulation is either manual by the owner or automatic with a sun sensor or clock. In direct-gain spaces, the manipulation is usually manual. Movable insulation must seal to the south-facing glazing to avoid reverse circulation between the glazing and the insulation. For thermal-storage-wall (TSW) systems, an automatic system for positioning night insulation seems most appropriate, since the owner may often forget to lower the insulating shade if it is not visible from inside. Most movable insulation controls contain (1) a sun sensor which is positioned in an unobstructed location south of the glazing area and (2) a small motor which is used to roll and unroll the movable insulation. The function of this system is shown shortly in an example system design. The details of movable insulation and its controls are given in Chap. 4.

Heat dumping in summer is accomplished by opening operable glazing or by positioning the movable insulation in the down position.

Example Passive Space-Heating System

The Gunnison County Airport located in western Colorado is one of the first large commercial structures heated by a combination of passive systems in the West. This building serves a small community with several general-aviation airlines and has approximately 10,000 ft^2 of gross floor area. The passive systems used on the building include a combination of direct-gain and thermal-storage-wall elements on the south facade as shown in Fig. 2.25. Additional direct gain and daylighting are provided by the clerestory. Active air circulation occurs between the thermal wall and the north zones, where the heat load is higher.

The building incorporates many energy-conservation features. The roof is insulated to R-30 and walls to R-20. All glazing is insulated to R-10 after sunset; R-10 perimeter insulation is used to prevent heat loss from the floor slab. Airlocks are used for all passenger entries and night setback to 55°F is used during periods when the terminal is closed. Infiltration is reduced by using zero building pressure provided by makeup air, and the low building profile further reduces wind infiltration. Thermal-buffer spaces on the north side of the building are maintained at only 45°F. No nocturnal ventilation is used when the facility is closed. The average installed lighting level is 1.5 W/ft^2. No mechanical cooling is used, and a reduced fenestration area of 9 percent of the floor area was specified. The aspect ratio of the building is exactly 2 to 1 to maximize winter solar gain and minimize summer gain. Stratification is reduced in the building by an air-circulation system between the roof peak and lower levels, thereby reducing roof heat loss normally present in multistory buildings.

Figure 2.26 is a section drawing of the thermal-storage wall. The double glazing is a standard manufactured window and is tightly sealed to the insulated walls at the top, bottom, and sides. A movable insulation curtain roll is positioned above the aperture and is controlled by a sun sensor placed in an unshaded location on the south wall. A 12-in masonry wall is used for collection and heat storage; the concrete is dyed dark-brown. Upper and lower wall vents, each having an area equal to approximately 2 percent of the total TSW area are used. The upper wall vent is connected to the heat-recovery system duct.

During winter, heavy snowfall is common in western Colorado, and the reflection from snow immediately south of the window is expected to improve performance by approximately 22 to 30 percent. Snow will not be removed from this area during the entire winter. The overhang serving the passive aperture is designed to block solar gains for approximately 4½ mo in summer. Although Gunnison does have a small space-heating load in summer, it is expected that this will be carried for the most part by internal heat sources, and the passive system will not serve as a heating system in summer. In the summer, the insulated louvers at the top of the passive wall are opened, and a natural circulation of air through them cools the thermal wall overnight and provides a small space-conditioning effect during the day. Large day/night temperature excursions of 30 to 35°F occur each summer night in this area, and the use of the TSW for cool storage will permit improved comfort in the building.

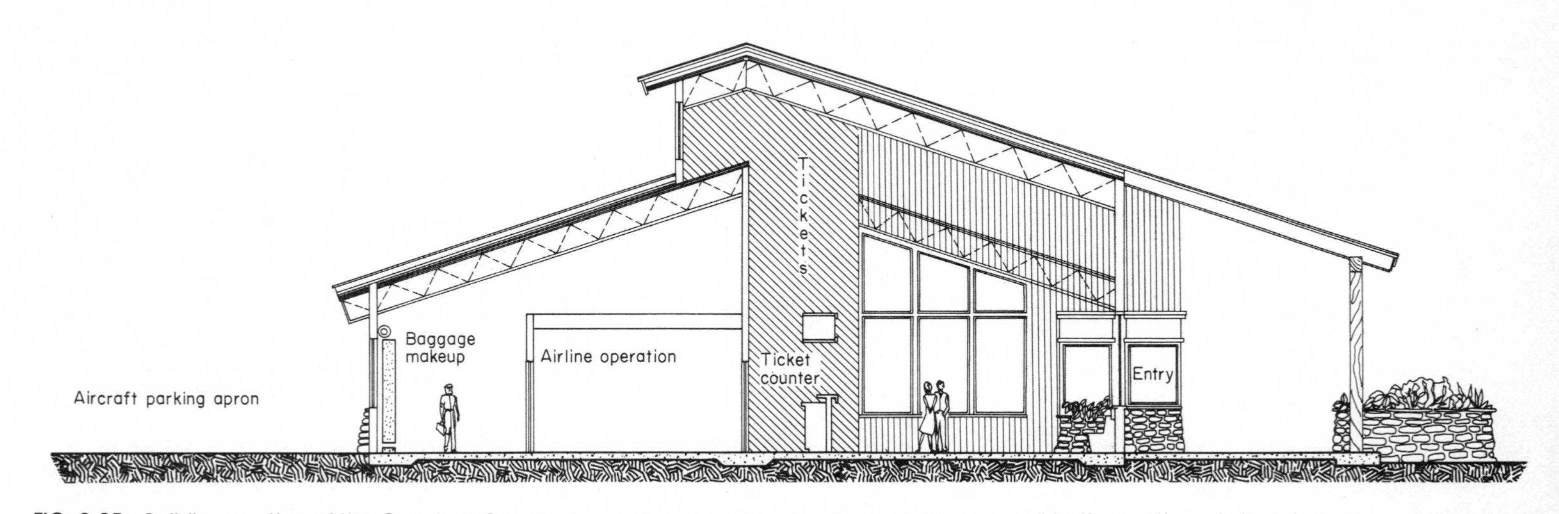

FIG. 2.25 Building section of the Gunnison County Airport Terminal showing thermal-storage wall to the south and clerestory to provide natural lighting as well as direct-gain heating. *(Courtesy of Associated Architects of Crested Butte, Colo.)*

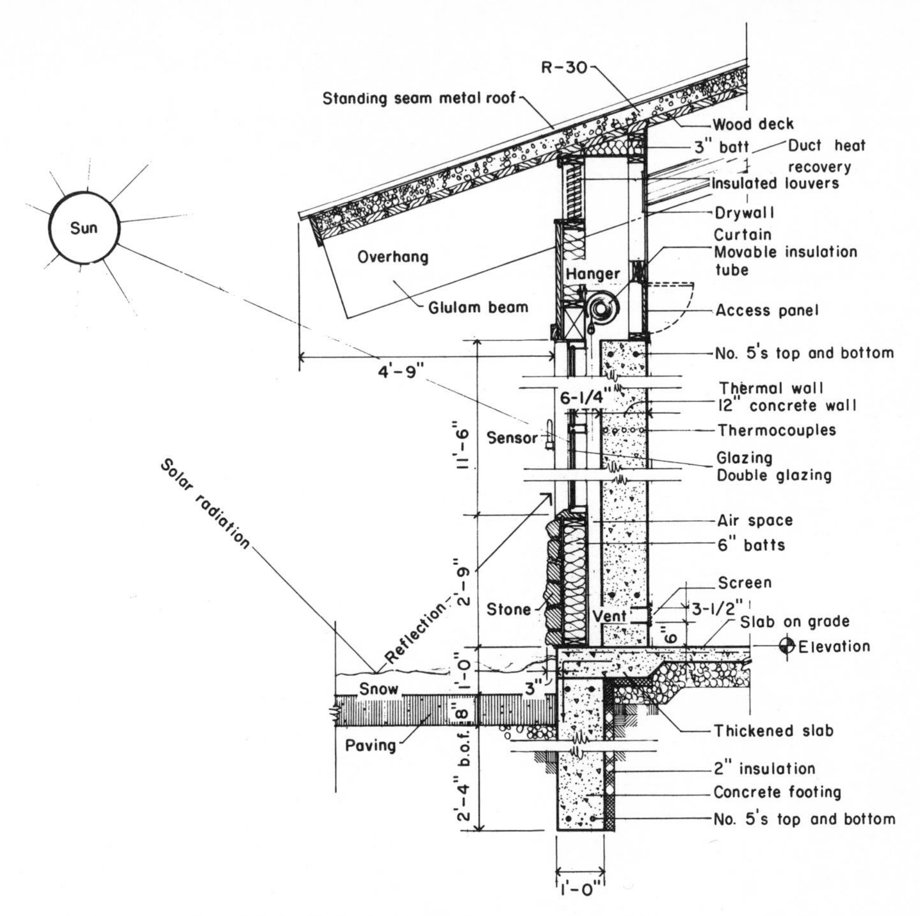

FIG. 2.26 Section of thermal-storage-wall section of the Gunnison County Airport showing aperture, storage wall, movable insulation, wall vent, and heat-recovery duct. *(Courtesy of Phil Tabb Architects, Boulder, Colo.)*

Operating Modes

Figure 2.27 is a schematic airflow diagram for this example project as well as a listing of the control sequences. Mode 1 is the mode by which the south zones are directly heated by radiation and convection from the thermal-storage wall. In this mode, blower B1 does not operate and the dampers are closed. This type of heating, which occurs whenever there is heat in the thermal-storage wall, is not subject to active control by the control system.

The second mode is active delivery of solar-heated air to the north zones of the building. This is accomplished by operating blower B1 and opening motorized damper MD1. The flow rate is 1 ft^3/min per square foot of TSW area. Airflow from the storage-wall cavity is directed to the north zonal distribution system. The thermal wall cavity is an air preheater; the final heating input is provided by the electric

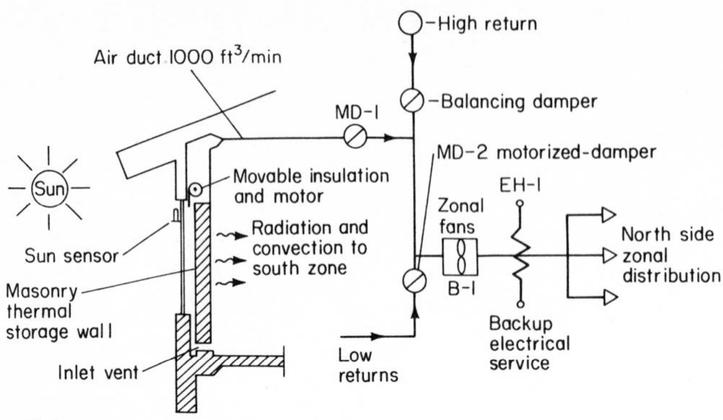

Control sequence

Mode	*B-1*	*MD-1*	*MD-2*	*EH-1*	*Zone thermostat*		*Movable insulation*
					Contact 1	*Contact 2*	
1. Direct wall heat south zones	Off	Closed	Closed	Off	Open	Open	Up if sun above setpoint level. Down otherwise; reverse in summer
2. Solar heat to north zone	On	Open	Closed	Off	Closed	Open	
3. Backup heating	On	Closed	Open	On	Closed	Closed	
4. No heat load	Off	Closed	Closed	Off	Open	Open	

FIG. 2.27 Energy flow diagram for Gunnison County Airport and table of operational modes. *(Courtesy of JFK Associates, Boulder, Colo.)*

heating element EH1, which consists of several steps of resistance heat activated by the control system.

When no heat is needed, blower B1 is off, and the damper MD1 is closed. Movable insulation is controlled by the level of sunlight independently of the heating control system. Whenever movable insulation is in the down position, motorized damper MD1 is interlocked closed. The air-distribution system also includes a high return from the clerestory shown in Fig. 2.27 to reduce stratification. This passive solar system is expected to provide 50 percent of the annual heating requirement for the building. Additional energy savings are realized by the natural daylighting via the clerestory.

O **Thermal Admittance.** Temperature control in passive heating systems may not be as precise as in active systems, since the release of heat from thermal-storage-wall surfaces is determined by the thermal characteristics of the wall as well as the environment. Calculation of temperature swings in passively heated spaces is done by application of the concept of thermal admittance. The thermal admittance of a wall is a measure of the wall's ability to absorb and store heat during part of a thermal cycle and then to release that heat through the same surface during the second part of the cycle. For example, in a direct-gain space, heat production occurs during daylight hours. Some of this heat is stored to be released during nighttime hours.

Quantitatively, thermal admittance depends upon thermal conductivity, density, and specific heat, and upon the period of the temperature swing occurring at the surface of the wall. The admittance is the ratio of (1) the heat stored during the heating cycle to (2) the temperature swing ΔT which occurs in the space during the heating cycle. If an allowable ΔT for the room air is stipulated, the amount of heat which can be stored is then known from the value of the thermal admittance. Details of thermal-admittance calculations are contained in Chap. 16 of Ref. 9.

ENERGY-CONSERVATION OVERVIEW

Any building which is to incorporate a significant amount of heating provided by an expensive energy source such as solar heat must be carefully designed to reduce heat loss by use of all practical energy-conservation methods. Since the payback period and life-cycle cost for energy-conservation features of a building are usually less than for a solar heating system, most of the common energy-conservation features must be included in the building design. The details of energy conservation in buildings are covered in a number of textbooks, for example Ref. 14, but are not the subject of this book. Control of heat lost through transmission and infiltration, proper design of lighting levels, and mitigation of microclimate effects are the principal areas of energy conservation in buildings.

The extent of energy-conservation features to be included in a building depends upon relative cost and benefit. For example, adding 3 in of insulation to a wall already incorporating 12 in of insulation would not be as cost-effective as providing double glazing on a single-glazed window. One of the principal heat transmission losses from buildings occurs through windows. It is more cost-effective to reduce that loss by 50 percent than to reduce the loss through an already well-insulated wall by 20 percent, since the transmission loss may account for less than one-third of the total heat loss. Energy-conservation economics is considered in Ref. 1, where recommendations are made about the best mix of various energy-conservation strategies.

"Superinsulated" residences carry energy conservation to the extreme. Heat losses are so small that internal gains provide most heat required. Although cost-effectiveness is not yet clearly established and very low fresh-air exchange rates may cause problems, the concept is worth considering. Shurcliff (64) presents the best treatment of this new concept.

The use of thermal mass within a building may also be considered an energy-conservation feature. Studies at the National Bureau of Standards have shown that thermal mass does not have much impact on the annual heating requirement but can reduce the peak heating load significantly. In many locations, peaking energy is more expensive than baseload energy, particularly if electricity is used for heating by means of a heat pump or resistance coils. Thermal mass may consist of materials similar to those discussed above for thermal storage in passive heating systems and may be located in the floors, ceilings and walls of buildings.

The orientation of buildings is obviously important. In addition to the south-facing requirement for all solar buildings, orientation must also respond to the microclimate, particularly regarding the direction of prevailing winds and the time of principal sunlight.

Load Calculation

The calculation of the benefits accruing from energy-conservation features can be done approximately by using heat-load calculation methods designed for steady state. However, the effect of mass, for example, is not present in simple heat-loss calculation methods such as bin method or the degree-day method and must be evaluated by more sophisticated analyses. Computer models of buildings are useful in determining trade-offs in energy-conservation features. Two of the most efficient programs for heat-load calculations in buildings are known as BLAST and DOE-2. These models use hourly weather data and sunshine information to calculate peak, annual, and monthly heating and cooling loads for an arbitrary number of zones of arbitrary occupancy and usage in any building. The computational algorithms in these programs result in short-run times, and a number of optional configurations may be evaluated for thermal performance and economics without great expense. These computer models are written in the Fortran computer language and are available to the public at nominal charge.

Many other computer programs for buildings have been developed; however, most are proprietary but can be used for a fee on a time-sharing basis; however, if the examination of many optional combinations of energy-conservation features is to be done, the cost of private programs may be excessive. Computer modeling of commercial buildings is now the standard method of calculating heating loads as well as calculating the performance of solar heating systems for space and DHW heating.

Investments in energy-conservation systems and solar systems may place the two in a competitive position if a limited budget is available. The amount of money to be invested in solar systems and energy-conservation systems can be determined from marginal-cost and benefit analysis. A simple overview of this method is contained in Ref. 15. If a detailed life-cycle-cost picture for a specific facility is to be developed these two sources of conventional-energy-use reduction should be studied together. Since energy conservation may take many different forms, it is not possible to calculate the best mix of solar heat and energy conservation by means of simple rules of thumb. The building computer models described above are best suited for this purpose. In the design of small commercial or residential buildings, such an exercise is normally beyond the available design budget and the best rule is to incorporate all standard energy-conservation features to their practical limit and then add solar heating, if economical, to replace most of the fossil-fuel heating loads remaining.

Energy Codes. Energy-conservation standards have been adopted in many states. The reference standard for many of these states standards is ASHRAE Standard 90, which undergoes revision periodically. This standard is a prescriptive standard giving the maximum allowable heat-loss coefficient according to the annual degree-day total for the location. Various types of buildings including residential, multiple-family, and commercial buildings as well as hotels and motels are identified in the standard. The *U* value for roofs, ceilings, floors over unheated spaces, and slab-on-grade insulation are each specified as a function of annual heating degree-days. Recommendations are provided for piping insulation levels for both heating and cooling fluids, and the function of control systems for energy conservation is described. Illumination levels to be used are specified as a function of the room-cavity ratio and the types of tasks involved.

The standard does not specify the method by which a given overall *U* value is to be achieved. Therefore, for example, any combination of wall insulation and double or triple glazing may be used to provide the required wall *U* value. Figure 2.28 shows the maximum permitted wall thermal conductance as a function of annual heating degree-days. If a location has an annual degree-day total of approximately 7200, the wall *U* value is to be 0.20. The *U* value for double glazing is approximately 0.60; therefore, the reduction of overall *U* value to the 0.20 level is

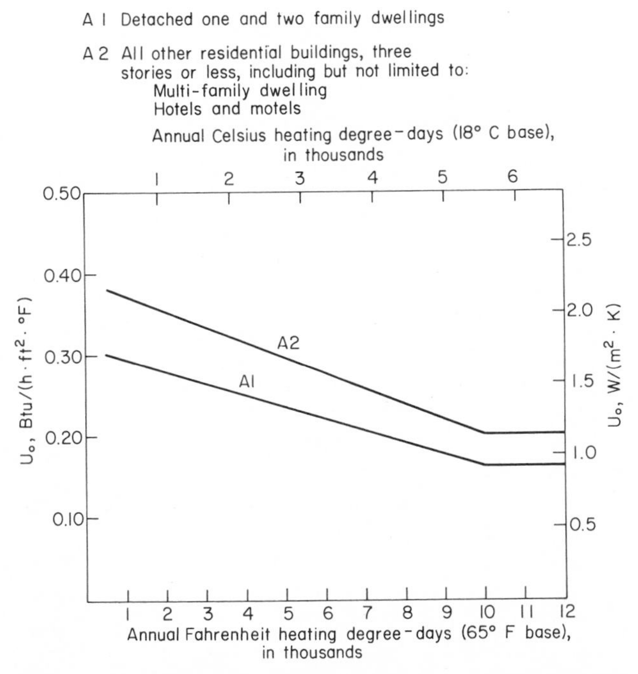

FIG. 2.28 Prescriptive U_o value curve from ASHRAE Standard 90-75 showing the average wall *U* value required for two types of buildings as a function of annual heating degree-days. *(Reprinted with permission from ASHRAE Standard 90-75, Energy Conservation in New Building Design.)*

accomplished by adding insulation to the opaque sections of wall or by triple glazing.

The amount of wall insulation required by the ASHRAE standard is normally below that which would be suggested by an economic analysis. Therefore, the use of the standard is primarily directed toward providing an energy code which can be adopted by localities or states. A simplified economic analysis will indicate where further energy-conservation features are appropriate, and the values given in the standard should not be considered as optimum.

An additional feature of energy-conserving design is the possible compromising of comfort levels in a building in order to save on cash outlays for fuels. If the comfort level of an office building, for example, is reduced beyond the point at which the workers in that office can operate efficiently, the possible costs of reduced performance far outweigh the small benefit of a reduced heating bill. For example, a worker earning \$20,000 per year and using 100 ft^2 of office space costs the employer \$200 per square foot of space per year in direct labor plus fringe benefits. A relatively inefficient building may consume 100,000 Btu/($ft^2 \cdot yr$). Heating energy costs approximately \$8 to \$10 per million British thermal units, therefore, the cost of energy per square foot per year is approximately \$1. If energy-conservation features are used to reduce the heat load by 50 percent, the saving will be approximately 50 cents per square foot per year. If these energy-conservation features reduce the efficiency of the workers by only 0.25 percent, the entire effect of energy conservation on the profit-loss statement of the firm will have been obliterated! Energy-conservation features affecting comfort include reduced thermostat settings in winter, reduced ventilation rates in summer and winter, and excessively high humidity and temperature levels permitted in air-conditioned buildings in summer. As in most economic analyses, it is important to consider all impacts in a systems context rather than isolating one budget item for particular emphasis. With imaginative design, however, most energy-conservation features can be incorporated without any deleterious effect on worker performance. In buildings with long heating seasons, *conservation and solar heat save more energy, economically, than either alone.*

INTRODUCTION TO SOLAR ECONOMICS

Many of the preceding sections have emphasized the importance of economics in solar energy system selection. Solar systems are fundamentally different from fossil or conventional heating systems in that the initial investment in the solar system is 10 to 20 times larger than that in a nonsolar system. The nonsolar system, of course, will require significant outlays year by year for fuel, whereas the solar system requires much smaller payment for fuels. It is necessary to examine these two fundamentally different types of cash-flow situations on the same basis. The method is described in engineering economics texts and goes by the name of *discounted-cash-flow analysis* (this methodology is used but not described in detail in this book). The key idea in discounted cash flows is that money has a time value. A measure of this time value of money is the interest or discount rate.

The use of the discount-rate permits future cash flows to be reduced to present dollars by dividing by the appropriate discounted-cash-flow factor. A very simple example would be a $100 investment at 10 percent. According to discounted-cash-flow assumptions, investors would be indifferent to whether they were offered $110 a year from now or $100 today. The two sums of money would have the same *present worth* since investors could invest the present $100 amount at 10 percent and have $110 available at year's end.

Discounted-cash-flow ideas are used in solar system analysis by applying present-worth factors to annual payments for the solar system. In some analyses, it is assumed that a sum of money is initially borrowed to pay for the solar system. Uniform annual payments are made to repay this loan over a period of years. The annual payments can be compared to the payments which would be made by the owner of the same building, if fuel heat were used instead of solar heat. If the total of discounted cash flows is less with the solar system than without, there is an economic incentive to use solar energy.

Another feature of solar energy systems is that they do not increase their benefit or output at a constant rate as size is increased. For example, if collector area is doubled, the amount of useful energy delivered will not be doubled. This effect is the result of reliance on a highly variable energy source—the sun. This performance characteristic is known as the law of diminishing returns.

The principal consequence of the law is that a solar system should rarely be sized to carry the entire building heating load since it would be uneconomical to do so. An economic analysis which balances the cost of additional solar capacity against the value of additional fuel savings can identify the "best" size of a solar system. This is the proper method for sizing any solar heating system. Rules of thumb relating collector area to heated floor area, for example, or to geographic location are inaccurate and can produce solar system designs which are far from the economic optimum. Solar heating systems are expensive, and errors in system sizing can greatly reduce potential benefits of solar heat use. Therefore, more care must be taken in the design of solar heating systems than has historically been taken in the design of nonsolar heating systems.

SOLAR RADIATION CALCULATIONS

The position of the sun and the amount of solar radiation striking a collector surface must be known in order to design a solar heating system. The details of solar energy calculations are given in Refs. 1 and 13. A brief summary of these calculations for the interested reader is provided in this section.

Solar Motion. For purposes of this book, the sun may be viewed as moving on an imaginary sphere called the celestial sphere centered at the center of the earth. Two angles are sufficient information to calculate the location of the sun. The two angles most commonly used are the solar altitude and azimuth angles, shown in Fig. 2.29. The altitude angle α is the angle between the sun's rays and the local horizontal plane; the azimuth angle a_s is the angle between due south and the projections of the sun's rays onto the horizontal plane.

The altitude and azimuth angles can be calculated from spherical trigonometry principles and are given in Eq. (2.21*a* and *b*) below.

$$\sin \alpha = \sin L \sin \delta_s + \cos L \cos \delta_s \cos h_s \qquad (2.21a)$$

$$\sin a_s = \cos \delta_s \sin h_s / \cos \alpha \qquad (2.21b)$$

In these equations, three fundamental angles are used—the latitude L, the solar declination δ_s, and the solar-hour angle h_s. The latitude is a measure of the site distance north or south of the equator. The solar declination relates the position of the sun at noon to the season of the year, and the solar-hour angle is a measure of the time. Solar declination can be read from Fig. 2.30 or from the more detailed declination values in Table A.1 in the appendix. The declination depends upon the month and is negative in winter and positive in summer. It is considered constant during any given day. Table A.3 in the appendix tabulates altitude and azimuth angles for several latitudes for each month.

The solar-hour angle is defined as zero at local solar noon with the 15°/h rate applied to times away from noon. For example, at 10 a.m. the hour angle is 30°, and at 4 p.m. the hour angle is 60°. The solar-hour angle is directly related to clock time with a correction of 4 min per degree of longitude required for those locations not on the standard meridians for United States time zones. The standard meridians for time zones are 75°, 90°, 105°, and 120° for the Eastern, Central, Mountain, and Pacific time zones respectively.

Altitude and azimuth angles for a specific latitude can be presented graphically in a construction called the sun-path diagram. Sun-path diagrams are described in Chap. 3 with reference to the assessment of shading problems during the programming phase of a solar system design.

The amount of radiation striking a fixed solar collector depends on the sun's position relative to the surface. The angle between the sun's rays and the perpen-

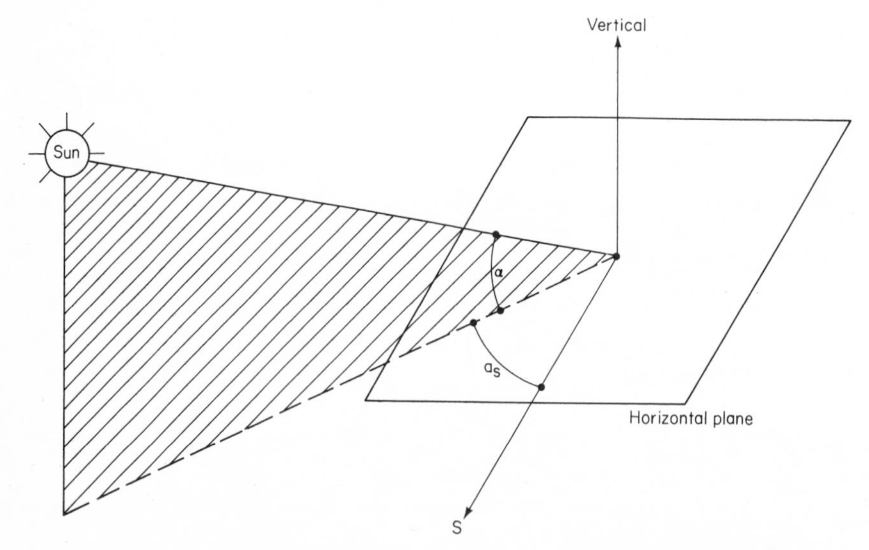

FIG. 2.29 Diagram showing solar altitude angle α and solar azimuth angle a_s.

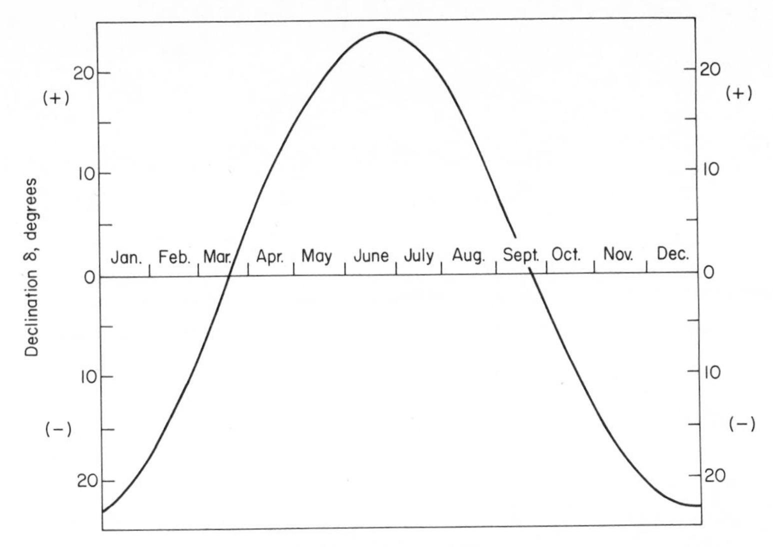

FIG. 2.30 Graph of solar declination in degrees for each month of the year.

dicular to the surface—the incidence angle—must also be known. The incidence angle i for a south-facing fixed, flat-plate collector surface is given by Eq. (2.22).

$$\cos i = \sin (L - \beta) \sin \delta_s + \cos (L - \beta) \cos \delta_s \cos h_s \tag{2.22}$$

The angle β is the collector tilt angle measured in degrees up from the horizontal. Incidence-angle equations for many other surface orientations and collector types are given in Ref. 1.

Sunrise and Sunset

The amount of solar radiation intercepted by a collector surface depends on the length of the day, the time between sunrise and sunset. The sunset time can be easily calculated from Eqs. (2.21*a*) and (2.22). When the sun sets, the solar altitude angle is zero and Eq. 2.21 can be solved for the hour angle h_{ss} $(\alpha = 0)$:

$$h_{ss} (\alpha = 0) = \cos^{-1} (-\tan L \tan \delta_s) \tag{2.23}$$

Between the spring and autumnal equinox, a day is longer than 12 h and the sun may pass "behind" the solar collector surface before it sets below the horizon. Therefore, the solar collection period is determined by collector cutoff and not by sunset. The collector-cutoff-hour angle can be calculated by setting $i = 90°$ in Eq. (2.22), the value 90° corresponding to grazing incidence of the sun's rays on the fixed flat-plate collector surface. This sunset angle is denoted by h_{ss} $(i = 90)$ and is given by Eq. (2.24).

$$h_{ss} (i = 90) = \cos^{-1} [-\tan (L - \beta) \tan \delta_s] \tag{2.24}$$

The useful collection period is the smaller of the two sunset angles given in Eqs. (2.23) and (2.24). In equation form, the sunset angle h_{ss} is given by

$$h_{ss} = \min [h_{ss} (\alpha = 0), h_{ss} (i = 90)] \tag{2.25}$$

Sunset-hour angles are used in the next section to calculate solar radiation on tilted surfaces.

Monthly Solar Radiation on Tilted Surfaces

Most solar data measured over long periods are on horizontal surfaces. In this book, we require solar radiation data, not on horizontal surfaces, but on tilted collector surfaces on a monthly time scale. Monthly solar radiation on vertical surfaces with overhangs is treated in the final part of this section.

The purpose of tilting a solar collector up from the horizontal is to favor the sun's location for the part of the year in which peak solar collection is desired. Beam radiation is the most effective form of solar radiation for heating. Therefore, solar collectors are oriented to maximize its capture.

Diffuse radiation is that radiation which issues from all directions in the sky dome other than the directions of the sun. It can also be used by solar collectors. The relative amounts of diffuse and beam radiation in the total radiation measured by the National Weather Service are not known from the published statistics. However, it has been shown (16) that their ratio can be related to a single meteorological variable called the clearness index $\overline{K}_T$. The clearness index is defined as the ratio of monthly horizontal solar radiation $\overline{H}_h$ to horizontal solar radiation on the same surface if no atmosphere were present $\overline{H}_o$:

$$\overline{K}_T = \frac{\overline{H}_h}{\overline{H}_o} \tag{2.26}$$

$\overline{H}_o$ is called the monthly averaged daily extraterrestrial radiation and can be calculated from known solar intensity just outside the atmosphere (1). Table 2.5 contains values of horizontal extraterrestrial radiation in both English and SI units.

The values for $\overline{K}_T$ and $\overline{H}_h$ for over 200 locations are given in Table A.4 in the appendix. The ratio of monthly horizontal diffuse radiation $\overline{D}_h$ to $\overline{H}_h$ can be calculated from the simple polynomial equation given below if $\overline{K}_T$ is known:

$$\frac{\overline{D}_h}{\overline{H}_h} = 1.390 - 3.909\overline{K}_T + 5.210\overline{K}_T^2 - 2.842\overline{K}_T^3 \tag{2.27}$$

Horizontal beam radiation $\overline{B}_h$ can now be calculated as the difference between the total and the diffuse component.

The next step in the calculation is to apply tilt factors to each of these radiation components. A tilt factor is the ratio of the value of a radiation component on a tilted surface divided by the value of the same component on a horizontal surface. The tilt factor for beam radiation $\overline{R}_b$ is given by Eq. (2.28).

TABLE 2.5 Average extraterrestrial radiation on a horizontal surface H_o in SI units and in English units based on a solar constant of 429 Btu/(h·ft²) or 1.353 kW/h²

Latitude, degrees	Jan.	Feb.	Mar.	Apr.	May	June	July	Aug.	Sep.	Oct.	Nov.	Dec.
					SI units, W · h/(m² · day)							
20	7415	8397	9552	10,422	10,801	10,868	10,794	10,499	9791	8686	7598	7076
25	6656	7769	9153	10,312	10,936	11,119	10,988	10,484	9494	8129	6871	6284
30	5861	7087	8686	10,127	11,001	11,303	11,114	10,395	9125	7513	6103	5463
35	5039	6359	8153	9869	10,995	11,422	11,172	10,233	8687	6845	5304	4621
40	4200	5591	7559	9540	10,922	11,478	11,165	10,002	8184	6129	4483	3771
45	3355	4791	6909	9145	10,786	11,477	11,099	9705	7620	5373	3648	2925
50	2519	3967	6207	8686	10,594	11,430	10,981	9347	6998	4583	2815	2100
55	1711	3132	5460	8171	10,358	11,352	10,825	8935	6325	3770	1999	1320
60	963	2299	4673	7608	10,097	11,276	10,657	8480	5605	2942	1227	623
65	334	1491	3855	7008	9852	11,279	10,531	8001	4846	2116	544	97
					English units, Btu/(ft² · day)							
20	2346	2656	3021	3297	3417	3438	3414	3321	3097	2748	2404	2238
25	2105	2458	2896	3262	3460	3517	3476	3316	3003	2571	2173	1988
30	1854	2242	2748	3204	3480	3576	3516	3288	2887	2377	1931	1728
35	1594	2012	2579	3122	3478	3613	3534	3237	2748	2165	1678	1462
40	1329	1769	2391	3018	3455	3631	3532	3164	2589	1939	1418	1193
45	1061	1515	2185	2893	3412	3631	3511	3070	2410	1700	1154	925
50	797	1255	1963	2748	3351	3616	3474	2957	2214	1450	890	664
55	541	991	1727	2585	3277	3591	3424	2826	2001	1192	632	417
60	305	727	1478	2407	3194	3567	3371	2683	1773	931	388	197
65	106	472	1219	2217	3116	3568	3331	2531	1533	670	172	31

$$\overline{R}_b = \frac{\cos (L - \beta) \cos \delta_s \sin h_{ss} + h_{ss} \sin (L - \beta) \sin \delta_s}{\cos L \cos \delta_s \sin h_{ss} (\alpha = 0) + h_{ss} (\alpha = 0) \sin L \sin \delta_s} \tag{2.28}$$

Values of $\overline{R}_b$ are also given in Table 2.6.

The tilt factor for diffuse radiation is difficult to calculate. However, if the assumption is made that the distribution of diffuse radiation is uniform over the sky dome, the calculation is greatly simplified. Then the tilt factor $\overline{R}_d$ is given by Eq. (2.29), where β is the tilt angle.

$$\overline{R}_d = \cos^2 \tfrac{1}{2}\beta \tag{2.29}$$

In addition to beam and diffuse components solar radiation can also be reflected from the collector foreground onto the surface of the collector. The amount of radiation which is reflected depends upon the reflectance of the foreground, which may vary from 0.7 for snow to 0.2 for concrete or gravel. A table of foreground

TABLE 2.6 Beam radiation tilt factor for various latitudes and collector tilts*

Latitude	*Jan.*	*Feb.*	*Mar.*	*Apr.*	*May*	*June*	*July*	*Aug.*	*Sep.*	*Oct.*	*Nov.*	*Dec.*
						(a) Tilt = latitude − 15°						
25	1.22	1.15	1.08	1.01	0.97	0.95	0.96	1.00	1.05	1.13	1.20	1.24
30	1.38	1.26	1.14	1.03	0.96	0.93	0.95	1.00	1.10	1.22	1.35	1.42
35	1.59	1.40	1.21	1.06	0.96	0.92	0.94	1.02	1.15	1.33	1.54	1.66
40	1.80	1.58	1.31	1.10	0.97	0.92	0.94	1.04	1.22	1.46	1.80	2.00
45	2.31	1.83	1.43	1.15	0.98	0.92	0.95	1.07	1.31	1.69	2.17	2.51
50	2.99	2.18	1.60	1.21	1.00	0.92	0.96	1.11	1.43	1.97	2.75	3.34
55	4.19	2.71	1.82	1.28	1.02	0.93	0.97	1.17	1.58	2.37	3.71	4.92
						(b) Tilt = latitude						
25	1.47	1.31	1.14	0.98	0.87	0.83	0.85	0.93	1.07	1.25	1.43	1.53
30	1.66	1.43	1.20	1.00	0.87	0.81	0.84	0.94	1.12	1.35	1.60	1.74
35	1.91	1.58	1.28	1.02	0.87	0.81	0.83	0.96	1.17	1.48	1.82	2.02
40	2.25	1.79	1.38	1.06	0.88	0.80	0.84	0.98	1.24	1.65	2.12	2.42
45	2.76	2.06	1.51	1.11	0.69	0.80	0.84	1.01	1.34	1.87	2.56	3.03
50	3.54	2.46	1.68	1.17	0.90	0.81	0.85	1.05	1.45	2.17	3.22	4.01
55	4.93	3.05	1.91	1.24	0.92	0.81	0.86	1.10	1.61	2.62	4.32	5.87
						(c) Tilt = latitude + 15°						
25	1.63	1.38	1.12	0.88	0.73	0.66	0.69	0.81	1.02	1.29	1.56	1.71
30	1.83	1.50	1.18	0.90	0.72	0.65	0.68	0.82	1.06	1.39	1.74	1.94
35	2.10	1.66	1.25	0.92	0.72	0.64	0.68	0.83	1.12	1.52	1.98	2.25
40	2.47	1.87	1.35	0.95	0.73	0.64	0.68	0.85	1.18	1.69	2.30	2.69
45	3.01	2.16	1.48	1.00	0.74	0.64	0.68	0.88	1.27	1.92	2.77	3.34
50	3.86	2.57	1.05	1.05	0.75	0.64	0.69	0.91	1.38	2.23	3.47	4.41
55	5.34	3.18	1.88	1.12	0.77	0.65	0.70	0.95	1.53	2.68	4.64	6.42

*The solar collector tilt factor is the ratio of monthly beam insolation on a tilted surface to monthly beam insolation on a horizontal surface.

reflectances is given in Ref. 1. The tilt factor for reflected radiation $\overline{R}_r$ is given by Eq. (2.30) in which ρ is the foreground reflectance.

$$\overline{R}_r = \rho \sin^2 \tfrac{1}{2}\beta \tag{2.30}$$

The total radiation on the collector plane is the sum of each tilt factor multiplied by its radiation component. This is expressed in equation form below.

$$\overline{I}_c = \overline{R}_b\overline{B}_h + \overline{D}_h \cos^2 \tfrac{1}{2}\beta + \overline{H}_h\, \rho \sin^2 \tfrac{1}{2}\beta \tag{2.31}$$

$\overline{I}_c$ is called the monthly averaged, daily collector-plane radiation and will be used in performance prediction procedures in subsequent chapters.

Example. Calculate the solar radiation on a surface tilted at 56° in Ogallala, Nebraska, for the month of February.

Data Required for the Calculation. To calculate the total collector-plane insolation, the horizontal terrestrial and extraterrestrial values must be known along with the three tilt factors. The extraterrestrial radiation interpolated from Table 2.5 for the latitude of Ogallala (41°) is 1730 Btu/(ft^2·day). Terrestrial horizontal radiation is 892 Btu/(ft^2·day).

The tilt factor $\overline{R}_b$ can be calculated from Eq. (2.28) or interpolated from Table 2.6. The value of h_{ss} must be known to use Eq. (2.28). Since the length of the day is less than 12 h in February, Eq. (2.23) applies and the sunrise-hour angle is 70.4° (the declination for the middle of February is −13°). The beam radiation tilt factor $\overline{R}_b$ is 1.94.

The diffuse tilt factor is calculated from Eq. (2.20) and is 0.78. The reflected radiation factor $\overline{R}_r$ is given by Eq. (2.30). If a reflectance of 0.5 is used, $\overline{R}_r = 0.11$. These numbers are the inputs to the five-step calculation of tilted-surface solar flux.

The Five-Step Calculation Sequence. First, calculate the clearness index $\overline{K}_T$ from Eq. (2.26). For the terrestrial radiation value of 892 Btu/(ft^2·day), $\overline{K}_T = 0.52$.

Second, find the diffuse-to-horizontal ratio from Eq. (2.27) using the preceding value of $\overline{K}_T$. The diffuse-to-horizontal ratio for February is 0.36. Third, calculate the horizontal beam and diffuse components. The diffuse component $\overline{D}_h = 0.36 \times 892 = 320$ Btu/(ft^2·day). The difference between total horizontal radiation and the diffuse component is the horizontal beam component $\overline{B}_h$, which is 572 Btu/(ft^2·day).

Fourth, evaluate the three tilt factors explained above. Fifth, multiply each radiation component by its tilt factor and then add the three together using Eq. (2.31). The tilted surface insolation value $\overline{I}_c = 1439$ Btu/(ft^2·day).

Solar Radiation on Vertical Surfaces with Overhangs

Many passive heating systems use vertical, south-facing apertures with overhangs to reduce summer solar gain through the glazing. Sizing of overhangs (see Fig. 2.31*d*) can be carried out by matching shading characteristics to heat demand.

Since overhangs are not designed to cut off solar gain completely on a particular day but rather over a period of weeks, the attenuation of solar flux must be calculated on an hour-by-hour basis during transition periods of the year. Results of these calculations carried out by Utzinger and Klein are shown in Fig. 2.31.

Figure 2.31 shows, for three values of latitude (35°N, 45°N, and 55°N) and three values of relative overhang width W, the monthly fraction $\bar{f}_i$ of vertical-surface beam radiation which reaches a vertical surface shaded by the specified overhang. Also, three values of relative width (overhang physical width divided by aperture height; see Fig. 2.31*d*)—1, 4, and 25—are plotted. The horizontal axis in Fig. 2.31 is the relative projection from the wall, P (distance of overhang edge from wall divided by aperture height).

As an example, consider the amount of beam radiation reaching a 5-ft-square window shaded by a 3-ft overhang in Albuquerque ($L = 35°N$) in January and June. For this example $W = 1$, $P = 0.6$. Then, from Fig. 2.31*a*, $\bar{f}_i = 0.72$ for January and $\bar{f}_i = 0.17$ for June.

The figures apply for horizontal overhangs whose bottom plane intersects the vertical wall at the top of the aperture as shown in Fig. 2.31*d*. The overhang has the same width as the window. Other geometries are covered in Ref. 49. If Fig. 2.31*d* applies, $\bar{f}_i$ for other values of latitude and W may be interpolated. The interpolation in latitude is linear, and in relative width W the interpolation is based on $1/W$, not on W.

The fraction $\bar{f}_i$ of *beam* radiation can be used in Eq. (2.31) to find the *total* vertical surface flux with an overhang:

$$\bar{I}_{c,\text{vert}} = \bar{R}_b \bar{B}_h \bar{f}_i + \bar{D}_h F_{rs} + \frac{\rho}{2} \bar{H}_h \qquad (2.32)$$

F_{rs} in the second term is the "view factor" between the vertical aperture and the sky. It is tabulated in Table 2.7 for the geometry of Fig. 2.31*d*; with no overhang $F_{rs} = \cos^2 \beta/2 = 0.5$ as in Eq. (2.31).

An important conclusion of the work in Ref. 49 is that overhangs do not block much more than 50 percent of the incident sunlight from a vertical surface in summer in sunny locations, even if the beam component is fully blocked, because of inability of overhangs to block diffuse and reflected radiation. Therefore, if solar gain is to be completely blocked from a passive system, external shutters or awnings must be used in summer. An overhang by itself is not enough.

Systems of Units Used in Solar Radiation Calculations. Four different sets of solar radiation units are in common use. In English units, the solar radiation is expressed in British thermal units per square foot per hour when an hourly time scale is used. In many solar radiation calculations, the langley is used. The langley is a unit named for the man who first made accurate solar radiation measurements for the Smithsonian Institution and is defined as a flux of 1 calorie per square centimeter. Solar flux data are usually expressed as langleys per hour or langleys per day. In the SI set of units, both kilowatts per square meter and kilojoules per square meter per hour are used. Most designers use English engineering units, and the

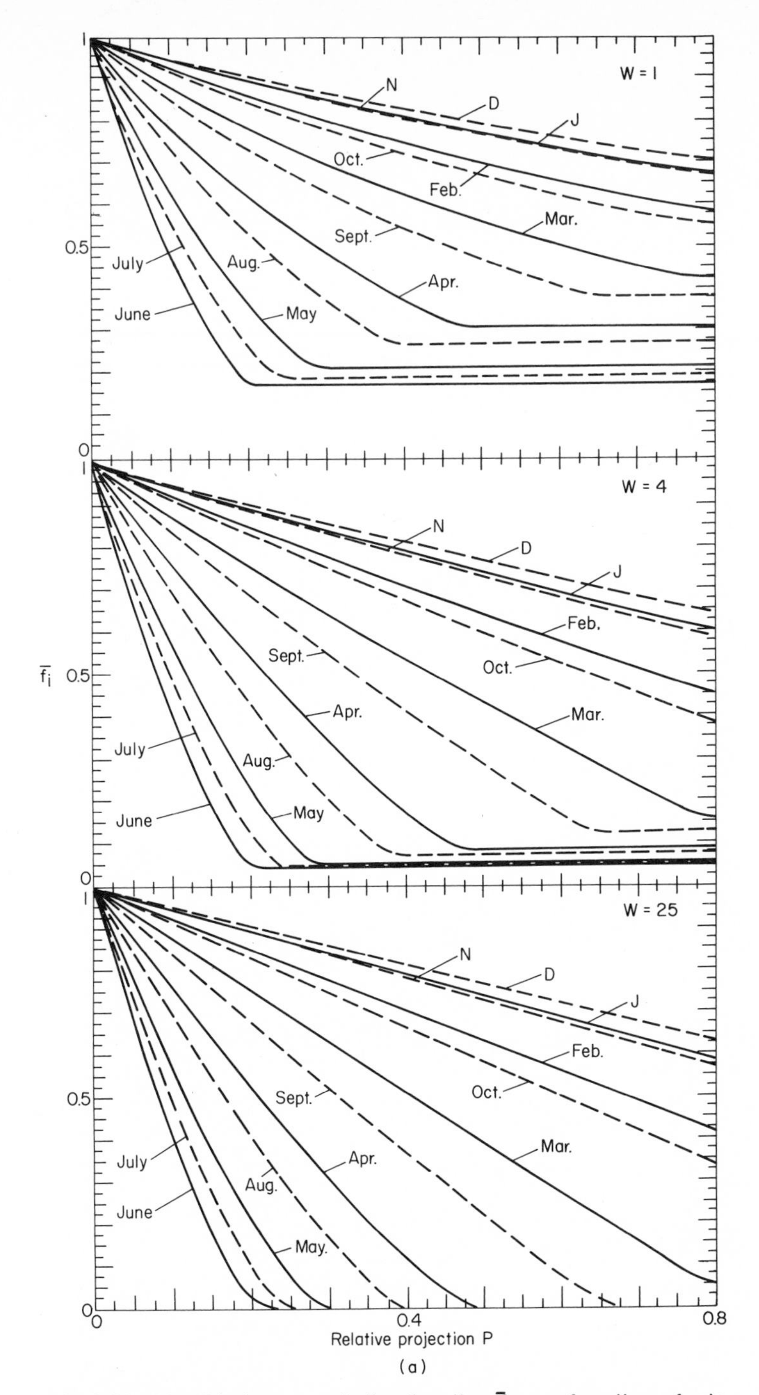

FIG. 2.31 Monthly beam radiation fraction $\bar{f}_i$ as a function of relative overhang projection P for three relative widths W: (*a*) 35°N latitude, (*b*) 45°N latitude, (*c*) 55°N latitude, (*d*) geometrical configuration. *(Reprinted with permission from Solar Energy Journal, D. M. Utzinger and S. A. Klein, Vol. 23, copyright 1979, Pergamon Press, Ltd.)*

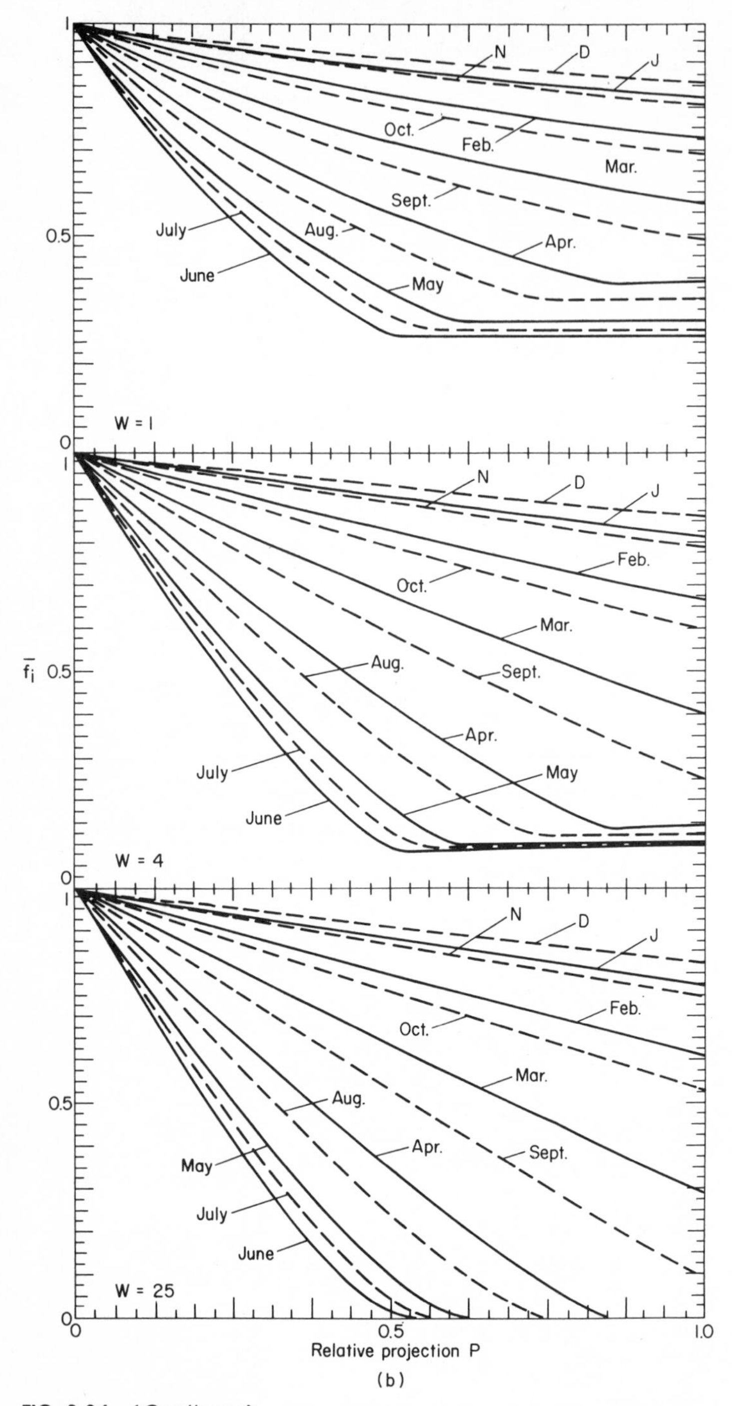

FIG. 2.31 *(Continued)*

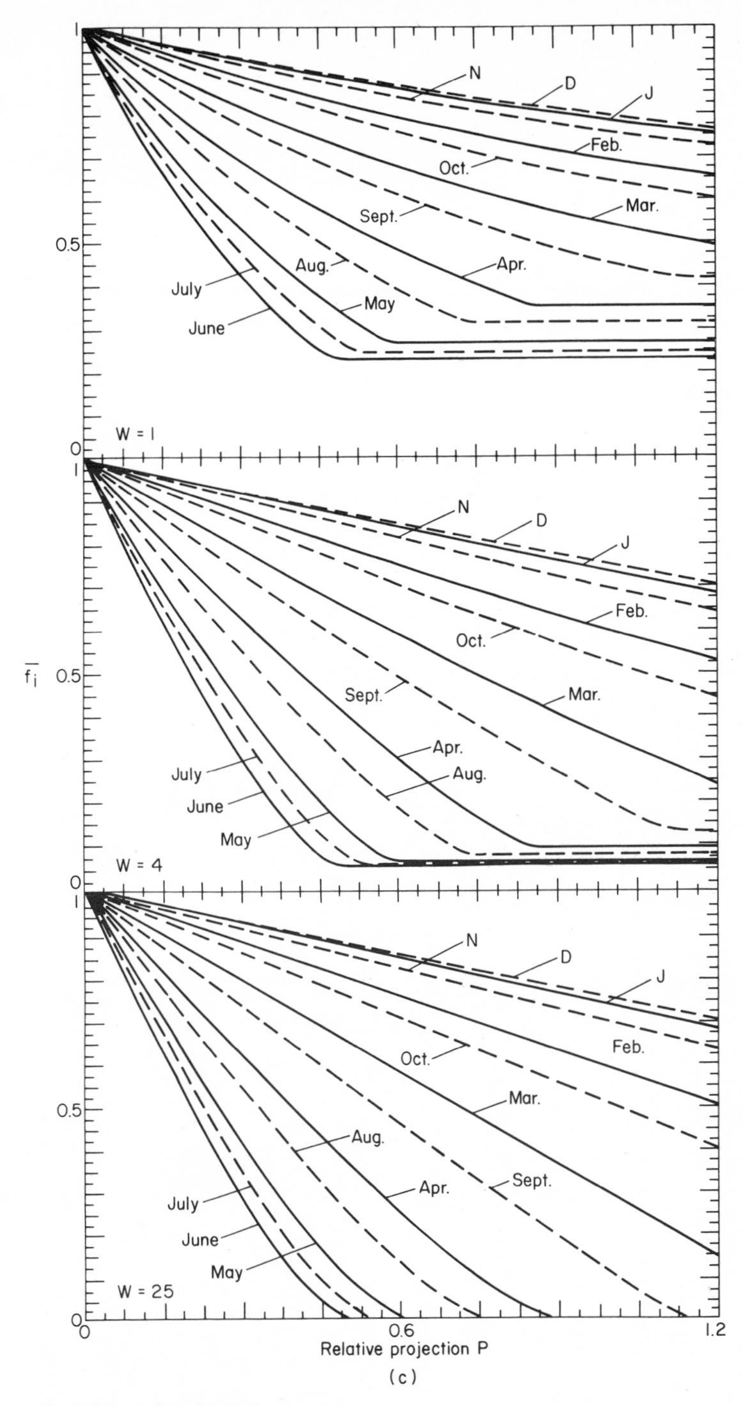

FIG. 2.31 *(Continued)*

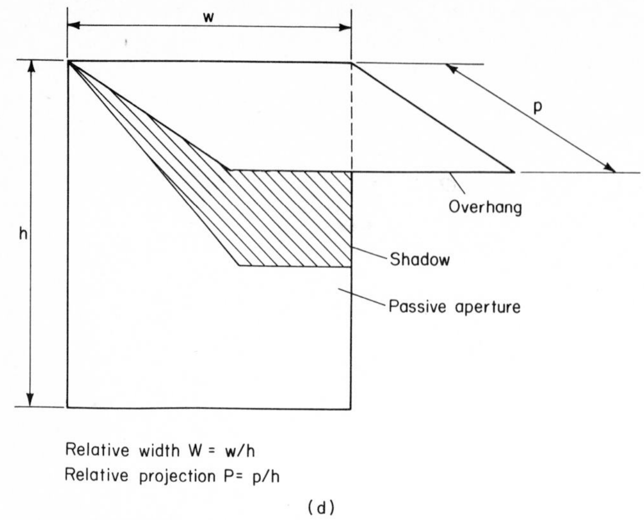

FIG. 2.31 (*Continued*)

only conversion normally required is between langleys and the English unit. To calculate British thermal units per square foot per day from langleys per day, multiply the latter by 3.687. Figure 2.32 is a nomograph which can be used to convert from one set of units to another. Place a horizontal straightedge at the value of solar flux in the known set of units, and read the value in any of the other three sets of units from the straightedge.

TABLE 2.7 View factor F_{rs} between a vertical receiver and the sky*

Relative projection P	*Relative width W*		
	1.0	*4.0*	*25.0*
0	0.50	0.50	0.50
0.1	0.46	0.46	0.45
0.2	0.42	0.41	0.41
0.3	0.40	0.38	0.37
0.4	0.37	0.35	0.34
0.5	0.35	0.32	0.31
0.75	0.32	0.27	0.25
1.00	0.30	0.23	0.21
1.50	0.28	0.19	0.15
2.00	0.27	0.16	0.12

*Reprinted with permission from *Solar Energy Journal*, D. M. Utzinger and S. A. Klein, Vol. 23, copyright 1979, Pergamon Press, Ltd.

Note: Refer to Fig. 2.31*d* for definition of *P* and *W*.

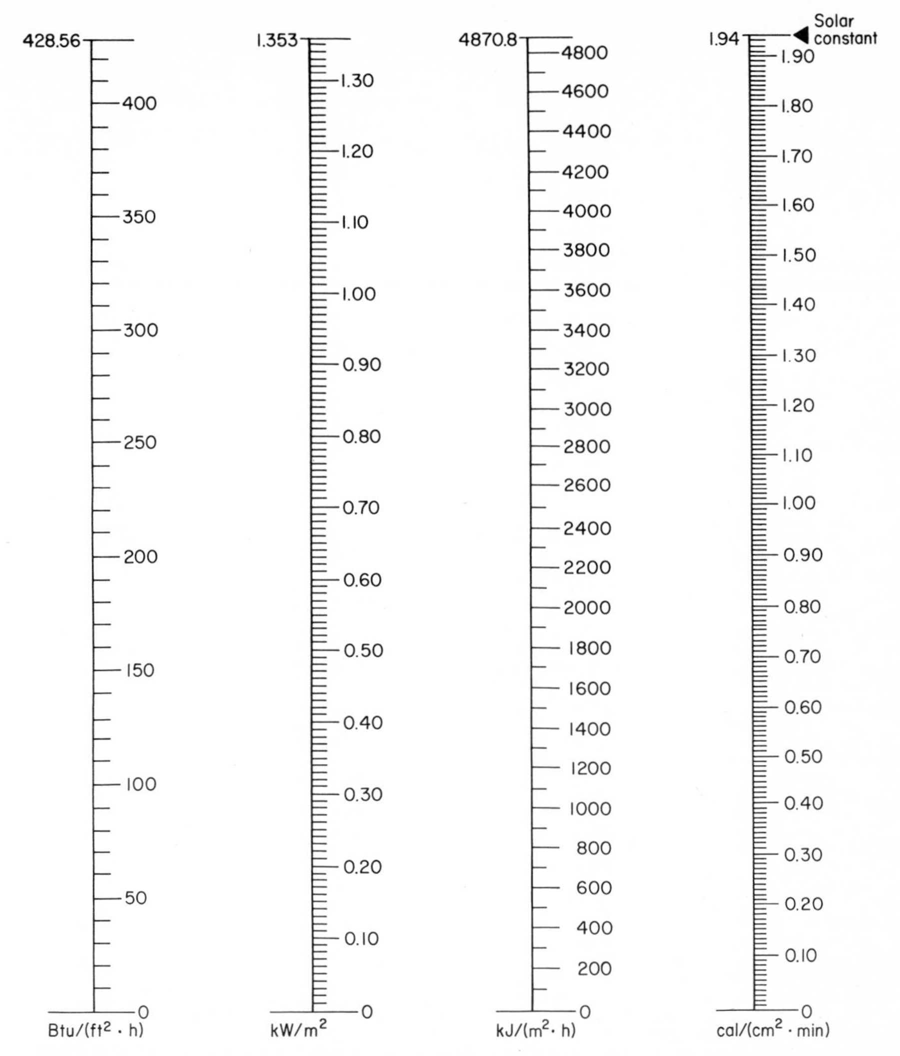

FIG. 2.32 Nomograph to convert from one set of solar radiation units to another. Conversion factors: kW/m^2 to $kJ/(m^2 \cdot h)$, $\times$ 3600; $kJ/(m^2 \cdot h)$ to kW/m^2, $\times$ 0.00028; kW/m^2 to langleys/min, $\times$ 1.432; langleys/min to kW/m^2, $\times$ 0.698; kW/m^2 to $Btu/(ft^2 \cdot h)$, $\times$ 316.75; $Btu/(ft^2 \cdot h)$ to kW/m^2 $\times$ 0.00316. *(Courtesy of Alpha Solarco.)*

SUMMARY

This chapter has summarized the configuration and operation of all solar heating systems to be discussed in this book. Subsequent chapters will deal with variations on these designs, rules for sizing the important components, economic analysis, and design checklists.

The brief overview of solar systems provided in this chapter did not emphasize the theoretical basis for the information given. Theoretical textbooks such as Ref. 1 can be consulted if the reader wishes to reexamine the assumptions and theoretical considerations leading to the material presented in summary form in this chapter.

In the remainder of this book, it is assumed that the reader is familiar with the operation of the solar heating systems presented in this chapter. *The design rules, specifications, and procedures to be presented deal with the design process and not with the physics underlying the performance of a given component of the solar energy system or of the system itself.*

3 PROGRAMMING AND SCHEMATIC DESIGN

One of the great maladies of
our time is the way sophistication
seems to be valued above common sense.

NORMAN COUSINS

The first two steps in the design of a solar building are described in this chapter. The first step—programming—defines the building and the conditions under which construction, occupancy, and use take place. The size of the building is identified, along with the expected budget and all site information. Schematic design is the first step in reducing the building requirements developed during programming to one or more building concepts represented by schematic drawings. Schematic drawings and diagrams show all major components of the building including architectural, structural, and mechanical items in their general relationship one to another. The end product of the schematic design phase is a set of schematic drawings and calculations, a general outline specification of the building, and data on its expected cost.

In this chapter, the conventional architectural process is briefly described first, with regard to both programming and schematic phases. Next, the solar aspects of the two phases are described in more detail. Methods of estimating solar system size, expected performance, and cost are presented.

PROGRAMMING

The American Institute of Architects *Handbook of Professional Practice* states: "Before any real progress can be made in the design of a building it is necessary to have a definite design program giving conditions precedent to construction and requirements to be met by the proposed project. The owner is expected to furnish all basic data required of the program. The architect formulates the program after gathering this basic information. . . ."*

**The American Institute of Architects Handbook of Professional Practice* (Washington, D.C.: AIA, 1969), Chap. 11, p. 6.

Prior to programming the owner will normally have acquired the land upon which the building is to be built. This is not always the case, and the architect may be asked to assist in land acquisition although this is not a normal part of programming and requires an additional fee. If land acquisition has proceeded normally, the site is stipulated for the building design process and is not subject to change. Another input from the owner which is typically not subject to significant change is the budget. The project budget is one of the first items furnished by the owner and is one of the key determinants of building type and size. Budget negotiation may be needed to match requirements to funds.

During programming, the first building variable to be identified is the type of building, that is, the purpose for which the building is to be constructed. Several building types can be identified, the first of which is residential. Residential buildings may be either single-family detached dwellings, multiple-family low-rise buildings, or multiple-family mid- to high-rise buildings. The next type—commercial buildings—includes small individual stores or shops, shopping centers, regional shopping centers, office buildings, either high-rise or low-rise, and office parks.

Industrial buildings frequently built for sale or lease may be of the light-industrial type, medium-industrial type including small fabricating businesses, or heavy industrial. Heavy-industrial buildings include process industries, large manufacturing plants, etc. The fourth class—institutional buildings—includes educational facilities such as colleges, universities, and primary and secondary schools, as well as special education facilities. Health care and correctional institutions are included in this category, and governmental administrative or military installations may also fall within the institutional classification.

The fifth category of building types includes recreational and other special-purpose buildings such as vacation facilities, sports complexes, libraries, and laundries.

Site Data

Data concerning conditions existing at the site are collected during programming. The initial site data collected will be contained in a topographic map which may be ordered from a land surveyor. The data must include boundaries of the property, rights of way, existing easements, and utility poles. Sewers, water and gas mains, and water taps should be located. Any interior roads, driveways, walkways, fences, large boulders, or walls are to be identified.

Additional information in the site survey includes the variety and location of trees within the site, and buildings or large trees on nearby properties which could shade solar collectors on the site. This information is contained in a solar horizon plot. Elevations of first-floor and basement levels of existing buildings should be identified since buried solar storage tanks may be specified at a later design step. Contour lines at specified intervals are on the site map, and a permanent benchmark should be established for use during construction.

Subsoil investigations may be carried out at this time, and the location of the local water table should be determined either by test drill or by contact with the county or city engineer. The location of the water table is important when assessing buried thermal storage. Seismic-zone type should be identified (Ref. 60 contains a seismic zone map for the United States).

Programming Investigations

In the standard architectural programming phase, the investigation of information furnished by the owner and by the land surveyor will lead to criteria determining the various types of buildings and facilities suitable for the required function. Additional investigations are required for solar system programming and are described in this section.

Programming studies include investigation of local and state laws, codes (see Ref. 65 for a model solar code), and customs which will impact building design. Existing land-use patterns should be determined by conference with the local planning commissions, and transportation features in the vicinity should be identified. The suitability of the site from a topographical viewpoint must be assessed, and the cost of land per unit area identified.

A particularly important investigation is the determination of applicable regulations. These include local, state, and federal regulations which control certain design features and construction. Zoning ordinances determine the construction type and height limits as well as the many other variables in the design process. Solar access zoning is particularly important.

Applicable building, plumbing, electrical, mechanical, and special codes must be identified. If an industrial building is to be designed, applicable safety codes and Occupational Safety and Health Administration (OSHA) requirements must be determined. If an elevator code applies to a high-rise building, it must be identified at this time. Federal government projects will have additional criteria.

Another investigation regards the determination of availability of utilities. The availability and connection plus unit costs of water, gas, electricity, district steam, sewers, and other required utilities must be determined at this time. If a gas hookup will require an application followed by a 1- or 2-year waiting period, as is common in many parts of the United States, the owner must be informed of this so that early application for gas can be made. The expected electrical service requirements and applicable rates should be identified at this time. If sewer discharge temperature is limited, this must be known, since it may affect certain aspects of the mechanical system. Programming investigation must also include the quantitative description of site conditions. The character and condition of nearby buildings must be known, and their landscaping and subsurface structure should be determined.

Particularly important in the design of an energy-conserving solar building is a quantitative assessment of local microclimatic conditions. The direction and speed of prevailing winds must be known. Other weather data—including minimum, maximum, and average wet- and dry-bulb temperature; diurnal temperature swing by season; heating degree-day totals; solar radiation levels; and seasonal snow and rain amounts—must be determined. Hourly data for all weather parameters is necessary if computer simulation is to be done. Other useful microclimatic data include expected hail and lightning, the nature and quantity of existing air pollution, and site orientation with respect to the due south direction.

Space Requirements and Size

A major programming activity identifies the activities and functions to be carried out in each area of the building. Floor-area requirements, occupancy schedule, and internal gain profiles will be studied. Mechanical requirements such as operating temperature range, night setback, acceptable humidity levels, and outside air ventilation rates will also be identified. Any specific space location requirements will be noted. For example, a certain area may require natural daylighting. Hence, it will be positioned on the north side of the building.

In industrial and some institutional applications, not only are the area requirements for process machinery considered, but also the flow of materials and people determining relative area placements is studied. One of the key determinants of the floor plan may be a smooth flow of materials from one processing step to the next. Circulation and functional relationships are key inputs to the programming effort and must be determined in conference with the owner.

In addition to continuously used spaces, service and intermittently used spaces must also be identified. These include storage and parking facilities, as well as special equipment spaces. Electrical and mechanical demands and diversity are identified, and the location and size of mechanical space is stipulated. As described shortly, the mechanical space for solar-heated buildings may be somewhat larger than that present in nonsolar buildings because of the additional mechanical equipment needed (for active systems). At the completion of this step the ratio of net to gross area for the building can be calculated. This parameter is useful later in estimating lighting and heating requirements.

Other space parameters determined during programming may include finishes of surfaces, colors, and interior furnishing needs. Safety, security, and maintenance facilities are identified, and expected foundation and structural requirements based on soil survey information are determined, in general. In addition to areas of spaces, their volumes and nonstandard ceiling heights should be specified. It is the task of the architect to determine these space and volume requirements from the owner's guidelines regarding function.

The best combination of spaces meeting the requirements of the owner may not be immediately obvious. Visits to similar facilities can be made during programming, and very rough scale models may be used to elucidate the process. Photographs of other installations may be useful.

Energy and Economics

Agreement must be reached on equipment sizing criteria and complexity and extent of energy studies. Parameters for life-cycle costs must be established. If an energy budget for the building envelope and processes is to be used, it must be established at this time in terms of either net or gross (or both) consumption per year. Peak electrical demand may also be stipulated.

Additional Programming Activities for Solar-Heated Buildings

In the design of many buildings, the use of solar energy will impose additional activities on this and every subsequent design phase. In the past *it has become obvious in later stages that all required data have not been quantified during programming. Much extra effort was required in schematic design and design development* as a result.

The key determinants to the performance of a solar heating system are the availability of sunlight during the heating season and the heating load. In this section, the additional program-level detail required to identify these matters is given quantitatively. Variables determining the availability of sunlight to an active or passive solar system include the building orientation, solar radiation intensity levels, and possible shading from structures and trees on the site or on adjacent sites.

Solar Availability. Solar radiation data for many sites in the United States are given in Table A.4 in the appendix. If the location under study is not listed in this table, data for a nearby location can be used. However, geographical distribution of solar energy can vary significantly on a relatively small scale of the order of tens of miles. Therefore, if the appendix data are to be used, no specific microclimatic effects causing a variation should exist. For example, fog or urban air pollution at a site may reduce solar radiation data versus those measured at an official National Weather Service (NWS) station located at a nearby airport outside the city.

The second determinant is building orientation, siting, and any possible shading which may be present during the heating season. Shading of solar collectors is a particular problem in the winter when sun angles are very low and shadows are very long. This is shown schematically in Fig. 3.1, in which the relative positions of the sun in summer and winter are indicated. Relatively longer shadows occur in winter, and nearby obstructions will have their greatest effect on collector arrays during this time of the year.

In addition to existing obstructions, the possibility of future obstructions must be considered. For example, a future tall building built south of the site may have more impact on solar availability than existing masses on the site. Zoning restrictions determine the height limit on future construction. In order to determine the shading impact of such future projects, a shadow map should be prepared (as described in the technical supplement section immediately following, "Sun-Path Diagrams and the Shadow-Angle Protractor") assessing the worst case conditions of height and location.

One way of assessing shading problems uses the "solar window" idea. Figure 3.2a shows an example of a solar window construction. Its upper and lower boundaries are determined by the sun's paths on the longest and shortest days of the year, June 21 and December 21. The right and left boundaries of the solar window are

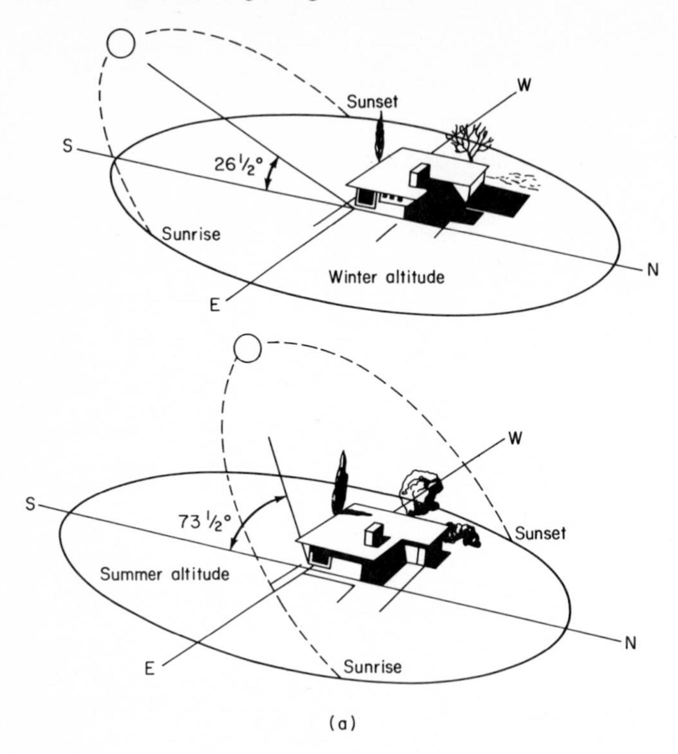

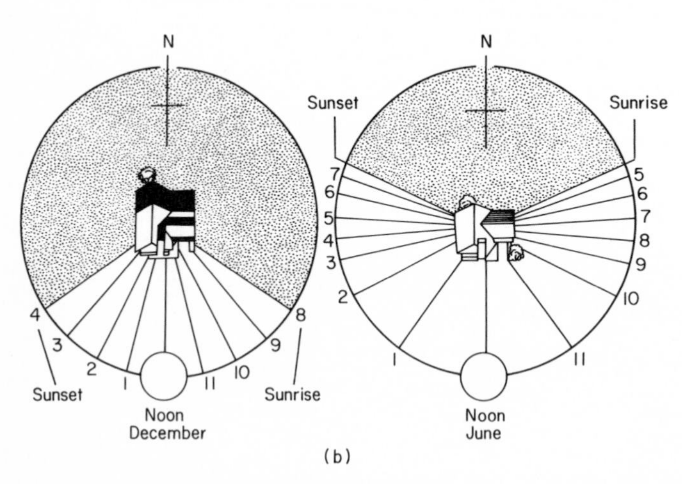

FIG. 3.1 Sun angles (40°N): (*a*) summer and winter solar paths showing noontime solar altitude angles, (*b*) plan view of solar motion in summer and winter.

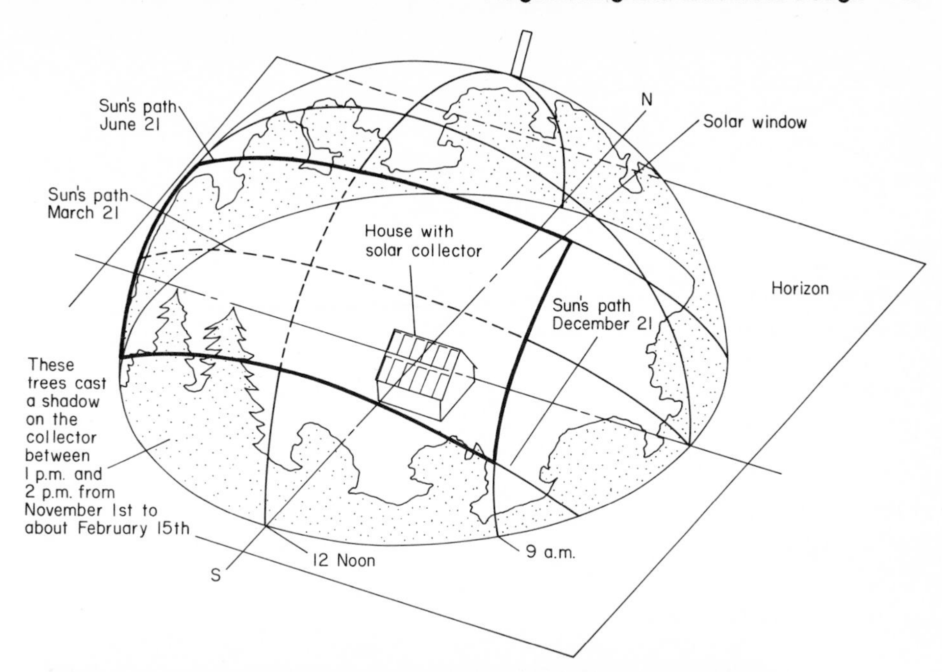

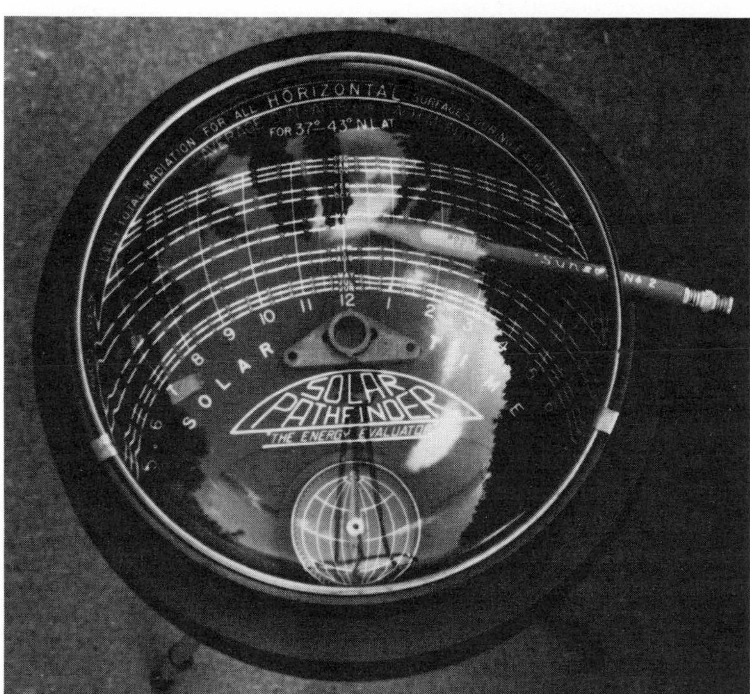

FIG. 3.2 (*a*) Solar "window" construction showing the relationship of the position of the sun to a solar building during the central part of the day when solar heat collection is most efficient. (*b*) Solar shading estimator manufactured by Solar Pathways, Inc. *(Courtesy of Solar Pathways, Inc., Glenwood Springs, Colo.)*

determined by the central hours of the day during which most solar collection takes place. Although the sun may rise at 8 a.m. in a specific location, the little energy available during early morning may not be collectable. Therefore, as shown in Fig. 3.2*a*, collection will not start in winter until approximately 9 a.m. on a sunny day.

Any obstruction penetrating into the solar window will cast a shadow on the collector. Although the impact of this shading cannot be directly determined by means of the solar window construction, at least the problems can be identified. Specific hours and months when shading occurs can be determined by use of sun-path diagrams described in the section "Sun-Path Diagrams and the Shadow-Angle Protractor."

The glass-dome device shown in Fig. 3.2*b* affords a method of studying the solar window quickly in the field. When the dome is properly positioned, its reflection will show the time of the year and hours of the day during which shading is caused by any obstruction in the vicinity. The device uses the sun-path-diagram idea.

Solar Collector Sizing—Programming Phase

After building areas have been determined by the architect, the total floor area of the building can be determined. Using average heat load and internal gain values for similar buildings of the same type, the net annual heat load may be estimated. Finally, on the basis of average solar performance for buildings in the same area, an estimate of optimum collector area can be made.

It is incorrect to base collector area estimates on building floor area only. This becomes clear after a little thought. Consider, for example, identical office buildings located in Minneapolis and in Albuquerque. Precisely the same building has vastly different heat loads in the two locations. As a result, the optimum collector area will be vastly different. The proper method of estimating collector area for programming is to use building floor area to identify expected heat gains and losses. The net heat load can be used to find the collector area by the method delineated below.

The heat load on buildings which are surface dominated is about 6 to 8 Btu/(degree-day$\cdot$ft^2) (2 to 5 for well-insulated buildings), depending upon insulation levels and other energy-conservation features. For ventilation-dominated buildings, the analogous number is easily calculated. The total annual load on the building can be determined by multiplying this unit heat load by the number of degree-days per year [to the proper base, accounting for internal gains and night setback; refer to Eq. (2.20) and Table A.4] multiplied by the heated floor area, assuming each day of the week uses the same schedule. The product gives an estimate of the annual heating load in millions of British thermal units per year. The range of values of this number will be between 40,000 and 200,000 Btu/(ft$^2\cdot$yr), depending on many factors. (Building Energy Performance Standards to be promulgated in 1983 will specify this goal according to location and building type.) The annual energy demand for water heating can be determined by the product of expected daily use per person (from Table 2.4) times the number of persons times the expected temperature rise times the number of use days in the year. This water-heating load plus standby losses added to the total space-heating load represents the total heating load on the building.

An approximation to the required collector area can be made by dividing the total load by the expected *useful* energy delivery per square foot of collector for a location similar to that being analyzed. For example, if a heating load is calculated above to be 1000×10^6 Btu/yr net of internal gains and a solar system on the average delivers 200,000 Btu/($ft^2 \cdot yr$), the estimated collector area is 5000 ft^2. As will be seen in subsequent chapters, the amount of useful energy delivery per square foot of collector decreases with increasing solar fraction. However, if an arbitrary solar fraction is picked—say 50 or 75 percent—it is possible to calculate the useful energy delivery per square foot of a typical collector for average loads. Then this value and the annual load can be used to estimate collector area. Table 3.1 lists expected energy deliveries for an arbitrary solar heating fraction of 50 percent for a typical system. Of course, the final design value of solar heating fraction may be 80 percent as determined in subsequent chapters.

Another approach to programmatic sizing uses the map in Fig. 3.3. The area to provide about 75 percent of the annual demand is calculated by dividing the estimated unit load by the number from the map for the location in question. For example, a 10,000 Btu/degree-day building in Albuquerque would require roughly 10,000/30 or 330 ft^2 of collector. The map is most suitable for skin- or ventilation-dominated buildings.

Once the size of the collector array has been roughly estimated, the availability of space for this array on the building can be studied. If the array is to be placed on

TABLE 3.1 Useful solar heat delivery for a typical active system (50 percent solar heating fraction)*

City	*Useful energy delivery, British thermal units (per net square foot of collector per year)*
Boston	147,000
New York	132,000
Baltimore	181,000
Rochester	161,000
Pittsburgh	158,000
Detroit	145,000
Memphis	204,000
Minneapolis	166,000
Des Moines	174,000
Kansas City	192,000
Great Falls	229,000
Salt Lake City	215,000
Seattle	122,000
San Francisco	276,000
Montreal	129,000
Vancouver, B.C.	100,000

*From Ref. 18 with permission.

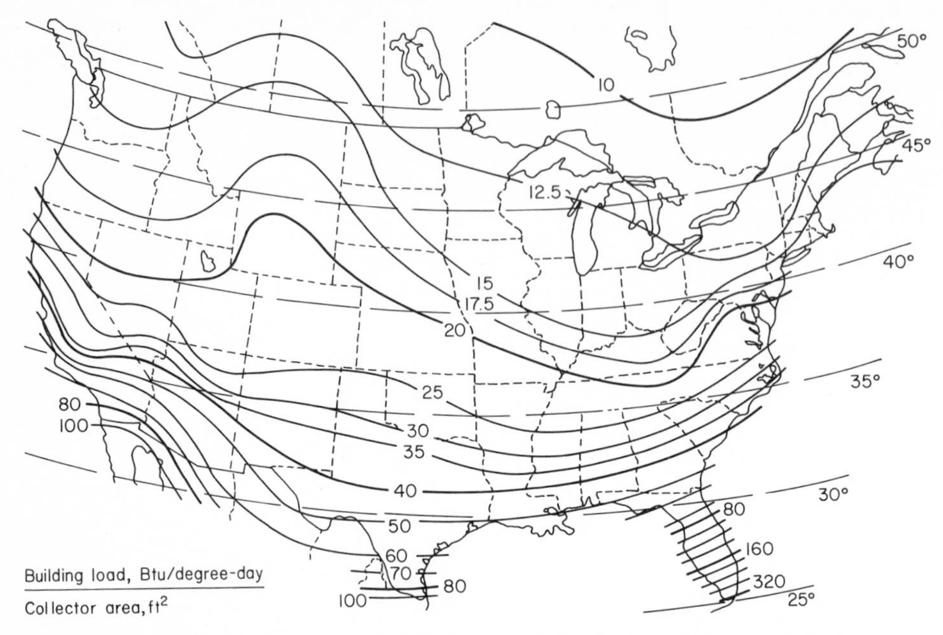

FIG. 3.3 Map showing the ratio of unit heat load (Btu/degree-day) to collector area required to provide about 75 percent of total annual space-heating demand. *(From Ref. 58 with permission.)*

the flat roof of a building, the general rule is that the area of the flat roof must be 3 times the area of the active collector array. This rule is developed from solar angle considerations described later.

The preceding method applies for both active and passive systems. Therefore, the rough amount of south-facing wall required for a passive system is also known.

Consultants

Since many architectural and engineering firms are not presently familiar with the design of energy-conservation and solar heating systems, the programming phase is the time at which a solar consultant first provides expert input. Although the amount of time such a consultant spends on programming will be small, the expert input is important in determining the best use of energy conservation and active and passive solar systems. The consultant will later prepare an estimate of solar system cost.

Absence of expert assistance from the program phase onward typically results in sunk costs in later design phases. The work not done cannot be done later to find optimum energy conservation and solar utilization.

Such things as mechanical-room sizing, location of solar storage, building length-to-width ratios, roof profiles, amount of glazing, and building mass and axis all affect both solar performance and design efforts for each discipline. These factors must be clearly understood at the program phase if rational fees, work scope, and interface points are to be established. The money will be spent—having input from experts will assure it is spent to obtain maximum benefit.

The consultant may also suggest methods of financing the solar system from sources normally not known to the owner or developer. These may include federal grant monies, investment tax credits on both a local and state scale, or other sources of public financing.

Preliminary Cost Estimation and Payback Period

Since the selected collector area is a good measure of the size of all components in a solar system, it can be the basis of a system cost estimate. The estimated solar system cost will depend upon system complexity and function, but similar systems built in the same area can be used as a basis for this preliminary costing.

In the early 1980s, active solar space-heating systems were built for a cost of approximately \$35 to \$50 per square foot of collector in the West and Midwest and along the Eastern seaboard. These costs are based on commercial systems installed by contractors and not on systems built in place by the building owner. This cost, when used for estimates in subsequent years, should be inflated by the local building cost index.

The economic feasibility of solar heating can be roughly estimated at the programming level but should be calculated carefully at the schematic design level. At the programming level, the simple payback period for a solar investment or the return on a solar investment can be calculated using the local fuel prices. The payback period is defined as the time required to pay off an initial investment in a solar system by savings in conventional fuel. A solar system does not pay itself off by replacing the conventional heating system but only by displacing the fuel which would have otherwise been used. For commercial analyses, a discounted payback period may be used, but several definitions of this term are widely used.

Example. A solar heating system is expected to deliver 125×10^6 Btu/yr. If the collector area is 500 ft^2 and its cost is \$35 per square foot, what is the simple payback period?

Solution. The value of energy saved by the solar system is the unit cost of backup energy multiplied by the number of units of energy saved. If backup energy costs \$10 per million British thermal units, the annual energy saving is \$1250. The initial cost of the solar system is estimated to be \$17,500. Therefore, the payback period is 14 yr. Since the cost of conventional fuels increases year by year, the value of solar energy savings also increases year by year into the future. If this inflation effect is taken into account, the payback period for a solar system decreases significantly. For example, if the fuel price in this example were to inflate at 12 percent per year, the payback period would be 8.7 yr instead of 14 yr.

Summary

In the programming phase, a building is quantified and budgeted with regard to its space and volume requirements as well as numerous characteristics of the site and of the building systems. This programming information can be used to estimate the size and cost of the required solar system. Also, area required for installation of the solar system is calculated.

The economics of solar heat can be roughly estimated if the cost and expected inflation rate of fuels is known. At the end of programming, the decision whether or not to use solar heat normally will not have been made. During schematic design sufficient additional detail is generated to allow a more careful feasibility study of solar heat use to be made. In most cases the solar feasibility question is answered at the schematic design level. The final step in the programming activity is the approval of the program by the owner. It should be reviewed in detail and approval in writing should be acquired. Programming approval is normally also taken to be approval to begin schematic design.

O **Sun-Path Diagrams and the Shadow-Angle Protractor.** During programming, the effect of shading on the viability of solar heating for a building must be studied. The sun-path diagram is used for this purpose. Briefly, the sun-path diagram is a plot of solar altitude and azimuth angles in a polar coordinate system. Figure 3.4 is an example sun-path diagram showing azimuth angles around the circumference of the circle and altitude angles along the radius. The roughly vertical lines in the diagram correspond to various times of the day. The roughly horizontal lines correspond to different times of year. The altitude and azimuth angles were shown earlier (Chap. 2) to depend upon the latitude, the declination, and the hour angle. Since only two of these three angles can be plotted on a two-dimensional graph, separate sun-path diagrams are prepared for each latitude. The diagram shown in Fig. 3.4 is for 30° north latitude. Additional diagrams are given in the appendix.

The diagram can be used to locate the sun on any day of the year. For example, the position of the sun at 10 a.m. on March 8 would be determined as follows. The 10 a.m. line is the second vertical line to the right of the center line of the diagram. The table above and to the right of the figure shows declination values for various dates of the year. The table shows that on March 8 solar declination is −5°. The −5° sun path intersects the 10 a.m. hour line at the point shown by the heavy dot in the figure. The intersection is then used to read off azimuth and altitude angles. If a line is constructed between the center of the figure and the point of intersection and then extended to the circumference, an azimuth angle of 45° is read. Likewise, the altitude angle, read from the radial scale, is approximately 45°.

A second geometrical construction called the shadow-angle protractor is used to assess shading problems. The shadow-angle protractor shown in Fig. 3.5*b* is a plot of profile angle P versus solar azimuth angle. The profile angle shown in Fig. 3.5*a* is solar altitude angle α projected onto arbitrarily positioned plane *ABCD*. In Fig. 3.5*a* the plane *ABCD* is the cross-sectional plane through the window overhang. The profile angle can be evaluated geometrically or from Eq. (3.1) below.

$$\tan P = \sec a \tan \alpha \qquad (3.1)$$

P varies with hour angle, declination, and latitude. The method of using the

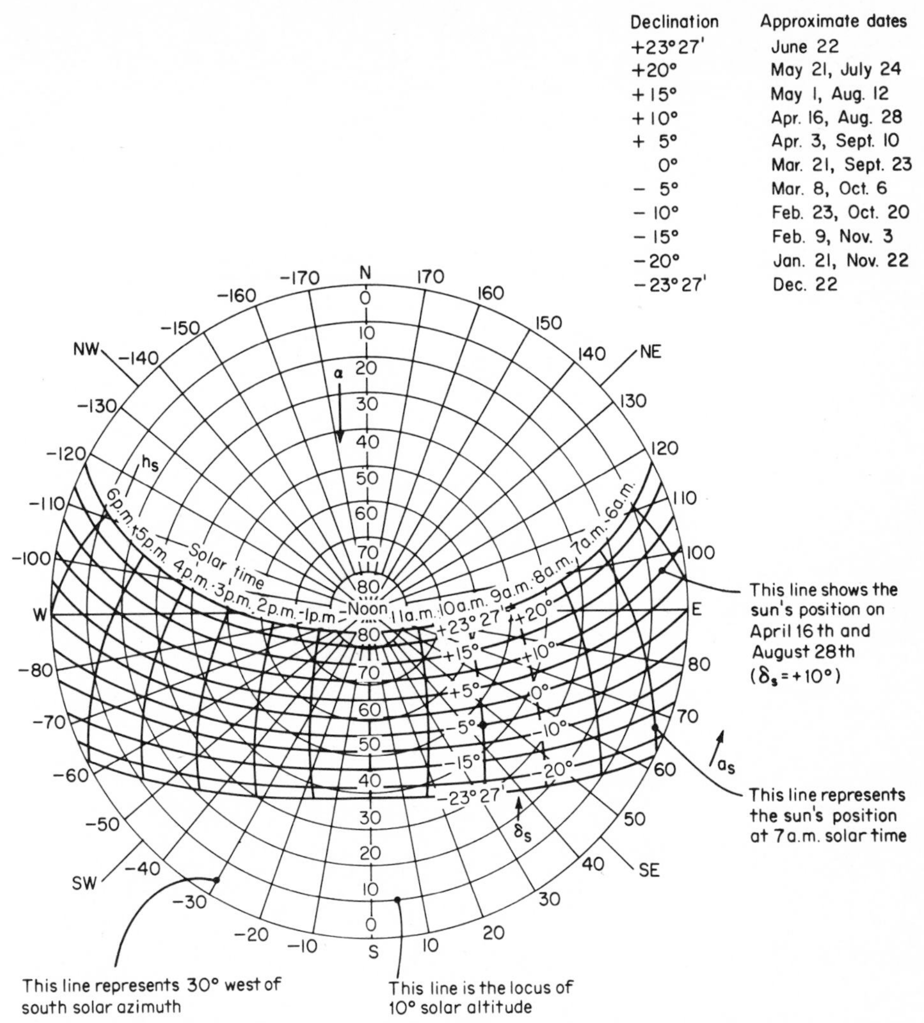

Declination	Approximate dates
+23°27'	June 22
+20°	May 21, July 24
+15°	May 1, Aug. 12
+10°	Apr. 16, Aug. 28
+ 5°	Apr. 3, Sept. 10
0°	Mar. 21, Sept. 23
− 5°	Mar. 8, Oct. 6
− 10°	Feb. 23, Oct. 20
− 15°	Feb. 9, Nov. 3
−20°	Jan. 21, Nov. 22
−23°27'	Dec. 22

FIG. 3.4 Sun-path diagram for 30° north latitude showing altitude and azimuth angles as a function of solar time and declination δ_s.

shadow-angle protractor along with the sun-path diagram to assess shading difficulties is made clearer by an example.

Example. A proposed solar building with a south-facing collector is sited to the north of a tall existing building. Prepare a shadow map to determine during what months of the year and what times of the day the lower edge of the collector at point *C* is shaded by the existing building. Section and plan views are shown in Fig. 3.6.

Solution. The controlling value of profile angle as shown in the sketch is 40° above the horizontal. The limits of azimuth angle *a* are seen to be 45°

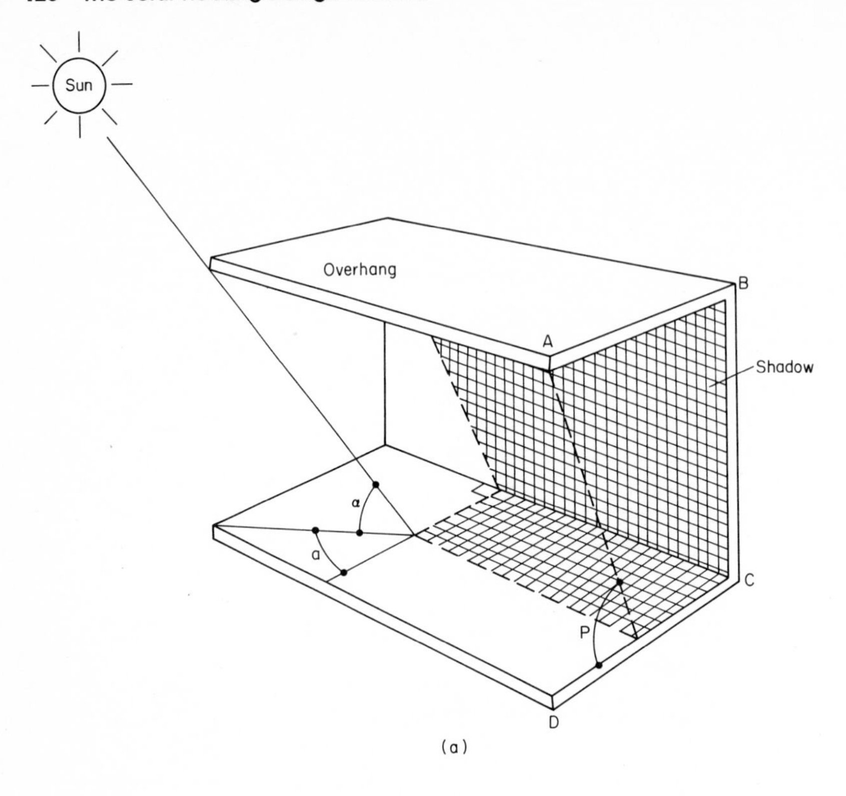

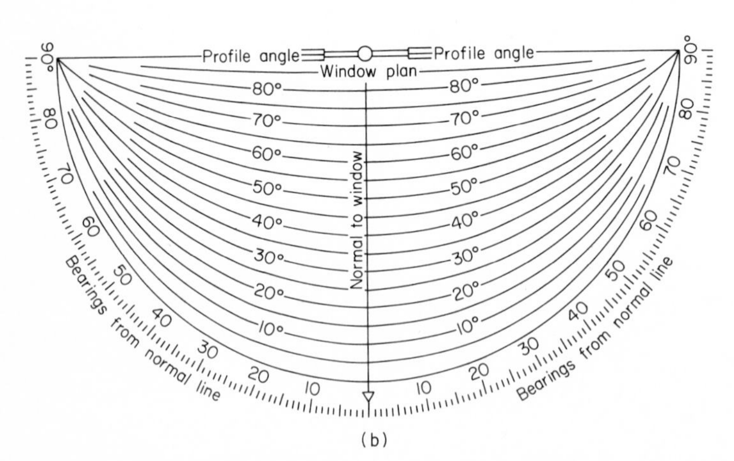

FIG. 3.5 The shadow angle protractor (*b*) used to calculate shading periods. The profile angle *P* on the protractor is defined in (*a*). *(Courtesy of Libbey-Owens-Ford Co.)*

east and 10° west. These values are plotted directly onto the shadow-angle protractor shown in Fig. 3.7*a*. When the profile angle is less than 40°, shading occurs; therefore this region of the shadow-angle protractor is crosshatched.

The shadow map in Fig. 3.7*a* is superimposed on the sun-path diagram which applies to the particular location in question. When the shadow map is superimposed on the sun-path diagram, one can read the time of the year and the hours of the day when shading will occur.

According to the overlay in Fig. 3.7*b*, shading occurs for those months of the year when the declination is less than approximately −10°. The summary declination table in Fig. 3.4 shows that −10° or less corresponds to the months of October, November, December, January, and February—approximately 4 mo in the center of the heating season. The hours of the day in which shading occurs in these months can also be read from Fig. 3.7*b*. Shading begins at approximately 9:30 a.m. and ends at approximately 12:30 p.m. Therefore, point *C* is shaded for 3 h for approximately 4 mo during the heating season.

This extensive amount of shading would not be acceptable in the design of an efficient solar heating system. One way to solve this problem is to position the solar building farther north of the existing building; or the solar collector could be repositioned toward the top of the sloped roof and not along the lower edge where point *C* is located.

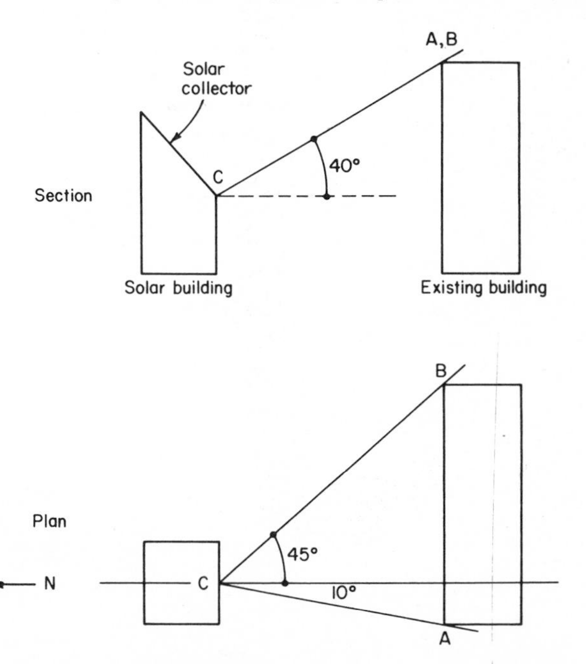

FIG. 3.6 Shading problem section and plan views relative to point *C* on solar collector. Points *A* and *B* are used to plot the shadow map, Fig. 3.7*a*.

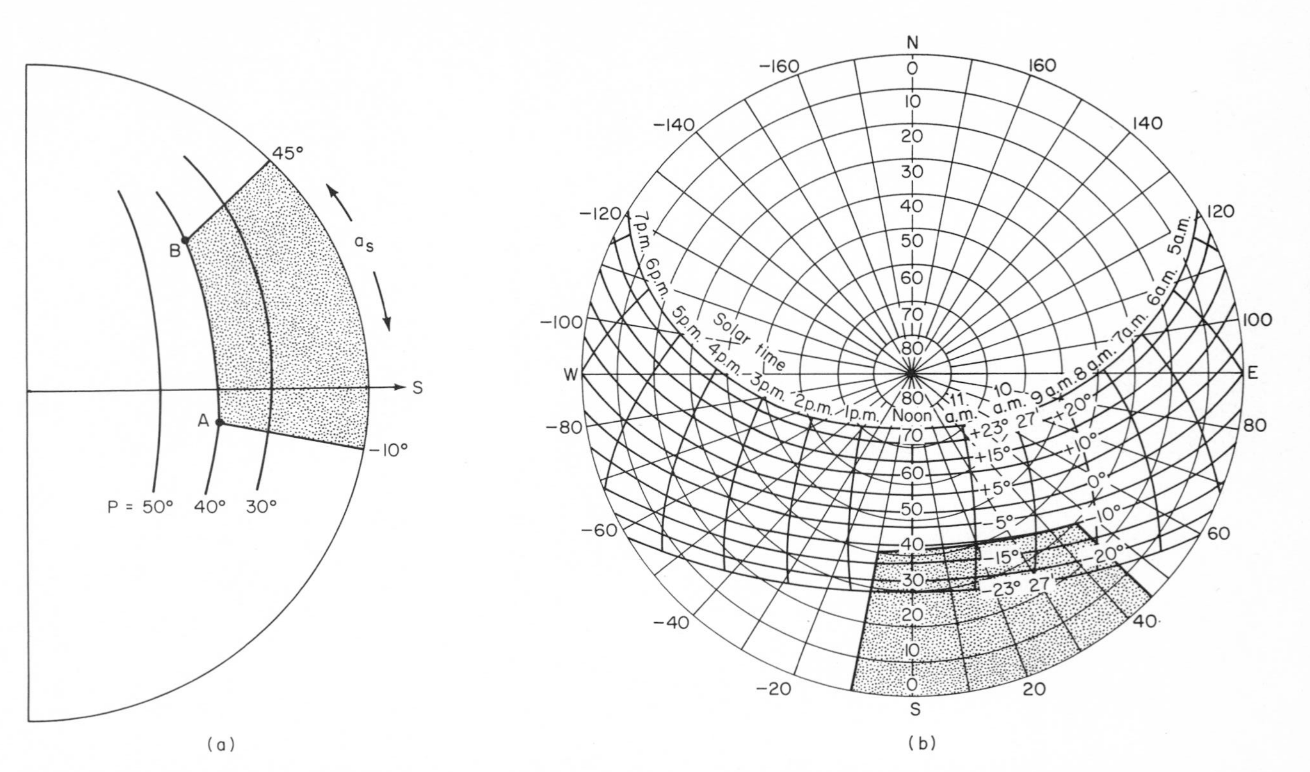

FIG. 3.7 (*a*) Shadow map for example showing limit of azimuth angle 45° east and 10° west and controlling value of profile angle *P* equal to 40°; (*b*) shadow map superimposed on sun-path diagram for site location.

The sun-path diagram method is a rapid graphical approach useful in assessing the severity of potential shading problems. It should be used at the programming level to determine whether present or worst-case future shading is a problem. If so, the location on the site and orientation must be corrected at the programming phase. At schematic design, it is too late to do this type of study since the approximate building orientation and location on the site have already been approved by the owner. Larger sun-path diagrams and shadow-angle protractors are available; for example, an 8-in-diameter version is sold by the Libbey-Owens-Ford Glass Co. (Toledo, Ohio) for a few dollars. This sun-angle calculator kit also includes other constructions prepared on transparent overlays which can be used to assess the amount of daylight transmitted by a window of known size and orientation.

PROGRAMMING PHASE CHECKLIST

The following checklist summarizes the work required during programming for a solar-heated building. The level of detail required depends upon building type, size, and usage. The emphasis is on heating and ventilating systems, not cooling or dehumidification systems. Refer to preceding text for details on each entry.

Architectural

Acquire building program from owner; to include project size, usage by zone, applicable codes, minimum property standards (HUD), etc. (see AIA *Handbook of Professional Practice*) ☐

Acquire or prepare thermal zone program—schedule, usage, temperature, etc. ☐

Identify total building energy goal [$Btu/(ft^2 \cdot yr)$] ☐

Identify project budget, dollars ☐

Identify project size—gross ft^2 ☐
—gross ft^3 ☐

Determine Building System Types to Be Considered

Heating ☐
Ventilation ☐
DHW ☐
Humidification ☐
Illumination ☐
Active Solar ☐
Passive Solar ☐

Acquire Site Information Location and Zip Code		——, ——, ——,
All conditions precedent to construction	☐	
Latitude, longitude	☐	
Site plan	☐	
Site topographical survey	☐	
Site solar survey	☐	
Special diurnal climatic characteristics	☐	
Identify solar shading/access problems	☐	
True north direction	☐	
Views to be preserved	☐	
Trees and other flora to be preserved	☐	
Site elevation, feet above mean sea level	☐	
Water table depth	☐	
Water chemistry and hardness	☐	
Seismic zone type	☐	
Soil resistivity (Ω·cm) if buried storage is to be considered	☐	
Zoning—height limits, solar access	☐	
Climatic extremes—hail, wind, snow, expected hail size	☐	

Identify Available Conventional Energy Sources and Prices		
Natural gas + available rate structures	☐	$____/CCF (hundred cubic feet)
LP gas	☐	$____/gal
Fuel oil	☐	$____/gal
Coal	☐	$____/ton
Electricity + available rate structures	☐	$____/kWh
Steam	☐	$______/1000 lb
Wood	☐	$____/cord
Other	☐	$____/unit

Fuel Heating Values		
Natural gas	☐	____ Btu/CCF
LP gas	☐	____ Btu/gal
Fuel oil	☐	____ Btu/gal
Coal	☐	____ Btu/lb × 2000 lb/ton
Electricity	☐	3413 Btu/kWh

Steam	☐	____ Btu/1000 lb
Wood	☐	____ Btu/cord
Other	☐	____ Btu/unit
Fuel cost inflation rates	☐	____ % per year
Utility hookup fees; permit application procedures	☐	

Characterize and Screen Site Energy Potentials

Solar	☐
Wind	☐
Earth temperature	☐
Diurnal ΔT	☐
Low wet bulb	☐
Other useful microclimatic characteristics	☐
Architectural +/− list for above	☐

Acquire Climatic Data (Hourly, Monthly as Required for Calculation Methods to Be Used)

Ambient dry-bulb temperature	☐	
Ambient wet-bulb temperature	☐	
Ambient wind speed	☐	
Ambient wind direction	☐	
Cloud cover	☐	
Solar flux	☐	
Heating degree-days to various bases	☐	
ASHRAE or local energy code peak design conditions		
Dry bulb—summer	☐	____ °F
Wet bulb—summer	☐	____ °F
Dry bulb—winter	☐	____ °F
Wet bulb—winter	☐	____ °F
Wind load	☐	____ mi/h
Monthly precipitation—rain	☐	
—snow	☐	
—ice	☐	
Air pollutant species and concentrations	☐	

Quantify Building Zonal Design Conditions ("Thermal Program")

Interior dry-bulb temperature	☐
Interior wet-bulb temperature	☐
Ventilation rates	☐
Schedules—people and/or materials	☐

Setbacks ☐
Internal gains ☐
Natural daylight required ☐
Peak allowable electrical demand ☐

Quantify Energy Needs Using Simplified Methods by Month

Heating ☐
(Cooling) ☐
DHW ☐
Lighting ☐
Ventilation ☐
(Dehumidification) ☐
Process ☐
Humidification ☐
People—latent ☐
—sensible ☐
Annual total [Btu/(ft^2·yr)] ☐
Consider monthly plots by end use for client review ☐

Determine Owner's Economic Criteria

Discount rate ☐ ____%
Minimum acceptable ROI ☐ ____%
Minimum acceptable payback period ☐ ____yr
Accounting procedure ☐ ____
Tax bracket—federal ☐ ____%
—state ☐ ____%
Profit/nonprofit ☐
Public/private ☐

Contract for Consultants and Establish Full Design Fees for Project

Solar ☐
Energy conservation ☐
Economist ☐
Standard architectural consultants ☐
Legal ☐

Energy Supply Estimates—Percent of Connected Load by Source

	Space Htg.	*DHW*
Gas	☐	☐
LP gas	☐	☐
Fuel oil	☐	☐
Active solar	☐	☐
Passive solar	☐	☐
Wood	☐	☐

Coal	☐	☐
Steam	☐	☐
Other	☐	☐
Energy Economics Estimates		
Solar feasibility—payback, ROI, life-cycle cost/benefit rates, etc.	☐	
Solar vs. nonsolar comparison	☐	
Acquire Program Approval		
Architectural	☐	
Economic	☐	
Thermal	☐	
Solar	☐	
Authorization to proceed to SD	☐	

SCHEMATIC DESIGN

The objective of schematic design is to help the client understand the program, to illustrate possible budget-allowable solutions within the shortest possible time, and to assist the client in determining the feasibility of the project. More than one solution to the design problem may be presented, and the goal is then to secure the owners' approval of the best solution and its budget. Results of the schematic design studies will include a site plan showing relationships between the building and its surroundings. Selected vertical sections will be used to identify important relationships within the building, and floor plans are presented. A general written description of the project is included in the schematic design document, and finally a statement of probable construction cost is prepared.

In this section, the schematic design process is first outlined and then the specific inputs to schematics required by the additional requirement of solar heating are given. Methods of assessing solar feasibility at the schematic level are presented in detail. This represents a new activity not present in nonsolar building designs. In addition, a table of approximate sizing rules for all components of active and passive solar systems is presented to enable the designer to quickly develop a schematic design for the system once the optimal collector area is known from an SD-level economic study.

Summary of the Schematic Design Phase

The following summary is abstracted from the American Institute of Architects *Handbook of Professional Practice*. The initial step in schematic design is the attempt to find one or more solutions to the design problem identified during programming. One of the most common methods of arriving at a solution is to prepare functional space diagrams that illustrate the program requirements. Each plan is drawn at a relatively small scale selected to illustrate only building story heights and orientations. The diagram is then integrated into an overall whole to assure proper role relationships. This step primarily consists of constructing a diagram of the information developed during programming.

The next step in SD represents the documentation of solutions in more detail. A series of sketches is normally made at this time so that various practical and economical solutions to the problem can be studied. The limits of the budget must be recognized in preparation of these solutions. Each sketch must include all important features of the area as well as passive and active solar impacts. These sketches can serve as a record of the SD process and should be considered part of the file for the specific job. The owners will review various solutions during the course of SD, and their inputs will be important determinants of the ultimate solution.

In addition to sketches, computations are made during SD. For example, the expected floor area, heat load, and energy consumption for the building may be calculated as well as expected solar system size and cost. The rough optimum sizes of various solar features are determined. These matters are described in the next section.

Another document prepared during SD is a statement of probable construction cost. Preparation of such a cost estimate requires a listing of major equipment as well as building element characteristics. Sufficient data for an accurate estimate are normally not available during schematic design, and as a result the cost estimate must be considered only preliminary. The estimated total construction cost may exceed the budget, and the designer must account for the discrepancy. If the owners' approval of schematics is dependent upon strict adherence to a budget, this problem is best solved at this time, and scarce capital must be optimally allocated among all potential end users. This is done in such a way that the last increment of investment in all energy-conservation and solar features, for example, provides equal rates of return.

Depending upon their sophistication, the owners may or may not be able to understand architectural sketches made during this phase. As an aid to the owner in understanding the project, small-scale models are frequently prepared during schematic design. In addition, perspective drawings and interior renderings will help elucidate the expected relationship of building elements. Models are particularly useful during solar design in determining the appropriate relationship of active and passive elements to the building mass and in determining the possibility of solar-aperture shading by nearby building components. If the project is part of a larger development, the relationship of the present building to the overall plan should be considered and indicated by means of a site plan and additional renderings if necessary.

Solar Activities in Schematic Design/Drawings

Several types of drawings or sketches are prepared as part of SD—site plans, floor plans, and building sections. Each of these drawings will show certain solar energy aspects of the building and help quantify the solar impact on the building design.

Before the addition of active solar heating to a building, all cost-effective energy-conservation features must be included. These will be represented on the drawings by specific building orientation, wall and roof thickness, insulation amounts shown in the sections, and efficient relation of one space to another on the floor plans. Following the incorporation of energy-conservation features, the inclusion of passive solar heating systems should be considered. Passive features are most orientation-sensitive and are usually somewhat lower in cost than active ones. The third consideration is the addition of an active system to a building which already incorporates proper energy-conservation and passive features.

Site Plan. In addition to the information normally shown on the site plan, the solar site plan must include an accurate determination of the due south direction. This is normally done during programming by the surveyor. If not, the due south direction must be established either using the North Star or a compass. The correction for magnetic declination must be made as described in Ref. 1. The variation between true north and magnetic north may be as much as 20° in certain parts of the United States. A 20° mislocation error can produce large solar performance penalties. The site plan will also show possible locations for collector siting either on the building or on the ground adjacent to the building. Possible locations for remote storage siting and connecting piping will be identified on the site plan. Shading obstructions and the worst-case shading configuration should be shown on the site plan.

Figure 3.8 is a typical schematic level site plan with solar and microclimate effects emphasized. The direction of the prevailing winter winds and trees used to provide a wind screen from the north are shown. In addition, an annual wind rose is given depicting prevailing wind direction and magnitude. The dashed shadow line shows the worst shading case which could exist on this site, given the existing zoning height limit of 35 ft. The shadow line is determined by graphically positioning the maximum allowable future building on the lot to the south at its north buildable limit, and then geometrically laying out the resulting shadow line. In order to ensure that no collector shading takes place, the collectors (active or passive) on the new building must fall outside the worst-case shadow line.

Roof and Floor Plans. Schematic-level floor plans of solar buildings are normally not much different from nonsolar ones. The floor plan must reflect initial estimates of extra space required for mechanical components of an active system. Active systems will require a larger mechanical space since thermal-storage tanks, heat exchangers, pumps, and expansion tanks may be present. The location of the mechanical space within the building must be determined in part by the requirement that solar piping runs be as short as possible. Long pipe runs result in excessive heat loss from the pipe and large requirements for pumping power. The mechanical space on the ground floor is centrally located. Figure 3.9 shows an example SD-level floor plan.

If a flat-roofed building is used, the floor and roof plans are closely related and the former can be used to determine the maximum amount of area available for

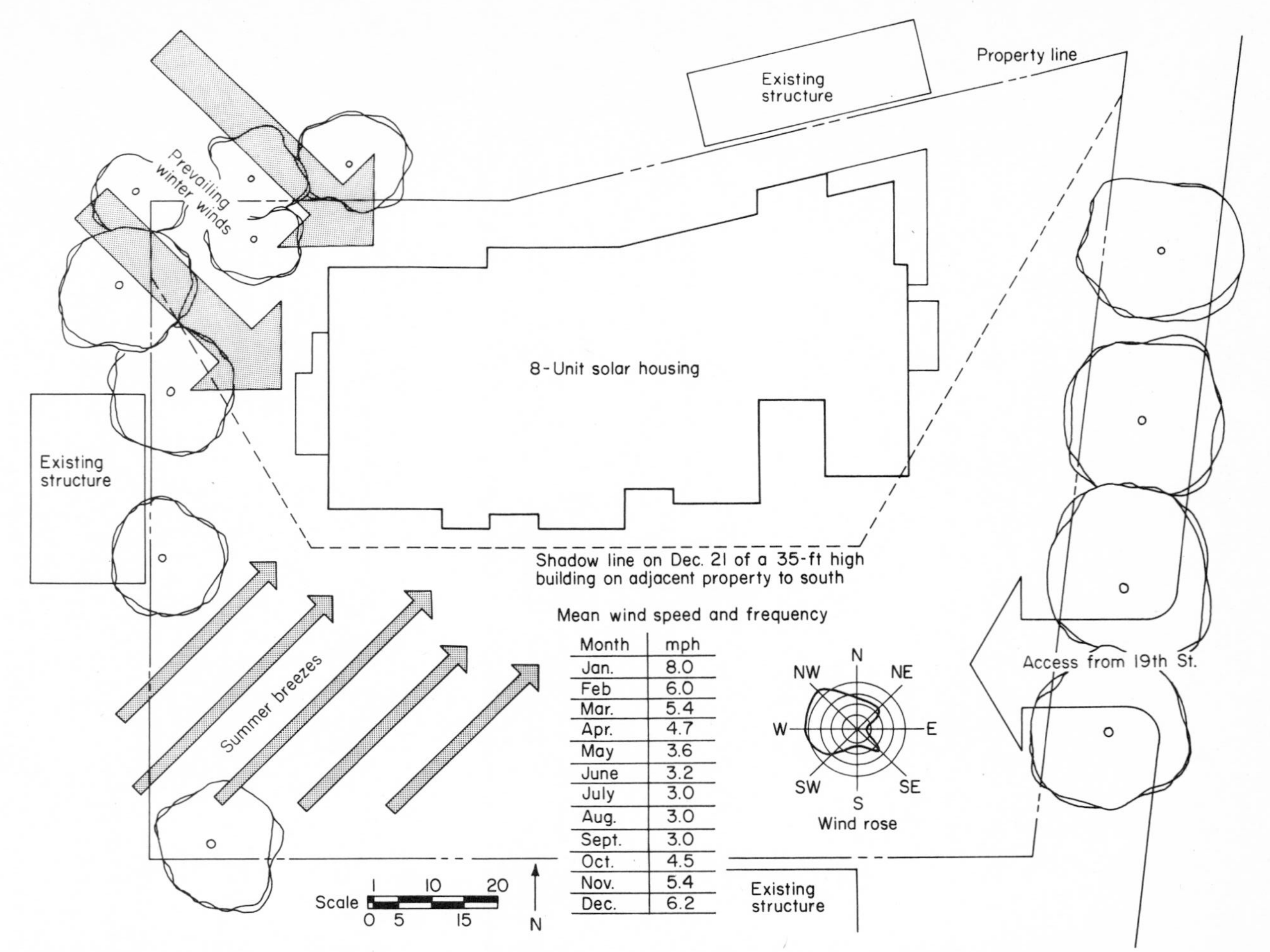

FIG. 3.8 Typical schematic-level site plan showing building orientation climatic information, and worst-case shadow line. *(Courtesy of Phil Tabb Architects, Boulder, Colo.)*

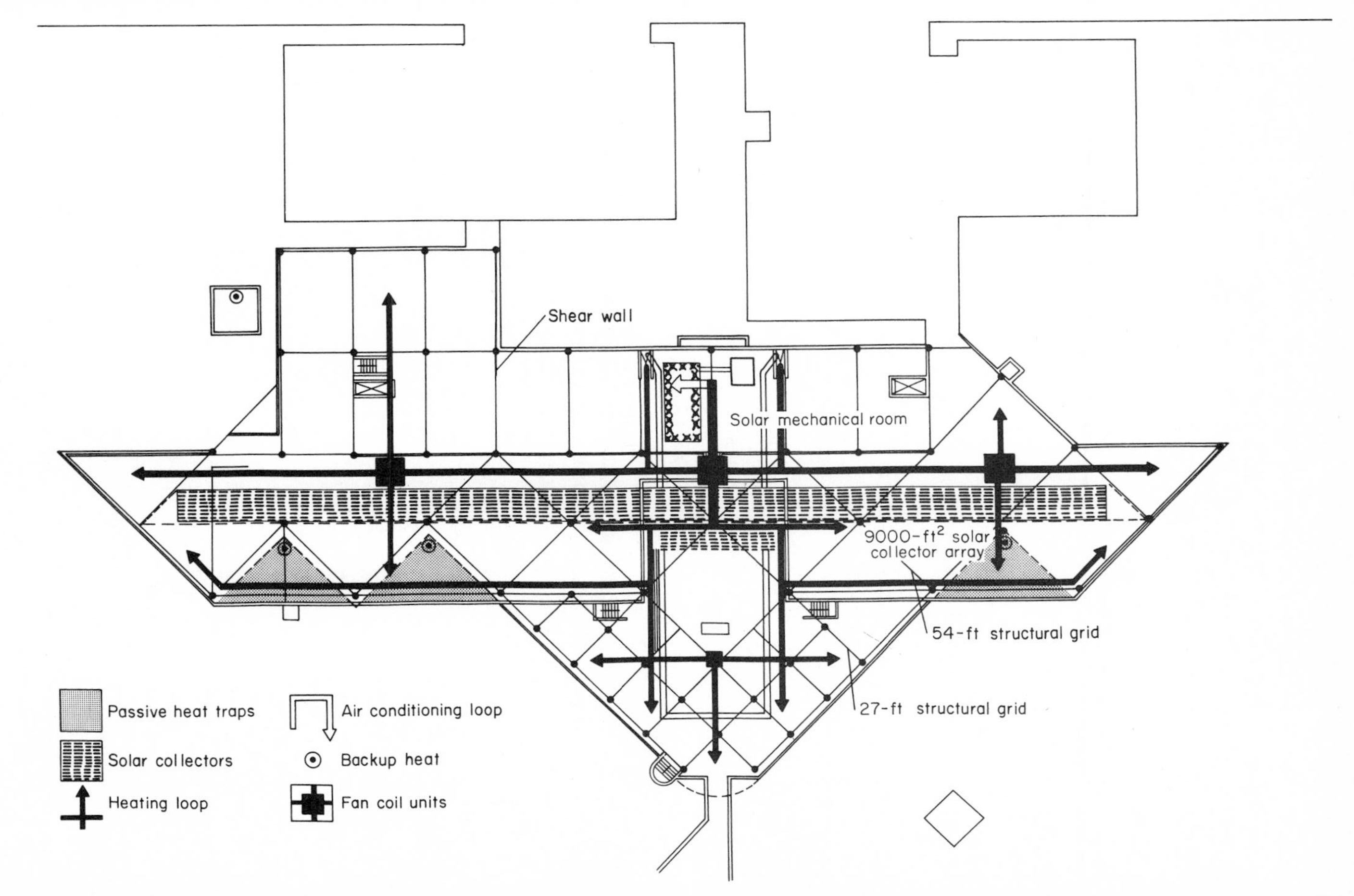

FIG. 3.9 Schematic-level building plan showing major space relationships and mechanical system flows. *(Courtesy of Phil Tabb Architects, Boulder, Colo.)*

collector roof mounting. As noted in the section on programming, the amount of collector mountable on a flat roof is approximately one-third the net flat-roof area. The floor plan can therefore be used to quickly determine whether on-roof or off-roof mounting of the collector array should be used.

Cross sections of the building can be used to determine several important characteristics from a solar viewpoint. For example, if roof mounting is to be used, the building section will depict the sawtooth collector array arrangement. This arrangement determines whether a certain row of the array is shaded by the row immediately to the south, as shown in Fig. 3.10. The sawtooth arrangement in Fig. 3.10 shows the approximate relationship to scale of a solar collector of face length L_c. The spacing D between two collector rows is determined by the collector length, the collector tilt angle, and the governing solar profile angle P. It is possible to stipulate that no collector shading whatever shall be tolerated on the shortest day of the year. However, in early morning and late afternoon in late December, shadows are very long and strict adherence to this requirement would cause collector rows to be spaced very far apart. A practical rule which seems to work well is that the collectors be spaced sufficiently far apart to allow no shading to occur between the hours of 10 a.m. and 2 p.m. on the shortest day of the year. This requirement will result in good annual solar collection without excessive collector spacing. Since the sun spends little of its time at the winter solstice position, a requirement for a longer period of collection during this time is often not practical.

The governing profile angle P can be calculated from Eq. (3.1), given above, or can be determined from the solar position tables given in appendix Table A.3. For a typical midlatitude location in the United States, a collector tilt of 55° will be used and a governing profile angle of approximately 18 to 20° therefore applies. In that case, the collector spacing D is approximately 3 times the collector face length.

Sections—Glare and Easement. Building sections are also useful in the assessment of any potential glare problems from large collector arrays. As shown in Chap. 2, the amount of light reflected from solar collectors can be sizable if the sun

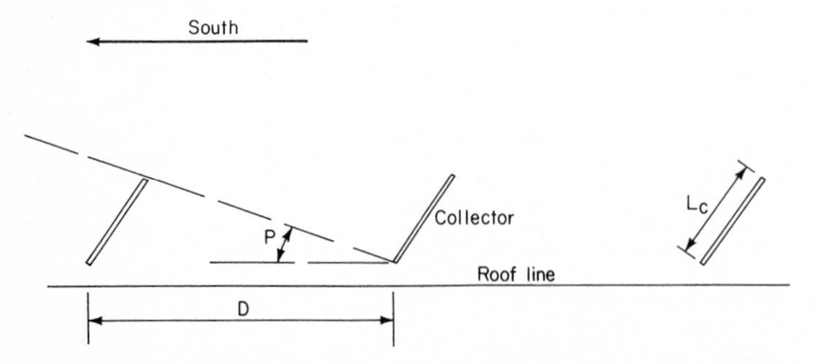

FIG. 3.10 Section of sawtooth collector arrangement used on flat roofs, showing collector spacing D related to collector length L_c and profile angle P.

strikes the collector surface at grazing incidence. The potential impacts of glare upon surface traffic, air traffic, or adjacent buildings should be assessed. The air and surface traffic impacts are particularly important from a hazard viewpoint. Reflection onto adjacent buildings should be minimized in order that glare and additional air-conditioning requirements not be created. The assessment of glare problems is relatively simple, since the angle of reflection is equal to the angle of incidence of sunlight on the reflecting surface. Using a building section, the incidence angles may be laid out and the reflection onto the ground or into the adjacent air space can be determined. Since the reflectance is high only at off-normal incidence ($i > 50°$), these are the angles which should be studied. When the sun shines onto a collector from within approximately 50° of the collector normal, little reflection takes place—typically less than 10 percent of the incident sunlight. At incidence angles beyond 70°, however, the reflection can approach 50 percent, and reflected glare will exist.

Building sections and shading analyses can determine whether a solar easement ensuring perpetual access to sunlight should be purchased over adjacent properties. The question of solar easements and their formulation is beyond the scope of this book; however, Refs. 17 and 9 contain information on these matters. The cost of a solar easement will vary significantly depending on the present or expected commercial value of property to the south of the solar building being analyzed.

Solar Schematic Design Activities—System Configuration and Sizing

During schematic design, important solar system characteristics are defined. For example, the types of passive heating systems and their expected performance should be established. Likewise, the appropriate types of active systems, if any, should be identified. Characteristics of solar systems which differentiate each from others include the working fluid (water, steam, or air), the energy delivery system (forced air, hydronic pumps, or heat pumps), the type of solar service to be provided (domestic hot-water heating, space heating, air conditioning, or a combination), and the type of backup system to be used (gas, electricity, fuel oil, wood, or other).

Sizing Rules. The collector area itself is determined in most designs by an economic optimization process (see pages 137 to 153), although it could be arbitrarily specified as that area corresponding to a specific percentage of the annual heating load, for example, 75 percent. In this section, a summary of sizing rules is presented which relates the size of all other components to the collector area. The rules are based both on computer simulations of solar systems and on performance measurements made in the field.

The principal components of active and passive systems include the following: collector, pumps or fans, heat exchangers, control valves, terminal load devices, and storage systems. In passive systems, some of these components are not specifically separable one from another. Table 3.2 is a summary of schematic sizing rules for various types of systems discussed in this book.

TABLE 3.2 Schematic design sizing rules

Component characteristic	*Active water heating*	*Active space heating—liquid*	*Active space heating—air*	*Passive space heating*
Storage capacity	1.5–2.0 gal/ ft_c^2	1.5–2.0 gal/ft_c^2	½–¾ ft^3/ft_c^2	8–12 in masonry or water equivalent
Heat-exchanger effectiveness	$E_{hx} \geq 0.8$	$E_{hx} \geq 0.8$		
Pump or fan flow*	0.02 gal/ $(min \cdot ft_c^2)$	0.02 gal/ $(min \cdot ft_c^2)$	2.5 ft^3/ $(min \cdot ft_c^2)$	1 $ft^3/(min \cdot ft^2)$ if hybrid
Load heat-exchanger effectiveness		$E_{hx} \geq 0.5$		
Collector tilt angle	Latitude	Latitude +15°	Latitude +15°	Normally vertical for TSW; to favor winter sun for others
Collector azimuth angle	Within 25° of due south for all systems			
Collector area	0.8–1 $ft_c^2/(gal \cdot day)$	*There is no rule of thumb for collector area!*†		

*Liquid flow rates are based on water. If water is not used, make the proper adjustments for density and specific heat. For air, flow rates are at sea level; make proper adjustment for reduced density at high altitude.

†See performance prediction and economic analysis section, pages 137–153.

Additional Sizing Information for Passive Schematic Designs

The schematic design of passive heating systems cannot be isolated from the overall building design. Therefore, certain design criteria for passive systems must be included in the building design as well as in the heating system design. In this section, additional information on direct gain, thermal-storage wall, and greenhouse passive heating systems is provided.

Direct Gain. Direct-gain systems consist of three generic types—south vertical walls, clerestories, and skylights. South-facing glazing is the most common direct-gain aperture. In order for direct gain to heat the adjacent space directly, the depth of that space should be no more than 2½ times the height of the south-facing window. This rule is based on the position of the sun in winter. Of course, south-facing direct-gain apertures must be confined to azimuth angle ranges between southeast and southwest, with due south being preferred.

One particularly troublesome feature of direct-gain systems is the likelihood of objectionable glare and overheating within the space during a sunny day. This is particularly a problem in offices where high internal gains already exist in the daytime. By proper choice of materials, placement of furnishings, and so forth, the impact of glare can be reduced somewhat, but never eliminated. The person choosing materials exposed to sunlight in a direct-gain space must take into account the tendency for intense sunlight to fade dyed fabrics and to deteriorate plastic materials, particularly if blue and near-ultraviolet rays are present.

Other types of direct-gain systems include skylights and clerestories. A clerestory used for both natural daylighting and space heating is shown in the example passive system described in Chap. 2 (see Fig. 2.25). Skylights and clerestories can provide heat to a space if they reflect both heat and light to a lower level. Clerestories may consist of one large vertical section or may be arranged in a sawtooth configuration. The spacing of sawtooth clerestories is governed by the same considerations that apply to flat-plate collectors on a flat roof (Fig. 3.10). Their performance can be enhanced by a reflecting surface of approximately the same area, placed horizontally to the south of the vertical window.

Skylights are apertures in the roof plane which are used for both natural daylighting and heating. Skylights do not have an ideal orientation for space heating since the sun is relatively low in the sky in winter and an aperture at typical roof angles intersects relatively little sunlight during the deep heating season. In addition, the large solar incidence angle results in significant reflective losses. The performance of a skylight can be enhanced by placing a nearly vertical (or adjustable) reflecting surface above and to the north of the skylight to direct additional sunlight through its aperture. One difficulty with skylights is their orientation ensuring large heat gains in summer when the sun is high. Therefore, any skylight design must include some method of summer shading. The reflector mentioned above, if shaped properly, can be used as a summer skylight cover. It can permit a small amount of sunlight to enter, giving natural daylighting in summer.

The amount of storage recommended for direct-gain systems in Ref. 19 is as follows. Interior walls and floor of solid masonry should be 8 in thick. If water storage is used, approximately 2 to 3 ft^3 of water are required in an interior water wall for each square foot of direct-gain glazing. The larger the amount of storage, the smaller the day-to-night temperature excursion. Insulation should be exterior to all indirectly illuminated storage components.

Indirect-Gain Thermal-Storage-Wall Systems. Thermal-storage-wall systems are used to provide a more controlled passive space-heating behavior than is possible with direct-gain systems. In addition, the glare problem and effects of directly illuminating furnishings and carpets are avoided by means of a wall. The thermal-storage wall need not be a solid masonry element but can be pierced with window openings as desired. Part of the heat delivered to the space by a thermal-storage wall is by thermal radiation in the far infrared. Therefore, the depth of the room to the north of a thermal-storage wall is limited to approximately 20 ft. Beyond this distance, the effectiveness of thermal circulation and radiation heating are reduced. This constraint by itself results in a fairly linear building design with an aspect ratio of 2 or 3 to 1, favoring a long south wall.

The amount of passive wall aperture area for a TSW system may be in a range similar to that for direct gain—approximately ¼ to ⅜ ft^2 of aperture per square foot of heated floor area.

Storage in TSW systems is by a solid masonry wall between 8 and 12 in thick or by a water wall 6 in or more thick. The amount of storage is determined by the time lag desired between initial heating of the wall and release of the heat to the space after sunset. If a high-conductance material is used for storage, the wall must be thicker to provide a given time-lag effect. Water and masonry have been most commonly used because of their availability and low cost.

Another determinant of wall thickness is comfort. Although TSW systems with 8- and 12-in walls may produce the same heating energy per year, the 12-in wall will have reduced temperature savings vis-à-vis the 8-in wall. Thinner walls may feel too warm on sunny, mild days and too cool on overcast, cold days.

Wall outer surface color is important. Darker colors with an absorption coefficient above 0.60 should be used. Table A.12 in the appendix contains absorptance values for many passive wall materials and coatings.

Passive solar systems carrying a relatively small percent of the load produce most of their heating effect during the day; storage is often not fully charged. For these systems, the thermal circulation vents are relatively large (2 to 3 percent of wall area) to pemit rapid heat up during the day. Passive systems carrying most of the heating load use smaller vents (1 to 2 percent of wall area). Small vents protect against overheating during the day, a potential problem if large TSW vents are used. Details of vent sizing are presented in Chap. 4.

Greenhouses. Greenhouses are essentially enlarged thermal-storage-wall cavities placed to the south of storage. Since greenhouses have a larger glazing area per square foot of wall area than a TSW system, heat losses will be greater and solar gains higher. Overall, the energy delivery through the storage wall to the building space to be heated will be less than for a pure TSW system. Reference 19 suggests that approximately 50 percent more greenhouse aperture area is required to heat the same space for the same percent of time as a TSW system.

Sloped greenhouse glazing presents many problems vis-à-vis vertical glazing:

Higher cost
More leaks
More collected dirt
More summer heat gain
Reduced load-carrying capacity
Increased snow load
More difficulty in construction

Heat storage for an attached greenhouse is provided in the common wall between the greenhouse and the building. The amount of storage to be used is roughly the same as for the TSW system described above. If the greenhouse is used in a free-standing mode and not as an attached greenhouse, the amount of thermal mass may need to be larger. Storage suitable for a free-standing configuration can easily be determined by experiment. Detailed sizing rules do not yet exist. Storage is added until the day-to-night temperature swing is within the range which can be tolerated by plants in the greenhouse—about 20°F or so.

Other design guidelines for greenhouses are similar to those for systems described above. For example, since solar gains in winter are much larger on a south face than on an east or west surface, the south wall will be the largest and a linear building configuration will be the most efficient. East and west windows are net winter heat losers. Insulated walls work better. Summer solar-gain controls must be used to prevent overheating; heat rejection via high vents should be included. Greenhouse design studies have lagged behind those for direct-gain and TSW configurations; therefore, the design guidelines are less clearcut.

Schematic System Diagram

The component sizes shown in Table 3.2 are expressed on a unit-collector-area basis. Once the collector area has been estimated, the sizes of all other components in the solar heating system are, therefore, known. These sizes are used in preparation of the schematic diagram of the solar heating system. The schematic diagram shows the main features and interrelationships of all system components. The diagram shows all fluid-flow loops and control sequences used in the design recommended at the end of SD. Figure 3.11 is an example schematic diagram of an active liquid-based system. The diagram shows the sizes of the collector, pump, heat-exchanger and storage backup system, and terminal delivery device. The domestic water-preheating system is also shown.

Major component sizes shown on the schematic diagram are the basis of the SD cost estimate. This becomes a part of the schematic-design-phase cost statement and will be approved by the owner if solar heating is to be used. The determination of the economic viability of solar heating must be made by means of a feasibility study. The feasibility study will determine the relative cost of solar heating equipment and operation costs vs. the savings in conventional fuels.

A second determinant beyond that of cost competitiveness frequently enters into the heating system selection process. This constraint is the availability of initial funding for a solar system. Although the investment may be shown to be economically viable on a year-by-year basis, initial budget constraints may make the initial, relatively large investment impossible. At this point the solar consultant can assist by suggesting to the owner possible tax credits and sources for funding including low-interest loans from the Solar Development Bank.

Building models are often prepared as a part of schematic design. Figure 3.12 shows an example of an SD scale model useful for client review, shading studies, and daylighting experiments.

PERFORMANCE PREDICTION AND ECONOMIC OPTIMIZATION AT THE SCHEMATIC DESIGN LEVEL

The sizing of a solar system is determined by trade-offs of projected fuel savings vs. the cost of the associated solar system. This type of economic study is not required for conventional heating systems, which always provide 100 percent of the annual heating load. Solar systems are almost never sized at the 100 percent level. At the schematic design level, an approximate economic analysis is made

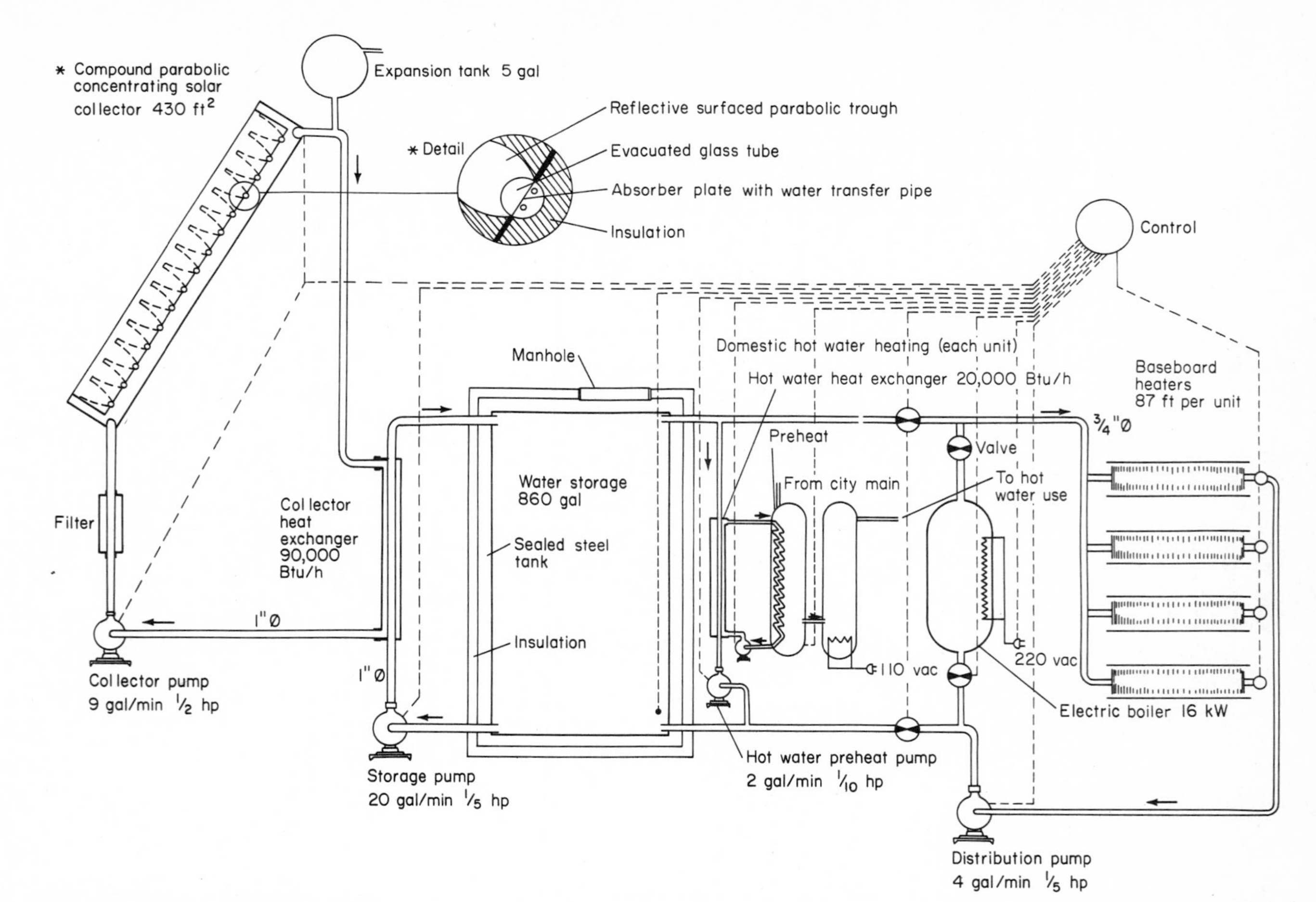

FIG. 3.11 Example schematic-level solar heating system diagram showing major components and their sizes. *(Courtesy of Phil Tabb Architects, Boulder, Colo.)*

FIG. 3.12 SD-level building model showing major building masses and collection features to scale. *(Courtesy of Phil Tabb Architects, Boulder, Colo.)*

based on only the major variables determining system performance. In this section, the methodology for determining the *optimum system size for schematic purposes* is described.

The cost of heating a solar building consists of two parts. The first part is the annualized cost of the solar system itself, including the capital investment, any interest charges, operating cost, etc. The second component of the heating-cost equation is the cost of auxiliary fuel which must be purchased, since a solar system has less than 100 percent capacity. Since the cost of backup energy is expected to inflate over the lifetime of the solar system, its inflation rate is to be included in the SD feasibility study.

Both components of total heating cost depend on system size. The cost of the solar system obviously depends upon the amount of collector area and associated hardware. The total cost of backup fuel also depends upon collector area since the larger the solar system, the smaller the amount of backup energy needed. Inasmuch as useful solar heat production does not increase linearly with system size (as a result of the law of diminishing returns), doubling the size of the solar system does not double the energy savings and economic benefit. This nonlinear effect of system size is not present in standard heating systems.

Figure 3.13 shows diagrammatically the relationship of backup fuel costs and solar system costs to the size of the solar system. In this book, *the index used to size a solar system is the active or passive collector area*. In Fig. 3.13, the total cost shown in the upper curve is the arithmetic total of the fuel cost and the solar cost.

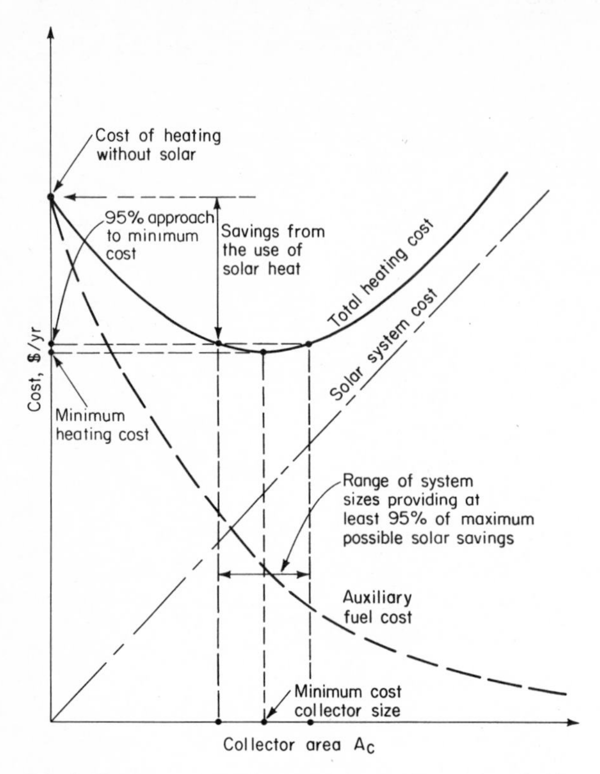

FIG. 3.13 Annual heating cost for various sizes of solar system as measured by collector area A_c. Total cost is the sum of fuel and solar costs.

The total cost curve will normally show a minimum at a specific collector size. This minimum point represents the size of the solar system resulting in the minimum heating cost. This system size is selected by the building designer as the basis for SD. However, variations of ± 15 percent from the cost-optimal system size have no statistically significant effect on life-cycle costs. Examples in Chap. 4 illustrate this.

The calculation of the total cost requires knowledge of the annual amount of energy delivered by solar systems of various sizes. This calculation is the first step in solar-system sizing. The second step in the sizing exercise is finding the costs of the various systems for which performance has been calculated. The ratio of this cost, when amortized over the expected life of the solar system, to the energy delivery is the *average cost* of solar energy expressed in dollars per million British thermal units.

Passive System Performance Prediction

The first step in the construction of the total cost curve is the calculation of expected solar system performance. This calculation is very complicated and is described in Ref. 1. However, it has been shown that the performance of a solar system, once calculated in precise detail, can be used to predict the performance of other solar systems in other locations if the results are used properly.

Passive predictions are based on a method developed by Arney, Seward, and Kreider (56). This method, called the *P-chart*™,* uses only the building heat load. The building heat load is expressed in British thermal units per degree Fahrenheit per day (British thermal units per heating degree-day) as described in Chap. 2. The correlation procedure developed in Ref. 56 is based on a month-by-month procedure called the SLR method and described in Chap. 4.

The P-chart will specify the solar fraction and optimum size of three types of passive systems: direct gain, masonry storage wall, and water storage wall. The following five steps should be followed to find the annual solar fraction $\bar{f}_s$. The solar fraction is the percent of annual heating load delivered by the passive system. The quantity $(1 - \bar{f}_s)$ is the fraction of the annual load which must be provided by the backup system. This fraction, when multiplied by the annual heat load and the cost of auxiliary fuel, is used to plot the auxiliary-fuel-cost curve shown in Fig. 3.13. The five P-chart steps are as follows.

Step 1. Determine the building *unit heat load* (excluding the passive aperture area) in British thermal units per degree-day. For a well-insulated residential building it will be about 5 to 7 Btu/(degree-day·ft²).

Step 2. From appendix Table A.4, find the total annual degree-days (base 65°F) for the site nearest the location to be studied. Multiply the unit load from step 1 by the degree-day total and divide by 1 million to find the *annual heat load L* in millions of British thermal units per year.

Step 3. Select one of the three types of passive systems for application to the building, decide whether it will use night insulation, and then *read constants A and B* from the P-chart table in appendix Table A.9 for the location.

Step 4. Calculate the annual solar fraction $\bar{f}_s$ from Eq. (3.2) for a specific aperture area A_c.

$$\bar{f}_s = A \ln [1 + B(A_c/L)] \tag{3.2}$$

The load L is in million British thermal units per year and the collector area A_c is in square feet.

Step 5. Repeat step 4 for a range of aperture of collector areas A_c. Then plot an auxiliary-fuel-cost curve like that shown in Fig. 3.13 and find the minimum cost point.

As shown later, the P-chart method permits direct calculation of the optimum area associated with minimum heating cost (bottom point of total cost curve in Fig. 3.13). As a result, the repetitive calculations in steps 4 and 5 above can be circumvented since the proper area does not need to be found by trial and error, which other passive sizing methods use.

*P-chart is a trademark of the Solar Energy Design Corporation of America, P.O. Box 67, Fort Collins, Colorado.

Example. Use the P-chart method to find the annual percent of heating provided to a 16,000 Btu/degree-day building in Grand Junction, Colorado. A thermal-storage wall with night insulation is to be used. What is the total cost of backup fuel required for two sizes of solar systems—350 ft^2 and 700 ft^2—if the net unit cost of fuel is \$10 per million British thermal units?

Solution. From appendix Table A.9, $A = 0.496$, $B = 0.317$, and from appendix Table A.4, the annual total degree-days are 5605. Therefore, the annual heat load is 89.7×10^6 Btu. From Eq. (3.2), solar fractions for each system size can be calculated. For 350 ft^2,

$$\bar{f}_s = 0.496 \ln (1 + 0.317 \times 350/89.7) = 0.40$$

and for 700 ft^2,

$$\bar{f}_s = 0.496 \ln (1.0 + 0.317 \times 700/89.7) = 0.62$$

Therefore the smaller system delivers 40 percent of the 89.7×10^6 Btu load or 36×10^6 Btu/yr, and the larger system provides 56×10^6 Btu/yr. The 350-ft^2 system requires 54×10^6 Btu of backup heat costing \$540. The larger system uses 34×10^6 Btu costing \$340. The law of diminishing returns is evident, since doubling the size of the passive system did not double the amount of useful energy delivered but only increased it by 55 percent.

The P-chart is a major simplification of the SLR passive prediction method developed by scientists at Los Alamos Scientific Laboratory. It predicts passive system performance within 1.5 percent rms of the SLR method for all cities in the United States in a small fraction of time needed by the SLR method. It applies for standard design features of the three types of systems. Included are 45 Btu/(°F·ft^2) thermal storage, masonry wall thermocirculation vent storage amounting to 3 percent of the wall area, double glazing, and, optionally, night insulation (R-9 if used).

In using the P-chart, it is essential to note that the load L must be calculated exactly as stated in steps 1 and 2. Therefore, degree-days to base 65°F must be used (this excludes night setback, for example) to find the load. No other load-calculation method can be used. If other degree-day bases or load-calculation methods are required, the SLR method described in Chap. 4 or an hourly simulation method must be used to find $\bar{f}_s$.

Performance Prediction for Active Systems

SD performance prediction for active systems uses a method called the *G-chart*™ (21).* The G-chart is based on a correlation of many hundreds of computer simulations for solar heating systems of both air and liquid types. The only inputs required to predict the annual performance of an active system are the slope and

*G-chart is a trademark of the Solar Energy Design Corporation, P.O. Box 67, Fort Collins, Colorado.

efficiency of the collector curve and the expected annual heat load. The performance is calculated for a range of collector areas as in the preceding example.

Procedure for Use of the G-Chart. The G-chart can best be used with a 10-step approach as described below. Note that SI units are required, and conversions are given below.

Step 1. Select an air or liquid flat-plate collector and determine the magnitude of the slope $F_R U_c$ and intercept $F_R \tau\alpha$ of its efficiency curve as shown in Chap. 2.

If the efficiency curve is in English engineering units, convert the $F_R U_c$ value to SI units by multiplying it by 5.67. Typical values of $F_R \tau\alpha$ are between 0.5 and 0.8; values of $F_R U_c$ are in the range 2 to 8 W/(m$^2 \cdot$ °C) or 0.5 to 1.5 Btu/(hr $\cdot$ ft$^2 \cdot$ °F).

Step 2. If there is a heat exchanger between the collector and storage, calculate $F_R' U_c$ and $F_R' \tau\alpha$ as follows:

$$F_R' U_c = (F_R U_c) F_{hx} \tag{3.3}$$

$$F_R' \tau\alpha = (F_R \tau\alpha) F_{hx} \tag{3.4}$$

A typical value of heat-exchanger penalty factor F_{hx} (see Chap. 2) is 0.97 for liquid systems. For air systems or liquid systems without a heat exchanger, $F_{hx} = 1.0$.

Step 3. Determine the building unit heat loss (including transmission and air-change heat losses) UA expressed in kilojoules per degree Celsius per hour. If the heat loss is known in English units (British thermal units per degree Fahrenheit per hour), multiply its value by 1.9 to convert it to SI units; if the heat loss is expressed in watts per degree Celsius, multiply its value by 3.6.

Step 4. Determine the annual heat load L in gigajoules per year using the degree-day method:

$$L = \left(24 \times \frac{UA}{1{,}000{,}000} \right) \times \left(\frac{\text{degree C-days}}{\text{year}} \right) \tag{3.5}$$

If available data are in degree F-days per year, divide by 1.8 to get degree C-days per year. (See appendix Table A.4.) If degree-days to a base other than 65°F are used, the G-chart may be less accurate. In this case, the F-chart (Chap. 4) should be used.

If a combined space- and water-heating system is considered, add the annual water-heating load to L above. The G-chart method assumes a daily water usage, typical for a residence, of 80 gal/day or 300 L/day corresponding to 22.4 GJ/yr. However, any usage rate can be specified.

Step 5. Select a collector area A_c in square meters. If the collector area is expressed in square feet, divide its value by 10.76 to convert to square meters. For an initial guess at A_c, multiply L from step 4 by ⅔ to get the calculations started for residences. Other collector areas are selected later for further calculations.

Step 6. Calculate parameter Y for a liquid system from

$$Y = \left(\frac{F_R'\tau\alpha}{0.75}\right)\left(\frac{L_o}{L}\right) A_c \tag{3.6a}$$

For an air system use

$$Y = \left(\frac{F_R'\tau\alpha}{0.50}\right)\left(\frac{L_o}{L}\right) A_c \tag{3.6b}$$

$F_R'\tau\alpha$ is from step 2, L is from step 4, A_c is from step 5, and L_o is from Table A.6 in the appendix for the location nearest the solar building site.

Step 7. Calculate parameter X for both air and liquid systems from

$$X = \left(\frac{F_R'U_c}{F_R'\tau\alpha}\right) - 8.0 \tag{3.7}$$

$F_R'U_c$ and $F_R'\tau\alpha$ are both from step 2.

Step 8. Calculate parameter R from

$$R = A + X(B + C \cdot X) \tag{3.8}$$

Values of A, B, and C are read from appendix Table A.6 (air systems) or from Table A.7 (liquid systems). (The E notation used in Tables A.6 and A.7 denotes powers of 10. Thus 0.15E-02 is the same as 0.15×10^{-2} or 0.0015.)

Step 9. Calculate parameter S from

$$S = D + X(E + F \cdot X) \tag{3.9}$$

Values of D, E, and F are read from Table A.6 (air systems) or from Table A.7 (liquid systems).

Step 10. Calculate the useful solar heat delivered Q_u (gigajoules per year) from:

$$Q_u = L[1 - e^{-(RY+SY^2)}] \tag{3.10}$$

To express Q_u in million British thermal units per year, multiply the results of Eq. (3.10) by 0.95. Do not use the G-chart for G/L (i.e., $\bar{f}_s$) ratios greater than 0.9 or less than 0.15. To repeat the calculations for other collector areas, repeat only two steps—6 and 10.

Assumptions Underlying the Use of the G-Chart. During the development of the G-chart, certain assumptions were made. Included among these are the orientation of the collector. The collector was assumed to face due south and to be tilted up

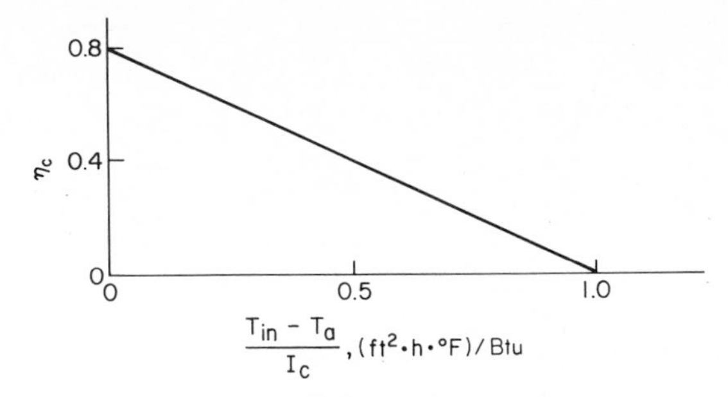

FIG. 3.14 Liquid collector curve for G-chart example.

from the horizontal at an angle equal to the local latitude. This tilt angle is slightly less than that often used in solar-heated buildings (latitude plus 15°), but has been selected for the G-chart in the interest of reducing architectural impact and structural cost.

The G-chart applies only to solar heating systems of the type shown in Figs. 2.16 and 2.19; it does not apply to other configurations which include features not present in these systems. The G-chart has been tested against more detailed calculation methods and found to agree within 2 percent for several dozen arbitrarily selected cities in the United States. The G-chart does not apply for systems that supply only domestic hot water. However, the G-chart for liquid-based systems can be used to estimate DHW performance to within about 10 percent.

Example. Find the annual solar energy delivery Q_u for a 1000-ft^2, liquid-based solar collector system on a residential building with a heat load of 1000 Btu/(h·°F) and the standard 80-gal/day water demand. The building is located in Bismarck, North Dakota, and experiences 8850 degree F-days per year.

Solution

Step 1. Find the collector efficiency curve slope $F_R U_c$ and intercept $F_R \tau\alpha$ from the curve in Fig. 3.14.

Intercept $F_R \tau\alpha = 0.8$
Slope $F_R U_c = 0.8/1.0 = 0.8$ Btu/(h·ft^2·°F)
Converted to SI units, $F_R U_c = 0.8 \times 5.67 = 4.5$ W/(m^2·°C)

Step 2. Find $F_R' U_c$ and $F_R' \tau\alpha$. Take the value of F_{hx} to be 0.97.

$F_R' U_c = 4.5 \times 0.97 = 4.37$ W/(m^2·°C)
$F_R' \tau\alpha = 0.8 \times 0.97 = 0.78$

Step 3. The given building unit heat loss UA is 1000 Btu/(h·°F). In SI units, UA is $1.9 \times 1000 = 1900$ kJ/(h·°C).

Step 4. Find L, the annual heat demand, including 22.4 GJ/yr for water heating, from Eq. (3.5). The 8850 degree F-day total is equivalent to 4917 degree C-days.

$$L = 24 \times (1900 \div 1{,}000{,}000) \times 4917 \text{ degree C-day/yr} + 22.4$$
$$L = 247 \text{ GJ/yr}$$

Step 5. The given collector area A_c is 1000 ft^2. Divide by 10.76 to convert to square meters; the collector area is 93 m^2.

Step 6. Calculate parameter Y from Eq. (3.6*a*) using L_o = 163.13 from Table A.6.

$$Y = \left(\frac{0.78}{0.75}\right)\left(\frac{163.13}{247}\right) \times 93 = 64 \text{ m}^2$$

Step 7. Calculate parameter X from Eq. (3.7).

$$X = \frac{4.37}{0.78} - 8$$
$$X = -2.4$$

Step 8. Read parameters A, B, and C from Table A.7, and calculate parameter R from Eq. (3.8).

$$A = 0.1867 \text{ E-01}$$
$$B = -0.1162 \text{ E-02}$$
$$C = -0.1103 \text{ E-04}$$
$$R = 0.0214$$

Step 9. Read parameters D, E, and F from Table A.7, and calculate parameter S from Eq. (3.9).

$$D = -0.4728 \text{ E-04}$$
$$E = 0.2469 \text{ E-05}$$
$$F = 0.7172 \text{ E-06}$$
$$S = -4.91 \times 10^{-5}$$

Step 10. Calculate the annual solar delivery Q_u using parameters R and S from steps 8 and 9 and parameter Y from step 6.

$$Q_u = 170 \text{ GJ/yr}$$

Comments on the Example. The load to be used in a G-chart calculation is the total heating load on the building less the input of any passive heating systems. Therefore, *L in the example and in the G-chart method is the net heating load to be met only by the active system.* If other collector areas are to be studied, use the same values of X, R, and S. Only the value of Y in step 6 changes. The solar fraction $\bar{f}_s$ is defined as Q_u/L, and in the example it is equal to 0.69.

System Size Selection

After the calculation of solar performance over a range of sizes is complete, the specific size of the system for schematic design purposes can be identified. This may be done in either of two ways. The owner may arbitrarily decide that the solar heating system is to provide a given annual solar fraction, for example, 75 or 80 percent. If this is the manner of system sizing, the P-chart or G-chart results directly give the collector area. Then, the sizes of the remaining components of the system are calculated from the rules summarized in Table 3.2 and the subsequent discussion.

The other system sizing method results in the lowest heating bill over the life cycle for the given building. This method was shown in Fig. 3.13. Calculation of the *auxiliary-fuel-cost curve* has been described above (page 139). If fuel price inflation is to be considered, another step in the calculation is required and is described shortly. The calculation of the *solar system cost curve* will be described later in this section.

Initial Cost. During schematic design the cost of a solar system can be based upon past experience with systems of a similar type or upon telephone quotations. In order to conduct the optimization study, costs for each component at several sizes are needed.

Table 3.3 summarizes typical costs of components of active systems used in commercial installations. The collector and supports compose approximately one-third of the system cost and storage and heat exchangers about 20 percent. The balance is devoted to piping, controls, and installation. A relatively wide range of collector costs is shown; the cost depends primarily upon the type of collector and the sophistication with which it is built. Built-in-place collectors may cost as little as one-half as much as commercially available equivalent. Controls and electrical components will vary widely by application, the figure shown in Table 3.3 applying for commercial installations. In residential installations, the cost is approximately $1000.

If time for acquiring quotes or at least telephone estimates is not available during SD, the rough figures in Table 3.3 can be used. However, one of the deliverables for schematic design is a cost statement. *If at all possible, the SD cost statement should be based upon estimates reflecting local labor and material prices, since the decision about whether solar will be used is normally made during schematic design.* The amount of money involved in purchasing a solar system is substantial; hence a reliable cost estimate is important.

Schematic design may be undertaken more than a year before the planned beginning of building construction. Purchase of solar components may take place 6 mo after construction has started. The 1- to 2-yr interval between schematics and system purchase and installation must be accounted for in the cost estimate by applying an inflation figure. The uniform construction index (UCI) or Means' cost index (44) can be used to estimate inflation. Although inflation is not known for future years, projections can be made based on immediately preceding years. Of the 16 sections of the UCI, solar design will be most concerned with Sec. 10 (specialties), Sec. 11 (equipment), Sec. 13 (special construction), Sec. 15 (mechanical), and Sec. 16 (electrical). The schematic design cost statement should specify inflation rate, contingency, profit, and overhead rates.

Initial solar system cost estimated in this section will be used by the owner while securing financing. For a residential application, the Federal Home Loan Mortgage Corporation will finance solar systems up to the 90 percent level. For commercial buildings, no specific guidelines are available, but the extra cost of the solar system must be known.

Annualized Costs. The *annualized solar heating cost* must be less than the annual cost of heating without solar energy for solar heat to be economically viable. This section describes the calculation of the annualized solar cost using the capital recovery factor idea. The annualized cost for schematic design purposes does not include details such as tax deductions for interest, salvage value, etc. These matters are treated in the more detailed DD-level economic analysis section in Chap. 4.

TABLE 3.3 Active solar system cost estimates for commercial buildings useful in schematic design*

Subsystem costs	*Percent of total solar cost*
Collectors and supports	35
Storage and heat exchangers	20
Piping, controls, electrical, and installation	45
Subsystem component	***Cost per square foot of collector***
Collectors	\$5–\$20
1. Nonselective	\$5–\$10
2. Selective	\$10–\$20
3. Collector support	\$3–\$10
Heat exchangers	\$0.40–\$0.80
Collector fluid	\$0.15–\$0.20
Storage tank and insulation	\$2–\$5
Piping, insulation, expansion tanks, valves	\$3–\$6
Pumps	\$0.40–\$1.00
Controls and electrical	\$3000–\$5000 per installation
Type of system	***Installed cost per square foot of collector***
Building service hot water (BSHW) only	\$20–\$35
Space and BSHW heating	\$25–\$50
Space heating and cooling	\$35–\$65

*1978 dollars, from Ref. 23.

TABLE 3.4 Capital recovery factors*†

	Interest rates *i*									
n	*0%*	*2%*	*4%*	*6%*	*8%*	*10%*	*12%*	*15%*	*20%*	*25%*
1	1.00000	1.02000	1.04000	1.06000	1.08000	1.10000	1.12000	1.15000	1.20000	1.25000
2	0.50000	0.51505	0.53020	0.54544	0.56077	0.57619	0.59170	0.61512	0.65455	0.69444
3	0.33333	0.34675	0.36035	0.37411	0.38803	0.40211	0.41635	0.43798	0.47473	0.51230
4	0.25000	0.26262	0.27549	0.28859	0.30192	0.31547	0.32923	0.35027	0.38629	0.42344
5	0.20000	0.21216	0.22463	0.23740	0.25046	0.26380	0.27741	0.29832	0.33438	0.37184
6	0.16667	0.17853	0.19076	0.20336	0.21632	0.22961	0.24323	0.26424	0.30071	0.33882
7	0.14286	0.15451	0.16661	0.17914	0.19207	0.20541	0.21912	0.24036	0.27742	0.31634
8	0.12500	0.13651	0.14853	0.16101	0.17401	0.18744	0.20130	0.22285	0.26061	0.30040
9	0.11111	0.12252	0.13449	0.14702	0.16008	0.17364	0.18768	0.20957	0.24808	0.28876
10	0.10000	0.11133	0.12329	0.13587	0.14903	0.16275	0.17698	0.19925	0.23852	0.28007
11	0.09091	0.10218	0.11415	0.12679	0.14008	0.15396	0.16842	0.19107	0.23110	0.27349
12	0.08333	0.09156	0.10655	0.11928	0.13270	0.14676	0.16144	0.18148	0.22526	0.26845
13	0.07692	0.08812	0.10014	0.11296	0.12652	0.14078	0.15568	0.17911	0.22062	0.26454
14	0.07143	0.08260	0.09467	0.10758	0.12130	0.13575	0.15087	0.17469	0.21689	0.26150
15	0.06667	0.07783	0.08994	0.10296	0.11683	0.13147	0.14682	0.17102	0.21388	0.25912
16	0.06250	0.07365	0.08582	0.09895	0.11298	0.12782	0.14339	0.16795	0.21144	0.25724
17	0.05882	0.06997	0.08220	0.09544	0.10963	0.12466	0.14046	0.16537	0.20944	0.25576
18	0.05556	0.06670	0.07899	0.09236	0.10670	0.12193	0.13794	0.16319	0.20781	0.25459
19	0.05263	0.06378	0.07614	0.08962	0.10413	0.11955	0.13576	0.16134	0.20646	0.25366
20	0.05000	0.06116	0.07358	0.08718	0.10185	0.11746	0.13388	0.15976	0.20536	0.25292

*From Ref. 1 with permission.

†For interest rates *i* of from 0 to 25 percent and for periods of analysis *n* of from 1 to 20 yr.

Engineering economics texts show that an initial amount invested in a solar system can be converted to a series of uniform or *level* annual payments by multiplying the initial cost by the capital recovery factor. The capital recovery factor includes the interest rate and the duration of the economic period of analysis. This period may be selected as equal either to the mortgage period or to a period representing the expected lifetime of the solar system. Capital recovery factors are given in Table 3.4. The interest rate is shown along the top of the table and the period of analysis—life-cycle period—is shown to the left.

Consider the annual payment for a solar heating system costing $10,000 after tax credits. Money is available at 10 percent interest to be paid off in 20 yr. The capital recovery factor from Table 3.4 is 0.11746. The product of the capital recovery factor and the initial cost is $1175. This amount is paid annually over 20 yr to repay both loan principal and interest; $1175 paid annually for 20 yr is a total payment of $23,500. The initial system cost was only $10,000. Therefore, the interest payment was $13,500. It will be shown in Chap. 4 that approximately one-third of this interest payment, depending on the owners' tax bracket, will be returned as a tax deduction. However, these details need not be considered during schematic design since property taxes, maintenance, and operating costs tend to offset the tax saving.

The capital recovery factor CRF(i,N) is given by Eq. (3.11).

$$\mathrm{CRF}(i,N) = \frac{i}{1 - (1 + i)^{-N}} \tag{3.11}$$

In this equation i is the annual, nominal interest rate expressed as a decimal, and N is the life-cycle period of analysis expressed in years. CRF calculations can be done on a monthly basis instead of an annual basis, but the difference is very small.

Effect of Inflation on Fuel Prices. During the 1970s, the price of conventional fuels increased more rapidly than the consumer price index (CPI), which is one measure of the general inflation rate of the economy. For example, in 1979–1980, fuel oil costs increased from 40 cents per gallon to 80 cents per gallon, an annual rate of increase of 100 percent. During this same period, the CPI increased by approximately 13 percent.

The effect of fuel differential inflation is to increase the value of the future fuel savings. For example, 100 million British thermal units saved may be worth $1000 during the first year of solar system operation. However, if the price of fuel were to double, the value of the same 100 million British thermal unit energy saving will be $2000. The fuel price inflation rate is one of the key variables in an economic analysis of a solar system. Since this inflation rate will not be known accurately until after the system has been in operation, it is recommended that a reasonable range of values be used during the SD economic study. If solar heat is feasible for all reasonable values of inflation rate, then the decision regarding solar system adoption can be made with confidence.

Inflation in fuel prices has been estimated by the U.S. Department of Energy (24). The table below shows the Department of Energy differential inflation guidelines which are to be used in assessing the feasibility of solar energy in federal facilities. *The differential inflation rate is the rate of inflation above the general inflation rate measured by the CPI.* Therefore, the total fuel price inflation will be that shown in the table plus the CPI.

Type of energy	*Percentage*
Coal	5
Fuel oil	8
Natural gas	10
Electricity	5.6–7.5

Fuel price inflation over the life cycle of a solar system can be expressed in what is called the *levelized fuel cost.* The levelized cost $\overline{C}_f$ is the cost of fuel $C_{f,o}$ during the base year multiplied by a ratio of capital recovery factors as shown in Eq. (3.12).

$$\overline{C}_f = C_{f,o} \frac{\mathrm{CRF}(i,N)}{\mathrm{CRF}(i',N)} \tag{3.12}$$

The capital recovery factor in the numerator is evaluated from Table 3.4 or from Eq. (3.11) using the expected interest rate on the solar loan and the life-cycle period

N. The capital recovery factor in the denominator is based on an inflation-modified interest rate i'; i' is equal to the difference between the interest rate and the inflation rate divided by the quantity 1 plus the inflation rate j—i.e., $i' = (i - j)/(1 + j)$.

Example. If the cost of oil $C_{f,o}$ is $40 per barrel today, what is the levelized cost over 15 yr if the discount rate $i = 10$ percent and the price of oil increases uniformly at $j = 5$ percent per year?

Solution. Find i' and the capital recovery factors from Eq. (3.11).

$$i' = (0.1 - 0.05)/1.05 = 0.0476$$
$$\text{CRF}(0.1, 15) = 0.1315$$
$$\text{CRF}(0.0476, 15) = 0.0947$$

Then from Eq. (3.12) the levelized cost

$$\overline{C}_f = \$40 \times 0.1315/0.0947 = \$55.54 \text{ per barrel}$$

Total Annual Cost. The total annual heating cost is the sum of the annual cost of the solar system and the annual cost of backup fuel. The annualized cost of the solar system is added to the *levelized fuel cost.* The levelized fuel cost is the levelized unit cost of fuel multiplied by the amount of backup energy required, i.e., $(1 - \bar{f}_s)$ times the annual load L.

This total cost C_{tot} of heating a solar building is

$$C_{tot} = [C_{s,tot}\text{CRF}(i,N) + \overline{C}_f(1 - \bar{f}_s)L] \tag{3.13}$$

where $C_{s,tot}$ = initial total cost of the solar system net of tax credits
$\text{CRF}(i,N)$ = capital recovery factor
$\overline{C}_f$ = levelized unit fuel cost
$\bar{f}_s$ = annual solar heating fraction
L = annual heating load for space and/or water heating.

The minimum total cost identifies the optimum solar system size. The two terms in the equation are represented by the two dashed curves in Fig. 3.13. The solid curve is C_{tot}.

Example. Find the optimum size of a passive solar heating system whose performance is given in Table 3.5. Assume that the levelized cost of fuel is $20 per million British thermal units and the passive system costs $18 per square foot and is purchased over a 20-yr period at 10 percent interest. The annual heat load is 88.38 British thermal units.

Solution. The capital recovery factor is 0.11746; therefore, the passive system costs 0.11746 × 18 = $2.11 per square foot per year. This cost is multiplied by collector areas in Table 3.5 in order to find the solar system cost per year. This is added to the levelized auxiliary fuel cost to find the total cost. The calculations are summarized in Table 3.5.

TABLE 3.5 Solar, backup fuel, and total heating costs for example passive system

Area, ft²	*Solar cost, dollars per year*	$\bar{f}_s$	*Fuel cost, dollars per year*	*Total cost, dollars per year*
53	112	0.1	1591	1703
112	236	0.2	1414	1650
176	371	0.3	1237	1608
246	519	0.4	1061	1580 ← least cost
333	703	0.5	884	1587
444	937	0.6	707	1644
593	1251	0.7	530	1781
842	1777	0.8	354	2131
1230	2595	0.9	177	2772

The minimum cost is for a collector area between 246 and 333 ft² or a solar fraction $\bar{f}_s$ of about 50 percent. The exact minimum point can be found by numerical interpolation.

Direct Sizing of Passive Systems. As noted above, the P-chart has the particularly attractive feature that the roundabout selection method used in the preceding example can be avoided. That is, the optimal area can be found explicitly rather than indirectly by considering a whole series of collector aperture sizes. For all P-chart systems, the optimal, least-cost size is given by

$$A_{c,opt} = L\left[\frac{A \times \overline{C}_f}{CRF \times c_s} - \frac{1}{B}\right] \quad \text{ft}^2 \tag{3.14}$$

All terms are described above and c_s is the solar system *cost per square foot*. Equation (3.14) is derived from Eqs. (3.2) and (3.13). If the right side of Eq. (3.14) is negative, solar heating is not feasible.

An alternative to Eq. (3.14) is the P-chart nomograph shown in Fig. 3.15. Start at the upper right quadrant at axis 1 with the value of CRF × c_s (extra passive cost per square foot times capital recovery factor). Move up to a curved line with the value of $\overline{C}_f$ (levelized unit fuel cost). Then move left to the straight line labeled with the value of P-chart constant A. Finally, drop down to the lower left quadrant of the P-chart.

Now start at axis 2 at the lower right with the value of P-chart constant *B*. Move up to the curved line and then left and intersect the line constructed down from the upper left quadrant. The intersection represents the ratio of optimum passive collector area $A_{c,opt}$ to annual load *L*. Multiply the ratio by *L* from step 2 of the method (page 141).

Summary

The method of selecting the optimum system size presented in the preceding sections is based upon the cost of solar energy relative to the cost of nonsolar energy. This is the only rational system sizing method in a free economy. Arbitrary rules of thumb advocated by some designers basing collector area on building floor area do not reflect the cost of competing fuels, the effect of geography on solar system performance, nor the level of energy-conservation features in the given building. Therefore, these *rules of thumb must not be used since they will result almost always in an erroneously sized system with far from optimum economic performance.* The calculation methods presented above are simple and are within the scope of a schematic design effort. The added effort in making these calculations is commensurate with the increased fee which an architect or engineer will receive owing to the additional cost of the solar system beyond that for a conventional building.

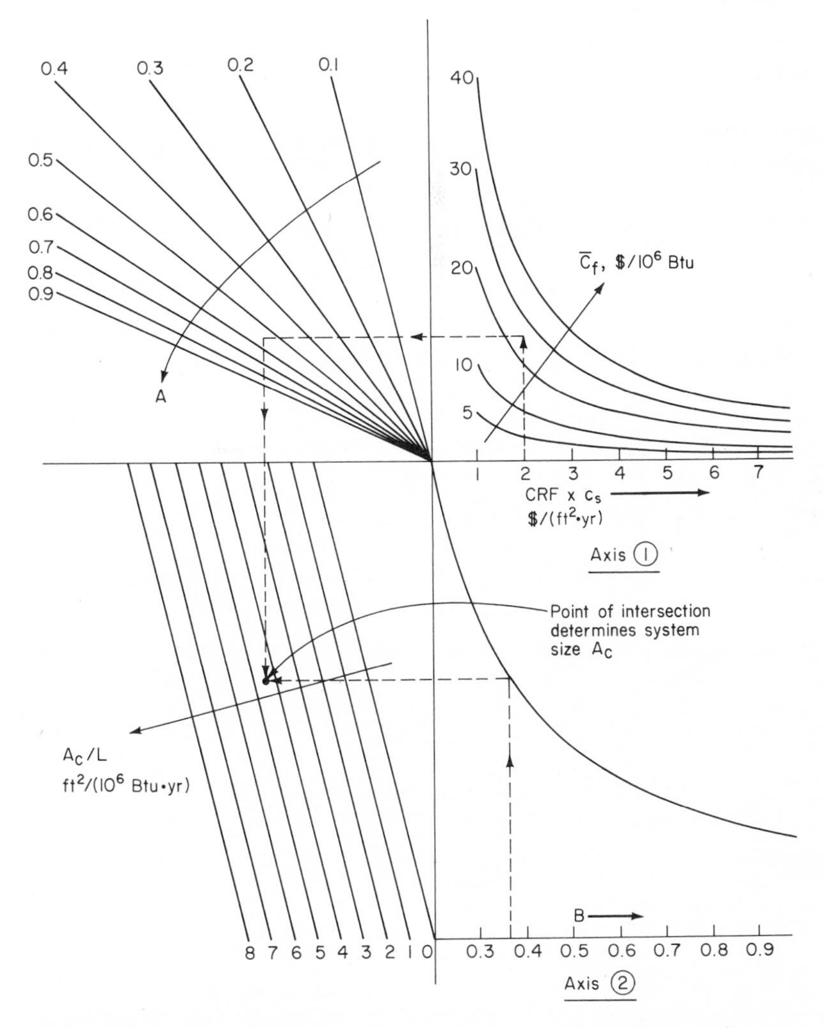

FIG. 3.15 The P-chart for finding the optimum size of passive heating systems. P-chart is a trademark of the Solar Energy Design Corporation of America, Fort Collins, Colo.

CLIENT REVIEW OF SCHEMATIC DESIGN DOCUMENTS

The owner will review all schematic diagrams, SD calculations, outline specifications, and other material prior to requests for either modification or approval of the schematic design phase. Presumably no changes will be required.

Feasibility Studies

The SD economic analysis is often summarized in a formal solar feasibility report indicating the economic viability of solar heating at the SD level. A more detailed study may also be done during DD with more refined costs. The SD study should list all thermal and economic assumptions, methods of calculation, and *expected accuracy of the results*. A statement regarding solar feasibility or infeasibility is the conclusion of the report supported by thermal and economic calculations and graphical displays of both.

In order to conduct an SD feasibility study, the following information is needed at a minimum (more detailed economic data can be used if the accuracy of the performance prediction method so warrants):

- Thermal information
  - Space heating load and schedule
  - Water heating load and schedule, water source and delivery temperature
  - Night setback, if any
  - Types of solar system to be considered—any limits to component sizes
  - Type of auxiliary system and fuel
  - Internal heat-gain estimates
- Economic information
  - System cost
  - Discount rate
  - Net fuel cost
  - Fuel inflation rate
  - Length of period of analysis
  - Applicable tax credits

This information should be available at the end of schematics if the owner has been involved in the design process in the proper manner. An additional deliverable item of the schematic design phase is the building model. The building model should accurately represent the expected impacts of solar energy including the proper collector area, orientation, tilt, and location.

SCHEMATIC DESIGN PHASE CHECKLIST

The checklist below summarizes work required to carry out solar aspects of SD. The various major steps used to organize the list are:

Load calculation
Solar system performance
Economic analysis

System sizing and component SD specifications
SD deliverables
Refer to preceding text for details on each entry.

SD Solution Selection—Overview

- Organize solar design team ☐
- Identify and document feasible solutions to program ☐
- Examine architecturally using sketches, etc., technically using models, etc., economically ☐
- Select solution ☐
- Prepare SD deliverables ☐

Building Energy Needs for Heating

- Select method of heating and DHW load calculation ☐
- Identify or specify all parameters needed for method ☐
- Calculate zonal heating and DHW loads varying important parameters—fenestration, R values, aspect ratios, colors, mechanical systems—monthly time scale ☐
- Estimate operating cost for nonsolar heating ☐
  - Space—demand plus energy charges ☐ $______/yr
  - DHW—demand plus energy charges ☐ $______/yr
  - Other—demand plus energy charges ☐ $______/yr
- Estimate nonsolar mechanical system cost for all loads ☐
- Carry out above steps for all proposed solutions ☐

Examine Solar Systems to Address Loads

- Passive DHW ☐
- Passive space heating ☐
- Active DHW (net of passive contribution) ☐
- Active space heating—liquid (net of passive contribution) ☐
- Active space heating—air (net of passive contribution) ☐
- Architectural constraints on system types ☐
- Site (shading orientation, etc.) constraints on system type ☐

Programmatic constraints on system type	☐	
Economic/budget constraints on system type	☐	
Code constraints on system type	☐	
Community (glare, etc.) constraints on system type	☐	
Calculate annual auxiliary fuel savings for a wide size range of each system or hybrids thereof using P-chart and G-chart	☐	

Economic Calculations for Solar Systems

Determine

Installed unit costs of all systems to be studied	☐	\$_______/ft_c^2
Interest rate	☐	_______%
Tax rates	☐	_______%
Life-cycle (LC) period	☐	_______ yr
Tax credits	☐	_______%
Solar system cost inflation until construction date	☐	_______
Fuel cost	☐	\$_______/million Btu
Fuel-cost inflation rate	☐	_______%/yr
Auxiliary system efficiency or COP	☐	_______%
Calculate LC savings (net of system cost) for each size of each solar system considered using P-chart and G-chart	☐	
Find size of each system including energy-conservation features with maximum LC savings (same as minimum LC cost)	☐	
Select least-cost combination of above systems subject to architectural, program, code, etc., constraints	☐	

System Sizing

Size components of cost optimal passive system(s) using SD rules based on cost-optimal aperture area	☐	
Vent area	☐	
Storage—masonry, water, etc.	☐	

- Position storage for desired winter illumination ☐
- Size storage area for proper heat transfer ☐
- Storage mass exterior surface colors ☐
- Night insulation ☐
- Overhangs or other solar control ☐
- Glare control for direct gain systems ☐
- Reflectors ☐
- Heat distribution system if any ☐
- Size components of cost-optimal active system(s) using SD rules of thumb based on cost-optimal aperture area ☐
- Storage ☐
- Pumps/fans ☐
- Pipes/ducts ☐
- Insulation of components ☐
- Collector heat exchanger ☐
- DHW heat exchanger ☐
- Load heat exchanger ☐
- Collector tilt ☐
- Collector azimuth angle ☐
- Expansion tanks ☐
- Heat rejector ☐
- Reflectors ☐

SD Solar Deliverables

Drawings

- Site plan ☐
- Floor plan ☐
- Roof plan ☐
- Sections adequate to show all solar features ☐
- Renderings ☐

Solar Features to Be Shown

- Collector—active/passive ☐
- Mechanical space requirements for storage, pumps, heat exchangers ☐
- Heat rejector (roof plan, site plan) ☐
- Internal space for passive storage mass, night insulation, heat distribution ☐

- Active system distribution system ☐
- One-line solar system diagram with notes and control operation ☐
- Summary of SD solar and load calculations and feasibility report including annual energy usage and cost compared to program targets ☐
- General outline performance specification ☐
- Statement of probable solar system cost ☐
- Building model with solar features shown ☐
- Secure SD approval and authorization to proceed ☐

4 DESIGN DEVELOPMENT

A good design is a simple design.
RAYMOND LOEWY

Design development (DD) is the heart of the solar heating design process. During this phase all architectural and engineering questions are formulated and resolved. The purpose of the design development documents is to display the size and character of the project in a formal manner. Design development documents are prepared from approved schematic design studies and when approved form the basis for construction documents described in Chap. 5.

All solar questions must be addressed and answered in great detail because of the lack of familiarity of builders with this new technology. Improper designs of solar systems can set back the widespread implementation of solar heating by decades. It is therefore of great importance that the design process be carried out with attention to detail.

Most of the detailed information in this book is contained in this chapter. This chapter first treats each component in detail. Next, system considerations not specific to any component are delineated. Finally the method of economically sizing the solar heating system is treated.

OVERVIEW OF DESIGN DEVELOPMENT

Design development documents are prepared from approved schematics. During schematic design more than one type of system may have been analyzed. At the end of SD one system will have been selected.

The DD documents include a site plan indicating the location of the project and the nature of required site improvements. A full set of building plans, elevations, and sections is included. Architectural, structural, mechanical, and electrical systems are described fully. An outline specification is included. Finally, a statement of probable construction cost is presented. During DD the solar components and system will be optimized. The output of this optimization process will be the least-cost solar system.

Before the beginning of any solar design activity during DD, a detailed heat-load calculation is needed. Loads are refined on the basis of improved estimates of ventilation and infiltration rates; expected window areas; walls, roof, and floor; and building orientation. Internal heat loads must be known. Loads for nonresidential buildings should be analyzed by computer, including all energy-conservation features and energy-management programs. This book does not treat building load calculations, but Ref. 14 can be studied for a summary of approaches to building energy analysis.

Since solar DD includes an economic optimization, it will be necessary to collect costs for a range of solar system sizes. In other words, the cost of an undersized system, a nominally sized system, and an oversized system—relative to the final SD size—should be determined. The optimum size is then established and a one-line mechanical diagram and an outline specification are prepared on the basis of this optimum size. These diagrams and specifications become part of the DD package.

SOLAR COLLECTORS

The information in this section applies to both air and liquid collectors. It is divided into two parts—the first part on construction and the second part on the operation of the collector in the field.

Collector Design

The solar collector materials must be selected with care, as described in Chap. 2. The collector environment is one of the most severe in a building.

Absorber Plate. Proper fluid flow is important if heat removal is to be efficient. Liquid piping should be in the reverse-return mode. For example, fluid could enter at the lower left-hand collector corner and exit at the upper right-hand corner. Reverse return assures that the path length for all fluid is the same. However, reverse return does *not* assure that the flow through each riser will be the same since the withdrawal of fluid from the inlet header affects the header pressure distribution. Header design is beyond the scope of this book, but most manufacturers have done this work properly. A useful rule requires that 90 percent of the collector pressure drop occur in the risers and 10 percent in the headers. If the manufacturer has adhered to this rule, the rise flow balance will be within approximately 3 to 4 percent.

The pressure drop across a solar collector should not be very small. If it is below 6 to 8 in of water in a liquid collector, the collector array will be very difficult to balance. It is better to have some pressure drop in the collector since balancing is easier.

If liquid risers are simply soldered to holes drilled in headers, leaks are likely. During large temperature swings, absorber plate expansion and contraction stresses this joint. Risers may penetrate into the header and block the flow distribution, causing fluid starvation in one or more risers. A better method uses a specific fitting at the end of each riser. It is soldered into the header. Solder used in a collector must maintain its strength to above 400°F. Silver or a 95/5 tin-antimony mixture is best for copper.

Internally manifolded collectors require less labor and materials. In addition, collector header insulation is eliminated. In this design the collector header serves as the collector array header. Therefore, it will be larger than if it were only a collector header. Typically, pipe sizes in excess of 1 in are used. The manufacturer must specify the number of such collectors which can be internally connected in parallel. If too many were connected, the pressure drop in the internal manifold could approach that in the collector risers. Then flow balance in the array would be lost. Since the manifold is internal to the collector, there is no method of correcting the imbalance. For example, several internally manifolded collectors available in the United States may be connected together in arrays of not more than six panels. Internal manifolding saves money and should be used where possible.

Flow balancing is as important in air collectors as in liquid collectors. If air collectors do not have internal manifolds, each inlet and outlet must be connected to a header sized so that its pressure drops no more than 10 percent of the collector drop. Many air collectors have small pressure drops; therefore, the duct connecting them may be larger and have lower velocity than would be expected from conventional HVAC designs. A few air collectors can be connected without intervening ducts. The resulting cost reductions are significant, since duct insulation, duct hangers, and the ducts themselves are all eliminated.

Solar Collector Operational Considerations

Supports. Attachment of a collector to a building requires a mounting structure. The structure may be either part of the roof or a separate structure above the roof. Structures must be designed to carry all live and dead loads—wind, snow, hail, the collector, its fluid, and the structure itself. Liquid collectors are heaviest; the maximum unit weight is approximately 11 pounds per square foot of collector. Local design loads for wind and snow are given in structural handbooks.

Roof snow loads are affected by collector arrays on flat roofs. If no collectors were on a roof, snow would be distributed uniformly. However, if collectors are on the roof, snow will normally slide from the collector and gather in drifts at the base of the collector row, thereby presenting a locally higher load than it otherwise would. In addition, the area north of most sawtooth array rows is shaded all winter. Therefore, snow or ice buildup will persist for the entire winter. These effects must be accounted for in the structural design.

Reflectors. If a reflector is used to enhance solar performance, the loads on the reflector must be considered. Most heating reflectors are approximately horizontal if the collector is tilted at an angle between the latitude and the latitude plus 20°. Although horizontal, the reflector is still subject to wind and snow loads. Lift on a flat reflector must be calculated so that hold-downs are adequate for the large reflector area. Most reflecting materials do not have much structural strength, and the substrate must carry the design snow load. Otherwise, the mirror could deform.

Collector dimensions should be considered in the selection of a manufacturer. Large collectors, although more bulky, normally cost less to install than an equivalent area of smaller collectors. The number of pipe fittings and the amount of pipe insulation required is obviously reduced. In addition, the large collector will make better use of a crane which is normally on site for locating collectors. More square feet of collector can be positioned per load and the installation of a given number of square feet accomplished more rapidly. Large collectors also have a larger ratio of net to gross area. Figure 4.1 shows a typical collector structure used to mount 16-ft-long collector units.

Array support materials must be selected with care. Metal and wood are most commonly used. If a metal structure is used to mount collectors with metal housings, it is important that the two metals not have dissimilar chemical characteristics. For example, an aluminum collector mounted on a steel frame will corrode in a few years. If dissimilar metals are used, electric isolation is essential. Sheets and washers of Teflon or other polymers have been used successfully for this purpose.

Wood collector mounts should be considered with care. Properly cured, preservative-impregnated wood may be the most economical material in some cases. However, wood which has not been fully cured will warp. In addition, ambient air

FIG. 4.1 Steel collector supports for 16-ft-long collector modules used on a large commercial building.

FIG. 4.2 Example wind break and appearance facade for the north side of a sawtooth collector array mounted on a flat roof.

pollutants can affect the long-term durability of wood in urban installations or at industrial sites where stack effluents are present.

Collector structures also support piping runs. The designer must specify the type of piping hanger to be used and the length between pipe hangers to assure that excessive stresses are not developed in the piping.

The back of solar collectors in a sawtooth arrangement is exposed to the environment. If the solar collectors are used in a location with relatively high winds in winter, a windbreak for the north side of sawtooth collector arrays may be useful. This need not be expensive, as shown by the example in Fig. 4.2. The windbreak can also improve the appearance of a sawtooth collector array since piping can be concealed behind it.

The expansion and contraction of large collector arrays must be accounted for as noted in Chap. 2. Figure 4.3 shows an example of a piping arrangement which can be used to absorb differential expansion in both directions in a large collector array. (Of course, before system startup the piping must be fully insulated.)

Collector Piping and Ducting. The fluid flow rate is suggested by the manufacturer in most cases. Large variations from these recommendations should not be made either by trying to improve performance with higher flows or by reducing the price of pumps with lower flows. If flow rates are increased, the performance improvement is relatively small as described in Chap. 2, and small fluid conduits can be eroded. Flow rates significantly less than manufacturer's recommendation cause a performance penalty since the fluid remains in the collector too long and

FIG. 4.3 Example of piping loop to accommodate expansion and contraction in a long manifold on a commercial building.

becomes too hot, with greater heat loss from the absorber. Collector flow recommendations are based on a trade-off of this heat loss and excessive pumping power and erosion problems.

Leaks in collectors and manifolds should be avoided. Depending upon the location of leaks in air systems, performance may actually improve if air is entrained into the system (but not leaked out of the system) near the collector inlet. If the air collector is below atmospheric pressure, the flow will be into the collector. Of course, leaks of hot air from the collector or ducting must be avoided even if the leaks are to the building interior since they represent an uncontrolled heat gain to the structure.

Leaks in liquid collectors can be more serious than in air collectors, although they are more easily located and controlled. In addition to the expense of replacing costly antifreeze fluids, other problems will arise from liquid leaks. Glycol-based antifreezes can severely damage asphalt roof materials. Also, leaks within a building can cause obvious damage to building finishes and structural components. Glycol solutions have relatively low surface tension and are more prone to leak than water. Therefore, additional attention is required in the design of piping and the use of appropriate pipe fittings. Flare-type and compression fittings on copper piping can cause problems.

Other Details. Collector arrays may be the highest point on flat-roofed buildings. Therefore, lightning arresters are advisable. The highest point of the lightning rod must be above the highest point of the collectors.

The piping between liquid collectors is most often metal. Since water is an electrolyte, the possibility for galvanic corrosion exists. Table 4.1 shows the relative electromotive relationship of common materials present in solar systems. The three most common materials are aluminum, copper, and iron. These three have significant electromotive differences. The greater the difference, the greater the corrosion rate. Corrosion may also occur within piping systems of a single metal if soldering flux is present. The most common example is the corrosion of copper conduits by the flux used to solder them together. A sacrificial anode of higher electromotive potential than any other metal in the system inhibits corrosion. For example, in a system consisting of copper collectors with steel piping, an anode of magnesium would go into solution rather than either of the two other metals. Magnesium anodes are commonly used in domestic water heaters for this purpose. Aluminum presents a more difficult problem. Although magnesium is above it in the electromotive series, it does not afford proper protection; other chemical methods are required.

Materials for piping and for insulation of piping must tolerate high-temperature stagnation periods. During stagnation, not only will the collector absorber plate become very hot, but the connecting metal piping will also be well above normal operating temperatures. The piping insulation must be able to remain intact at temperatures up to 400°F.

Collector seals are also exposed to high temperature both during stagnation and during initial installation if piping is soldered by an acetylene torch. The seals must be protected during high-temperature soldering. They are designed to withstand stagnation conditions.

The amount of flat roof available in buildings of more than three stories may be insufficient to hold all collectors. By use of the method shown in Fig. 4.4, some additional collectors can be added to the array beyond those fitting the usual sawtooth arrangement. Of course, additional structure is required, and building height

TABLE 4.1 Electromotive series for solar system metals*

Element	*Oxidation potential in 1-molal solution, V*
Sodium	+2.71
Magnesium	2.37
Aluminum	1.66
Zinc	0.76
Iron (to ferric ion)	0.44
Cobalt	0.28
Nickel	0.25
Tin	0.14
Lead	0.13
Hydrogen (reference)	0.00
Copper (to cupric ion)	−0.34
Bromide ion (to Br_2 gas)	1.09
Chloride ion (to Cl_2 gas)	1.36

*Adapted from *Solar Heating and Cooling* by J. F. Kreider and F. Kreith. Copyright 1977 by Hemisphere Publishing Corporation. Used with the permission of Hemisphere Publishing Corporation.

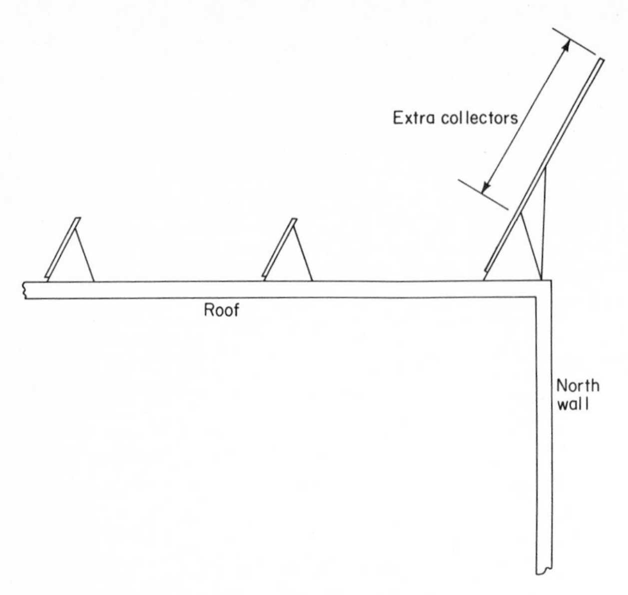

FIG. 4.4 Sketch of a method for adding additional collectors to a roof-limited collector array.

limits and buildable property line limits must be considered. If still more collectors are needed, remote collectors arrays can be used.

Collector Sizing

Since the collector is the key component of heating systems, its size is the index of the size of all other components. Therefore, "system size" can often be interpreted as collector size. In DD, the final size of the collector and all other components is determined. A greater level of accuracy is required in these calculations than in SD calculations. The method of precise economic calculations, taking into account all important variables, is described in subsequent sections in this chapter. In addition, more versatile DD methods of predicting performance are outlined. This method, based on the F-chart, can be used for air- and liquid-based active heating systems. The analogous SLR method for passive systems is used for DD passive analysis.

SD system sizes will be the starting point for detailed DD sizing. The cost-optimal system selected during DD will be that used to prepare DD documents and construction documents. During the CD phase, only minor alterations to collector area could be expected, possibly to accommodate structural or support limitations.

Collector Selection Criteria

The information presented in this section and in Chap. 2 is sufficient to specify collector design and size. In addition to the physical criteria mentioned herein, a minimum efficiency curve is specified. Collectors to be considered must perform better than this curve.

According to Chaps. 2 and 3, the selection of the collector is not based on initial cost alone. Energy delivered relative to cost is the selection criterion. This method of selection accounts for the higher cost of more efficient collectors vis-à-vis the lower cost of less efficient collectors. The final collector selection is made during the bidding process described in Chap. 5.

Acceptable collector performance is important, but the manufacturer must also be considered. In the United States, there are presently several hundred companies manufacturing solar collectors. In the past 5 yr, many have come and gone. A company which is out of business will not honor a warranty no matter how impressive the warranty appeared during purchase of the collectors. Therefore, the designer must consider the probability that the collector manufacturer will be in business a decade or two after the system is built. Accordingly, it is the responsibility of the designer to protect the owner by being circumspect in the selection of bidders. An example specification including this information is given in Chap. 5.

THERMAL STORAGE FOR SOLAR HEATING SYSTEMS

The amount of storage to be used is subject to the law of diminishing returns as shown in Fig. 4.5. This basic economic and physical law applies to all components in a solar system including storage pumps, heat exchanger, collector area, and pipe insulation. For all other inputs held fixed, the solar load fraction increases at a decreasing rate with increasing storage size. The rate of increase decreases because

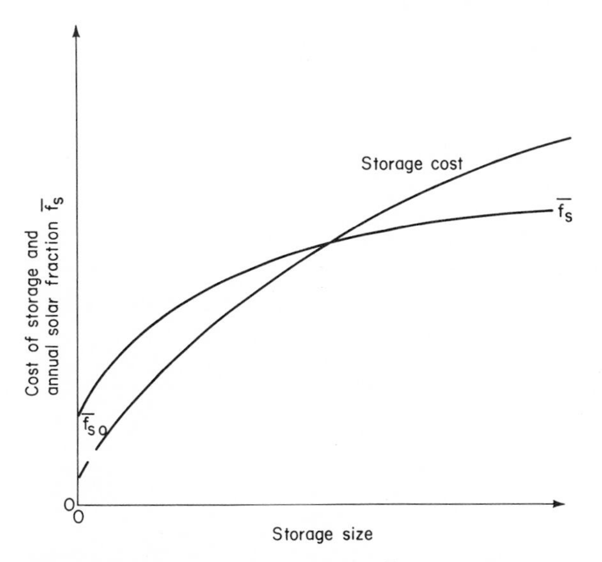

FIG. 4.5 The annual solar fraction $\overline{f}_s$ and cost for various sizes of thermal storage on a specific building in a specific location.

of the saturation of storage and the limits to the amount of heat which can be delivered to a given load by an every-increasing storage size. For example, large storage may be able to carry a given heating system through an extended period of bad weather. During the remainder of the heating season, the storage system will be oversized and will not be fully used. However, the storage system will be paid for during the entire heating season even if it is only fully utilized once.

Figure 4.5 shows that, as storage size increases, storage cost continues to increase, even though energy delivery is relatively flat for large storage. The economic consequence is that an optimum storage size exists. This optimum will be described later, but is established by the trade-off of storage cost with the load factor.

If no storage is present, solar energy can be used only during daytime; the corresponding solar fraction is $\bar{f}_{s,o}$. As storage is added, solar heating can be extended into the night and storage to accomplish this will be used each day. That is, storage goes through one temperature cycle per day. As storage size is increased still more, the frequency at which it is fully used decreases. This extra storage then becomes an additional cost without significant benefit. The precise sizing of storage is not critical. The performance prediction procedure described later can be used to evaluate the effect of various storage sizes on seasonal performance.

Storage Container Design for Liquid Systems

In this and the following six sections, the design of storage for liquid and air systems is described. Then storage materials used in passive systems are discussed.

Tanks are usually considered to be one of the simplest components in any mechanical system and are usually not treated in detail as a result. This has resulted in poor solar performance for both air and liquid systems. More attention should, therefore, be paid to the design of tanks.

Tank Connection Design. One of the key characteristics of a tank is the method of introducing heat from the solar loop and removing it from the tank to the heat-

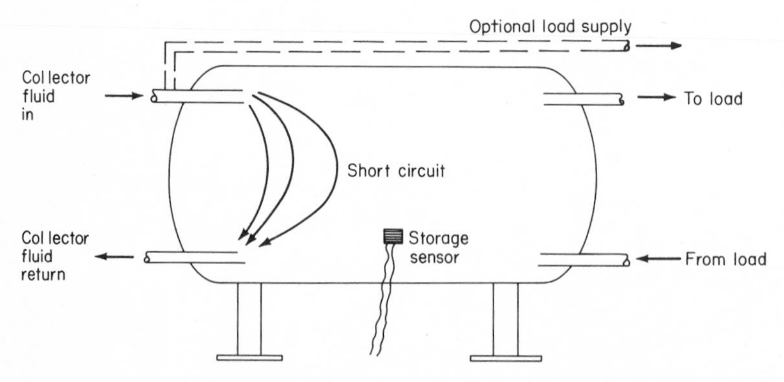

FIG. 4.6 *Incorrect* storage-tank plumbing arrangement in liquid-based solar heating systems.

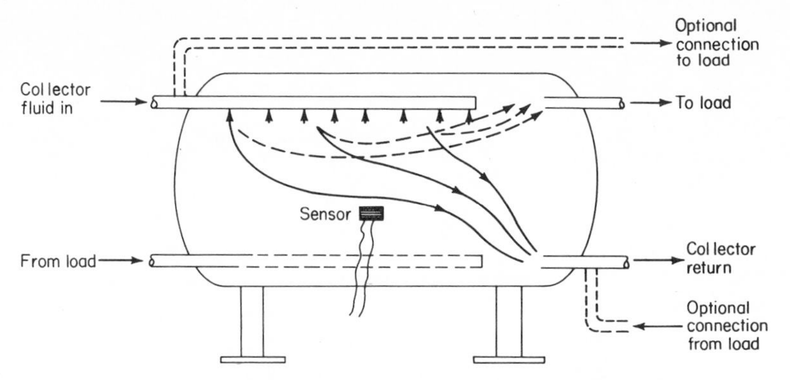

FIG. 4.7 Schematic diagram of proper method of plumbing storage tanks in liquid systems to ensure a cross flow of fluid in the collector loop and in the load loop for full use of the entire tank volume.

delivery loop. The location of these connections and their penetration into the tank must be considered.

Figure 4.6 shows the wrong way to connect the collector and load loops to storage. This method has often been used since it is the simplest, the collector connections being made to one end and the load connections to the other. However, in this type of configuration, collector fluid will short-circuit between the tank inlet and outlet at the tank left end. As a result, only a fraction of the tank is used for storage. Small storage causes the collector to operate at progressively higher temperatures with progressively reduced efficiency. The heat delivered to load is also relatively small when this piping arrangement is used. The load will short-circuit fluid in the right end of the tank, quickly exhausting any solar heat present. The backup system will come on and the use of solar heat will be terminated. The storage temperature sensor shown in Fig. 4.6 will get an erroneous tank temperature signal, and the controller will be unable to make proper decisions regarding collector pump operation.

A better method of connecting collector and load fluid loops to a storage tank is shown in Fig. 4.7. Collector-heated fluid enters at the upper left-hand corner, replacing fluid drawn off from the lower right-hand corner. This cross flow ensures that the entire volume of the tank can be solar heated. In addition, the cross-flow load connections are also used. Fluid sent to the load is taken from the top of the tank (the region of highest temperature). Alternatively, the load connection can be made directly to the collector fluid supply to storage (Fig. 4.7). The diffusers shown for collector inlet and outlet are optional but do enhance the stratification of fluid, resulting in somewhat lower collector inlet temperature and greater collector efficiency. Reference 70 gives a single method for diffuser hole sizing.

In addition, the flow arrangement shown in Fig. 4.7 results in the hottest possible fluid at the load. If both collector heat exchanger pump and load pump are operating, fluid heated in the collector loop can be drawn from the top of the tank. The storage-tank sensor in Fig. 4.7 also has an improved signal compared with that in Fig. 4.6. Storage-tank sensors are usually located in the lower third of the tank in a position where an average signal exists.

Increased stratification can be achieved by using tanks in series, although the cost is rarely justified unless a single large tank will not fit available space. Another way to enhance stratification in vertical tanks is to introduce solar-heated water into the bottom of a perforated vertical cylinder (large pipe) running from top to bottom in the center of the tank. The warm fluid released at the bottom of the cylinder rises to a level where its temperature and the tank water temperatures are the same, thereby automatically achieving stratification. The perforated cylinder can be made from a length of large plastic or iron pipe with many drilled holes around the circumference.

The interior of storage tanks must be treated according to the type of circuit to be used. In an open storage system (normally not recommended), the inside coating of the storage tank is important. A clay or glass lining, similar to that in domestic water-heating tanks, is used. A sacrificial anode is also needed. Closed systems of one metal do not require such protection since the oxygen in a closed volume of water is very rapidly consumed by a small amount of corrosion.

A storage tank must allow access to the interior. A manhole 2 to 3 ft in diameter located near the top of the tank serves well. The manhole is used if repairs are required.

Fluid expansion and contraction are accommodated by an expansion tank. Expansion tanks are treated in a later section.

Tank Insulation. The exterior of any solar storage tank must be well insulated. Solar heat is expensive and must not be dissipated by parasitic heat losses to a mechanical space or to the outdoors. One of the best approaches seems to be the insulation of a tank at the fabricating plant if possible. The tanks may be foam-insulated by the manufacturer and delivered to the site fully insulated. The foam must be protected from moisture or physical damage by a layer of fiber glass or other material.

If the tank is to be insulated on site, the insulation may be either foam or fiber glass. Vertical tanks are easier to insulate on site than horizontal ones. Foam can be applied to a storage tank after it has been permanently located. Foam must be applied above 40°F for proper curing. If fiber glass batting is used for tank insulation, it must be applied without any gaps. Although gaps may appear to be small relative to the entire insulated area, cumulative heat losses can be large since they occur 24 hours a day for the entire heating season. Fiber glass bats must be properly taped or, better, enclosed in a metal jacket to ensure that all areas are insulated permanently. Polystyrene insulation is unsatisfactory for solar storage tanks since this material distorts if overheated. The designer must specify insulation details fully.

Insulation must be protected from water. Fiber glass can absorb water and lose much of its insulating value. Some foams in the presence of water can form a weak acid which will attack a steel tank. Tanks located outside buildings are described below. Great care must be exercised to prevent long-term water exposure.

Insulation of the tank itself must be supplemented with piping insulation and support insulation. The thermal break between supports and the floor presents a problem since the weight of water storage is large. Most insulations are not capable of carrying such a load unless it is distributed over a relatively large area. One material not requiring very large support plates is Foamglass™. This material is capable of supporting 100 lb/m^2 dead loads and has a thermal conductivity of 0.025 to 0.035 Btu/(h·ft·°F). This insulation can be placed either between the tank and the saddles or between the base of the support saddle and the floor. A tank having insulated supports and a completely insulated surface area is thermally isolated from the environment; therefore, completely controlled heat addition and extraction can take place. Any other design will result in parasitic heat losses and unwanted heating of the building.

The amount of insulation to be used on a storage tank is determined by an economic trade-off of the cost of insulation relative to the value of the energy saved by adding that insulation. The results of such a calculation usually result in an insulation level between R-25 and R-30. Equation (4.1) below can be used to calculate the optimum thickness t (in feet) (22).

$$t_{opt} = \frac{\overline{C}_f \, \Delta T \, \Delta t \, k}{CRF(i,N) b} - R_0 k \qquad (4.1)$$

where t_{opt} = insulation thickness, ft
$\overline{C}_f$ = levelized cost of fuel including inflation effects, dollars per British thermal unit
ΔT = average temperature difference across the tank insulation, °F
Δt = number of hours per year during which heat loss occurs
k = thermal conductivity, Btu/(h·ft·°F)
$CRF(i,N)$ = capital recovery factor
b = cost of insulation, dollars per foot per square foot
R_0 = thermal resistance for heat loss if no insulation is present, $[Btu/(h \cdot ft^2 \cdot °F)]^{-1}$

The following example indicates how Eq. (4.1) is used to select the proper insulation thickness on a storage tank.

Example. Find the optimum insulation thickness on a storage tank whose average temperature is 125°F in a 65°F mechanical room. The solar system operates for 4500 h/yr, and the cost of fiber glass insulation [k = 0.025 Btu/(h·ft·°F)] is 5 cents per board foot. (A board foot is a 1-in-thick, 1-ft-square volume.) The cost of fuel is $12 per million British thermal units, and the interest charge on the solar system is 10 percent for 20 yr. Without insulation the thermal resistance R_0 is 1.0 $[Btu/h \cdot ft^2 \cdot °F)]^{-1}$.

Solution. Evaluate the parameters in Eq. (4.1).

ΔT = 60°F
Δt = 4500 h

$$b = \$0.05/(\text{in}\cdot\text{ft}^2) \times 12 \text{ in/ft} = \$0.60/(\text{ft}\cdot\text{ft}^2)$$
$$\text{CRF} = 0.1175$$

Then

$$t_{opt} = \frac{(12 \times 10^{-6})(60)(4500)(0.025)}{0.1175 \times 0.60} - 1 \times 0.025$$
$$t_{opt} = 1.12 \text{ ft} = 13.5 \text{ in}$$

This corresponds to an R value of 45.

Even with insulation as described above, a hot storage tank in a building in summer is undesirable. Solar heat is used in summer only for water heating, and the storage of heat in the main storage tank is not necessary. In summer, the large tank can be bypassed.

An alternative solution to summer heat control confines the storage tank in a separate insulated structure within a building. This housing can be opened to the environment in summer to avoid building overheating. This requires some additional space and cost but is quite effective in isolating overheated storage. The tank itself need not be insulated. Storage tanks located outside a building are not a summer problem.

Tank Location. The location of storage is determined by the availability of space either within or outside the building. An interior location is convenient; on-site insulation is easy. The tank can be inspected periodically for leaks and routine maintenance. Buried storage tanks, on the other hand, are difficult to service and may have significant heat losses. It is particularly difficult to adequately insulate a buried storage tank since moisture will almost always be present periodically in the immediate environment. The extra cost and insulation required by buried tanks must be considered in the selection of the tank site. A sight glass and access pit for it are also needed to check the fluid level. The pit must be warmed or insulated so that the sight glass does not freeze.

Other problems are associated with buried storage tanks. If the water table is above the bottom of the storage tank, the empty tank may tend to float. To prevent this, the tank must be tied down. The straps must be thermally isolated from the tank. All insulation on buried tanks must maintain waterproof integrity for an extended period in a severe environment with no possibility of repair or replacement.

The potential for corrosion between a buried metal tank and the surrounding soil should be examined. If the soil is corrosive (resistivity about 500 $\Omega\cdot$cm according to Ref. 60), a sacrificial anode is required; it must be replaced occasionally. The resistivity is measured during a soil test.

Buried tanks should not be located close to building foundations in a retrofit installation. They can undermine the existing foundation. Pipe connections to buried tanks should allow for possible tank settling. Settling can break rigid pipe connections entering the tank from the side.

Storage can be located outside and above ground. The problems of buried, inaccessible storage are avoided, but additional insulation and weatherproofing are required. The amount of insulation must be increased since the storage tank will be exposed to a cold environment and wind, conditions which are not present in other storage locations. The seismic overturning moment must be evaluated and provided for in tank anchor design.

The weight of storage in either air- or liquid-based systems is significant. For example, a 700-ft^2 liquid-based solar system would use approximately 1400 gal of water weighing approximately 6 tons, with several hundred pounds for the storage tank. This load is concentrated in a relatively small area and the support must be designed appropriately. For tanks located within a mechanical space, extra-thick floor slabs must be designed by the structural engineer with allowance for the dead load as well as seismic restraint. The slab loading for storage tanks is the weight of the fluid and tank divided by the support area. This loading determines slab thickness. The structural consultant determines the area of supports required and number of supports according to the span of the tank and the weight of material in it. In several early systems, steel tanks cracked when inadequate or too widely spaced supports were used.

Heat Rejection. During periods of excess heat collection, it may be necessary to reject heat from storage, using a P/T safety valve set at the design pressure and temperature. Fluid purged from storage may steam; a pipe conveys it to a drain or the outdoors. When the system cools, storage must be refilled to replace the rejected fluid. In most designs, heat rejection is done in the solar collector loop and not in the storage loop. This preferred method of heat rejection is described below.

If an open tank must be used, heat is rejected by boiling. The method shown in Fig. 4.8 permits boiling but controls any evaporation and convection losses at lower temperatures. The flapper valve is held closed against the overflow pipe seat by a counterweight. If pressure builds up within the tank, the flapper valve will open and release only that volume of water vapor required to maintain the tank slightly above atmospheric pressure.

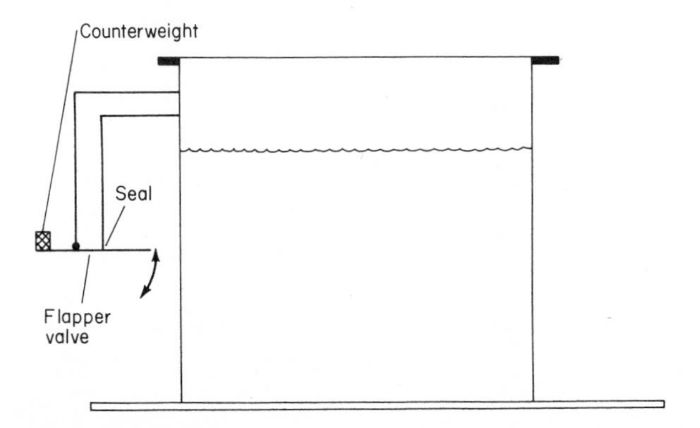

FIG. 4.8 Method for controlling evaporation and convection losses from an open storage tank.

TABLE 4.2 Material multiplier and scale-up exponent for solar storage tanks*

Material	*Multiplier* F_m	*Tank type*	*Scale-up exponent n*
Carbon steel	1.00	Atmospheric–steel	0.46
Aluminum	1.40	Steel	
Rubber-lined	1.48	100–250 kPa (15–35 psi)	0.47
Stainless steel	3.20	250–850 kPa (35–120 psi)	0.49
Glass-lined	4.25	Rubber-lined	0.57
		Aluminum	0.61
		Atmospheric—stainless steel	0.50–0.54
		Glass-lined†	0.43

*From *Chemical Engineering Reprints* by J. D. Chase. Copyright 1970 McGraw-Hill Book Co. *Chemical Engineering Reprints* by K. M. Guthrie. Copyright 1969 McGraw-Hill Book Co. Used with the permission of McGraw-Hill Book Co.

†Limited to 38 m³ (10,000 gal).

Cost. The cost of closed storage tanks can be estimated from the information given in Table 4.2 and Eq. (4.2) below.

$$\text{Cost} = \text{base cost} \times F_m \times \left(\frac{\text{volume}}{\text{base volume}}\right)^n \tag{4.2}$$

The "base cost" is that for a tank of specific material and of given reference volume. Equation (4.2) is useful when the cost of only one or two tanks is known but a range of costs is needed during optimization. Tank cost must include the cost of insulation (which is roughly proportional to the tank volume to the 0.667 power) and installation labor.

Storage Tank Selection. Tank selection is based upon cost, size, compatibility with other system components, and life-cycle economics. In summary, steel tanks are readily available, relatively light in weight, and relatively easy to pipe. They can be pressurized, are of relatively low cost, and can be purchased in almost any size or shape. Aluminum tanks are also light in weight and easily piped, but may have severe corrosion problems unless lined; they are unsuitable, as are bare steel tanks, for use in corrosive soils.

Concrete tanks are easy to build in many shapes and sizes. They are corrosion-resistant and have high strength. The tank itself provides additional storage capacity if proper external insulation is used. There are two difficulties with concrete tanks: (1) supporting their relatively large weight and (2) sophisticated workmanship to avoid leaks. Internal sealants and careful joint fabrication are required.

Plastic tanks are light and watertight but do not have sufficient temperature capability to handle elevated temperatures present in space-heating systems. Wood tanks are not used in many liquid systems but are common in air systems, as described in the next section. Wood tanks must not be pressurized. An algicide or other means of controlling biological growth must be used.

Storage Design for Air Systems

Rock is universally used in air systems. It should be stable and fairly smooth. More importantly, the rock size must be uniform. Typical sizes are 1 to 1½ in average diameter; that is, the rocks must pass through 1½-in screen and not pass through 1-in screen. If widely varying pebble sizes are used, the pressure drop in the bed is large, requiring excessive fan horsepower.

The rock must be cleaned by careful washing. This assures proper heat transfer. It is impossible to clean rock once in place; therefore, cleaning and sizing are done in advance of storage filling. The designer must specify all rock-bed details.

The rock container is most often fabricated from wood or concrete. In residences either material is low-cost. A concrete housing can be poured as part of the basement. If the box is poured as part of the floor slab, the rock must be insulated from the slab and from the concrete walls. This is best done by rigid foam insulation with adequate compressive strength. The insulation should be placed *inside* the rock box, ensuring a thermal barrier between the warm pebble bed and the relatively cold base (the floor slab) and walls of the box.

Rock storage units can be fabricated using 2 × 4-board and plywood construction. Wood boxes must be insulated—a fireproof insulation (or gypsum-board inner lining) of adequate compressive strength is essential. Overlapped plywood sheets help reduce leaks.

Figure 4.9 shows a typical residential pebble-bed heat-storage unit. From bottom to top, the unit consists of the base of the container, the lower plenum, steel mesh support for the pebbles, the pebbles, the upper plenum, and the cover. Air inlets and outlets are shown in the side walls although they may enter through top and bottom panels. A typical depth (dimension in the direction of airflow) is about 5 ft.

If more storage is required than can be accommodated in a cubical conformation, the storage unit should be a rectangular prism maintaining the 5- to 6-ft flow dimension. The shape in the plan view can be square or rectangular depending on available space.

Pressure Drop. The inlet and outlet plenums to a pebble-bed heat-storage unit must be sized to provide adequate flow to the most remote section of the bed. In Fig. 4.9 the plenums are approximately 6 to 8 in deep for both the upper and lower areas. If the lower plenum includes large bed supports, its depth must be increased accordingly. The longer the bed, the deeper the required plenum if only one inlet and outlet are used. Of course, multiple outlets can be used and the 6- to 8-in dimension retained even for very large pebble beds. As a design guide—one 16-in diameter duct carries air for about 250 ft^3 of rock storage. The key sizing consideration is that the plenum pressure drop must be smaller than the pressure drop through the rock. A minimum bed drop of 0.15 in water gauge is suggested. Calculation of pressure drop is described below in a technical supplementary section. The plenum pressure drop corresponds to one velocity head (6.2×10^{-8} FPM^2 = velocity head in inches of water for entering velocity in feet per minute). The frictional pressure drop through the rock bed should be 2 to 3 times the velocity head.

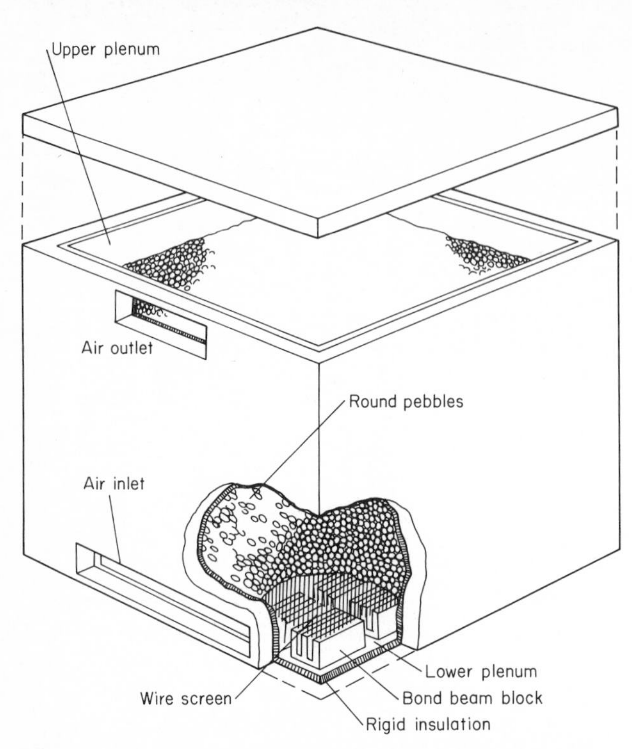

FIG. 4.9 Cutaway drawing of residential pebble-bed heat-storage unit used with air-based heating systems. *(Courtesy of Solaron Corp., Englewood, Colo.)*

If excessive pressure drop occurs in the plenum, flow channeling will result. This is to be avoided, since the rock bed is only partly used. For example, in Fig. 4.9 if the plenum areas were smaller than suggested above, a flow short circuit would exist between the air inlet and outlet. Then only the left end of the bed would be used effectively. If flow is not uniform through a rock bed, the stratification which is absolutely essential to the efficient operation of an air-heating system will not occur and system performance will be poor.

Rock Support. The lower plenum contains supports for the rock. It is frequently supported by a wire mesh, e.g., "expanded metal." This material is specified based on the depth of the bed. For example, support for a 5-ft-deep bed containing rock weighing 100 lb/ft^3 (including air spaces) would need to support 500 lb/ft^2. This loading dictates the thickness of the screen and the spacing of the supports below it. If available expanded metals are too coarse, the pebbles can pass through into the lower plenum. This is avoided by placing a layer of hardware cloth above the expanded metal. Support blocks below expanded metal are spaced on approximately 1-ft centers.

Seals. Proper design of seals for all bed openings must be addressed in the DD phase. Wood containers are prone to leakage over a long period of time as a result of warping of the wood. Leaks can occur at the inlet and outlet fittings and around the perimeter of the bed cover. Standard sheet metal ducting and flanging is adequate at duct fittings. Silicone caulking (e.g., Dow No. 732) works well.

Cover sealing presents more of a problem. The designer should prepare a detailed drawing of the cover and include all necessary features for sealing—lag bolts to secure the cover to the box, sealing systems and materials, angle-iron locating supports, handles for positioning the cover, material of construction, and spacing and size of main structural elements.

The location of the storage temperature sensor is to be unequivocally specified. This sensor is located just above the lower plenum, that is, in the cold zone of storage. The best location is described in the section on controls; however, the designer must indicate by a drawing note that care is to be exercised in filling the bed so that the sensor will not be damaged. One method of protecting the storage sensor is to place it within a PVC tube or conduit between the sensor location and the edge of the box. The tube locates and protects the sensor after the rock bed has been filled. Also the sensor can be replaced if necessary in the future. Of course, the end of the tube outside the box must be sealed.

Some designers are concerned about moisture buildup in pebble beds. Since storage is warm, it is an ideal environment for growth and multiplication of bacteria, spores, or fungi. It has been suggested that drains be installed to avoid this problem. However, installation of a drain in the base of a storage container is not necessary. No instances of bacterial growth in properly designed rock storage have been reported.

Installation of a drain in storage may present more problems than it solves. In order to keep a drain functioning properly, the trap must not dry out; therefore, it must be occasionally primed, a difficult process for a drain located in the base of a pebble bed. If the trap should dry out, sewer gases would be released into the heating system and distributed through the building. Therefore, the installation of a drain in storage is not recommended. The preferable approach to inhibiting growth in a storage tank is to keep the medium dry at all times. The designer should position the air inlet above the basement floor by several inches to keep water out.

Large Storage. Pebble-bed storage for larger buildings is designed according to the same principles described above for residential units. However, the large size can introduce flow-balancing difficulties. It is easy to balance flow if several storage units contained in one large housing are used. This design eliminates deep plenums required for long runs. Each subunit within the main storage unit is separated by vertical partitions. The spacing of partitions is determined by pressure drop rules given above. In principle, there is no difficulty in accommodating large beds in this fashion. The largest storage unit in the world contains 96,000 ft^3 of rock. It uses the compartmentalized storage idea.

The deeper the storage bed, the greater the pressure drop and the greater the fan operating cost. In large rock beds the flow path should be less than 8 ft long.

Horizontal Beds. In some applications the space available within a building for a pebble bed is small. Some designers have suggested placing pebble beds in an

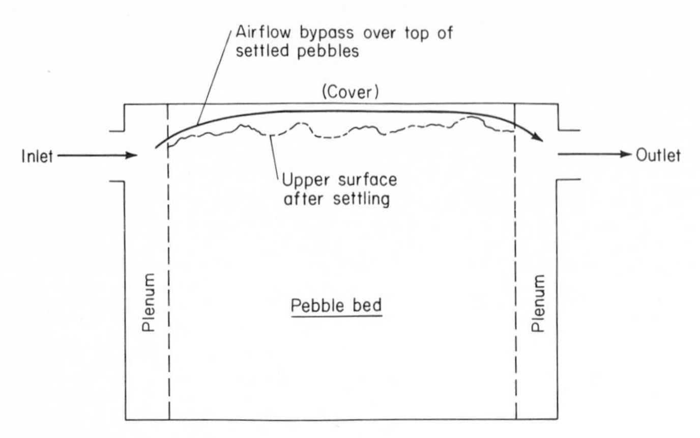

FIG. 4.10 Horizontal pebble bed showing possibility of airflow bypass after settling has taken place.

unused crawl space below the inhabited level of the building. This normally requires the use of horizontal flow in the bed. Horizontal flow has two difficulties. The first is the problem of maintaining a proper vertical interface between hot and cold zones of storage. Since hot air tends to rise, the flow in the pebble bed may become two-dimensional. Instead of moving in plug flow through the pebble bed, the air may rise toward the outlet plenum.

The second difficulty with horizontal flow is shown schematically in Fig. 4.10. After the storage bed has been filled, the material may settle. This will cause a gap at the top of the box. Then, an air short circuit above the pebble bed from inlet to outlet occurs. Consequently, very little heat storage occurs.

If horizontal storage must be used, flow baffles or seals are needed between the cover and the top of the pebbles. For example, a high-temperature compressible foam several inches thick could be attached to the cover when the cover is clamped in place. The foam is compressed against the pebbles and offers an effective barrier for airflow. Vertical flow through the storage container avoids these problems.

○ **Flow and Heat Transfer in Pebble Beds.** Pressure drops and heat-transfer coefficients in pebble beds have been measured by several investigators (1). One method for calculating the pressure drop through a pebble bed is given in Eq. (4.3).

$$\Delta p = \left(\rho V_s^2\right)\left(\frac{L}{D_s}\right)\left[\frac{(1-\epsilon_v)^2}{\mathrm{Re}\,\epsilon_v^3}\right]\left(1.24\frac{\mathrm{Re}}{1-\epsilon_v} + 368\right) \tag{4.3}$$

where $D_s = \left(\dfrac{6}{\pi}\,\dfrac{\text{net total volume of particles}}{\text{number of particles}}\right)^{1/3}$ = effective particle diameter

ρ = fluid density

$\dot{m}$ = mass flow rate

$V_s = \dot{m}/\rho A_b$

A_b = bed volume/bed length

L = bed length

ϵ_v = void fraction, usually 0.35 to 0.5
Re = Reynolds number based on the superficial mass velocity v_s and the particle diameter D_s

Equation (4.3) is valid for Reynolds numbers between 100 and 13,000.

Figure 4.11*a* shows the pressure drop measured in rock beds for various pebble sizes and superficial velocities. Pressure drop is expressed as inches of water per foot of bed length, that is, $\Delta p/L$. The figure shows that velocities are usually relatively small.

The heat-transfer coefficient in pebble beds between air and rock is shown in Fig. 4.11*b* in English units. The heat-transfer coefficient is on a volumetric basis rather than the usual area basis. In equation form, the volumetric heat-transfer coefficient in a pebble bed is given by Eq. (4.4), from Ref. 1.

$$\bar{h}_c = 0.79 \left(\frac{G}{D_s}\right)^{0.7} \tag{4.4}$$

where h_c = heat-transfer coefficient, Btu/(h·ft²)
G = the superficial mass flow rate (based on bed cross section area), lb/(h·ft²)

The total particle surface area A_p is given by Eq. (4.5) below.

$$A_p = \frac{6(1 - \epsilon_v)A_b L}{D_s} \tag{4.5}$$

In the DD rock bed design, the heat-transfer coefficient is normally not required information, but it is included here for specialized designs which may require its calculation. However, the pressure drop Δp is always required in DD designs in order to size air movers. Figure 4.11 can be used for preliminary estimates, but final design should be based on the use of Eq. (4.3).

Design of Storage for Passive Heating Systems

The two most common materials used for passive storage in passive systems are masonry and water. Storage requirements per unit collector area in passive systems are greater than in active systems by a factor of 2. Typical passive-system storage is between 30 and 40 Btu/(°F·ft²). The larger storage is necessitated by the slower exchange of heat with storage in passive systems.

Storage in most passive systems is located in intimate contact with the collector surface; the sun-facing side of the storage is the collecting surface. In direct-gain systems, the storage is frequently a massive floor or a south-facing wall positioned several window heights behind the aperture. Storage surfaces are dark in color ($\alpha > 0.6$) and are made of masonry, plastic, or metal (which may be selectively coated). It is possible to decouple storage from the collector in passive water-storage systems. The heated water can be pumped from the collector wall to holding tanks for later use. This hybrid method affords more heat delivery control. This is useful in spring and fall; afternoon overheating can thereby be avoided. Most passive systems, however, have closely coupled collector, storage, and delivery surfaces.

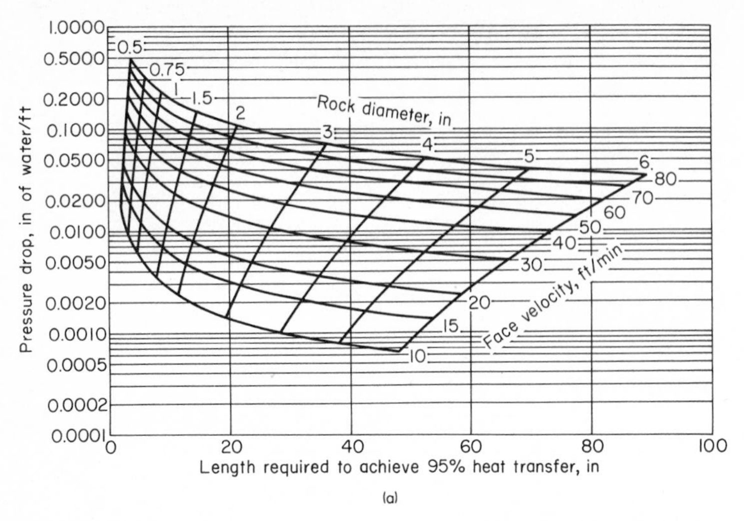

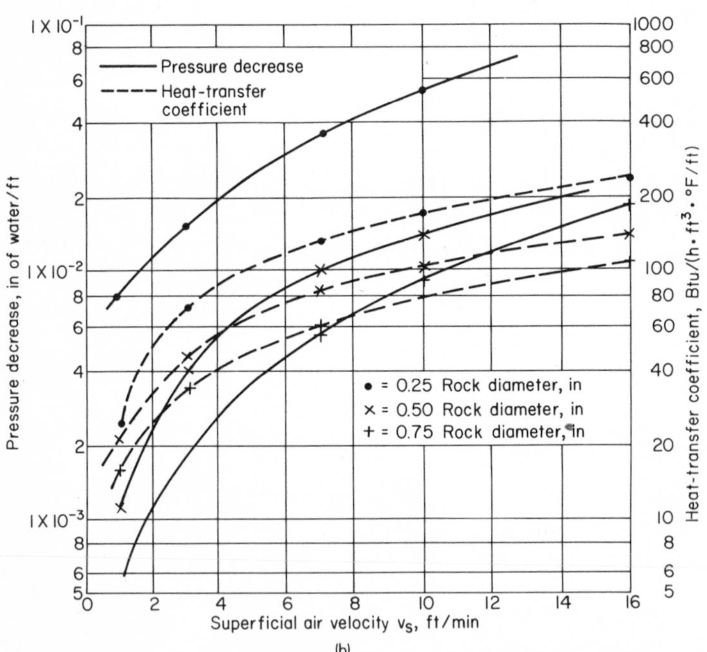

FIG. 4.11 (*a*) Rock-bed performance map showing pressure drop and air path length for various rock sizes and air velocities for sea-level conditions. *(From Ref. 57 with permission.)* (*b*) Heat-transfer coefficients and pressure drops per foot of depth for pebble beds of various size. *(From Ref. 28 with permission.)*

Absorptance. The color of conventional passive storage is determined in part by thermal requirements but more importantly by appearance. A dark color, of course, is essential; however, the color need not be black. A green or gray dye or paint may have an absorptance of 80 percent—10 percent less than black paint—but may be more acceptable from an appearance viewpoint. Slate colors have an 89 percent absorptance; red paint has between 75 and 80 percent absorptance. Since these absorptances are below the value of black paint, a somewhat larger aperture area is required to achieve the same thermal performance.

The absorptance of unfinished concrete is only 60 percent and is probably too low for an economical passive system. See appendix Table A.12 for a tabulation of absorptances.

The designer should specify the finish to be used on both sides of storage walls. It is not essential that the surface of masonry walls be perfectly flat. Since the flow over these walls is laminar, the pressure drop is not sensitive to the microcharacteristics of the surface. Therefore, an attractive texturing of these surfaces is possible with little thermal performance effect. This is particularly noteworthy regarding the interior wall surface. A decorative design can even be cast into the wall as it is poured.

The interior of TSW's should not present a barrier to heat transfer. At most, only paint should be applied. No paneling, insulation, or other heat-flow obstructions are permitted on the inner surface of the storage element. Further, the DD drawings should contain a note specifying that no large furniture or other obstructions to natural air circulation or radiation be placed immediately adjacent to the wall. Of course, baseboard heaters should be placed on walls other than those used for storage.

Weight. Thermal storage for passive systems is heavy. The design of the structure and foundation to support storage is the responsibility of the structural engineer, who must know the storage mass to be used before the foundation can be designed. Soil properties collected during the programming phase will be used by the structural engineer to size and shape the foundations below passive storage walls. In locations with expansive soils, the foundation may be rather complex and the cost high. This cost must be included in the economic analysis of solar heating described later in this chapter.

Wall Construction. The designer must consider the method by which the storage wall will be assembled. Windows in the wall require structural support above them. The size and shape of storage-wall vents has structural implications. If a relatively wide but narrow vent is used, the structural impact is great since the remaining relatively small nonvent area must support the entire wall mass above the lower vent. Preferably, use a relatively tall vent with reduced width achieving the same flow area.

The designer must also include instructions on the drawings requiring the concrete contractor to properly fill the wall area below the vent. The vents will normally be formed by 2 × 4 knockouts. It is important that concrete fill the entire area below the vents—approximately 6 to 8 in above the floor elevation. Proper vibration of concrete ensures that no entrained air is trapped in the wall.

Wall Air Gap. The space between the sun-facing surface of thermal storage walls and inner glazing of passive windows is normally 3 to 4 in wide. This dimension is not critical and may be adjusted to accommodate specific designs of movable insulation and windows. The window-to-wall gap should not be made excessively large, since a thermal circulation loop could be formed between the relatively cool glass and the wall. Heated air rising along the surface of the storage wall loses heat to the glass. The cooled air sinks to the base of the wall. Useful heat flow into the space is thereby reduced.

Insulation of passive walls is described on pages 240 to 242.

Storage Sizing

Active Systems. Most active heating systems of standard design have storage volumes independent of geographical area. For liquid systems, the storage amount is between 1.5 and 2 gal of water per square foot of net collector aperture area. For air systems, ½ to ¾ ft^3 of rock per square foot of collector area is normally used. These are not critical values, and variations of 10 or 15 percent will have almost no effect on thermal performance.

The key to proper storage sizing is to use a volume that is not below the knee of the performance curve (e.g., Fig. 4.5). For example, liquid systems with a storage volume below 1.2 gal/ft^2 will have poor performance. However, doubling liquid storage will result in only a 5 to 10 percent increase in annual energy delivery. Oversized storage is expensive and will rarely pay itself off in increased fuel savings. Seasonal storage is not economical in small installations and is not considered in this book. For larger district installations, the inverse heat-loss relation to size may result in economies of scale sufficient to justify the method. Further, seasonal storage may be appropriate in certain severe climates where sunlight is rarely available in winter if a very large storage volume can be charged in summer.

Passive Systems. The sizing of passive system storage is a trade-off between comfort and thermal performance. For example, a TSW system will have roughly the same thermal performance for an 8-in wall as for a 15-in one. The designer would select an 8-in wall because of its 50 percent lower cost. However, relatively thin walls will have large temperature excursions at the inner surface. These temperature swings affect the mean radiant temperature and hence thermal comfort.

An increase in wall thickness to 12 in is recommended to maintain good comfort levels. The 50 percent volume increase is determined by comfort and not by thermal performance criteria. The amount of storage in water walls is approximately the same, on a thermal basis, as in masonry walls. Between 30 and 40 British thermal units per degree Fahrenheit per square foot of aperture area are required. This is equivalent to 4 to 5 gal of water per square foot of net aperture. If storage is increased beyond these nominal levels, little performance gain is achieved and costs rise because of larger container costs and larger foundation requirements to support the increased storage mass.

The amount of storage surface area required in direct-gain systems using indirectly illuminated heated storage is approximately 3 times that for directly illuminated storage (25).

HEAT EXCHANGERS

The solar designer must minimize the detrimental effects of heat exchangers, subject to the constraint of cost effectiveness. In large system designs, this is done during the DD economic study. The designer of smaller systems can use information in this section to size the heat exchanger without going through a detailed economic analysis. The conclusions below are based on the trade-off of heat-exchanger cost and the value of energy penalty incurred by the heat exchanger.

This section will describe three types of heat exchangers used in active systems: (1) between the collector loop and the storage tank in liquid systems or between the storage tank and a domestic water-preheat tank; (2) between storage and the heated space; (3) between solar-heated air and the domestic water preheat system, as shown in Chap. 2. Performance characteristics for heat exchangers are presented in Chap. 2 and are not repeated here.

Liquid-to-Liquid Heat Exchangers

Liquid-to-liquid heat exchangers used in solar heating systems are of the tube-and-shell configuration. Since the effectiveness of the heat exchanger, as described in Chap. 2, must be relatively high for solar applications, counterflow designs are most often specified. Counterflow heat exchangers have the largest effectiveness for a given number of transfer units (NTU) and given capacitance ratio $(\dot{m}c_p)_{min}/(\dot{m}c_p)_{max}$. Immersed-coil heat exchangers have low heat-transfer coefficients, requiring relatively large heat-transfer areas to achieve a given UA product. However, submerged coils are less expensive than shell-and-tube exchangers and they may be appropriate for some applications. The design of submerged-coil heat exchangers is a mixture of art and science since the convection cells set up within the storage tank surrounding an immersed coil have not been carefully measured. Therefore, the free-convection heat transfer is not well represented by empirical equations useful to the designer. An understanding of the physical processes accompanying free convection in containers is useful to the designer, who must specify coil diameter, location, and size.

In solar designs, the heat exchangers are specified by their effectiveness. Heat-exchanger manufacturers do not use the effectiveness concept in their computerized design programs but rather use the log-mean-temperature method. It is the responsibility of the solar designer to translate an effectiveness specification into a form which can be used by the heat-exchanger manufacturer. The steps of the method are relatively straightforward. The designer selects an effectiveness, usually 70 to 80 percent for liquid-to-liquid heat exchangers. This effectiveness and the heat rate can be used at a specified operating temperature to calculate heat-exchanger inlet and outlet temperatures for both fluid streams. The heat rate, the flow rates, and temperatures can be used by vendors to select a specific heat exchanger. Since solar systems operate over a very wide range of temperatures, the operating point selected for specifications should be in the middle of the expected solar system operating range. The effectiveness of a liquid-to-liquid heat exchanger is not affected in any major way by operating temperature. Therefore, the specifiation of an 80 percent effectiveness does not require an additional specification of the temperature at which that effectiveness is to apply.

The log-mean-temperature difference for solar liquid-to-liquid heat exchangers is in the vicinity of 5 to 7°F. This is somewhat smaller than used in many commercial applications with which a heat-exchanger vendor will be familiar. The vendor may suggest that a smaller heat exchanger be used on the basis of experience with nonsolar applications. Larger heat exchangers with smaller mean-temperature differences are economical since they result in improved energy delivery in solar systems.

In addition to specifying heat rates, flow rates, and temperatures, allowable pressure drops in both shell and tube sides of the heat exchanger are to be noted. These pressure drops can be calculated from empirical correlations, but it is suggested that the manufacturer's computer program be used instead. For given flow rates, the manufacturer will select the heat exchanger and inform the designer of the pressure drops. In addition to pressure differences, the designer must specify the operating pressure. In solar applications, the collector-to-storage heat exchanger will rarely be subjected to pressures above 150 lb/in^2. In the storage-to-preheat tank exchanger, the tube side will be exposed to city-water pressures which may reach levels of 125 lb/in^2 or above. These pressure requirements are to be included in the specifications.

The following example illustrates the method for converting effectiveness ratings to the information required by the heat-exchanger manufacturer. This procedure must be followed in the design of any heat exchanger for a solar system.

Example. Convert an 80 percent effectiveness rating to fluid stream temperatures and flows required by a heat-exchanger manufacturer for a 500-ft^2 solar system in Denver. Also calculate the log-mean-temperature difference (LMTD).

Solution. The maximum heat rate Q must be known. This will occur on a sunny day in fall or spring when solar incidence angles are small near noon. Under these conditions, the collector is found to operate at 53 percent efficiency. Therefore,

$$Q = 0.53 \times 310 \times 500 \text{ ft}^2 = 82{,}150 \text{ Btu/h}$$

Clear-sky solar data [in this case 310 Btu/(h·ft^2)] for various latitudes are in appendix Table A.3.

The next step is to pick an average storage temperature which becomes the tube-side inlet temperature. Here we will use 130°F. Recall that the heat rate Q, from Eq. (2.13), is as follows:

$$Q = E_{hx}(\dot{m}c)_{\min}(T_{h,i} - T_{c,i}) \tag{4.6}$$

For a liquid system, the minimum capacitance rate is through the collector since storage-side flow is usually 50 percent to 100 percent above that in the shell side to improve the effectiveness. Using the 0.02-gal/min rule for collector flow (50 percent glycol) and glycol properties,

$$(\dot{m}c_p)_{\min} = [(0.02 \times 500) \times 0.9] \times 500\ [\text{Btu}/(\text{h}\cdot{}^\circ\text{F})]/(\text{gal/min}) = 4500\ \text{Btu}/(\text{h}\cdot{}^\circ\text{F})$$

From Eq. (4.6), the hot fluid inlet temperature is

$$T_{h,i} = \frac{Q}{E_{hx}(\dot{m}c)_{\min}} + T_{c,i} = 152.8^\circ\text{F}$$

The final information for the manufacturer is the tube-side flow rate. Using the 50 percent rule above,

Tube flow = 1.5 × (0.02 × 500) = 15 gal/min

In summary, the thermal data for the manufacturer is then

Heat rate = 82,150 Btu/h
Shell EWT = 152.8°F
Tube EWT = 130.0°F
Shell fluid and flow = 50 percent glycol, 10 gal/min
Tube fluid and flow = water, 15 gal/min
Type = counterflow

To find the LMTD, the shell and tube outlet temperatures must be found. Using a heat balance on the shell side, we write

$$Q = (\dot{m}c)_h(T_{h,i} - T_{h,o}) \tag{4.7}$$

Solving for $T_{h,o}$, we find

$$T_{h,o} = T_{h,i} - \frac{Q}{(\dot{m}c)_h} = 152.8 - \frac{82{,}150}{4500} = 134.5^\circ\text{F}$$

Likewise for the tube side,

$$T_{c,o} = T_{c,i} + \frac{Q}{(\dot{m}c)_c} = 130.0 + \frac{82{,}150}{15 \times 500} = 140.9^\circ\text{F} \tag{4.8}$$

The LMTD is

$$\text{LMTD} = \frac{\Delta T_1 - \Delta T_2}{\ln(\Delta T_1/\Delta T_2)} \tag{4.9}$$

Here,

$$\Delta T_1 = 152.8 - 140.9 = 11.9°\text{F}$$
$$\Delta T_2 = 134.5°\text{F} - 130 = 4.5°\text{F}$$
$$\text{LMTD} = \frac{11.9 - 4.5}{\ln (11.9/4.5)} = 7.6°\text{F}$$

The effectiveness value of 80 percent used in the preceding example has been selected based on experience. If the designer wishes to optimize the effectiveness on the basis of the cost of the heat exchanger and the value of energy saved with progressively larger and larger heat exchangers, the effectiveness values in Table 2.3*a* can be used for this purpose. In this table, the effectiveness is expressed as a function of NTU which in turn is related to heat-exchanger area. Heat-exchanger manufacturers can provide costs for various heat-exchanger sizes, and these may be used to optimize the selection. An example of such an optimization is contained in Ref. 1.

○ **Submerged-Coil Heat Exchangers.** Submerged coils in storage tanks can be used to transfer heat from a solar-collector loop to storage. A reliable equation for calculating the free-convection heat transfer from the outer surface of a helical tube to a the tank does not presently exist. However, an approximation of the heat-transfer coefficient [Btu/(hr·ft²·°F)] is given by Eq. (4.10).

$$h_{coil} = 56(\Delta T/D_{tube})^{1/4} \tag{4.10}$$

where ΔT = temperature difference between the coil outer surface and the storage water temperature, degrees Fahrenheit

D_{tube} = tube outside diameter, feet

In the preceding example, an average ΔT of 7.6°F was calculated. If this value is used in Eq. (4.10) for a 1-in-diameter tube, the heat-transfer coefficient is approximately 175 Btu/(h·ft²). Combining this coefficient with a typical coefficient inside the tube yields a U value for a submerged coil of approximately 100 Btu/(h·ft²). In a shell-and-tube heat exchanger with forced convection, the U value will typically be 300 Btu/(h·ft²) without fouling factors. Therefore, for a coil, the effective heat-transfer area must be approximately 3 times that for a shell-and-tube heat exchanger.

The effectiveness of a submerged coil heat exchanger can be calculated from Eq. (4.11).

$$E_{hx} = 1 - e^{-NTU} \tag{4.11}$$

where NTU = $(UA/\dot{m}c_{coil})$.

In addition to reduced heat-transfer rates, submerged coils present other difficulties in liquid-based heating systems. Storage tanks with coil heat exchangers tend to stratify into two zones—one above the coil, the other below. The design of the controller for the collector pump requires proper positioning of the storage-tank sensor. If the tank sensor is located within either one of the two zones, the controller must be programmed to take this effect into account. The sensor will usually be located in the lower zone since the temperature here represents to some extent the inlet temperature to the collector. However, when the load pump is operated stratification is destroyed. The controller will then sense a temperature regime fundamentally different from that present when the coil is transferring solar heat to the tank with the load pump off. The 25 to 30°F stratification for which the controller is designed does not exist when the load pump is on, and improper control function results. No simple solution has been found to this problem without adding additional relays and sensors to the control system designed for external shell-and-tube heat exchangers.

The principal difficulty with the control of the submerged coil as described above is that larger stratification can occur within the storage tank than is developed across the collector during morning or afternoon solar conditions. That is, 30°F stratification in the tank causes the collector pump to operate. If the fluid temperature rises only 10°F across the collector, it is returned to the tank at a temperature lower than that already present in the top of the tank.

Other Design Details. Heat exchangers must be insulated to control heat loss. The amount of insulation is subject to the same criteria described earlier for storage tanks. Equation (4.1) can be used to calculate the optimum insulation thickness for a heat exchanger approximately. However, Eq. (4.1) is based upon heat transfer through a surface with a large radius of curvature, such as a storage tank. The heat exchanger has a much smaller radius of curvature. However, the result from Eq. (4.1) will be approximately correct and can be used as a starting point for a more detailed calculation taking into account the effect of curvature using Eq. (2.12).

The flow velocity through shell-and-tube heat exchangers is constrained above by the tendency for heat-exchanger tubes to rattle. The rattling is caused by fluid impingement at oblique angles on the tubes. Most heat-exchanger manufacturers have guidelines for selecting flow rates below the rattle level in relatively large heat exchangers. This should be checked by the designer. Heat-exchanger rattle must be avoided since tube fatigue permanently damages the exchanger.

Double-wall heat exchangers are to be included in the specifications. Double-wall heat exchangers are most commonly needed in potable-DHW-heating systems. In space-heating systems using boiler-treated water, a double-wall heat exchanger will also be necessary. Double-wall heat exchangers have reduced effectiveness for a given heat-transfer area and are more expensive than single-wall devices.

The designer should select heat exchangers which are standard units. The required exchanger may be nonstandard, but the extra cost of special heat exchangers should be avoided. Therefore, use the closest standard units. Two or three standard exchangers may be connected in series or parallel. As in most solar-system sizing calculations, a variation of a few percent from the nominal design will have minimal impact on overall system life-cycle economics. The cost of acquiring a special exchanger, the lead time required, and its associated cost will not be offset by the slight performance improvement associated with the exact size calculated.

Air-to-Liquid Heat Exchangers

Load devices in liquid-based solar systems require the transport of heat from a liquid stream to air which heats the space. Solar air systems require an air-to-liquid heat exchanger for domestic water preheating. The design of these heat exchangers can be carried out by application of the effectiveness concept. The effectiveness-NTU relationship for cross-flow air-to-liquid heat exchangers is contained in Table 2.3*b*.

Fin-Tube Exchangers. If a fin-tube coil is used as the terminal load device in a forced-air system, the ratio shown in Eq. (4.12) can be used to estimate the required size of the heat exchanger:

$$\frac{E_L(\dot{m}c_p)_{air}}{(UA)_{bldg}} \approx 2.0 \tag{4.12}$$

where E_L = load heat-exchanger effectiveness
$(\dot{m}c_p)_{air}$ = airflow capacitance rate over the fin tube
$(UA)_{bldg}$ = unit heat loss from the building, Btu/(h · °F) or W/°C

The building load $(UA)_{bldg}$ includes both transmission and infiltration losses. The airstream capacitance rate will be determined by the flow required for building heating in the forced-air distribution system. In order to achieve the effectiveness specified by Eq. (4.12) without excessive pressure drop through the coil, the flow rate may need to be increased.

Solar systems, like heat-pump heating systems, may deliver energy to the space at a temperature below that normally found in a gas-fired or oil-fired system. These reduced air temperatures may feel cool to the occupants of the building but are above room temperature so that a heating effect does take place. If reduced temperatures or higher velocities than normal are present, the designer must ensure that proper diffusers are used so that these airstreams do not affect the building occupants more than absolutely necessary. The example calculation shown below shows how Eq. (4.12) is used to size the terminal device in a liquid heating system.

Example. What is the heat-exchanger effectiveness required for a load heat exchanger on a building with a 150,000 Btu/h load under design conditions with indoor-to-outdoor ΔT = 75°F?

Solution. To use Eq. (4.12), find UA.

UA = 150,000/75 = 2000 Btu/(h · °F)

If the heating system is designed to carry the *design* heat load with an air temperature rise across the heating system (return grilles to supply grilles) of 30°F,

$$(\dot{m}c_p)_{air} = \frac{150{,}000}{30\,°F} = 5000 \text{ Btu/(h} \cdot °\text{F)}$$

and from Eq. (4.12),

$$E_L \sim 2 \times 2000/5000 = 0.8$$

The effectiveness can be converted to manufacturer's parlance using the approach of the previous example. The liquid-side capacitance rate is usually greater than that for the air side. For example, if a 15°F drop in water temperature is to be used in the terminal device, the water-side capacitance rate is 150,000/15°F = 10,000 Btu/(h·°F), twice that of the air side.

Baseboard Exchangers. Baseboard heating systems are also liquid-to-air terminal devices and can be analyzed using the effectiveness idea. Most baseboard heating systems are designed to operate at temperatures well above those available in solar systems, for example, 180 to 200°F. In order to achieve the effectiveness required by Eq. (4.12), greater lengths of baseboard are used than would be present in a nonsolar heating system for the same building.

Figure 4.12 shows the performance of a typical baseboard system presented as a function of fluid inlet temperature. The data for this figure are based upon a pipe diameter of ¾ in and a flow rate of 1 gal/min. Flow rate is not a critical variable in the selection of baseboard systems, since most of the resistance to heat transfer occurs between the outer surface of the baseboard fins and the room air. The heat-transfer resistance between the water and the pipe inner wall is small by comparison. The heat rate in Fig. 4.12 is expressed as British thermal units per hour per foot of radiation. The standard method of sizing a baseboard heating system divides the heat rate into the design building load. This determines the length of baseboard required to carry the design day heat load. For example, if the building heat load is 150,000 Btu/h, a baseboard system operating at 180°F would require 246 ft of hydronic baseboard to carry the load. This length is based on a heat rate of 610 Btu/(h·ft) taken from Fig. 4.12 at 180°F.

The linearity of the data shown in Fig. 4.12 indicates that a relatively constant value of $E_L(\dot{m}c)_{\text{air}}$ applies. From the data in the figure and additional information from the manufacturer, it can be shown that the numerator of Eq. (4.12) is equal to approximately 6 Btu/(h·°F) per foot of baseboard. The length of baseboard can immediately be determined, therefore, from Eq. (4.12). In the preceding example, a *UA* value for the building was calculated as 2000 Btu/(h·°F). To satisfy Eq. (4.12), the numerator must be 4000 Btu/(h·°F). Dividing this quantity by 6 results in a required baseboard length of 670 ft. This is approximately 2½ times the length of baseboard required in the nonsolar application described above.

Additional baseboard heater lengths may require more wall space than is available. Two solutions can be considered. The first is to use a double-tube baseboard system which will transfer more heat per linear foot than the single tube used above. Second, interior walls can be used for baseboard heater mounting.

Radiant Panels. Terminal devices of a third type can be used for space heating. These are *radiant* devices and consist of large, relatively low-temperature surfaces

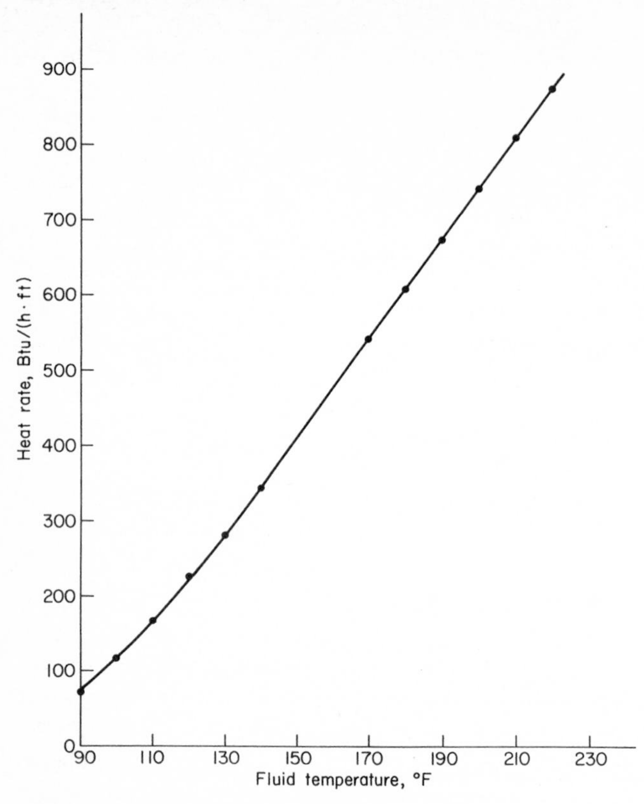

FIG. 4.12 Heat delivered in British thermal units per lineal foot for baseboard heating systems as a function of fluid temperature for ¾-in-diameter pipe and 1 gal/min flow rate.

which heat by radiation and convection rather than by convection alone. Radiant heating surfaces are composed of a conducting material containing a network of heating fluid pipes. One difficulty with radiant heating is the relatively large thermal mass involved and the slow response time. Difficulties occur in attempting to bring a space up to temperature quickly on a winter morning and then controlling temperature through the day as the outdoor temperature and sunlight levels increase. The mass in the radiant heater contains a significant amount of heat which may be released even when no heat is required.

From a reliability point of view, the corrosion and erosion of piping in a radiant system, its expense, and its repair difficulty mitigate against its selection for use in the solar system. However, the good low-temperature effectiveness is attractive for solar applications.

Humidifiers. Humidifiers are also terminal load devices. Heating outdoor air reduces the relative humidity below the comfort point. Indoor humidity should be about 20 to 40 percent in winter. Water evaporation maintaining the recommended RH level imposes an additional energy requirement on the heating system. The design of humidifiers for solar and nonsolar systems is the same. It is treated in detail in Ref. 6.

DHW Preheaters. The air-to-liquid domestic-water-preheat exchanger in solar air systems can be sized using the information in Table 2.3*b*. Typical effectiveness values for this heat exchanger are approximately 0.5 to 0.6. The airflow is the collector airflow. Hence, the minimum capacitance rate is usually on the air side. The face area of the heat exchanger is determined by the nominal face velocity specified by the coil manufacturer and the total collector array airflow. A typical face velocity is 500 ft/min. The following example illustrates how the preheat heat exchanger can be sized.

Example. Size the preheat coil for a system using 700 ft^2 of air collector.

For the air side,

$$(\dot{m}c)_{air} = 700 \times 2.5 \text{ ft}^3/(\text{min}\cdot\text{ft}^2) \times 1.08$$
$$= 1890 \text{ Btu}/(\text{h}\cdot{}^\circ\text{F}) \qquad \text{at sea level}$$

On a sunny day, the collector outlet temperature may be 160°F. Since efficiency is related to exchanger effectiveness, an iterative method must be used to find the proper exchanger area. In this example, however, an estimated value of efficiency will be used—40 percent.

Then on a sunny day

$$Q = 0.4 \times 310 \times 700 \text{ ft}^2$$
$$= 86{,}800 \text{ Btu/h}$$

Suppose 10 percent of the total heat rate is devoted to water heating. Then the air temperature drop across the coil is

$$\Delta T_{air} = 0.1 \times Q/(\dot{m}c_p)_{air} = 4.6^\circ\text{F}$$

which is a typical value for an air system.

Coil sizing may be done in another way by first specifying an effectiveness, then selecting a coil with the proper heat-transfer capacity for given stream temperatures. The calculated exchanger heat rate may be less than the delivery of the collector. Then the collector return temperature must be increased in the calculation. Its efficiency is therefore lower. Two or three iterations on the calculation are usually sufficient.

PUMPS, FLUIDS, AND PIPES

Energy is moved from one point to another in solar systems by pipes and pumps. The parasitic losses from pipes, including both heat losses and pumping power, are described in Chap. 2. The method of selecting pumps in various liquid loops is a two-step process. The piping layout and flow rates are prescribed by other considerations; these determine the pressure drop. The pressure drop and flow rate can then be superimposed on a series of pump curves until the pump with the best efficiency, given cost constraints, is selected.

Pipe Sizing

Figure 4.13 shows the pressure drop in smooth pipes (in turbulent flow) as a function of flow rate in gallons per minute. If the flow rate is outside the limits on the graph, both scales of the graph may be changed by powers of 10, the only requirement being that the Reynolds number be greater than 2500 to assure turbulent flow.

Figure 4.13 is illustrated in the following example. A 700-ft^2 collector array with a flow rate of 0.025 gal/(min·ft^2) has a total flow rate of 17.5 gal/min. Moving up along the 17.5 gal/min line in Fig. 4.13, we can select a series of pipe sizes. The larger the pipe size, the lower the pressure drop. A velocity of 4 to 6 ft/s can be used for preliminary selection. If the flow rate is 17.5 gal/min, a velocity of 6 ft/s corresponding to 1-in pipe should be selected.

A detailed economic analysis of pipe-size trade-offs vs. pump horsepower is rarely done for small solar systems. Large heating systems in commercial buildings may warrant an optimization study, however. The trade-off considered in the economic study is the reduced cost of pumps and pumping power, resulting from larger pipe sizes, vs. the cost of larger piping and insulation.* The calculation would be carried out for a range of pipe sizes limited below by the maximum allowable fluid velocity and above by the minimum velocity assuring proper entrainment of air for later removal. The cost of pipe and insulation is sensitive to pipe diameter. The pressure drop depends on the square of the flow rate and the inverse fifth power of the diameter. The horsepower is proportional to the product of flow rate and pressure drop. It determines the amount of electric power consumed.

Piping Materials

Pipes used in liquid solar systems are most commonly steel or copper. Plastic piping of the polyvinyl chloride (PVC) variety is not usable above 140°F, although high-temperature chlorinated polyvinyl chloride (CPVC) at 3 times the cost may be used up to 185°F. Galvanized piping is not recommended, since the electromotive relationship of steel and the galvanizing coating reverse above 130°F and the steel, no longer protected by the galvanizing, goes into solution.

Copper piping should be soldered with 95/5 or silver solder capable of withstanding elevated temperatures present at stagnation conditions. The velocity in

*Under certain conditions increasing pipe insulation thickness t may increase heat loss since the outer surface from which heat loss occurs increases in size more rapidly than the added insulation thickness reduces the loss. The critical insulation outer radius r_{crit} below which this problem occurs is given by $r_{crit} = k/\bar{h}_o$ where k is the insulation conductivity and $\bar{h}_o$ is the convection coefficient from the outer surface of the insulation. Note that aluminum insulation jackets must not contact steel pipe rests or shields, to prevent electrolytic corrosion.

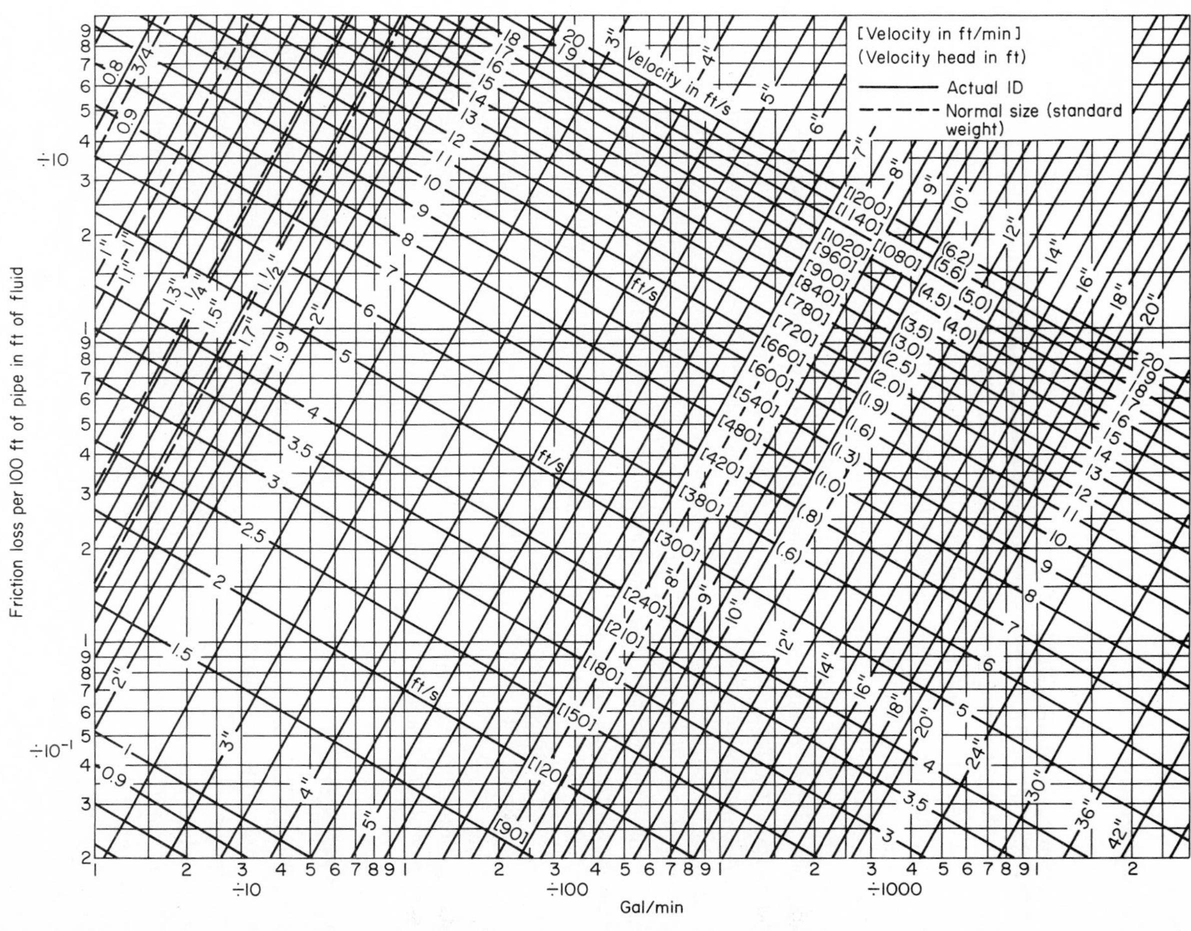

FIG. 4.13 Pipe pressure drop as a function of flow rate. *(From Ref. 48 with permission.)*

copper piping should be maintained below 6 ft/s to avoid erosion but above 2 ft/s to guarantee proper entrainment of air. Copper piping can use screw fittings if Teflon™ pipe dope is used.

Piping Details

Pipes must be provided with supports and anchors to ensure reliable operation. The spacing of piping supports is determined by the size of the pipe and the length of the span as well as by the weight. The spacing of piping supports is a standard HVAC calculation and is described in Ref. 6. (Typical values are 5 ft maximum for cast iron, 8 ft maximum for 1 in or smaller steel, 10 ft maximum for over 1-in steel pipe, 6 ft maximum for copper 1¼ in or smaller, 10 ft for over 1¼ in copper, 4 ft maximum for plastic pipe.)

Since solar-collector piping headers must drain, it is important for the designer to specify the slope angle of piping, including the effect of pipe sag between pipe supports. These details must be called out on drawings in more detail than normally required for piping systems, since plumbing personnel installing the system will undoubtedly be unfamiliar with the important features of the piping network. Expansion joints for piping should be of the preinsulated type covered with a metal jacket, and the piping must be properly anchored so that the expansion joints work in the intended fashion. The anchor mounting should be shown in detail on the drawings with adequate notes to specify all possible questions.

All solar piping is to be insulated, including all elbows and fittings connecting the collector manifolds to the collectors themselves. The insulation has been discussed in Chap. 2. It must be protected from the environment by means of a durable jacket. Expensive metal jackets seem to be the most reliable for long-term installation in severe environments; plastic PVC jackets may deteriorate in the presence of sustained ultraviolet exposure.

Pipe Fittings

Pipe fittings have associated pressure drops which can exceed the pressure drop of the pipes connecting them. Figure 4.14 is a nomograph which shows the effective *equivalent length* of standard pipe fittings for various diameters. For example, the dashed line shows that a standard elbow of 6-in pipe is equivalent to 17 feet of straight pipe. The equivalent pipe lengths for all pipe fittings are added to the physical pipe length in order to calculate the total pressure drop. The total pressure drop is given by the friction loss per 100 ft of pipe from Fig. 4.13 multiplied by the total of the physical and equivalent pipe lengths.

Other Sources of Pressure Drops

In addition to pipes and pipe fittings, loops have additional components that cause pressure drops. Figure 4.15 shows schematically some of the pressure drops encountered in a collector loop. Starting from exchanger outlet, the fluid passes through the fill-and-drain valve. Its pressure drop can be determined from Fig. 4.14. The fluid then enters an air scoop which is used to separate entrained air from the fluid. The air scoop serves as the connection to the expansion tank. Air scoop pressure drops can be of the order of several feet of fluid; the manufacturer must be consulted. After the air scoop, the fluid passes through the check valve. The pressure drops for check valves are given in Fig. 4.14.

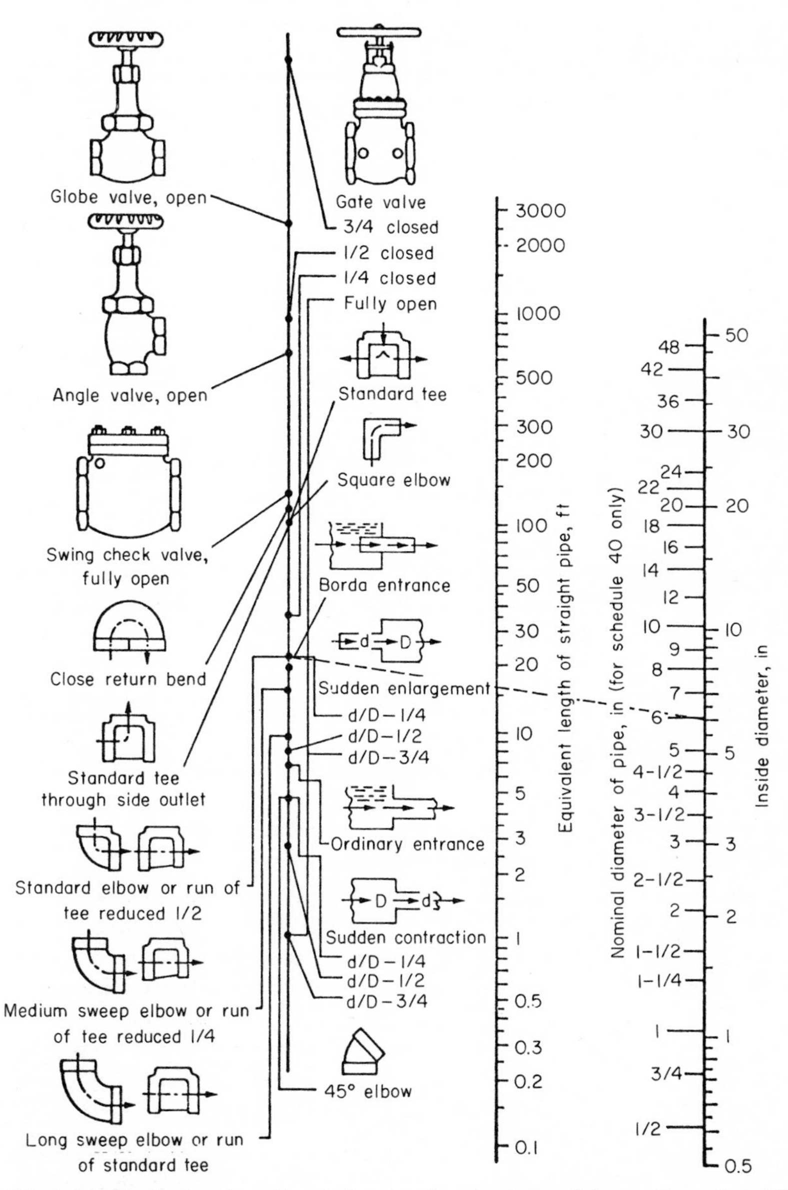

FIG. 4.14 Equivalent lengths of pipe for standard pipe fittings. *(From Ref. 63 with permission.)*

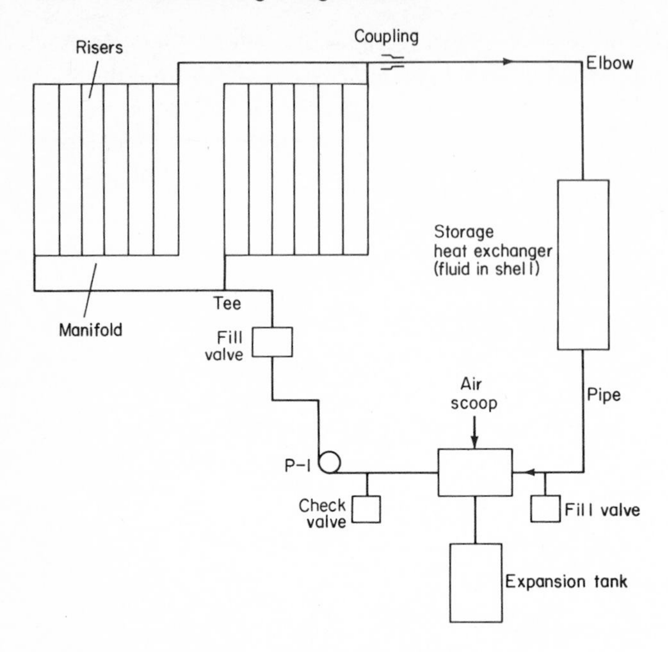

FIG. 4.15 Schematic diagram of liquid-based solar collection loop showing all sources of pressure drop which must be considered in the selection of the collector pump P1.

After the check valve is the second drain valve in the loop. Two fill-and-drain valves are required in collector loops, since fluid above a check valve cannot be drained from a drain below the check valve. Fluid then passes from the inlet piping to the collector manifold. The pressure drop in the manifold cannot be calculated as precisely as that in a corresponding simple length of pipe and Eq. (4.16) must be used for this purpose.

Fluid then passes through the collector array. Its pressure drop is determined from manufacturer's information. The outlet collector manifold pressure drop can be calculated from the same expression as for the inlet manifold. After returning to the mechanical space, fluid finally flows into the storage heat exchanger. The pressure drop across heat exchangers is determined from the manufacturer's data. The total of all preceding pressure drops composes the total system pressure drop.

In order to make an accurate calculation of pressure drop, the location of collectors on the roof, the length of piping runs, and the location of the mechanical space must be known. A checklist helps avoid overlooking any pressure drops. Pressure drops at flow rates other than the nominal rate can be estimated if it is assumed that the pressure drop is proportional to the square of the volumetric flow rate through the system.

A method of reducing pressure drops and operating costs in the solar collector loop while maintaining a uniform outlet temperature uses a variable flow rate in the collector pump. Use of such a pump for this purpose is not recommended.

Reduced flow rate means reduced heat-removal factor F_R and increased thermal losses from the surface of the collector. In addition, the added complexity of control systems is probably not warranted by the limited usefulness of fixed fluid temperature. In heating systems it is the storage temperature and not the collector outlet temperature which determines the temperature of fluid delivered to the load.

Pumps

The pump horsepower required to move fluid at a rate of Q (gallons per minute) against a pressure head Δp (feet of fluid) is given by Eq. (4.13).

$$\text{hp} = Q \times \Delta p/(3960 \times E_p) \tag{4.13}$$

The efficiency E_p depends upon the operating point. Peak efficiencies are about 50 or 60 percent.

In the final pump selection, characteristic curves must be acquired from manufacturers. An example curve is shown in Fig. 4.16. The performance is given for two rotational speeds, 3500 and 2940 r/min. In the upper part of the figure, the pump curve is plotted as the total head vs. capacity. Several pump curves are shown for different impeller diameters, as indicated to the left. The downward-sloping

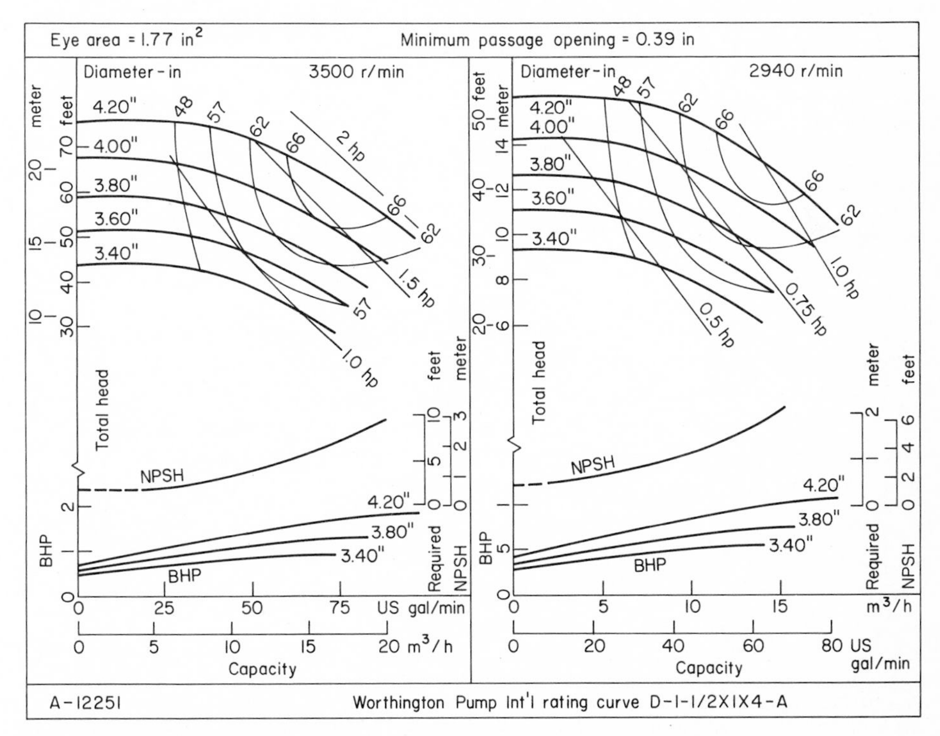

FIG. 4.16 Example pump curves for centrifugal pump with 1½-in inlet and 1-in outlet with nominal 4-in-diameter impeller. *(Courtesy of Worthington Pump Corp.)*

lighter solid lines represent the motor size required. The upwardly concave solid curves are pump efficiency curves used to select the most efficient pump operating point.

As shown by the left half of Fig. 4.16, this pump can flow 75 gal/min at 60 ft of head. The efficiency is above 66 percent and the power required is 1¾ hp if a 4.2-in-diameter impeller is used. The same pump operating at the same speed is also capable of delivering 25 gal/min at 78 ft of head. However, the efficiency is below 48 percent.

Complete specification of pumps includes other design variables. The pipe sizes must be specified. The pump whose curve is shown in Fig. 4.16 has an inlet pipe diameter of 1½ in and an outlet diameter of 1 in. These must be compatible with the remainder of the system piping. Reducers can be used if necessary.

Cavitation will occur unless adequate net positive suction head (NPSH) is present at the pump inlet. Required NPSH is shown in the pump curve. For example, in Fig. 4.16, at 3500 r/min and 75-gal/min flow rate, a NPSH of 7 ft of fluid (approximately 3 lb/in^2) is needed. Available NPSH is determined by the initial pressurization and temperature of the system. Adequate NPSH is assured by pressurizing the system.

The net positive suction head of a fluid depends both on pressure and temperature. Therefore, *available* NPSH is greatest at the highest-pressure, lowest-temperature point in a given fluid loop. For example, in a solar collection loop the pump should be located on the outlet side of the storage heat exchanger since temperatures at this point are the lowest in the system (see Fig. 4.17).

Impeller materials are to be included in pump specifications. A cast-iron

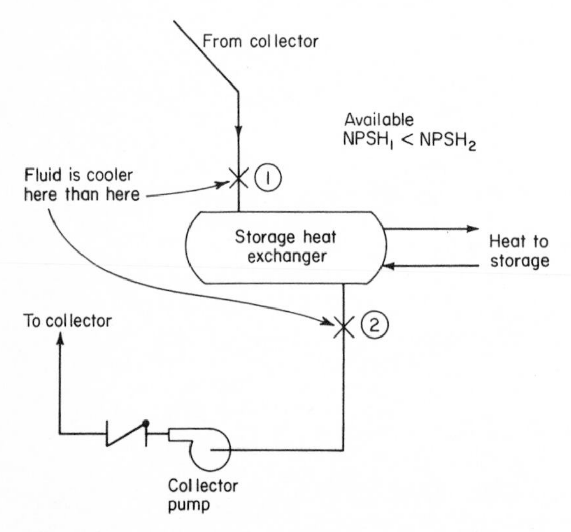

FIG. 4.17 Diagram showing location of solar collector pump at lowest temperature point in the collector loop to reduce the NPSH requirement when heat-exchanger shell-side pressure drop is small.

impeller is satisfactory in dead water. However, if air is dissolved in the fluid a bronze or stainless-steel impeller should be specified to avoid corrosion and erosion. Galvanized pumps are incompatible with glycol solutions. Polyethylene and polypropylene seals are incompatible with petroleum-based fluids.

Nonpriming pumps must be located in such a way that a prime is always maintained. This problem only occurs in open systems. A positive pressure at the pump inlet in an open system is achieved by placing the pump inlet below all other points in the system. A less reliable method of maintaining fluid at the pump inlet is by use of a check valve when the pump must be positioned above the lowest point in the liquid system.

In summary, the selection of pumps and pipes for liquid systems must consider the operating cost, the initial cost, the possibility of erosion at high fluid velocities, the optimum efficiency point of the pump, possible impact on flow balancing, air entrainment if fluid velocities are too low, and costs of insulation and piping. The design of piping systems is a standard HVAC exercise and is not described in further detail in this book. The solar impact on piping design is in (1) the specification of flow rates, described in Chap. 2, and (2) the existence of temperature excursions and the effect on fluid viscosity.

Fluids

In systems using antifreeze, the amount to be mixed with water is determined by the required freezing-point protection. Figure 4.18 shows the freezing-point depression of five common types of glycols used in solar systems. The desired freeze point is to be below the lowest temperature expected at the site. For example, if that temperature were −20°F, 40 percent by weight of ethylene glycol would afford adequate freeze protection. Beyond approximately 60 percent glycol by weight, little freeze protection improvement is achieved, but fluid viscosity is higher at low temperatures. Therefore, a glycol concentration greater than 50 percent is not used.

The type of glycol selected may be determined by both cost and toxicity problems. Ethylene glycol is more toxic than propylene glycol and costs are similar. Propylene glycol provides only slightly less freeze protection at a given concentration, and therefore the nontoxic fluid will usually be specified.

Fluids other than glycols are used for freeze protection. Organic and silicone heat-transfer fluids from industry have been used. These fluids have very low surface tensions and leak easily. The potential cost of a leak in a circuit using a silicone fluid is significant. It is not clear that the extra cost required to fabricate a leakproof system, plus the additional cost for the fluid, are paid off in added reliability over the life cycle of the system. Oils do not degrade at high temperatures as does glycol. Therefore, collectors may stagnate when no solar heat is needed. Glycol solutions deteriorate and must be replaced occasionally. In addition, the pH must be monitored and kept slightly basic by the addition of inhibitors.

Before a pump is selected to handle a specific fluid, its compatibility with pump seals must be established. An alternative to mechanical seals is to use a magnetic

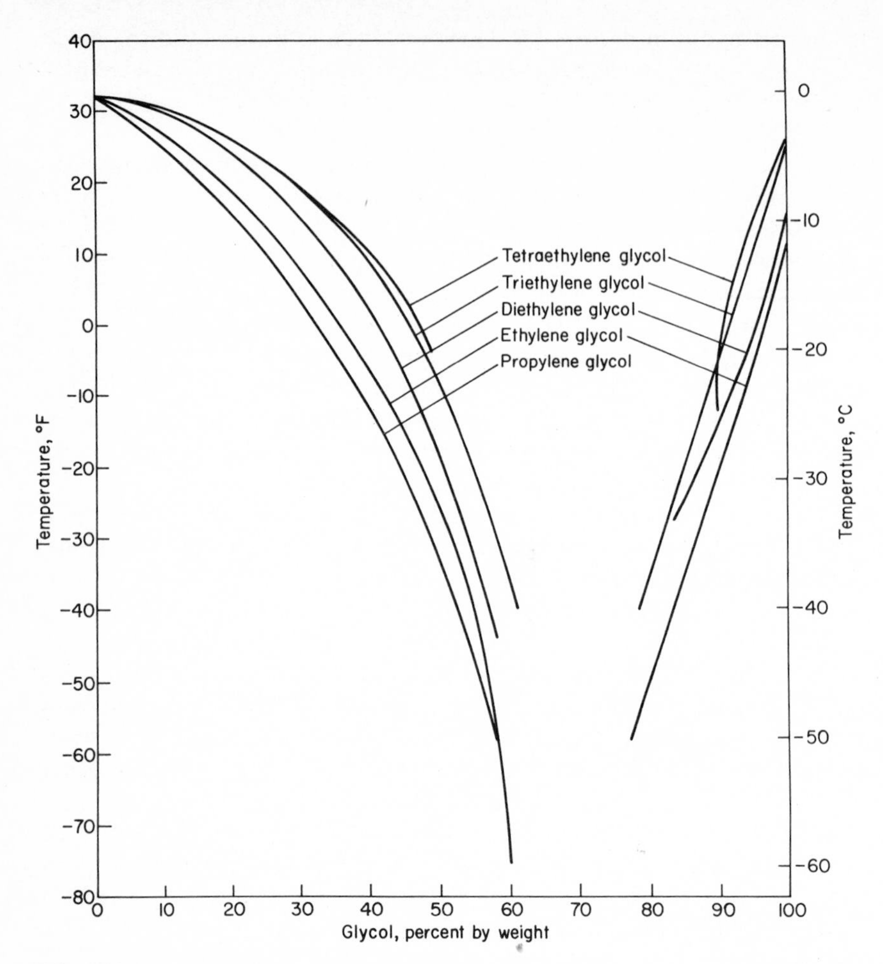

FIG. 4.18 Freezing-point depression of aqueous solutions of common glycol anti-freezes. *(From Ref. 32 with permission.)*

drive pump which requires no seals. Fluid leaks in storage and delivery loops are less of a problem, since the surface tension of water is greater than that of glycol solutions or heat-transfer oils. Heat-transfer oils have another problem—a high viscosity at low temperature.

In addition to freeze protection, fluids must provide efficient heat transfer over a wide range of temperatures from −20°F to over 250°F. The fluids must be chemically stable at high temperatures. Most glycols decompose above 300°F by changing to organic acids. Therefore, it is important that stagnation not occur in glycol systems. Glycol converts to glycolic acid and ultimately to oxalic acid; both corrode piping. Control systems are designed to avoid stagnation conditions.

Corrosion control in aqueous solutions should be considered by the designer. The type of corrosion protection required depends on the fluid. Glycol solutions designed for use in solar collectors include inhibitors which protect the collector loop metals. This is not the normal function of inhibitors in automotive-type antifreezes. There, the inhibitor is designed to protect the fluid, not the metal, from oxidation. However, this protection is not necessary in solar loops since no oxygen sources exist in closed loops.

Dichromate inhibitors are incompatible with glycols. They promote a glycol oxidation reaction. The accumulation of sludge in a number of solar systems has been attributed to the misuse of dichromate inhibitors. Dichromate is an inhibitor for water only, not for water-glycol solutions.

The volume of fluid in a collector loop is determined by the total volumes of all components in the loop. After piping and plumbing layouts are completed at the DD level, the designer can calculate the fluid volume and include the fluid cost in the DD cost estimate. The total volume includes the volume of collector inlet and outlet piping, collector manifolds, the shell side of the heat exchanger, the solar collectors, the expansion tank, and the pump. Fluid volume in a heat exchanger can be calculated easily if it is assumed that the exchanger is a cylinder of known diameter and length. The expansion tank may be considered to be one-third charged, and the fluid content of the collector can be determined from the manufacturer. Collector fluid volume will usually be less than 0.04 gal/ft^2. The total cost of antifreeze can be determined from the unit cost of glycol and the percent by weight to be used. For example, a glycol costing $5 per gallon and used in a 40 percent concentration results in a cost of $2 per gallon of fluid in the collector loop. Organic oils cost approximately $20 per gallon and are not diluted. Therefore, the effective cost vs. a glycol is approximately 10 times greater.

In summary, the important properties of a working fluid for collector loops include the following: cost, freeze point, pH, surface tension, pour point, transport properties, and corrosion inhibitors. If an aqueous solution is to be used, the hardness of the water should be known since hard water can precipitate scale on the collector tubes and reduce the heat-transfer effectiveness. In locations with very hard water, it may be necessary to use distilled water in the collector loop.

FANS AND DUCTS

Fans and ducts are the air-system analogs of pumps and pipes in liquid systems. Many of the same fluid-mechanics principles apply.

Pressure Drop

The pressure drop through ducts, for example, is proportional to the square of the flow rate and the length of the duct. Figure 4.19 shows the pressure drop in ducts as a function of flow rate. The inclusion of duct bends, turning vanes, and other fittings is analogous to that for pipe fittings in Fig. 4.14. An equivalent length of all fittings is determined, and this length is added to the physical length of duct and finally multiplied by the friction loss shown on the horizontal axis in Fig. 4.19.

The design of ducts is treated in Ref. 29, where three basic design methods are described—velocity reduction, equal friction, and static regain. The velocity-reduction method is useful only in the hands of a skillful designer with long experience.

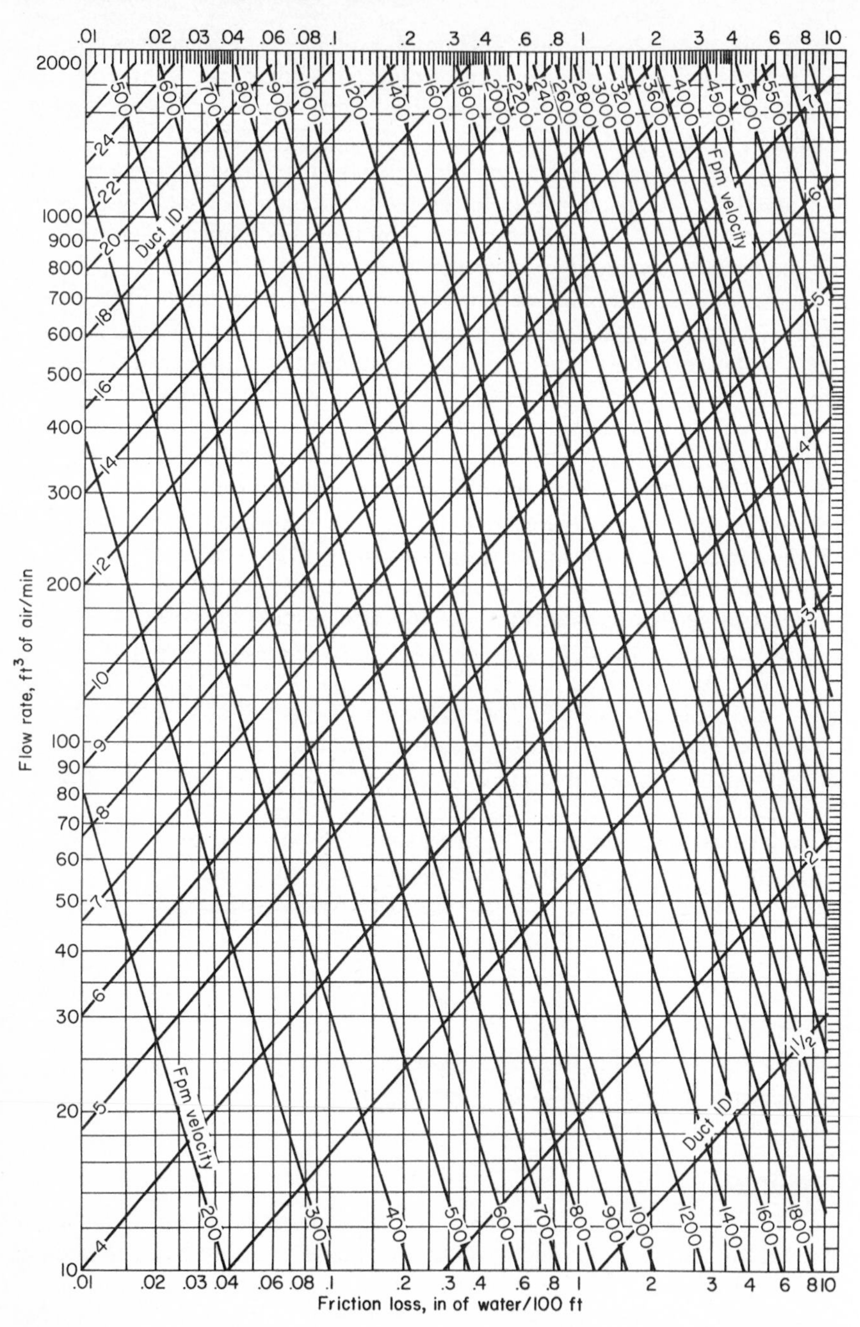

FIG. 4.19 Pressure drop of air flowing in round, smooth ducts. The equivalent pressure-drop diameter for rectangular x-by-y ducts is given by $\frac{1.30(xy)^{0.625}}{(x+y)^{0.25}}$. *(Reprinted with permission from the 1976 Fundamentals Volume, ASHRAE Handbook & Product Directory.)*

The equal-friction method established a constant pressure loss per foot throughout the system. It is the most popular method and makes the system almost self-balancing if the lengths of all runs are essentially the same. The static-regain idea uses the principle that static pressure gain resulting from fluid velocity reduction at a point of duct branching can be used to offset the frictional pressure drop through the next branch of the ducting. The downstream duct size determines the frictional pressure drop, and its size can be selected to give the proper balance of static-pressure rise and frictional pressure drop in each branch of the fluid system.

As described in Chap. 2, air systems operate in four distinct modes—collector to load, collector to storage, storage to load, and summer water heating. In each mode, the length of duct and the number of fittings and dampers is different. Therefore, a separate calculation of pressure drop must be made for each mode and the fan sized for the worst-case situation. It is desirable to have relatively equal flows in all modes. In the collector-to-storage mode, for example, the physical length of duct might be greater than in the storage-to-load mode. The summer water-preheating mode is not critical, since more heat is available than required. Therefore, this mode should not determine the sizing of ducts and fans.

Fan power is proportional to the product of pressure drop and flow rate and increases with each. One collector manufacturer recommends duct sizing based on a pressure drop of 0.08 in of water per 100 ft in residential applications. Duct design for larger systems will include an economic analysis including trade-offs of duct plus insulation cost vs. the fan initial and operating costs. Optimum flow rates and duct sizes can be identified in this manner. If proper values of airflow for the solar collection mode and the storage-to-load mode cannot be achieved with one fan, two fans must be used.

The fan horsepower required to move air at rate Q can be calculated from Eq. (4.14).

$$\text{hp} = Q \times \Delta p/(6356 \times E_f) \tag{4.14}$$

where Q = volumetric flow rate, cubic feet per minute
Δp = the system pressure drop, inches of water gauge
E_f = fan efficiency

This expression is analogous to Eq. (4.13) for pumps.

If collectors are connected in series, the pressure drop increases for two reasons, the first being the increased path length, and the second being the increase in airflow required. If a nominal airflow of 2 to 4 $ft^3/(min \cdot ft^2)$ is provided for each square foot of collector, the velocity through two series collectors will be twice that through the same two collectors in parallel. Higher flow rates in the series configuration result in higher pressure drops and heat-transfer coefficients.

Leaks

Other details of duct design concern the designer at the DD level. The control of leaks in air systems is a prerequisite for efficient operation. The drawings should show adequate detail and notes to ensure that all joints and fittings are properly sealed. The control of air noise is also important. High velocities should be avoided for this reason (in addition, high velocities cause high pressure drops). It may be necessary to isolate the blower from the ducting by a flexible connection. This eliminates blower motor noise telegraphing through the duct system. The designer must also specify the insulation to be present on all ducts whether inside the heated building or not. A thickness in excess of 1 in of fiber glass is usually applied to most ducts. The optimum thickness can be calculated from Eq. (4.1).

The designer must use an efficient arrangement of equipment in the mechanical room to avoid unnecessarily contorted ducting and excessive fittings and turning vanes. A logical flow of air from the blower either to the collector or the storage unit should be the basis for mechanical room layout.

The blower motor must be located where adequate cooling air is available. The allowable temperature rise of the motor must be considered. Standard blower motors should not be located in hot airstreams. This location was used in some early solar systems and caused continuous cycling of the blower motors throughout the collection period. As the fan began to operate, hot air was pulled across the motor, the motor temperature rose above the circuit breaker set point, and the motor turned off. The problem was most prevalent during the highest efficiency periods of the system when fluids were hottest. Repeated motor cycling is not desirable from a durability standpoint. Of course, high-temperature motors can be operated in the hot-air stream.

SOLAR HEATING CONTROL SYSTEMS

The control system is the key component in any solar heating system. The controls determine the operation of all mechanical components in the system and determine the overall efficiency of energy collection and use; they must be designed with the greatest care. Many solar heating systems designed properly from a mechanical standpoint in the past have not operated at the design performance level because of malfunction or poor design of the control system. In this section, the design of all control components is described in detail. Since the controls have been the cause of many malfunctions in the past, this section is particularly detailed.

The controller is designed after the mechanical configuration of the heating system has been finalized. Of course, during the design of the mechanical system, the type of control must be considered. The controller consists of three parts—the sensors which determine the state of the system; the control unit itself, which makes decisions based on preprogrammed criteria; and the actuators, which carry out the decisions made by the control unit. Actuators include pumps, valves, heat-rejection units, and other components operated by the controller.

In this section, the operation of the control system viewed as a system is treated first, and then the sensors and their location are presented. Details of system operation follow, and a section on special features of controls for nonstandard systems is presented thereafter. Controls for passive systems are described on page 217.

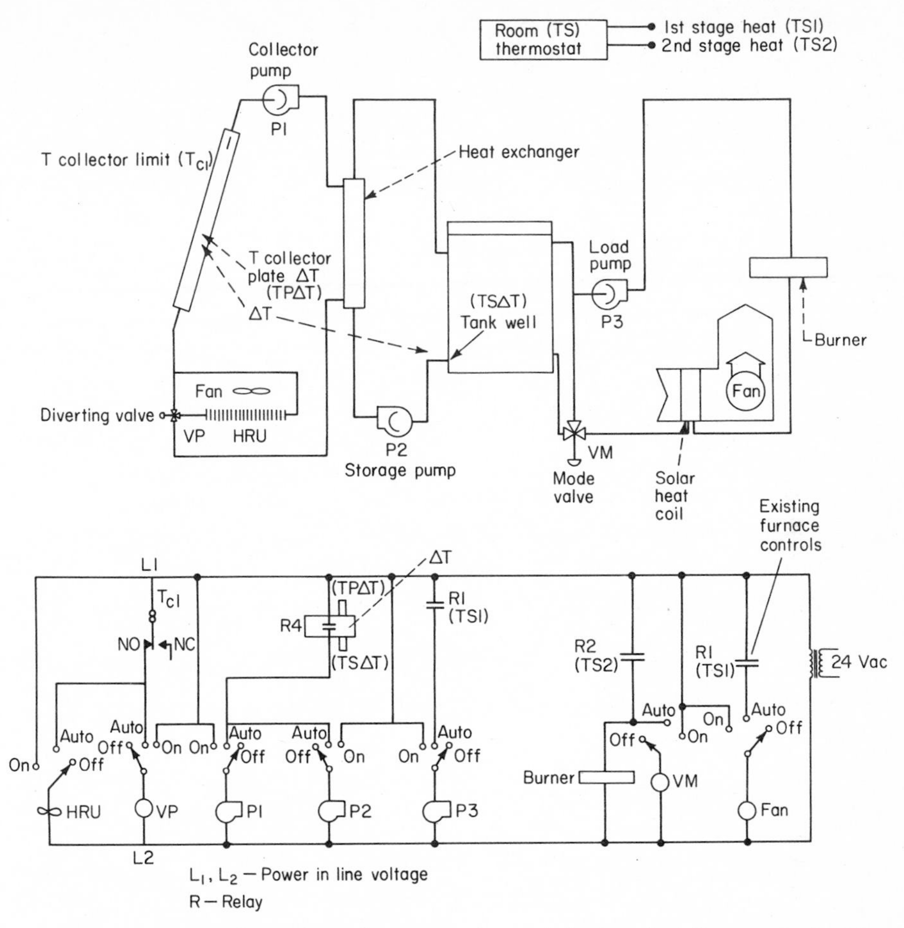

FIG. 4.20 Simplified solar-system diagram and associated control diagram for liquid system with parallel backup.

Control System Operation

The general operational modes of a solar heating controller have been described in Chap. 2. In this section, the details of operation are treated, and an example system configuration is presented. The controller operates the collector pump, the storage pump, and the pump delivering heat to the load. In addition, the backup system and heat-rejector are operated by the controller. Figure 4.20 is a schematic diagram of a solar heating system with the control diagram also presented. This system will operate in several modes:

Energy collection and delivery to storage
Heat delivery to load
Backup heat delivery to load
Summer heat rejection

Air systems operate in an analogous fashion, and the performance of air-system controls is similar to that of liquid systems. In this section, only the liquid-system controller is described. In a later section, the few differences between air and liquid controls are summarized.

Collection. The collector-loop-controller subsystem is where most design problems have occurred in the past. Therefore, the careful specification of this part of the controller is of prime importance and must be addressed by the designer.

Figure 4.21 illustrates one of the key operational characteristics of the collector-loop control subsystem. The decision whether pumps P1 and P2 in Fig. 4.20 operate is based on the collector temperature TP relative to the tank temperature TS. If the collector temperature is somewhat greater than TS, it is worthwhile operating pumps P1 and P2. These pumps will continue to operate until the collector temperature approaches the storage temperature, at which time operation is no longer worthwhile. A hysteresis characteristic is built into the controller to prevent cycling of the pumps.

The collector in Fig. 4.21 warms from point *a* to point *b* as sunlight first strikes it. The pump begins to operate at point *b*. As the pump operates, cool fluid from the collector inlet line is introduced into the collector. The collector sensor senses

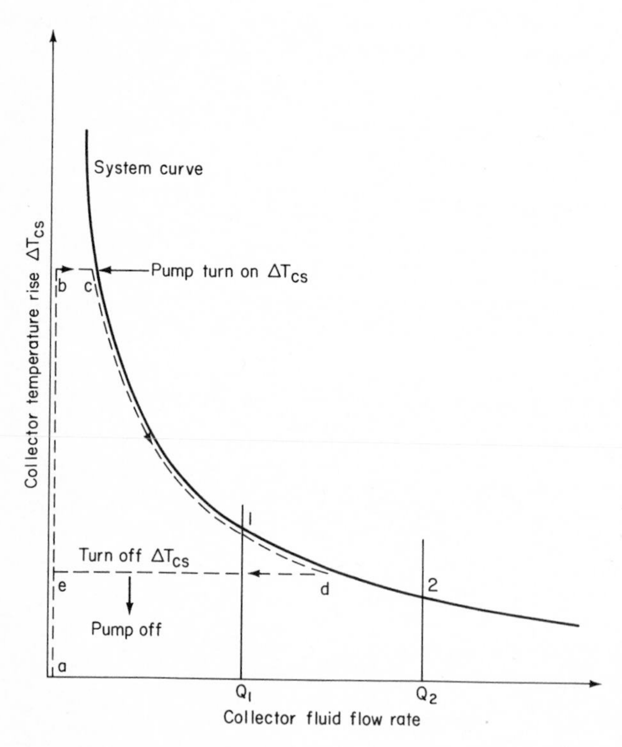

FIG. 4.21 Solar control characteristics superimposed on system start-up temperature curve.

a drop in temperature. This temperature drop is shown as the system curve in Fig. 4.21. As the pump continues to operate, the temperature drops to a steady-state operating point corresponding to pump flow rate Q_2. However, if flow rate Q_2 is too great, the pump turnoff temperature will be reached before temperature T_2 (corresponding to Q_2) is reached. If the pump should turn off, the collector temperature will again begin to rise from point e to point *b*. The pump again begins to operate with fluid temperature falling once again toward point 2. Then the cycle repeats.

The solution to this problem is to use a smaller flow rate Q_1. As the pump begins to operate and the temperature drops along the system curve, temperature T_1 will be encountered before point *d*, at which point pump turnoff occurs. Since point 1 represents a steady-state operating condition at collector flow rate Q_1, stable operation is assured and no pump cycling takes place. A typical value of collector-to-storage turnon temperature difference in liquid systems is 15 to 20 °F. The hysteresis band designed into most collector controllers continues pump operation until the collector-to-storage temperature difference drops to 3°F or so. An empirical rule which has been found useful is that the turnon temperature difference for liquid systems should be 5 to 7 times (1.5 to 3 times for air systems) the turnoff temperature difference.

Heat Rejection. The collector-loop controller also operates the heat-rejection unit shown in Fig. 4.20. The heat-rejection unit is used in summer when storage is fully charged and no additional heating of storage is desired. The storage temperature limit relates to a collector temperature limit T_{cl}. The collector sensor will activate the heat-rejection loop if the collector temperature exceeds T_{cl} (storage temperature rather than collector temperature can be used for this purpose). During heat rejection, flow is diverted through the heat rejector for dissipation to the atmosphere. This is done by blowing air over an air-to-liquid heat exchanger located outside the building. The overheat heat-rejection mode also includes pump P1 and divert valve VP operation. Pump P2 need not operate in the heat-rejection mode.

The heat-rejection unit can also be operated independently of the collector-loop controller. This can be done by installing a make-break switch on the collector absorber plate (or at the top of the storage tank). The switch will close at the design value of T_{cl} (about 220°F) and activate both the diverting valve and the heat-rejector fan. This switch also causes P1 operation so that fluid flow occurs in the collector loop. P1 may not operate during heat rejection, since the normal operation of pump P1 is based on a 15 to 20°F temperature excess above the storage-tank temperature.

Heat Delivery. The energy-delivery mode of the solar control system is quite reliable and uses standard HVAC controls. In the residential single-zone configuration, a single dual-point thermostat is used to determine whether heat is required within the heated space. As the first contact closes, a call for heat is sensed by the control system, and pump P3 and the heating unit fan shown in Fig. 4.20 are both activated. If sufficient heat at sufficient temperature is contained in storage to carry the heating load, the space temperature will rise, contact TS1 will open, and the pump and fan will cease operation when the heat load is satisfied.

If insufficient capacity exists in storage, the space temperature will continue to drop and the second-stage thermostat contact TS2 will close. The closure of the second contact activates the diverting valve causing fluid to flow not from storage but through the backup heating unit. The second-stage contact also activates the fuel valve and igniter, if present, or the heat-pump relay, initiating backup system operation. The anticipator on the thermostat is set so that backup continues until both contacts of the thermostat open.

In larger buildings with multiple heating zones, control is more complex. However, the same priorities apply—use solar heat first; then use backup heat.

Computer simulations and field studies have shown that series and parallel backups relative to storage do not differ much over a year. As noted in Chap. 2, the use of a series backup can permit backup heat to be inadvertently stored in the tank. In principle this can be avoided by adding features to the controller, but it can be automatically avoided by using a parallel backup.

Air-based solar systems using a heat-pump backup require two stages of backup heat. If solar capacity is inadequate, the storage fan is turned off and the motorized storage-to-space damper is closed. A third damper in a duct that bypasses storage is then opened and return room air is diverted through it directly to the heat-pump condenser. This heats the space if outdoor temperatures are above the heat-pump balance point. If not, either an outdoor thermostat or a third contact in the room thermostat closes (indicating inadequate heat-pump capacity) and the strip heaters are operated.

The design of the solar-control system is different for each system configuration. The designer should attempt to use the *simplest possible control system, but design this simple system with great care,* considering all operational modes and climatic conditions which may affect system operation. A simple system does not imply a simple and brief design exercise. Complex control systems, if they work, may increase the annual solar contribution to load over a year by a few percent. The designer must determine whether extra modes, which are rarely used but add expense to the system, are worthwhile economically. Additional control complexity makes future servicing more difficult and increases the possibility for control-system malfunction.

It has also been found that the use of one master control system for a building is better than adding a solar-control subsystem to a building heating-control subsystem. The integration of two dissimilar controls is always more difficult than the use of a single system designed to control all mechanical systems in the building.

Sensors for Solar-Control Systems

The decisions made in solar heating systems are based on temperature. For example, the collection of heat is determined by the relative temperatures of collector and storage. Devices called thermistors measure temperature. Table 4.3 shows the

TABLE 4.3 Temperature-resistance characteristic of example thermistors used in solar-control systems*

Temperature, °F	*Resistance, ohms*	*Temperature, °F*	*Resistance, ohms*
20	138,700	130	9,100
30	103,700	140	7,500
40	78,300	150	6,100
50	59,700	160	5,100
60	46,000	170	4,200
70	35,600	180	3,500
80	28,000	190	2,900
90	22,000	200	2,500
100	17,500	210	2,100
110	14,000	220	1,800
120	11,300		

*Courtesy of Solar Control Corp., Boulder, Colorado.

resistance of a typical solar-control thermistor. The resistance values indicate that the thermistor is nonlinear. At low temperature a greater resistance change occurs per unit temperature rise than at high temperature. During system installation and debugging, the thermistor resistance vs. temperature characteristic is very useful. A variable resistor can be substituted for either the collector or storage sensor. Then proper function of the controller can be checked by simulating values of temperature on a rheostat.

The location of sensors in solar heating systems is of crucial importance. In active systems, four temperatures are normally sensed—collector, ambient, storage, and heated space. The collector sensor must sense the collector temperature, not the temperature in the vicinity of the collector. The most reliable location for the collector sensor is *on the collector plate* attached by a bolt *within a few inches of the outlet header*. This sensor, furnished by the control contractor, should be put in place by the collector manufacturer.

If the collector sensor should fail at some point during the life of the solar system, it would be difficult to repair. Therefore, it is suggested that a spare sensor or two be installed with electrical leads run outside of the collector for possible future hookup. All sensor connections should be soldered and insulated in accordance with normal electrical installation procedures for these components.

Sensors cannot be mounted on the absorbers of evacuated-tube collectors and other nonflat plate designs. Satisfactory temperature signals are acquired as described in a later section (page 213).

Figure 4.22 shows typical thermistor-type sensors for both collector and storage. The collector sensors have a hole used for attachment to the absorber plate.

The storage sensor must be located with care. As described in the previous section on storage, it is possible for a storage sensor to measure a temperature which has little relationship to the temperature of the fluid entering the storage heat exchanger in liquid systems. The nominal location should be in the lower one-quarter of the storage tank and away from any tank fitting. If the sensor is located near

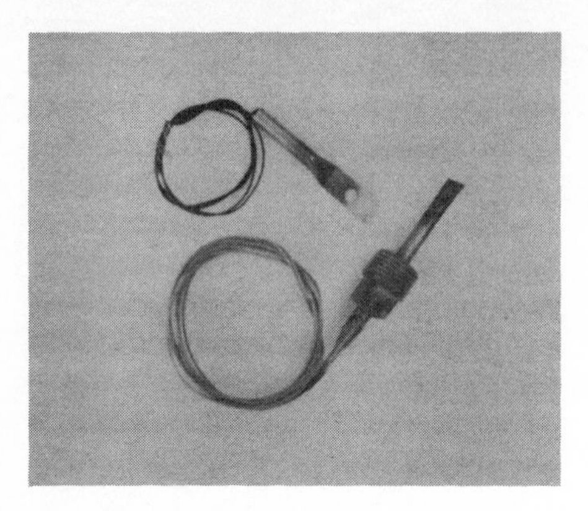

FIG. 4.22 Photograph of typical thermistor sensors used for solar control signals.

one of the tank connections, cool returning fluid from the load heat exchanger, for example, could present an artificially low temperature signal to the controller and cause the solar collector loop and the storage pump to be activated when no net heat can be accumulated. This is a frequent problem in domestic hot-water systems where low-temperature city water enters the bottom of the solar-preheat tank. If the storage sensor is located near the city-water inlet, a very low temperature will be measured. The controller will incorrectly decide that the proper differential for collection exists, and the pumps will begin to operate. It is possible in these situations for heat to be rejected through the collector if the storage tank is at a substantially higher temperature than the inlet water which was sensed by the storage sensor.

In rock-bed storage systems, the storage sensor should be located near the end of the cool zone of storage, normally the bottom of the storage container. The storage sensor must be protected during the rock-bed filling operation to avoid problems. Since the solar-storage sensor may not be accessible after the storage box is filled, it may be worthwhile to specify that two storage sensors be installed, one to be available for future connection if the original sensor should fail.

For both air and liquid systems, the storage sensor should be of the averaging type. That is, it should represent an average temperature in the lower or cooler zone of storage. A small sensor, if properly placed, may work satisfactorily, but a more reliable signal can be acquired from a sensor averaging over a horizontal foot or so of storage. These sensors are commercially available.

Thermistors are subject to degradation in the presence of humidity. Most controller manufacturers provide hermetically sealed sensors. If a home-built control unit is used, waterproof or glass-covered sensors must be used to ensure a reliable signal over a long period of time.

The replacement of sensors at some future time should be considered by the designer. Suggestions for methods of replacing collector and storage sensors have been noted above, and the slight extra investment in installing spares at the time of system buildup is certainly offset by the possibility of future high replacement cost. The liquid storage sensor should be placed in a well so that it can be replaced without draining the tank. In this case, a spare sensor is not necessary.

Conventional dual-point thermostats can be used to sense zone temperatures. The rules used in standard HVAC design apply to solar system design regarding the placement of thermostats. They should be located out of direct sunlight and away from any hot airflows or cool drafts. A position on an inside wall is common.

Other Control System Design Considerations

In this section, a collection of miscellaneous features of reliable control systems is presented. Some considerations apply only to liquid or air systems, and others to both.

Most actuators in solar-control systems are operated by a 24-V circuit. Twenty-four-volt power is provided by a transformer from 110-V line and is used to operate motor starters, dampers, control valves, and other components. The control-system designer should specify relay current capacity sufficient to operate all components which function during a particular mode. For example, in the heat-rejection mode in a liquid system the relay must be capable of operating the pump in the collector loop, the heat-rejection-motor starter, and the control valve. Or in an air system, the relays must be capable of operating collector dampers and the main fan plus the backup system fuel valve. The worst-case current flow situation should be identified and the relay sized for this purpose.

As described earlier, the stable function of control systems in the collector loop requires a hysteresis or dead band in order to avoid on-and-off cycling of the collector blower or pump. The size of the dead band has been estimated at 15 to 20°F for liquid systems and 40°F for air systems. This dead band should be adjustable. The precise values of the hysteresis and hysteresis ratio depend upon the amount of fluid contained in the liquid loop as well as the flow rate and collector heat capacity or time constant. Most solar controls are adjustable—in the form of either a potentiometer or a replaceable resistor in the circuit board. This adjustment is best made in the field, and the designer need only specify that the dead band be adjustable in the outline specification for DD.

As shown in the system schematic in Fig. 4.20, the controller should have a full manual override for all operational modes. This is important since the testing of a system will require the operation of pumps and fans during conditions when the controller in the automatic mode may not cause them to operate.

Pneumatic Systems. Pneumatic controls have been used for many years in HVAC systems. The previous discussion on controls has concentrated on electronic controls which are the suggested approach for solar systems. Pneumatic controls have the difficulty that the air which operates them can never be completely dried. The moisture remaining in pneumatic control system air can freeze in winter at the collector sensor location and cause the system to cease proper operation. The standard approach in HVAC systems is to use a small refrigeration system to remove as much humidity as possible, thereby drying the control system air. The dew point normally specified for these systems is about that which can be present on a winter night outdoors. If a pneumatic system must be used, the designer must specify a dew point below the minimum temperature ever expected to exist at the control site, allowing for any nighttime sky radiation effects.

Alternatively, hybrid electronic and pneumatic systems can be used. In these systems, the collector and storage sensors are thermistors, the electrical signal is converted to an air signal at a transducer, and the remainder of the control system can be pneumatic. However, the majority of solar control systems in use today are electronic and not pneumatic.

Power Failure. The control-system designer must also consider the possible consequences of a power failure. If the power failure occurs during a sunny day, the collector will boil. If glycol antifreeze is used, it will be degraded. If prolonged boiling occurs, the antifreeze may decompose to such an extent that sludge will block collector fluid conduits. The consequences of power failure in air systems during the day are less severe, since most good air collectors are designed to stagnate.

Power failures at night have undesirable consequences only in collector systems which are based on the drain-down method of freeze protection. Of course, power failure may result in a completely inoperable building heating system, in which case pipe freezing and other damage may occur.

One method of protecting a liquid system from daytime power-failure damage

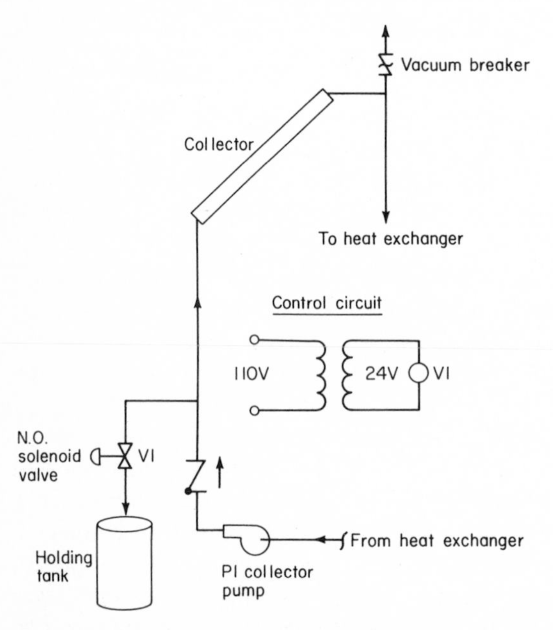

FIG. 4.23 Schematic diagram of NO system drain valve V1 to drain liquid system in the event of power failure.

is to install a normally open valve in the collector loop as shown in Fig. 4.23. If a power failure should occur, the valve will open and fluid will drain into a holding tank. The fluid may then be returned to the collectors after power is restored. Liquid-based collectors can stagnate, when empty, without damage.

Restarting a solar system on a sunny day after power failure must be done with care. If the drained fluid is simply pumped back into a 400°F collector, physical damage—including generation of steam and overpressure in the collector array, warping of the absorber plate, and possible glass breakage—is sure to result. It is suggested that a solar system not be restarted on a sunny day but after sunset. The restart of a solar system can be manual and not be programmed into the control system. If restart must be automatic, a high-temperature-collector pump lockout switch on a collector absorber plate can be used to prevent flow until the collector cools.

Power failure difficulties can be avoided with an auxiliary electric power supply for the collector and storage pumps. A small pump in parallel with pump P1 in Fig. 4.20 could be used to maintain the collector fluid below the boiling point. The design flow rate would not be required. Backup motor-generator (MG) sets are available for this purpose, or a battery-operated circulator could be used in both the collector and storage loops. The use of a battery system is probably most appropriate for a smaller installation, while an MG set would be appropriate for large commercial solar heating systems.

Display Panel. A display panel indicating operational modes and components currently in use can help monitor a solar system. The proper operation of the system can thereby be determined rapidly. Some controller manufacturers include a display panel as part of the system without a specific requirement being delineated by the designer.

Tubular Collector Sensors. The collector sensor cannot be mounted on the absorber of tubular collectors. Since temperature cannot be obtained by mounting a sensor on a collector header, another method is used. Some early installations did use a sensor mounted on the header as close to the collector as possible. Very unreliable system operation resulted since the flow of heat from a hot absorber to the cold header takes place only by conduction in the water. No convection can take place because of the check valves in the fluid circulation loop. In some instances, fluid boiled in the collector before the sensor temperature began to rise.

One solution to this problem uses a small fluid flow through the array during daylight hours, whether useful energy collection can take place or not. A small circulator, sized to provide only a small fraction of the design flow rate, will assure that a sensor mounted external to the collector will have a good signal. The sensor should be placed as close to the collector as possible.

In very large arrays of tubular collectors the trickle-flow pump need not circulate fluid through the entire collector array as shown in Fig. 4.24. Flow through one collector is enough. In this design, backflow of fluid through the small circulator when the collector pump operates must be prevented by a check valve. If only one

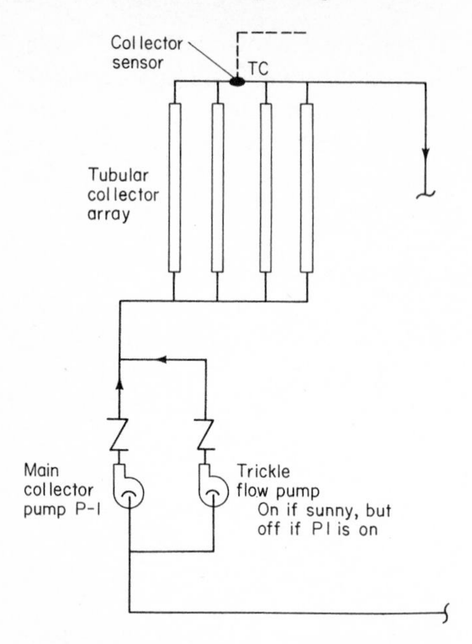

FIG. 4.24 Schematic representation of trickle-flow pump for tubular collector array to provide collector temperature signal to collector sensor when it cannot be mounted on the absorber surface.

collector is used, the circulator will normally be installed out of doors behind the selected collector. The pump and the control must be located in a weatherproof housing.

Thermal Transients. The volume of collector fluid contained in large liquid systems can be enormous. On a very cold winter night, this fluid will be well below zero. The following day, as the collector warms up in the presence of sunlight, the rising temperature eventually will cause the collector pump to begin operation in the normal manner. However, when the pump starts, a large amount of very cold fluid from the piping passes through the heat exchanger. Since this fluid has not been warmed, it could cool storage and even freeze the water in the storage heat exchanger.

One method of avoiding this problem is to heat all fluid in the collector loop with solar energy before the storage pump is turned on. This idea is shown schematically in Fig. 4.25. The fluid bypasses the storage heat exchanger until the entire fluid loop is 15 to 20°F above the storage-sensor temperature, at which time the storage pump P2 begins to operate.

Night Setback. Night setback is used to conserve energy during periods of low or no occupancy. Night setback affects solar system operation. If the night setback is the normal 10 to 15°F below the daytime temperature setting, the morning reset will cause the backup system to come on every morning whether solar heat is available in storage or not. This occurs since the night setback temperature is well below the daytime closing temperature of both contacts of the room thermostat. Figure 4.26 shows that the operation of the backup system causes a relatively rapid pickup of room temperature to the daytime setting. Since the backup system operates at a higher terminal temperature than a solar system, the recovery rate is faster than that shown for the solar system.

Solar heat can pick up the morning building heat load if a time-delay feature is used in the controller. This time delay should be based on the ability of the solar system to heat the building given a typical wintertime storage temperature. The solar recovery rate line shown in Fig. 4.26 will vary each day depending upon the residual heat in storage. In spring and fall, the slope of the curve may be as steep as that for the backup system; however, in winter the slope will be more shallow

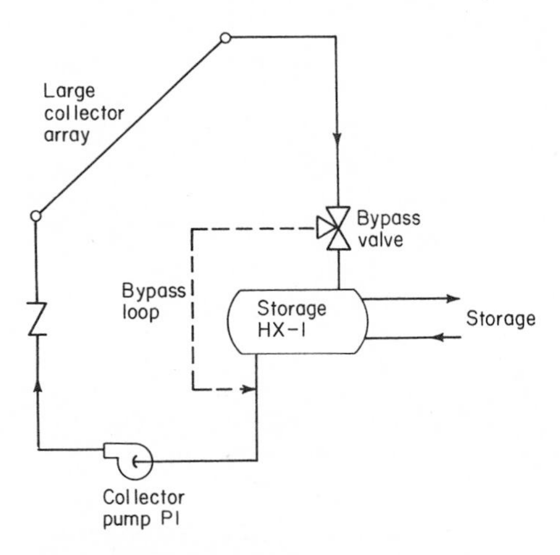

FIG. 4.25 Storage-bypass loop for use in large systems requiring collector fluid warm-up before activation of storage pump.

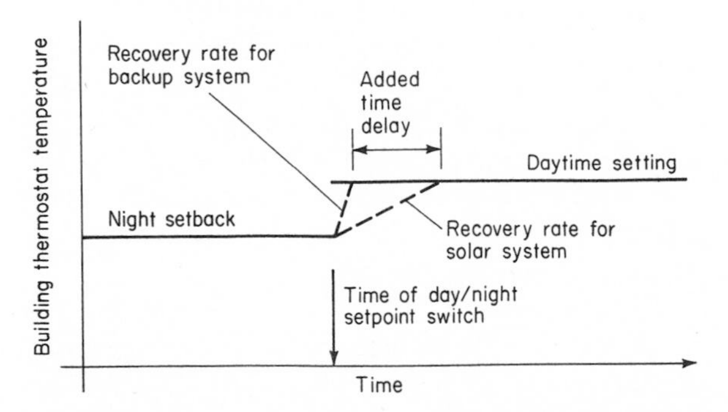

FIG. 4.26 Characteristic recovery curves for solar and nonsolar heating systems after night setback period.

as shown. The ability of solar storage to heat a building from the night setback temperature to the daytime setting can be calculated if the effectiveness of the load heat exchanger (described earlier) is known, along with a typical winter storage temperature. These temperatures will be in the vicinity of 100 to 110°F on a winter morning following a sunny winter day. One way of inserting the required time delay is via a lockout timer on the fuel valve of the backup heating system.

Controller Location. Solar control units use solid-state circuitry for the most part. These devices require a relatively restricted range of ambient temperature to operate properly. The designer should specify the location of the control unit so that it is not in either a very cold or a very warm location. Further, the temperature of the controller should not vary from season to season. A controller ambient temperature below 55°F may cause problems.

Additional Air System Control Features

Many of the preceding considerations apply to both liquid and air systems. The few applying only to air systems are described below. In the collector-to-load mode, any air temperature above room temperature will theoretically heat the space. However, air at 70 to 75°F is not comfortable in an occupied space. In some air systems, therefore, it is desirable to place a limit on the collector outlet temperature which can be used for space heating. A typical value is 90 to 100°F.

On a sunny day, the temperature rise through an air collector is 30 to 40°F. Given the fixed inlet temperature of 65 to 70°F, the delivery of air at a temperature below the comfort level will rarely be encountered during a morning start-up.

However, the delivery of cool air to the space can occur during the day or toward evening when the controller hysteresis effect applies and the system continues to operate until the collector temperature is within 10 to 15°F of the storage cool zone temperature. Since the cool zone of storage is 65 to 70°F, the delivery of air from the collector in the evening can be at a temperature of only 80 or 85°F. If this airflow is delivered to storage at a temperature of 80°F, the topmost stratum of storage will be 80°F. When a call for heat occurs after sunset, reverse airflow through storage in the storage-to-load mode occurs. The heat content of 75 to 80°F air leaving storage may not be sufficient to carry the heat load at night. As this low-temperature air is delivered, the space temperature will continue to fall, although most of the storage bed may be well above 75 to 80°F and could easily provide all heat required. This cycle can be repeated two or three times before sufficiently hot air is produced.

This problem can be avoided if the differential temperature for system cutoff is set sufficiently high. The designer should therefore, carefully specify the turnon and turnoff temperature differences in air systems, as well as in liquid systems. Since installed system function depends upon the interaction of the building with the solar system, final adjustments are made during the system installation phase described in the next chapter.

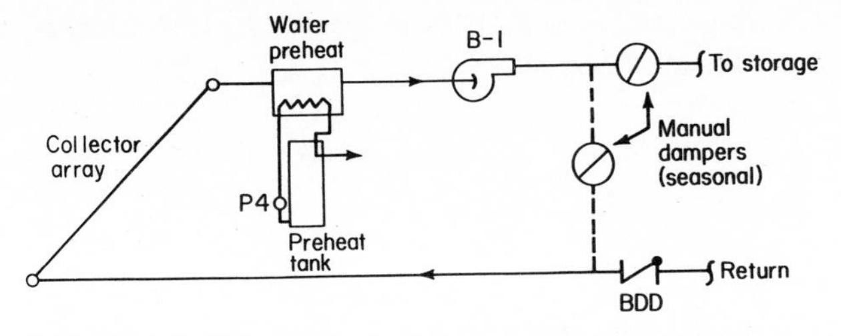

FIG. 4.27 Manual dampers used to bypass air-system storage in summer while still providing domestic water preheating.

Seasonal Changeover. An additional feature of air systems is the manual mode switch which takes place in spring and fall. The entire rock bed need not be heated in summer; therefore, a bypass around storage is activated during these periods. This bypass is shown in Fig. 4.27 by the dashed line.

The summer water-heating mode in an air system can be controlled simply by allowing the system to function in its winter sequence. Since the entire unheated storage bed will be at 70 or 80°F during the summer, the controller will turn on the blower B1 at a temperature of 110 to 120°F. The blower will continue to run until the collector temperature drops to within approximately 20°F of storage, or about 90°F. Since much more solar energy is available in summer than is required for only water heating, B1 may run for too many hours per day if controlled as above.

One method of solving this problem is to activate a slightly different control logic in summer by causing B1 only to be turned on by the normal temperature difference between the collector and storage. However, B1 is turned off by the same signal which turns off the water preheat circulator pump P4. This method of operation assures that both the blower and pump operate only when a water-heating load exists and not throughout every sunny hour of the summer.

Passive System Controls

The controls of passive systems are simpler than those for active systems. As described in Chap. 2, the only control device present in a fully passive system is that which operates the movable insulation. This can either be the occupant of a building or a very simple sun sensor which places the movable insulation in its down position if sunlight is below a certain level. A typical value is approximately 75 Btu/(h·ft^2). This threshold level for positioning the insulation should be adjustable, and the designer must specify this in the DD outline specification. The energy delivery from a passive system does not require control since it is determined by the aperture area, the size of TSW vents, and the thickness of the TSW.

In hybrid active/passive systems, some active controls are required. Air movers used to extract heat from the TSW cavity are actuated by space thermostats much as load devices are actuated in an active space-heating system. The thermostat-blower interface requires no special attention since it employs the simple arrangement described above involving a dual-point thermostat. The first contact closure will remove heat from the cavity, and the second contact closure starts the backup system.

Summary

In summary, this section has described most of the control design details which should be addressed at the DD level. The key considerations in the design of a successful control system are as follows:

- Proper collector-sensor location
- Proper storage-sensor location
- Proper relation of turnon temperature difference to turnoff temperature difference
- Hysteresis characteristics
- Linear hysteresis ratio over the expected operating range
- Proper signal to storage and collector sensors
- Careful consideration of each possible operational mode
- Summer-winter mode switch—usually manual

The design of a solar-control system is somewhat different than that for a conventional system, and the designer must devote additional thought to the controls. This section has emphasized residential controls. Large commercial systems use the same ideas but each is custom-designed.

VALVES IN LIQUID-BASED HEATING SYSTEMS

The control of flow in liquid systems is accomplished by valves. The following types of valves are used:

- Gate or ball valves
- Mixing valves
- Flow-control balancing valves
- Pressure-reducing valves
- Check valves
- Two- or three-way solenoid valves
- Safety pressure-relief valves

Gate and Ball Valves

Gate and ball valves are used to isolate components which may require removal for service. In addition, they are used for filling and draining the system, for bypass, etc. The key parameters of gate and ball valves are the material, the pressure at which the valve is designed to operate, the size of the valve, and the compatibility of the seals with the fluid which flows through the valves. The selection of gate and ball valves is a routine process in HVAC design.

The only difficulty with seals is their exposure to high-temperature solvents such as glycol solutions, organic oils, or silicone working fluids. The low surface tension and chemical reactivity of some of these components should be considered in specifying the seal design for the valve.

The components normally isolated by gate or ball valves include the collectors, pumps, heat exchangers, and storage tanks. A relief valve is needed in any branch of a solar system which can be isolated completely by valves and to which heat can be added. For example, a solar collector with closed valves at the inlet and outlet can rise very rapidly in pressure when exposed to the sunlight, and pressure relief is mandatory.

In small systems which are readily refilled, the use of valves to block off components is probably not warranted. Simple unions can be opened between components when the system is drained. Valves make the system-refill process easier but do not eliminate the necessity for air bleeding after component reinstallation.

Mixing Valves

Mixing valves are used in DHW heating systems. During summer or periods of low water demand (vacations, etc.), the temperature of water in preheat tanks in systems such as that shown in Fig. 2.15 can become quite high—180 to 190°F. If this water were to be used in the building, users could be scalded. Mixing valves mix cold city or well water with hot solar water. The blended output is then at a fixed outlet temperature. The amount of cold and warm water mixed in the outlet stream is determined by the valve set point. A typical valve is 140°F with some energy codes mandating a lower setting, typically 110 to 120°F. The location of the mixing valve is shown in Fig. 2.15.

Flow Control Valves

Flow control and balancing valves are present in each circuit of a solar system connected in parallel in which equal flows must be present. For example, in collector arrays flow balancing is required among all subarrays. If manifolds are designed as described in Chap. 2, balancing valves may not be necessary within subarrays.

Another application of balancing valves is shown in Fig. 4.28. In this circuit, two balancing valves are present to assure that equal flow occurs to the load heat exchanger if heat is extracted either from storage or from the backup heater. The two balancing valves are both adjusted at the design flow rate of the terminal device.

One method of setting balancing valves is based on the pressure drop across the valve. Most flow valves are equipped with inlet and outlet pressure taps. The flow can be determined by a calibration chart based on the valve pressure drop, the temperature, and the type of fluid in use. If the flow rate is not critical and erosion is not a concern, it is best to specify the maximum flow available. The installed

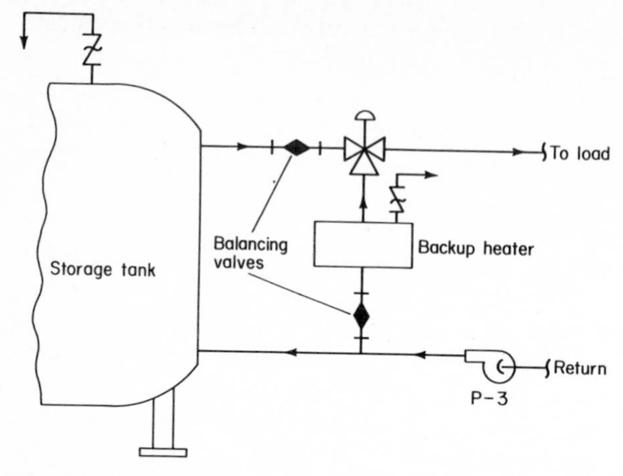

FIG. 4.28 Illustrative installation of balancing valves in two parallel fluid loops—storage-to-load loop and backup heater-to-load loop.

flow rate and head capacity of pumps selected will never precisely match the calculated characteristics of the fluid loop. If the highest possible flow rate is to be used, the balancing valve in the circuit with the greatest pressure drop should be fully opened. The installer can then adjust the balancing valves in other parallel circuits to give the same flow rate. For example, in Fig. 4.28 if the pressure drop through the backup heater is larger than that through the storage tank, the backup heater balancing valve would be opened fully and the storage valve would be adjusted at the same flow rate. Then the load device would operate at the same flow in either mode. Of course, different flows can be used in different modes if the design so dictates.

Flow rates in collector loops must not exceed the design flow rate, since erosion can occur in small fluid conduits within the collector. Therefore, some method of checking flow in a collector loop is suggested. Flowmeters should be permanently installed in large systems, but in small systems an orifice or Pitot tube is adequate for calibrating the fluid-loop flow. Safety valve protection of closed segments of fluid loops also applies to loops which can be isolated if a balancing valve is completely closed. For example, in Fig. 4.28 if the balancing valve were inadvertently closed and the three-way mode-control valve were in the storage-to-load mode, the backup heater would be completely isolated. If the heater were inadvertently actuated, pressure would build up in that loop unless a pressure-relief valve were used.

Safety Valves

Safety valves limit the pressure and/or temperature to which a component or fluid loop is exposed. The location of safety valves in a piping circuit must be such that its function cannot be deactivated by anything. For example, if a P/T valve were connected to a storage tank via a gate valve, closure of the gate valve would completely deactivate the safety relief valve for the tank. A safety valve can also be deactivated if it freezes. If the valve were to freeze shut, no protection would be afforded.

The sizing of safety valves is determined by the maximum possible rate of heat production in the subject loop. The design of safety valves is described in detail in Ref. 30.

During operation of pressure/temperature (P/T) valves, hot liquid or steam is released. It must be routed to a safe location. If toxic fluid is released from a collector loop, it should be retained in a holding tank for future loop recharge. Toxic fluids must not be dumped into city sewers.

The possible release of steam should be considered and the method of venting steam to the environment included in the design. Release of steam or hot liquid simply into the mechanical room is completely unacceptable, since injury and damage will result. (Note that P/T valves and pressure-relief valves are two completely different types of protective devices.)

Check Valves

Check valves are used to block reverse flow. Their selection is similar to that outlined above for gate valves. The size of the valve, its operating pressure, and its material must be specified. In addition, the seat, if not metallic, must be compatible with the highest temperature of the fluid. Two types of check valves can be used, one relying on a spring for closure and the other relying on gravity. The spring-type check valve will have a lower pressure drop but may be held open by small particles of dirt. Therefore, the check valve should be located downstream of a filter. Check valves relying on gravity to maintain closure should be located in a horizontal or vertical position, depending upon the design, so that gravity indeed does hold the valve closed. The exact position of the check valve should be specified carefully in the collector loop of solar systems since it is a critical component in these loops. In many nonsolar hydronic systems, check valves are held closed by fluids pressurized in the opposite flow direction by a pump. In solar systems, the reverse flow is not caused by a pump but by a relatively low pressure occurring from gravity flow. If absolutely certain backflow prevention is essential, a motorized ball valve with spring return can serve the purpose of a check valve.

In the solar collector loop the position of the check valve determines the drainability of the system. Figure 4.29 shows the proper location of a check valve to assure that both sides of the collector loop can be drained. If the system were not provided with a separate drain and a separate fill line on either side of the check valve, the riser to the collector loop from the collector pump could not be drained. The rule for locating check valves in this loop is to place them between the drain and the fill fitting as shown in the figure. Automatic fill systems are described below.

Codes require a backflow preventer or check valve to be present at city-water connections to avoid possible contamination of city water by water from the building.

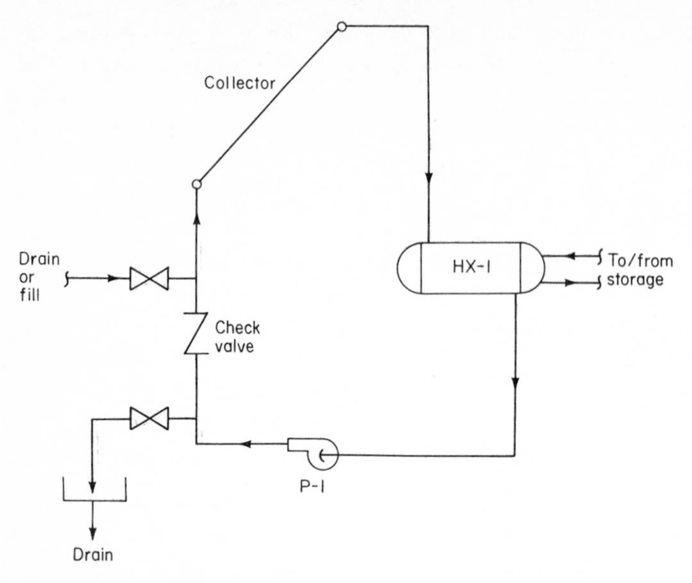

FIG. 4.29 Location of check valve in liquid circulation loop relative to drain and fill fittings.

Solenoid Valves

Solenoid valves are used to select modes in fluid loops controlled by the solar controller. For example, in Fig. 4.28 a three-way valve diverts flow to the load either from the backup heater or from the storage tank, depending upon controller decisions. Three-way valves are widely used in industrial and commercial applications, but the requirements of solar systems are somewhat more severe than in normal commercial situations. Problems arise in solar systems requiring a very tight closure of a three-way valve. Commercial-grade valves do not close completely, and some leakage can occur past the valve even when closed. In the circuit shown in Fig. 4.28 this causes no problem, and a commercial-grade three-way valve could be used. However, in systems relying on drain-down for freeze protection, leakage through the three-way valve (Fig. 2.15) could result in a collector freeze-up at night.

A type of three-way valve can be purchased which does not permit leakage. This is a butterfly design of the bubble-tight variety. These units are more expensive than ordinary commercial three-way valves but can be used in those circuits where a three-way valve is absolutely necessary.

Another method of avoiding problems which have occurred in many liquid solar systems with three-way valves is to use 2 two-port solenoid valves instead of 1 three-port valve. For example, in Fig. 4.28 the three-way mode-selector valve can be replaced with 2 valves as shown in Fig. 4.30.

The use of two-port valves to replace three-port valves introduces another possible difficulty which must be considered by the designer. In Fig. 4.30, if the sequence of operation of valves is not timed exactly, the pump P3 could have its outlet blocked momentarily if a mode switch is taking place between storage and backup. If V1 closes before V2 opens, all outlet streams to the pump are blocked and pump damage could occur. At the minimum, the pump circuit breaker will open and the services of maintenance personnel will be required. One method of avoiding this problem is to *mechanically link valves V1 and* V2. A single actuator is used, and whenever V1 closes V2 opens, not by means of an air signal to the diaphragm, but by a positive mechanical linkage. Therefore, if V2 should tend to move more slowly than V1, this mismatch is automatically compensated for by the mechanical linkage.

The use of solenoid valves can lead to a potential lack of long-term reliability. Therefore, the number of solenoid valves should be minimized by the substitution of check valves where possible. The use of solenoid valves is certain to involve increased maintenance cost as well as increased initial capital cost. Some solenoid valves, however, are unavoidable in most liquid systems.

DAMPERS FOR AIR-BASED HEATING SYSTEMS

The control of flows in air systems is accomplished not by valves but by dampers. Two types of dampers are commonly used—motorized dampers and backdraft dampers. Backdraft dampers are analogous to check valves in liquid systems and prevent flow from occurring through flow loops when not desired. The location of backdraft dampers is determined by the system flow logic. As described in Chap. 2, certain dampers in solar systems are critical to the proper function of the system. These dampers must be specified by the designer with extra care. Motorized dampers operate with 24-V actuators and are used to divert flow from one loop to another.

Two considerations determine the selection and design of motorized dampers—the leakage rate and the material. In conventional HVAC practice, a motorized damper is expected to leak from 5 to 30 percent of the airflow passing the damper. This leakage rate is completely unacceptable for solar systems, and higher-quality dampers are required. Even 5 percent leakage through a collector at night could

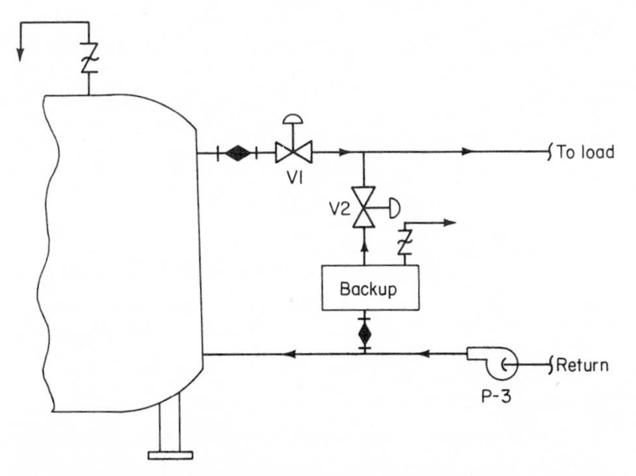

FIG. 4.30 Optional flow configuration replacing one 3-way valve with two 2-port valves. System function is the same as shown in Fig. 4.28.

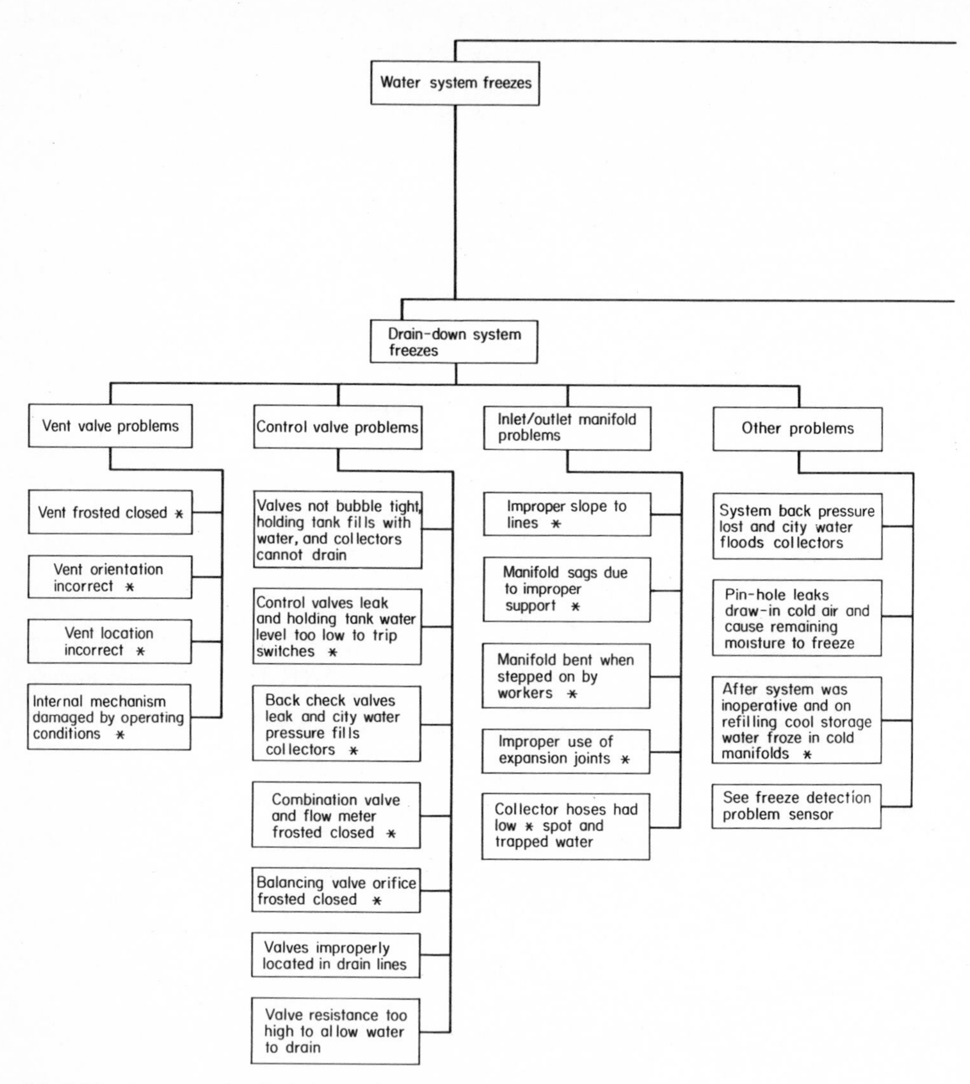

FIG. 4.31 Sources of solar system freeze-up. *(From Ref. 31.)*

freeze the domestic water-preheating coil. Butterfly valves with self-inflating seats limit leakage to approximately 1 percent. They should be used in critical locations in solar air heating systems.

Materials used in dampers are not critical for the most part; however, the seals and adhesives should be selected with care. Seals in air loops can be exposed to temperatures much higher than those present in standard HVAC systems. Seals may not function properly at elevated temperatures. The seal is the key to proper leakage control in air systems.

An example of a critical damper is shown in Fig. 4.27. The manual seasonal

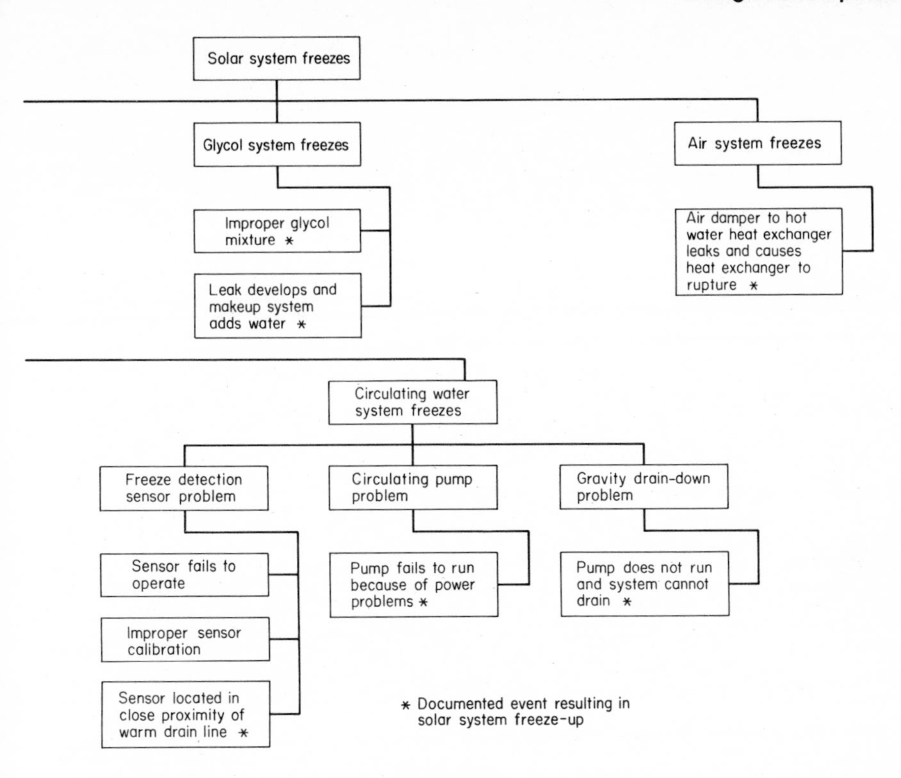

changeover damper in this flow configuration must seal tightly to avoid introducing very warm air into the heated space. In summer, the air in the collector loop, used only for water heating, could approach 200°F. The manual dampers must seal completely at this temperature. Any leakage would result in a significant cooling load within the space. The second critical damper in the air system is that which selects whether air is to flow through the collector or through storage for space heating at night. Any leakage past this damper will result in both deteriorated system performance caused by dilution of the hot airstream and possible freeze-up of the domestic water-heating coil in the collector outlet duct (see Chap. 2).

Dampers should have adjustable linkages, good-quality shafts, and shaft bearings. The failure of bearings can result in perpetual misadjustment problems. The use of poorly manufactured dampers in the two locations mentioned may result in minutely smaller initial cost with potential for significant system malfunction for many years.

Fire dampers are required in air system ducts at all fire-resistive separations such as walls, roofs, or fire partitions. These dampers require no special design for the solar application and are off-the-shelf items.

FREEZE PROTECTION

All space-heating systems used where a significant heat load exists will be exposed to freezing in winter. Therefore, a freeze-protection system that is 100 percent reliable must be used. In addition, solar water-heating systems used in locations where freezing is uncommon may still be exposed to infrequent freezing. Figure 4.31 shows graphically the sources of freezing which afflict the three types of active systems in common use—water, glycol, and air. This figure is based on field experience in many hundreds of solar heating systems in the 1970s. Four freeze-protection systems are shown—drain-down, circulating warm water, glycol, and air. Air systems do not of themselves freeze but leakage past the critical dampers can cause the hot-water heat exchanger to freeze.

Figure 4.31 vividly shows the potential for freezing in systems using water as the collector fluid, as opposed to those systems using glycol. For the glycol approach, two sources of freezing are identified in the figure. In water systems, 25 possible failure modes are identified. On the basis of this figure alone, it would seem imprudent to use water in the solar collector.

The problems with drain-down systems and fluid-circulation systems have previously been described in Chap. 2. Advocates of drain-down systems claim a slight advantage in warm-up time in the morning in addition to a saving in the cost of the heat exchanger. However, the amount of heat involved is small. The amount of fluid contained in a solar collector may be 0.02 to 0.03 gal/ft^2. Heating this fluid from an overnight temperature of 0°F to an operating temperature of 100°F requires approximately 20 Btu/ft^2. The total sunny day collection for a liquid system is in excess of 1000 Btu/(ft^2·day). Therefore, the amount of heat involved in warming the collector working fluid is only 2 percent of the total heat collected. In view of the improved reliability which can be achieved by glycol systems, the minute penalty is of little consequence. Drain-down systems have other problems in addition to those shown in Fig. 4.31, and the emphasis in this book is on glycol (or other organic fluid) freeze-protection systems because of their greater reliability. Therefore, the balance of this section deals with the design of glycol freeze-protection systems.

The freeze-protection mode which uses circulating warm water in winter cannot protect the collector in the case of a power failure. In addition, this approach rejects heat to the environment through the collector and can best be characterized as a simple, expedient method of freeze protection without much consideration for reliability or cost-effectiveness.

Two types of glycol makeup systems can be used—manual or automatic. One automatic system used in the past by some designers must not be used, however. This system uses an automatic fill valve connected to the city-water supply to maintain pressure within the collector loop. This expedient but inappropriate method is the cause for the second of the two glycol system failure modes shown in Fig. 4.31. If glycol were to leak from the collector loop, the fluid would be made up by city water. Eventually the freeze protection afforded by the glycol disappears as more and more water is added to a progressively more dilute glycol solution. The system ultimately freezes.

Figure 4.32 shows one example of both a manual fill system and an automatic fill system. In Fig. 4.32a, a glycol makeup tank is connected to the collector pump outlet. The normal flow of fluid is through the valve labeled NO. When glycol

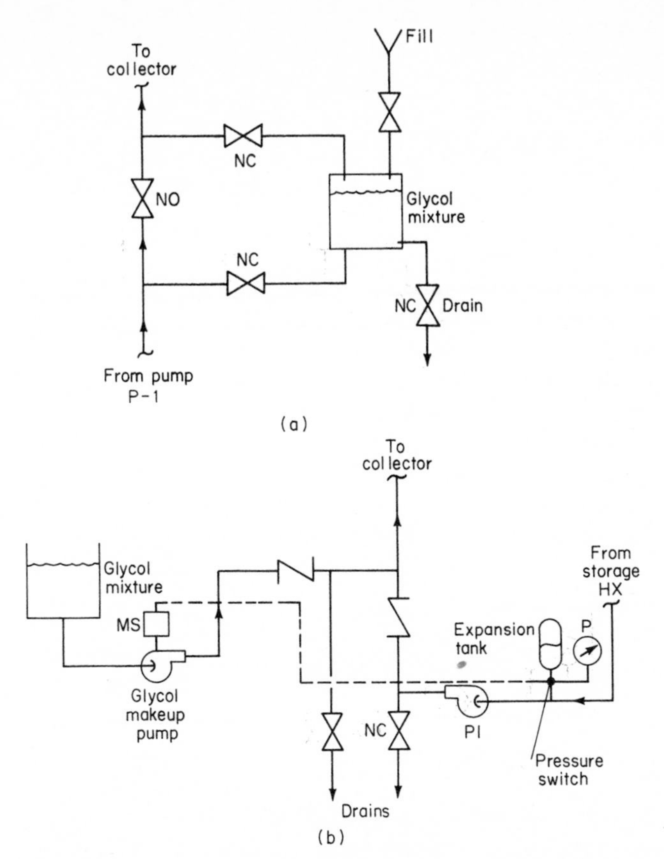

FIG. 4.32 (*a*) Manual glycol fill and makeup system for liquid heating systems; (*b*) automatic glycol fill subsystem using collector-loop pressure signal to operate glycol makeup pump. If system has a large leak, a fill pump timer or glycol tank float can activate an alarm.

makeup is required, the normally open valve is closed and the two normally closed (NC) valves are opened. Fluid is pumped from the collector loop through the glycol tank and back into the collector loop. This method can introduce any amount of glycol into the system. Additional glycol is added to the makeup tank by closing the two isolation valves and filling the tank from an atmospheric pressure drum as shown. The disadvantage to this system is that the glycol makeup tank must be capable of withstanding collector loop pressure. Pressurized tanks are costly.

Figure 4.32*b* shows an automatic glycol makeup system. The pressure within the collector loop is an indicator of the volume of fluid within the loop. When fluid leaks from the loop, the pressure at the expansion tank drops. The expansion tank is the pressure reference point for the entire system, and a pressure switch located at this tank can turn on the glycol makeup pump as shown. The pump will continue to add fluid to the collector loop until the pressure switch opens. The automatic system has the advantage of first, being automatic, requiring minimal owner input, and second, being refillable from an ambient-pressure makeup tank, which costs less than the pressurized tank used for the manual approach. Of course, the system shown in Fig. 4.32*b* can be used in a manual mode if the system operator monitors expansion tank pressure. When it drops, the makeup pump is turned on. An additional cost in the automatic system is the glycol makeup pump cost. This is a relatively low-flow-rate but high-head pump (e.g., a gear pump), since the static pressure to be overcome by the pump is equal to the system operating pressure, which may be 15 to 30 lb/in^2.

Other variations of glycol makeup and initial-fill subsystems can be used with equal effectiveness. However, *the expedient of using an automatic fill connected to city water main must be avoided.*

Occasionally it is necessary to freeze-protect entire solar systems located in

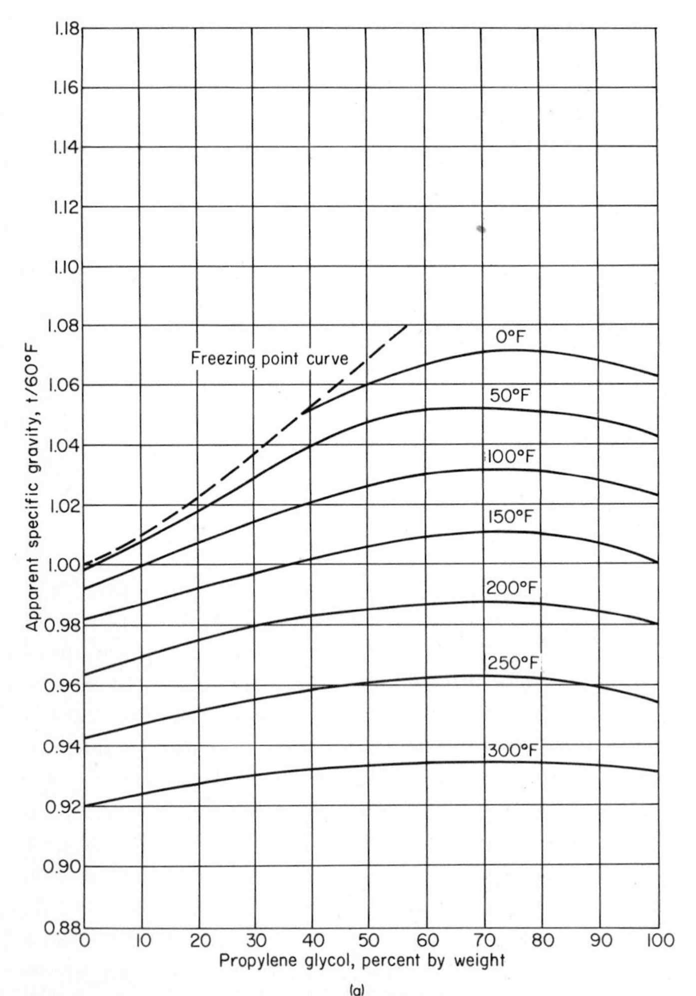

FIG. 4.33 Specific gravity of aqueous solutions of (*a*) propylene glycol and (*b*) ethylene glycol. *(From Ref. 32 with permission.)*

unheated spaces. A common situation is residential water heaters in unheated garages. First, insulate *all* components. Second, protect all components from drafts and cold air. Third, place a freeze-stat in a location such as the bottom connection of the collector-to-storage heat exchanger; this switch will operate collector and storage pumps briefly to raise pipe and collector temperatures above freezing. (This system, however, is not fail-safe during a power failure.)

SUBSIDIARY COMPONENTS OF ACTIVE SOLAR HEATING SYSTEMS

This section summarizes the design details which must be considered in the specification of subsidiary components of active systems. These components include expansion tanks, air-bleed fittings and vacuum breakers, heat-rejection units, auxiliary heating system, filters, instruments, and other small components.

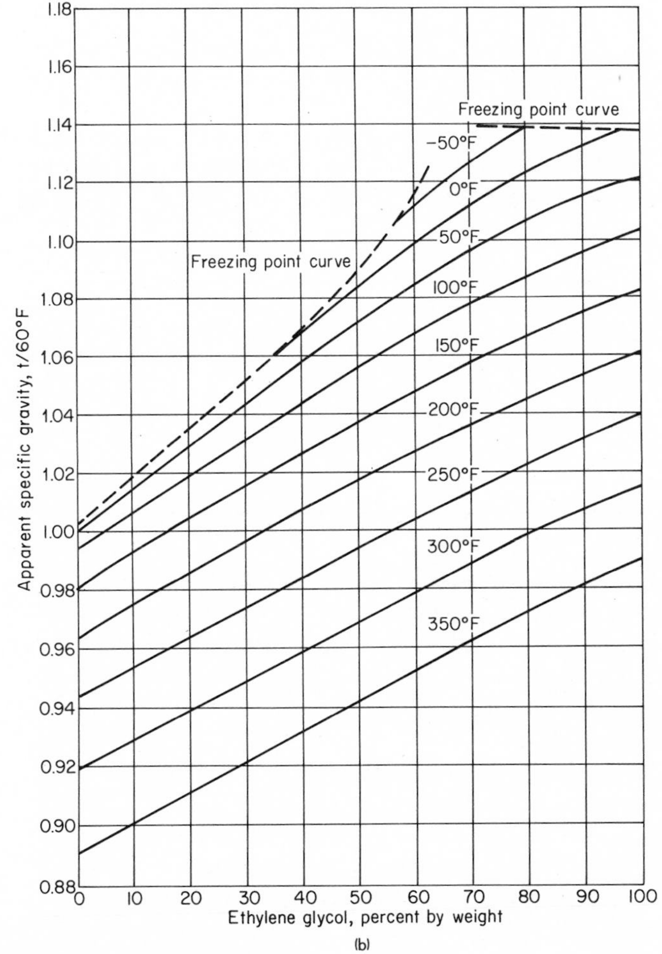

FIG. 4.33 *(Continued)*

Expansion Tanks

The volume change in closed fluid loops must be accommodated by an external expansion tank as described in Chap. 2. The sizing of this tank must account for the variation of density with temperature for the fluid used in the loop. Expansion tanks are required in the collector storage and delivery loops of liquid heating systems. The storage expansion tank can be sized from density-vs.-temperature characteristics for water, the normal storage medium. If liquid storage is part of the distribution loop, a single expansion tank can serve both loops. An example of this design is shown in Fig. 2.16.

The most common fluids used in liquid solar systems are glycols for freeze protection and water for heat delivery and storage. Figure 4.33 shows the specific gravity of solutions of ethylene and propylene glycol, the two most common glycols. Also included is the density of water as a function of temperature, shown along the left-hand vertical axes of the figure.

Example. What is the change in volume for a 50 percent ethylene glycol solution between 0°F and 350°F? Compare it to the corresponding change for water.

Solution. From Fig. 4.33*b* the following properties are read. The specific volume is inversely proportional to the specific gravity, and the change in volume can be calculated as shown.

Substance	*Temperature, °F*	*Specific gravity*	*1/specific gravity*	*Percent change in volume*
50% glycol	0	1.082	0.924	
	350	0.942	1.062	12.9
Water	32	1.002	0.998	
	350	0.891	1.122	12.2

The preceding example shows that water expands 12.2 percent over the prescribed temperature range, and a 50 percent solution of ethylene glycol would expand 12.9 percent. The use of empirical equations to size expansion tanks is not recommended for solar applications. Although Ref. 10 contains empirical curves and equations for expansion tank sizing, they do not apply to glycols nor do they apply to the wide temperature ranges which occur in solar heating systems.

Once the percent expansion and allowable pressure rise in the loop are known, the acceptance volume of the expansion tank can be calculated. The volume of the fluid present in the loop is first multiplied by the percent expansion to find the required acceptance volume of the expansion tank. For example, in the collector loop, the volume of the collector, the headers, the piping, the heat exchanger, and heat rejector unit are calculated and multiplied by the expected expansion. The resulting volume represents the *acceptance volume* of the expansion tank. The acceptance volume of the tank is usually one-third to one-quarter of the total volume of the tank. (Note that 100 percent acceptance can be achieved by using a separate air-pressure control circuit for air above the tank diaphragm. The loop pressure remains constant for all fluid temperatures in this approach.) To complete the specification of the expansion tank, the required operating pressure and the initial gas charge must be specified. The operating pressure of the collector loop is related to the elevation of the highest point in the collector loop relative to the inlet of the expansion tank. Once the pressure and expansion volume are known, expansion tank manufacturers may be consulted for selection of a model. The expansion tank should be of the prepressurized diaphragm type with an air separator and air float valve for purging air from the fluid loop. A pressure-relief valve set at the maximum system pressure should be attached to the expansion tank.

The expansion tank serves as the pressure reference point for the fluid loop in which it is located. Therefore, a pressure switch located at the expansion tank inlet can serve as a level alarm for the fluid loop.

Air Separators

Expansion tanks are frequently connected to liquid loops via an air separator. Air separators remove entrained air from the fluid and direct it to the expansion tank where it is removed from the circulating fluid. The expansion tank is connected to the pump inlet, since this represents the lowest pressure point. The pressure here is maintained above the required pump NPSH to avoid cavitation. This location is also ideal for the removal of entrained air.

Although closed expansion tanks are the most common in solar applications, open tanks can be used. In order to accommodate fluid expansion and still be connected to the pump inlet, these tanks must be located above the highest point in the fluid loop by several feet. The tank accommodates expansion and contraction of the fluid by a change of its level. In water loops, an open expansion tank can freeze if located in an unheated area. Therefore, the use of these tanks is restricted and uncommon in most liquid systems.

Air Bleeds and Vacuum Breakers

When a liquid system is filled, air must be removed from the system at the same rate liquid is added. Air removal is accomplished by positioning air bleed valves at high points in the system. These bleed valves are capable of passing air without passing an appreciable amount of liquid at the same time. They are operated either by a float or a wick. The float idea uses a small float located in the bleed valve tank. As air fills the cavity in which the float is located, the float drops, thereby opening the valve and permitting the air to escape. As liquid eventually enters the tank chamber, the float rises and the valve is held shut. The wick approach uses a valve operated by a wick material which changes volume as the amount of its water content changes.

Air bleed valves are also used after filling has been completed. As water is heated, air is forced from solution and is purged via an air bleed. In addition, when makeup fluid is later introduced into a fluid loop, it contains air which eventually comes out of solution and must be eliminated from the loop. Air bleed valves are required in the collector loop as well as in the energy-delivery-to-load loop. They should be positioned at all high points in the system as well as points of low velocity and high temperature where air entrainment may be difficult.

Automatic air valves have shown a tendency to leak when used with glycol solutions in collector loops. Therefore, the preferable design is an automatic air valve with a manual shutoff. Once air has been eliminated from the system, the air bleed valve can be isolated and its potential leak problem eliminated. Manual air bleed valves can also be used, but in large systems present a nearly insurmountable problem when system filling takes place. Many valves are needed in a large collector array. They all must be progressively closed and reopened as system filling takes place. The manual approach is acceptable, however, for small systems where only one or two bleed valves are required.

Air bleeds are frequently combined with vacuum breakers to permit entry of air into a system during fluid draining required in drain-down freeze-protection designs. In glycol systems, the draining of a collector occurs only when fluid is to be replaced or during a daytime power outage when fluid is drained from the collector to avoid chemical degradation. Vacuum breakers in collector loops must be located in nonfreezing locations if their function is to be assured throughout the entire heating season. This can be accomplished by insulating it or by placing it inside a building and running a small pipe to the topmost connection of the collector system.

Vacuum breakers are also needed in situations where a system can boil dry and then cool. If a system should boil most of the fluid away because of stagnation during the day, it will be filled only with vapor, which will condense overnight. During condensation, the pressure within the fluid loop will drop to almost a complete vacuum mitigated only by the vapor pressure (dependent on temperature) of the fluid trapped in the loop, since no air is present. Under vacuum conditions, some tanks may collapse, unless a vacuum breaker is present. In domestic hot-water loops, vacuum breakers are particularly important since no expansion tank is present in these systems. If they are heated to a maximum operating temperature and then the heat source is shut off for a long period, the drop in temperature is accompanied by a reduction in internal pressure. Atmospheric pressure can collapse hot-water tanks in low-pressure systems.

Heat-Rejection Units

Liquid heating systems require a method of rejecting excess heat collected during periods of relatively low load, which commonly occur during spring and fall when the solar collection level is high but the demand is relatively low. Several methods of heat rejection can be used, including a cooling tower, an air-to-liquid heat exchanger, or dumping of hot water from a solar-preheat tank. Rejection of heat through a solar collector at night uses the collector for a purpose it was not intended to meet. Collectors are heat traps and not heat rejecters.

The actuation of the heat-rejecter unit has been described in the preceding section on storage. The sizing of the heat-rejection unit is based on the peak energy collection rate. For example, if the allowable maximum temperature is 200°F in a collector, the collector efficiency curve can be used to calculate the efficiency at this operating point given the maximum solar radiation available during the heat-rejection season. The efficiency is multiplied by the peak insolation and the collector area to find heat collectable during peak heat-rejection conditions. The heat-rejection unit is sized to reject this amount of heat at a 200°F fluid inlet temperature. The heat-rejection specification includes the heat-rejection rate in British thermal units per hour, the incoming fluid temperature (200°F for this example), the fluid flow rate (the collector fluid flow rate), and the inlet air temperature and flow. The air temperature is selected corresponding to the warmest temperature anticipated during heat-rejection conditions. Figure 4.20 shows a typical heat-rejection-system installation in series with the collector.

Heat rejection from domestic hot-water heating systems can be done without an additional heat-rejection unit. Here warm water can be dumped from the top of the water-preheat tank as shown in Fig. 4.34. Cold water flows into the bottom of the tank. This cold water is immediately exposed to the collector fluid through the storage-to-collector heat exchanger and the temperature of the collector will drop. This method of heat rejection is clearly not appropriate in desert locations or where the water supply is limited. A dry heat-rejection unit described earlier would be used in these locations.

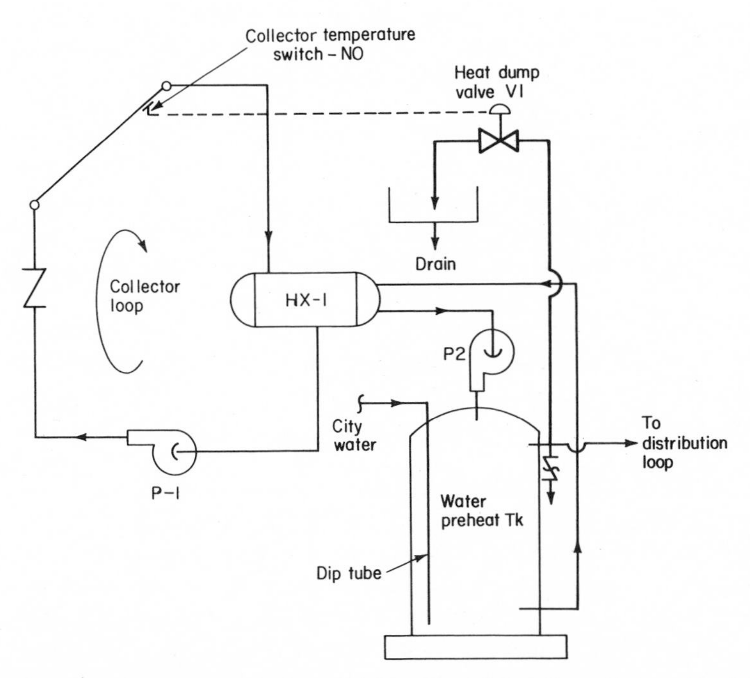

FIG. 4.34 Method of heat rejection from domestic water-preheat system.

Heat rejection from air systems is less of a problem than in liquid systems. It can be done by opening the inlet and outlet of the collector loop to the atmosphere and allowing a natural thermal circulation to take place. However, atmospheric dirt and dust can enter the collector. If this is a local problem, filtered air can be blown through the collector and exhausted to the environment.

Auxiliary Heat Sources

All solar systems considered in this book require an auxiliary system for economic reasons. As a result of the law of diminishing returns, very few solar systems will be economically optimal at 100 percent solar fraction. The backup is that system which would be present if no solar system were used to heat the building. The backup system is independent of the size of the solar system and is designed to carry the full system heat load (unless detailed building model studies show that the solar system reduces the peak load).

In many solar-heated buildings, the space-heating load has been reduced based on an economic analysis of energy conservation. In some residences, it has been found that the required heating rate for a furnace is smaller than the smallest available commercial unit. In these cases (superinsulated houses) a domestic hot-water heater has been used for the backup. Oversized systems cycle frequently and have increased transient loads, stack losses, standby losses, and reduced efficiency. In the past, oversizing did not have significant economic penalties since the cost of fuels was low; however, this is no longer the case.

The use of the same terminal load device for both backup and solar energy sources must be considered by the designer. The effectiveness of a solar terminal load device, when viewed as a heat exchanger, will be greater than that usually used for a nonsolar heat source since the nonsolar system operates a higher temperature. If the backup source operates at 200°F, for example, and the solar system has ceased operation at 100°F, the temperature of air leaving the terminal load device will rise rapidly to a high level when the solar-to-backup mode shift takes place. This can cause discomfort in the heated space and is best avoided.

Ambient reset of water temperature avoids temperature overshoot. This control feature modulates backup system output temperature as a function of ambient temperature. The colder the ambient temperature, the hotter the delivery temperature. Ambient reset is a standard feature in large commercial systems and can be incorporated into the boiler controls of residential systems as well without much additional cost.

Filters

Filters are required in both air and liquid systems to reduce dirt accumulation in solar-collector passageways. In liquid systems, the accumulation of dirt can result in flow blockage in liquid risers in the collector, overheating of the collector, and lower efficiency. The location of the filter in a liquid system should be upstream of the collector and the check valve. During start-up of a liquid system, the piping may contain substantial amounts of refuse, including pipe sealant and metal chips.

These are best removed from the system by means of a strainer which can be repeatedly emptied as the system cleans itself during start-up. Once the system is completely clean, a strainer is too coarse for proper filtration and a filter can be used. A typical size for filters used in liquid systems is approximately 150 μm.

The filters used in air systems are similar to those used in nonsolar systems. The filters keep both the rock bed and the solar-collector loop clean. The filter in air systems must be located so that air flows only in one direction through the filter. As shown in the air systems diagram in Chap. 2, air can flow in some solar system ducts in either direction, depending upon the mode. Filters should not be placed in these ducts.

Instrumentation

Instrumentation is required to monitor the proper operation of a solar heating system. If performance measurements are to be made, the instrumentation must be very complete and must be interfaced with a computerized data logger so that meaningful long-term results can be collected. For the purpose of assuring proper function of a system as opposed to monitoring its performance, relatively few instruments are required. The performance of pumps can be determined by measuring the pressure at the inlet and outlet. The pressure rise across the pump will be related directly to flow. A check with the pump curve determines whether the proper flow rate is present. The performance of heat exchangers and solar collectors can be determined by measuring inlet and outlet fluid-stream temperatures. The flow rate known from the pump performance curve multiplied by the temperature rise and the heat capacity will give the energy delivery of the system. This net energy delivery per square foot, divided by the local solar radiation, can be checked against the manufacturer's efficiency curve at the same conditions.

Operational modes can be indicated, as described in the control section above, by indicator lamps. A pressure gauge should be located at the expansion tank to assure that the system is operating at its design pressure.

If a direct measurement of flow rate is required, as opposed to the indirect measurement described above, flowmeters can be installed in long, straight pipe sections. A set of flow rates, pressure drops, and temperatures should be recorded when a system is new. These numbers can be used for comparison in future years to determine whether any degradation in performance of the collector or any plugging of fluid conduits has occurred. During initial system start-up, it is useful to temporarily install an hour meter across the pumps. In winter, a solar-collector-system pump such as P1 in Fig. 4.20 should operate between 6 and 8 h on a sunny day. Over a month, a typical operation of 150 to 200 h would be expected, depending on sunlight levels. If the pump hour meter gives a reading substantially different from this amount, it indicates a system malfunction.

On large systems, the installation of many thermometers and pressure gauges for each fluid stream may be prohibitively expensive. An alternative to permanent installation of these units is the substitution of pressure-temperature wells on all pipes or ducts of interest. These are small rubber diaphragms through which a pressure or temperature sensor can be temporarily inserted without any fluid leakage. The performance of a system can be checked by repeatedly reinserting the same instrument in several locations to measure either pressure or temperature.

In summary, pressure and temperature instrumentation should be inserted at each point in a fluid loop where either a pressure rise or heat addition takes place. Included are heat exchangers, collectors, storage tanks, terminal load devices, pumps, and heat-rejection loops.

Gaskets and Seals

If improper materials are used for seals and gaskets, they will have a very short lifetime since the operating conditions in solar collector loops are quite severe, especially if organic fluids at high temperatures are used. An excellent summary of the proper materials to be used for gaskets and seals in liquid-based solar systems is provided in Ref. 33. Most solar component manufacturers are aware of these restrictions, and the designer need only check these details.

DESIGN DEVELOPMENT CONSIDERATIONS FOR PASSIVE HEATING SYSTEMS

Most of the material in this chapter on design development has dealt with active systems for space and water heating. The reason for this is twofold. First, field experience with active systems far outweighs that for passive systems. For example, during the solar demonstrations sponsored by the U.S. Department of Housing and Urban Development in the mid- and late 1970s, hundreds of active systems and only a very few passive systems were built. It was only in the late 1970s and early 1980s that passive systems were explicitly funded for federal demonstration. Performance data from these systems operating over several years are not yet available. A second reason for the emphasis on active systems in the previous part of this chapter is that the design effort for active systems is hardware-oriented. That is, the specification of the system requires the specification of the performance of many components, each of which is used in a slightly different fashion than in a conventional heating system.

Passive systems are not hardware-oriented. Therefore, the specification of these systems is less lengthy than for active types. However, the design of passive systems is no simpler than for active systems. The added complexity comes not from hardware requirements and detailed control strategy design but rather from the proper interaction of passive-heating-system dynamics with building dynamics. This is an area in which much is yet to be learned and few design guidelines currently exist. In this section, the DD-level information required to specify systems of the TSW or direct-gain type as well as hybrid system is described. Since field experience on passive systems is so minimal, the coverage of passive design cannot be as exhaustive as the preceding coverage for active design.

Passive System Apertures

Passive system apertures are usually vertical in orientation, south-facing, and composed of one or two glass panes. The selection of the glass to be used must consider the loading on the glass, solar transmittance, and thermal properties. It is desired, within the constraints of cost, to have the highest transmission and the lowest absorption possible. This can be accomplished theoretically by using low-iron glass, although this material is frequently not available in the large sheets useful for passive glazing.

A more important consideration in the selection of a glazing material is the structural strength. Large south-facing glass surfaces near ground level present tempting targets for vandals; tempered glass is therefore to be specified. Tempered glass is also essential for the inner surface of double-glazed passive apertures. This glass in summer can reach temperatures in the vicinity of 200°F, and untempered glass is certain to break eventually from thermal stresses which occur at shadow lines. The designer should ascertain in addition that the edges of all glass panes have been smoothly polished prior to tempering. Any nicks can cause a stress concentration and fracturing of the glass. This is a particular problem during part-load periods of the year when the glazing will be partially in sun and partially in the shadow of the glazing overhang. The temperature difference between the two parts of the same sheet of glass can be considerable and any imperfection in the glass will tend to promote fracture.

It is essential that glazing be properly sealed to the building to avoid infiltration losses which can destroy the proper function of a thermal circulation loop. Seals must be selected with somewhat more care than conventional fenestration. Seals will be subjected to substantial levels of ultraviolet radiation with concurrent elevated temperatures. These two conditions, when combined, result in relatively rapid chemical reaction rates in the seals, and their deterioration can be rapid unless proper inhibitors are present. If operable vents for summer heat rejection are provided, these too must be sealed to withstand wide ranges of operating temperature during the course of a year.

Adjacent glazing panels are frequently overlaid with a common cap strip for appearance and sealing purposes. As with any other component of the glazing subsystem, these strips will be subject to significant expansion and contraction during the course of a year. Any holes, for example, used for mounting these strips to the passive aperture structure must be relatively large compared with the screw diameter to accommodate motion which is certain to occur.

The location of windows is another feature of passive-aperture design. For daylighting purposes, it is desirable to have windows located as high in the space as possible. However, from a heating standpoint it is desirable to have windows lower in the space so that stratification is minimized and lower as well as upper parts of the space are equally heated. In addition, the relative magnitude of direct-gain aperture vs. TSW aperture should be considered by the designer. These matters have not been quantified in sufficient detail to provide regional guidelines on the relative percentages of each. Experience in a few limited locations has indicated that approximately one-quarter to one-third of the total passive aperture should be direct gain to provide a quick warm-up in the morning. The balance of the passive aperture used for heating should be backed with storage elements to insert a phase lag between the capture of heat and its release to the space.

The design of daylighting systems is an ancient art which has been lost in recent times. The rediscovery of principles of daylighting and the size of aperture required to achieve a given footcandle level is presently being reestablished, although guidelines useful at the DD level do not presently exist. Reference 36 is an introduction to the use of natural daylighting in buildings.

The recommendation of Lam (36) is to prepare a scale model using proper materials with appropriate absorptance and reflectance values, and to carry out daylighting design and glare analysis in this fashion. The use of models is a typical architectural exercise but the use of daylighting models will require more care in the construction of internal spaces. A model is useful since it can be oriented in any direction. A set of daylighting readings taken inside the model at critical locations by a small photometer will indicate expected daylight levels in the final building.

Passive Storage Elements

The general characteristics of passive storage have been described earlier in this chapter in the section on thermal storage for passive and active systems (pages 179 to 182). In this section, specific details of the design of passive storage walls are presented.

Passive storage design determines the rate of release of solar heat to a space. The thickness of the wall, as described earlier, is determined more by comfort considerations than strictly by technical criteria such as the annual solar fraction $\bar{f}_s$. Since a TSW is a source of heat, it must be insulated from the environment. Insulation from the exterior during the day is achieved over the large aperture area by using double-glazed windows. The insulation at the bottom of the wall is more difficult but equally important. Since the mass of storage walls is very large, it is not possible in most cases to interpose a layer of structural insulation between the wall and the foundation. Heat loss control is best accomplished by adding insulation to the perimeter of the floor slab as well as underneath the slab in the vicinity of the TSW. A complete thermal break is rarely possible, but all sources of conduction from the wall to the environment should be identified and minimized by design during the DD phase.

Thermal storage for direct-gain systems is best distributed around the space to be heated to expose the maximum storage area possible (37). This storage can be thinner than that used in a directly illuminated TSW, but no thinner than 4 in. Performance of direct-gain systems increases as storage mass is added, the limit being the available surface of all walls in a direct-gain space. This is contrary to the performance of a TSW system, where an optimum amount of mass storage exists of the order of 30 to 40 Btu/(ft^2 · °F). For larger masses, the performance of a TSW system begins to fall off, whereas for a direct-gain system it continues to improve. Although an upper limit to the amount of storage for a direct-gain system does not exist from a technical point of view, economic limitations and the sheer physical size of direct-gain storage do place an upper limit on the storage mass.

The optimal amount of storage for most passive systems corresponds to diurnal storage—that is, storage adequate to carry the building through a night following a sunny day. Storage amounts capable of carrying the building through more than two or three sunless days are frequently uneconomical. This consideration is the basis of the rules suggested above for the size of storage.

Thermal-storage walls incorporate thermal circulation vents in most parts of the United States. The sizing of these vents is determined by the solar load fraction as described in Chap. 3. For small solar fractions, the vents are relatively large. Figure 4.35 shows the recommended vent size vs. solar heating fraction (37). The optimum solar heating fraction to be used depends upon aperture area and cost as described in Chap. 3. Once the cost-optimal solar fraction has been identified, the proper vent area can be read from the curve. Note that Fig. 4.35 applies for *one-story walls*; taller walls require more airflow and bigger vents to prevent excessive air overheating. No quantitative information is available.

Thermal circulation vents require backdraft dampers. Backdraft dampers are best placed at the inner surface of the upper vent as shown in Fig. 4.36. The upper vent is blocked by the damper when any tendency to backflow occurs. The same effect could be achieved by a backdraft damper placed at the outer surface of the lower vent, but repair of this damper would be impossible, and therefore the upper location is preferred.

The inlet to the lower vent should be located several inches above the floor to make certain that no obstructions are placed in front of it. In addition, a coarse

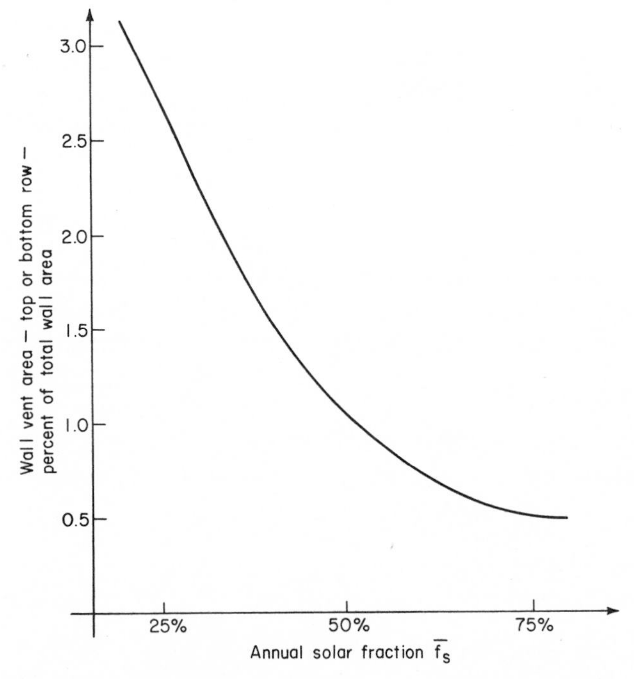

FIG. 4.35 Thermal-storage-wall vent size as it relates to annual solar heating fraction.

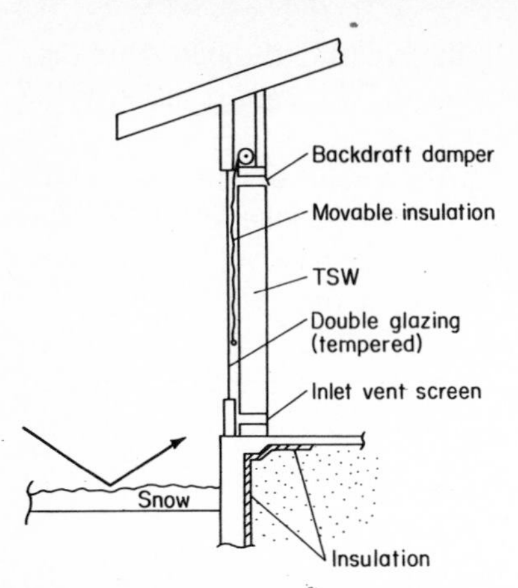

FIG. 4.36 TSW cross section showing vents with backdraft damper and inlet screen.

screen covers the opening. This prevents anything from being placed in the duct or in the wall cavity. A screen with ¼-in mesh is adequate for this purpose.

The potential for fire should be considered. In the design of the wall cavity and heat circulation path, the designer tries to optimize the free movement of air within the structure so that passive heat can contribute as much as possible to the heating load. However, passive airflow paths are excellent paths for fire propagation. Flames can move with astonishing speed through the same path as airflow for passive system. Spontaneous combustion can be a potential problem in passive solar buildings, especially in direct-gain spaces where exposed wood can become very dry. In addition, the insulation placed over the passive aperture should be selected with regard to fire prevention.

Another potential fire hazard occurs in the wall cavity in summer. If movable insulation is relied upon to serve as the only barrier for sunlight in summer, this insulation in the lowered position can become very hot since no thermal circulation will occur. The possibility of spontaneous combustion must be considered in this context as well. Use of an overhang or shutter for solar control can circumvent this particular difficulty.

Insulation for Passive Heating Systems

The optimum performance of a passive heating system requires insulation to retain solar heat in the building during nighttime and overcast cold periods. Without insulation, the performance of TSW and direct-gain systems is severely diminished and in many cases will be uneconomical, since large uninsulated glazing areas permit large heat losses to occur. In one of the early passive demonstration homes built in the United States in the mid-1970s, insulation was not used for cost reasons. As a result, in winter, the thermal-storage wall dropped to 20 or 30°F below the room temperature during long cold periods and was a source of great discomfort to the occupants. A surface at 40°F located in a room at 70°F acts as a large heat sink.

Several types of movable insulation have been developed. The insulation for a TSW passive system must be capable of being raised in the wall cavity. This can be accomplished by using a roll-up insulation of the type shown in Figs. 4.36 and 4.37 or some alteration in this design, or by using a system composed of particles of styrofoam which are blown into and out of the cavity between the two double-glazing panes. This approach has seen some difficulties because of the static charge which accumulates on styrofoam beads. Some beads remain stuck to the glazing as a result. For appearance purposes, many designers have rejected the use of beads. Movable insulation is currently available from only a very few manufacturers. It is assumed that other manufacturers will market new products in the next few years as a result of the federal incentive for passive-component design initiated in 1980.

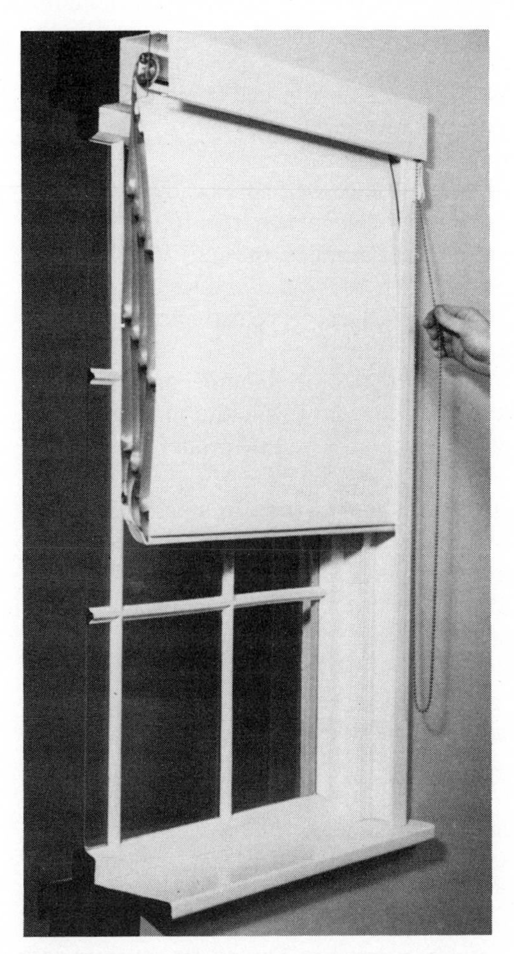

FIG. 4.37 Example of movable passive insulation. *(Courtesy of Insulating Shade Co.)*

A nominal value for insulation for passive TSW applications is approximately R-9. Other R values can be used, and the effect on solar heating fraction based on the R-9 level can be calculated from Eq. (4.15), which is from Ref. 37.

$$\frac{\bar{f}_{S,R} - \bar{f}_{S,R=0}}{\bar{f}_{S,R=9} - \bar{f}_{S,R=0}} = \frac{1 + R_0/9}{1 + R_0/R} \tag{4.15}$$

where $\bar{f}_{S,R}$ = the annual solar fraction with insulation level R
R_0 = the wall R value for no insulation; equal to 3.2 for water wall, 4.6 for TSW
R = wall resistance value other than R-9

Passive apertures can be insulated outside the wall glazing instead of inside the glazing. Although a less common approach than the internal glazing method, this approach is technically feasible. One of the practical difficulties involved is the moving of insulation after a snowfall has accumulated on it. If the building owner or manager is not present to continuously reposition insulation in response to changing weather, less than optimal control of heat flows will occur. The long-term durability of exterior insulations is also not known at this time.

Servicing of movable insulation in thermal-storage-wall applications presents a potential difficulty. The designer should provide adequate access to the thermal-storage-wall retraction mechanism and motor. For example, in Fig. 4.36 an opening above the upper TSW vent could be provided for removal of the insulation roller. Since the insulation is the full width of the TSW, the access panel must also be the full width and not be interrupted by structural supports.

Some systems have been designed using insulation on the inner surface of the thermal-storage wall in summer to prevent excessive heat gain into the space. This feature is not necessary since heat control can be achieved by proper use of movable insulation and overhangs.

Since movable insulation represents the only moving part in a passive system, alternative, more reliable methods for heat-loss control have been studied. Many owners reject movable insulation out of hand owing to its perceived poor reliability and human input for operation. An alternate concept uses a selective surface on the TSW. Computer studies have shown that heat losses from a selective wall surface are only slightly greater than from an R-9–insulated wall surface (57). Although promising from a technical point of view, few buildings have used this approach to date. If used, it is essential that a durable thermal bond between the wall and the selectively plated metal exist at all points. This is absolutely necessary to ensure proper transfer to storage.

Another method of reducing wall heat loss uses a "heat mirror" as part of the passive glazing system. A heat mirror is transparent to sunlight and cannot be seen by the naked eye. It is a reflector in the far infrared radiation regime, however, and does not permit low-temperature heat radiating from the TSW to be absorbed and reradiated by the glass. Heat mirrors have been used in Europe for several years and are beginning to become available in the United States.

Heat Rejection

The rejection of heat during low-heat-load periods of the year is as much a problem with passive systems as it is with active systems. The principal methods of heat control for vertical surfaces are shutters, louvers, and overhangs as shown in Fig. 4.38. Overhangs when properly sized can provide a complete cutoff of direct sunlight to a TSW. The size of overhang is determined by the period for which shad-

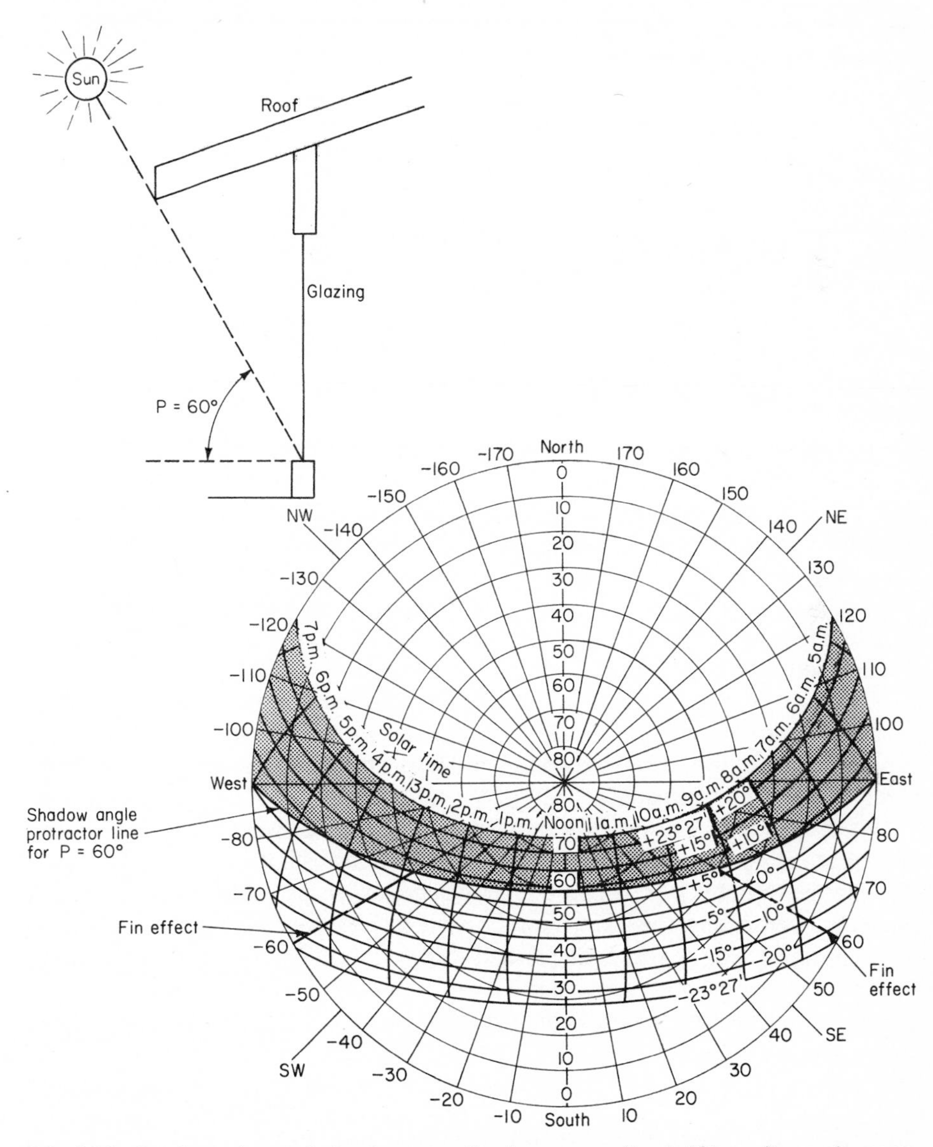

FIG. 4.38 Roof overhang design by sunpath-diagram method; 60° profile angle corresponds to heavy line; 60° azimuth fins add triangular zones shown by dashed lines.

ing is desired. The sunpath-diagram approach described in Chap. 3 can be used for quick design of these systems. For example, when the image of the overhang shown in Fig. 4.38 is superimposed on a sunpath diagram for 40° north latitude, it is clear that shading will occur from mid-April through the end of August. If the profile angle *P* shown (60°) results in either too long or too short a solar blockage period, the overhang can be adjusted accordingly. The sunpath diagram shows that the cutoff of sunlight does not completely occur on a given day but progressively during approximately half a month. The calculation of the shading effects on a month-by-month basis is done using the method described in the last section of Chap. 2. On one building, several overhang cutoff angles can be used to closely match solar gains to loads.

Passive heating systems must include overhangs. Without them, even though the solar incidence angles are fairly large, the amount of summer heat gain can be substantial. In one early residence built without an overhang, the thermal-storage-wall temperature was 85 to 90°F in summer, with resulting poor comfort levels within the space. On another project with a properly designed overhang and movable insulation, the interior space was consistently 15° below the outdoor ambient temperature without air conditioning.

The shadow map shown in Fig. 4.38 is for a very long overhang. If the overhang is of finite length, the extreme right- and left-hand edges of the shadow map are eliminated. In order to restore shading in early morning and late afternoon, vertical fins are useful. The construction of the fin effect on a sunpath diagram is quite simple; the terminator for the shadow map simply becomes a radial line corresponding to the azimuth angle of the fin as observed in a plan view. For example, if the overhang shown in Fig. 4.38 were equipped with vertical fins having a 60° azimuth angle cutoff east and west of due south, the shadow map would look exactly like that in Fig. 4.38 except for added radial lines corresponding to 60° and −60°. These are shown by the dashed lines. It is seen that fins reduce winter gains through south-facing apertures, but the cutoff is only in early morning and late afternoon when solar intensity levels are low. The improvement in heat rejection is usually more valuable, particularly in regions with significant cooling requirements. However, the cost of overhangs and fins is not a small item and sizing must also consider their ultimate cost.

Overheat control can also be a problem in greenhouses, where the same method of heat rejection has been used by many generations of horticulturalists, namely, opening the top of the greenhouse to let heated air escape. Greenhouses are more prone to overheating than are Trombe wall systems since they usually contain some upward-facing glazing components. Since the sun angle is high in summer, these glazing components capture significant amounts of solar radiation in summer when it is not needed. Any greenhouse must therefore include a method of very effective ventilation.

Reflectors and Draperies for Use in Passive Systems

The performance of a passive heating system can be significantly enhanced by placing a horizontal reflector immediately to the south of the aperture. The reflection effect can cause a 30 to 40 percent increase in the amount of solar radiation striking the passive collector. Practical difficulties occur with the use of reflectors, the most notable of which is the accumulation of snow and dirt. The reflection effect of clear snow can be nearly as significant as that for a reflecting surface such as weathered aluminum.

Direct-gain solar systems place severe requirements on fabrics and materials used in the space to be heated. These requirements are not present in TSW systems; this is one of the advantages of TSW systems. The method of selection of sun-resistant textiles is described in Ref. 38, summarized herein. The criteria which must be considered in the selection of fabrics exposed to sunlight—fabrics used on furniture or floors or in draperies—are as follows: (1) the possible reaction to sunlight including the fading of dyes and the deterioration of the fabric itself; (2) the effects of long exposure to heat, which can modify the reflectance of materials as well as thermal properties; (3) the reaction of fabrics to humidity, particularly in regard to dimensional changes.

All fibers are subject to a greater or lesser reaction to sunlight. This rate of deterioration varies according to fiber content, yarn and fabric construction, the finish applied to the textile, and the type of dyeing and printing process used. Some fabrics exposed to sunlight can deteriorate both in appearance and in physical strength. In extreme cases of sunlight exposure, fabrics can disintegrate and be destroyed during ordinary cleaning processes. The selection of fabrics is normally within purview of the interior decorator; however, the inputs of the solar designer must be considered regarding color and reflectance. The durability of fabrics in the presence of sunlight can be increased by the use of ultraviolet inhibitors and screens; artificial fabrics are less prone to degradation than are natural fabrics. In addition, fabrics designed for uses other than interiors are particularly prone to deterioration in heat and sunlight.

The thermal properties of carpets are important in the design of direct-gain systems. If the carpet acts as an insulator, the coupling between the solar heat and any storage mass below will be nonexistent. The insulative characteristics of a textile arrive from the dead air trapped in it. The amount of trapped air depends in turn on the fiber diameter and shape, the fiber length, the amount of twist, and the fabric construction. Fabrics with maximum insulative characteristics use crimped, irregularly shaped fibers with low-twist yarns made from crimped filament. Pile surfaces or open weaves along with expanded foam mats increase the insulative quality.

Recent developments in drapery materials permit them to function thermally. The reflectance of fabrics used on draperies can be significant, and in some cases an aluminized backing is available for summer solar gain control. Other designs include multilayer reflective insulation materials. The thermal durability of these materials is good and they can be used for movable insulation in direct-gain systems.

Reference 38 concludes with the following recommendations. Window treatments can use acrylics, fiber glass, nylon, and polyester fabrics. Upholstery exposed to sunlight can best use acrylics, cotton, and leather as well as linen, nylon, polyester, silk, vinyl, and wool. Leather and vinyl have negligible moisture absorbency,

TABLE 4.4 Use conditions for metals in contact with aqueous heat-transfer fluids in closed solar heating systems*

Generally unacceptable use conditions	*Generally acceptable use conditions*
	Aluminum
When in direct contact with untreated tap water with pH <5 or >9.	When in direct contact with distilled or deionized water which contains appropriate corrosion inhibitors.
When in direct contact with liquid containing copper, iron, or halide ions.	When in direct contact with stable anhydrous organic liquids.
When specified data regarding the behavior of a particular alloy are not available, the velocity of aqueous liquids shall not exceed 4 ft/s.	
	Copper
When in direct contact with an aqueous liquid having a velocity greater than 4 ft/s.	When in direct contact with untreated tap, distilled, or deionized water.
When in contact with chemicals that can form copper complexes such as ammonium compounds.	When in direct contact with stable anhydrous organic liquids.
	When in direct contact with aqueous liquids which do not form complexes with copper.
	Steel
When in direct contact with liquid having a velocity greater than 6 ft/s.	When in direct contact with untreated tap, distilled, or deionized water.
When in direct contact with untreated tap, distilled, or deionized water with pH <5 or >12.	When in direct contact with stable anhydrous organic liquids.
	When in direct contact with aqueous liquids of $5 < pH > 12$.
	Stainless steel
When the grade of stainless steel selected is not corrosion-resistant in the anticipated heat-transfer liquid.	When the grade of stainless steel selected is resistant to pitting, crevice corrosion, intergranular attack, and stress corrosion cracking in the anticipated use conditions.
When in direct contact with a liquid which is in contact with corrosive fluxes.	When in direct contact with stable anhydrous organic liquids.
	Galvanized steel
When in direct contact with water with pH <7 or >12.	When in contact with water of pH >7 but <12.
When in direct contact with an aqueous liquid with a temperature $>55°C$.	

TABLE 4.4 (*Continued*)

Brass and other copper alloys
Binary copper-zinc brass alloys (CDA 2XXX series) exhibit generally the same behavior as copper when exposed to the same conditions. However, the brass selected shall resist dezincification in the operating conditions anticipated. At zinc contents of 15 percent and greater, these alloys become increasingly susceptible to stress corrosion. Selection of brass with a zinc content below 15 percent is advised. There are a variety of other copper alloys available, notably copper-nickel alloys, which have been developed to provide improved corrosion performance in aqueous environments.

*From Ref. 33.

so comfort levels are reduced when compared with the other fabrics listed. Floor covering materials exposed to direct sunlight can include acrylic, nylon, polyester, and natural wool. Dyed wool is not recommended, since it has little resistance. The optimum properties of any fabric exposed to sunlight include minimal loss of tensile strength, good heat resistance, minimum moisture absorptance and dimensional change, and good resistance to air pollution. Carpet and fabric manufacturers have long been aware of the problems with textiles exposed to sunlight, and consultation with an informed vendor may be the best method of quickly acquiring this information.

OTHER SYSTEM DESIGN CONSIDERATIONS

The information presented in the preceding sections of this chapter covers most of the detailed component information required for active and passive system design. Of course, the general material to which these details refer was presented in Chaps. 2 and 3. In this remaining section, three additional systems considerations are described. The first is the selection of materials for use in liquid systems, the second is the effect of the law of diminishing returns on solar system performance, and the third is flow balancing in collector arrays.

In the selection of piping, storage, and collector materials, the relative electromotive activity of each must be considered. This has been described earlier. Additional considerations, however, are present in water-based systems because of special chemical interactions. A convenient summary of these data is contained in Table 4.4, in which the acceptable and unacceptable use conditions for five common metals used in closed liquid-based systems are given—aluminum, copper, steel, stainless steel, and galvanized steel.

Table 4.4 was presented in Ref. 33, the Minimum Properties Standards for solar heating systems. The most common materials used in liquid-based systems are copper and steel, which have few unacceptable use conditions in the table. Galvanized piping should not be used at temperatures above 140°F because of the reversal of the electromotive relationship between the galvanizing and the steel pipe above that temperature. Reference 33 contains a table for open systems similar to Table 4.4. Open systems are much less frequently used than closed systems because of the presence of reactive oxygen in the working fluid and the greatly enhanced possibility for corrosion. The selection of materials for piping in collector storage in delivery loops is more critical in open systems than in closed systems.

Diminishing Returns

The law of diminishing returns applies to solar systems but not to solar components. It is one of the key factors in solar system design since it determines economic viability, as described in the next section. The law of diminishing returns states that the output of a system increases at a decreasing rate as one component of the system is increased uniformly, all other components remaining unchanged. For example, if additional collector area is added to a solar heating system, the performance will not increase in a proportionate manner. Rather there will be an increase but it will be at a progressively decreasing rate.

The law of diminishing returns is shown schematically in Fig. 4.39 for an example system. This figure shows that a 100 percent increase in the size of one component of the system results in only a 33 percent increase in system output over the course of a year. A 100 percent increase in collector size, for example, causes a 100 percent increase in the cost of the collector array. This would not be repaid by a 100 percent increase in collector output. An economic analysis will determine the proper size of components subject to the law of diminishing returns.

Collector area, storage amount, collector fluid flow rate, heat-exchanger effectiveness, heat-exchanger flow rates, storage flow rate, load heat-exchanger effectiveness, load heat-exchanger flow rate, collector tilt, and even such details as the level of insulation on pipes in storage tanks are subject to this law. Although the

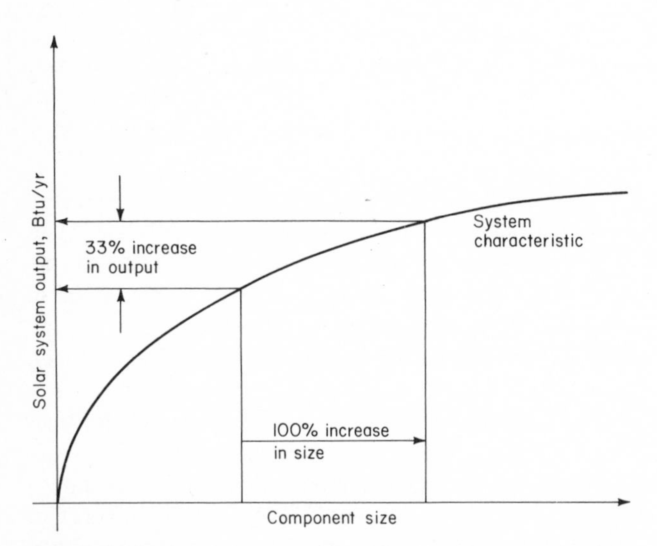

FIG. 4.39 Example law-of-diminishing-return characteristic curve for solar system showing diminished system output at progressively larger and larger component sizes.

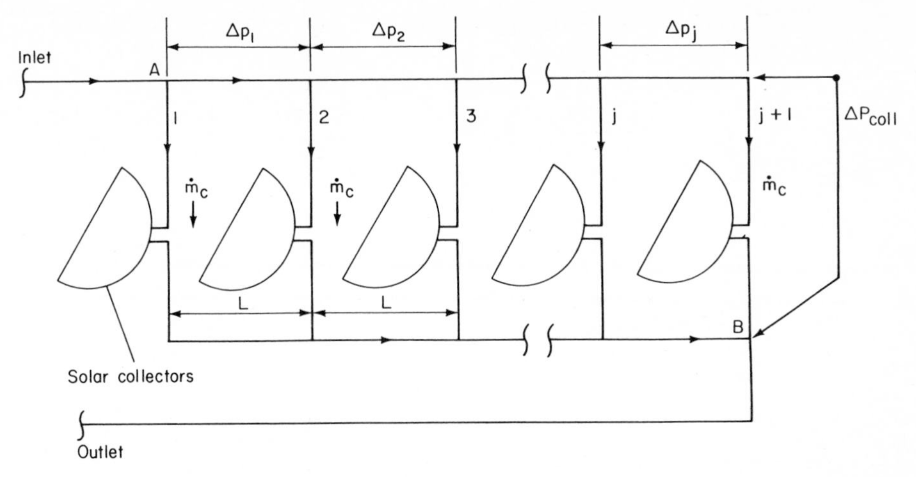

FIG. 4.40 Parallel flow arrangement for solar collector array showing collector spacing L and collector flow rate $\dot{m}_c$.

curve for each of these components is different when presented quantitatively, the qualitative shape is always the same as that shown in Fig. 4.39. For small amounts of the component, the increase in output is relatively large as extra amounts of the component are added. However, for relatively large sizes the increase in output becomes progressively smaller and smaller.

The law of diminishing returns has its basis in the physical characteristics of solar heating systems. In the case of collector area, for example, the law of diminishing returns prevails since the increase in performance of a progressively larger and larger collector cannot be used during progressively greater portions of the heating season. If a solar collector is designed to carry all the heating load in the worst month of the heating season, that collector will be oversized for the remainder of the year and heat will need to be discarded. Therefore, the amount of *useful* energy delivered by the collector per square foot will become smaller and smaller as additional collector area is added. Likewise, adding additional storage to a base line amount cannot increase energy delivery proportionately, since oversized storage would be used for only a few periods during the year, and its contribution to the year-long heating season output would therefore be relatively small. The same considerations apply to all the components listed in the paragraph above. In the next section, the impact of the law of diminishing return on system sizing is described in detail.

○ **Flow Balancing in Collector Arrays.** Pressure drop from the inlet of a collector array header to the array exit point cannot be calculated simply from friction factor and collector-pressure-drop information as is the case for simple piping runs. The pressure drop in the collector inlet header, for example, is determined both by frictional drops and by the effect of removal of fluid from the inlet header as shown in Fig. 4.40. The pressure drop in the inlet header of the figure is the total of the pressure drops in segments of the inlet header L units long. The total of ΔP_1, ΔP_2, ..., ΔP_j is the total pressure drop in the header. If the diameter of the header piping is uniform and the spacing between each collector connection into the header is also uniform, the pressure drop in the collector array can be calculated by a relatively simple analysis giving a simple, useful result. Equation (4.16) can be used to calculate the overall operating pressure drop (22).

$$\Delta P_{AB} = (K_f \dot{m}_c^2/6)\,[j(2j + 1)(j + 1)] + \Delta P_{coll} \tag{4.16}$$

where $j + 1$ = number of collectors connected in parallel
$\dot{m}_c$ = collector mass flow rate, lb/h
$K_f = 8\bar{f}L/(\pi^2 \rho D^5)$
ΔP_{coll} = collector pressure drop, assumed equal for all collectors in array; includes both the static and frictional pressure drop

The average friction factor $\bar{f}$ in the inlet or outlet header is evaluated at one-half the total array flow $(j + 1)\dot{m}_c/2$. The friction factor can be evaluated from Eq. (4.17).

$$f = 8\{(8/\text{Re})^{12} + (A + B)^{-3/2}\}^{1/12} \tag{4.17}$$

where $A = \{-2.456 \ln[(7/\text{Re})^{0.9} + 0.27\,\epsilon/D]\}^{16}$; ϵ = pipe roughness
$B = (37{,}530/\text{Re})^{16}$
$\text{Re} = \bar{u}D\rho/\mu$; D = pipe inside diameter
$\bar{u}$ = bulk velocity in the pipe

The requirement that the pressure drop through each collector be the same stipulated in the development of Eq. (4.16) is not automatically achieved. The so-called reverse-return method of piping shown in Fig. 4.40 is a first step toward accomplishing flow balancing in a set of parallel-connected collectors. The reverse-return idea is used to assure that the length of fluid paths through any branch of the collector array is the same from inlet A to outlet B. However, the change in pressure along the inlet and outlet headers imposes different pressure drops across each collector, thereby tending to cause different flows to each collector although the flow path lengths are the same.

In addition to reverse-return piping, an additional requirement must be met to assure equal flows through the collector. Flows can be automatically balanced to within a few percent by requiring that 90 percent of the overall collector-array pressure drop ΔP_{AB} occur across the collectors, the remaining 10 percent occurring in the header. This requirement will be met by using relatively large piping for the inlet and outlet headers and reducing the pressure drop in the headers as a result. Tubular collectors have high pressure drops, and arrays of them are easy to balance.

For large parallel-connected arrays, the pipe size may become rather large and its cost high. One method of reducing the cost of self-balancing headers to some extent is to use progressively smaller and smaller pipe as the flow rate decreases. For example, in the inlet header in Fig. 4.40, the header between collectors 1 and 2 would be relatively large, while the last element of the header between collector j and collector $j + 1$ would be relatively small. It is not possible to select the appropriate sequencing of pipe sizes in closed analytical form, and a numerical method such as the Hardy-Cross method can be used. A number of solar-collector manufacturers provide manifold sizing assistance as a service.

Another method of assuring proper flow through each collector in a large array is to insert a balancing valve in each branch. The valves can then be set by the balancing contractor to assure proper flow through the collectors. This method is not recommended, since the amount of labor and material cost is much higher than if a self-balancing approach using a properly sized header is used. Another effect of the balancing-valve method is to increase the horsepower requirement and consequently the operating cost and size of the collector pump. An additional reliability problem may occur since collector flows are small and the balancing valves may be nearly closed in some branches; dirt can easily plug nearly closed balancing valves.

The requirement that 90 percent of the pressure drop occur across the collectors in an array such as that shown in Fig. 4.40 is difficult to meet with those collectors which have a very small pressure drop. Some collectors originally designed for thermosiphon operation, where minimal pressure drop is essential, have such a low pressure drop that the 90 percent rule cannot be met with practical manifold pipe sizes. The method of solving this difficulty is to artificially introduce a collector pressure drop. This can be done by inserting in each collector branch an orifice providing a pressure drop which will permit flow balancing with headers of reasonable size. The pressure drop across such an orifice is typically 6 to 9 in of water gauge. Since flows through collectors are small, the orifice must also be small to produce this pressure drop. The introduction of orifices may eventually result in flow blocking unless filters are well maintained. The recommended approach uses a collector having the appropriate pressure drop by design.

The question of connecting collectors in series or parallel can be addressed by calculation of the required pressure drop and flow requirements for a pump in either approach. If a large number of collectors are connected in series, the pressure drop will be much higher than for the same number of collectors connected in parallel. Of course, the total flow rate in each case would be the same, since collector flows are specified on a flow rate per unit area basis. That is, 1000 ft^2 of liquid collectors would have a total flow rate of 20 to 25 gal/min whether connected in parallel or in series. The increased pressure drop in the series arrangement will cause higher initial pump cost and higher operating costs through the life cycle of the system. In addition, forcing the fluid for all collectors to pass through each collector causes relatively high velocities in the collector conduits. These velocities may be desirable from a heat-transfer standpoint, but in liquid collectors the likelihood of erosion is increased. The one advantage of a series connection is that the flow rate through each collector is automatically the proper flow rate, and flow balancing becomes moot. In addition, the need for piping headers is eliminated. A combination of series and parallel flow regimes is frequently used in large commercial arrays such as that shown in Fig. 2.18a.

SYSTEM SELECTION CRITERIA AND SIZING PROCEDURES

In the previous section, we discussed the effect of the law of diminishing returns on the physical output in energy units of any type of solar heating system. The consequence of this law for the designer is the requirement that a system be sized so that it is large enough to provide a significant amount of heating but not so large that it has an overwhelming cost relative to its output. In this section, we will address specifically how to calculate the economically optimum size of all components in solar heating systems.

The cost of solar energy is calculated as the ratio of the extra cost of the solar system to the owner per year divided by the useful energy delivered by that system per year as shown in Eq. (4.18). If the average cost of solar energy C_{SE} calculated from Eq. (4.18) is less than the cost of competing fuels, an economic incentive for the use of solar energy exists.

$$C_{SE} = \frac{\text{extra cost of system per year}}{\text{useful energy delivered per year}} \tag{4.18}$$

The Solar Energy Cost Equation

At the SD level, the only solar costs considered with the operation were the capital cost net of tax credits, the interest cost, the cost of backup energy, and the inflation in the cost of backup energy. During DD, a much more detailed cost analysis should be made for larger projects. For smaller projects, the effort involved in a detailed analysis is not worthwhile and SD economics (pages 137 to 154) is adequate.

The detailed DD cost analysis must include the following terms (the same equation is used for both active and passive systems):

Extra initial capital cost in solar system net of investment tax credit
Replacements of system components
Operating energy costs
Property taxes
Maintenance
Insurance
Interest charges

The negative costs include the salvage value and income tax deductions for interest payments. Profit from the sale of the system can also be considered.

Unfortunately, the inclusion of all costs to be considered during DD adds considerable complexity to the economic analysis. For commercial and revenue-producing buildings, this analysis is required. However, *in the residential context it is acceptable in many cases to use SD-level cost calculations with refined DD performance predictions*. This recommendation is based on the fact that the operating, maintenance, insurance, and replacement cash outflows are roughly offset by tax savings on interest, property taxes, and salvage value for residential systems. Therefore, for residences one may ignore all terms other than the initial cost of the solar system and the cost of backup fuel and its inflation.

Equation (4.19) represents the costs which are incurred by the owner of an active or passive heating system. Each term in the equation has been labeled, and the significance of the symbols in the equation is described in the list following the equation:

$$
\begin{aligned}
C_y = {} & [C_{s,tot} - \mathrm{ITC}]\mathrm{CRF}(i',N) \\
& \text{(Initial investment less tax credit and its interest)} \\
& - C_{s,salv}\mathrm{PWF}(i',N)\mathrm{CRF}(i',N)(1 - CT_{salv}) \\
& \text{(Salvage value)} \\
& + \left[\sum_{k=1}^{N} R_k\,\mathrm{PWF}(i',k)\right]\mathrm{CRF}(i',N)(1 - CT_{inc}) \\
& \text{(Replacements in years } k\text{)} \\
& + C_e\frac{\mathrm{CRF}(i',N)}{\mathrm{CRF}(i'',N)}(1 - CT_{inc}) \\
& \text{(Operating energy)} \\
& + T_{prop}C_{s,assess} \\
& \text{(Property tax)} \\
& + M(1 - CT_{inc}) \\
& \text{(Maintenance)} \\
& + I(1 - CT_{inc}) \\
& \text{(Insurance)} \\
& - T_{inc}T_{prop}C_{s,assess} \\
& \text{(Deduction for property tax)} \\
& - T_{inc}i_m\left[\sum_{k=1}^{N}\frac{P_k}{(1+i)^k}\right]\mathrm{CRF}(i',N) \\
& \text{(Deduction for interest)} \\
& - CT_{inc}\left[\sum_{k=1}^{N} D_k\,\mathrm{PWF}(i',k)\right]\mathrm{CRF}(i',N) \\
& \text{(Depreciation deduction)}
\end{aligned}
\tag{4.19}
$$

where $C_{s,tot}$ = total initial solar investment including the cost of all equipment installed, sales taxes, fees, and ancillary costs (the capital cost of solar components is described below)

$C_{s,salv}$ = solar system salvage value

$C_{s,assess}$ = assessed value of solar system

C_e = first-year energy cost to operate solar system

$\mathrm{CRF}(i',N)$ = capital recovery factor for N years at interest rate i'

$i' = (i - j)/(1 + j)$ = real discount rate

$i'' = (i - j_e)/(1 + j_e)$ = inflation-adjusted discount rate for energy

i = the interest or discount rate bounded below by the owner's rate of return foregone on the next-best alternative investment and bounded above by the cost of borrowing. The rate i is strictly the marginal discount rate, that is, the rate applied to the solar system owner's next set of investment decisions. It is frequently different from the average of discount rates on prior investments.

M = maintenance cost per year
j = general inflation rate (as measured by the consumer price index)
i_m = market mortgage rate (real mortgage rate plus general inflation rate)
j_e = total energy inflation rate per year
k = years at which replacement or repairs are done
I = insurance costs per year
R_k = replacements made in year k
$i_m P_k$ = interest charge in year k

P_k = principal in year k, given by $(C_{s,tot} - ITC)\left[(1 + i_m)^{k-1} + \frac{(1 + i_m)^{k-1} - 1}{(1 + i_m)^{-N} - 1}\right]$

$PWF(i',N)$ = present-worth factor for year N based on interest rate i'

T_{inc} = income tax rate, state tax rate + federal tax rate − state tax rate × federal tax rate, where the rates are based on the last dollar earned

[The summation term for interest deduction can be expressed in closed form as

$$i_m \sum \frac{P_k}{(1 + i)^k} = \left\{ \frac{CRF(i_m,N)}{CRF(i,N)} + \left(\frac{1}{1 + i_m} \right) \left[\frac{i_m - CRF(i_m,N)}{CRF\,[(i - i_m)/(1 + i_m),\, N]} \right] \right\} (C_{s,tot} - ITC)$$

If $i = i_m$,

$$i_m \sum \frac{P_k}{(1 + i)^k} = \left\{ 1 + \frac{N}{1 + i_m} [i_m - CRF(i_m,N)] \right\} (C_{s,tot} - ITC) \Big]$$

T_{salv} = tax rate applicable to salvage value
ITC = investment tax credit for initial solar system cost
C = commercial (1) or noncommercial (0) index; the commercial index applies if the solar system is used in an income-producing environment
D_k = depreciation amount in year k, depending upon type of depreciation schedule used

Equation (4.19) is based on cash flows in constant dollars, that is cash flows with the effect of general inflation factored out. Terms in the equation are expressed in dollars per year—annualized costs rather than total life-cycle costs. All cash flows are discounted by the *real* discount rate i'. The capital recovery factor (CRF) and present-worth factor (PWF) are tabulated in the appendix for various values of interest rate and life-cycle time N.

The tax deduction for mortgage interest depends upon the amount of interest paid in a given year. For a self-amortizing mortgage, the amount of interest in the annual mortgage payment becomes progressively smaller with time. Table 4.5 shows the interest fraction of the mortgage payment for each year of a 20-year mortgage for various interest rates between 7 and 12 percent. This table can be used to evaluate the ninth term in Eq. (4.19). The depreciation term applies only to income-producing owners. Depreciation schedules are discussed in Ref. 1 or in engineering economics textbooks.

The salvage value of a solar system enters solar cost consideration as a negative cash flow since the liquidation of this asset can result in a net benefit to the owner at the end of the life-cycle period or before. The resale is taxed at tax rate T_{salv}. The estimation of salvage value at the present time is difficult. To add a conservative bias to the calculations, the salvage value can be placed at zero, although a zero salvage value is unlikely for a well-maintained solar system. At the other extreme, salvage value could be equal to the assessed value of the solar system, denoting its possible resale value.

The cost of energy C_e required to operate the solar system is that consumed by motors for pumps and fans and by control operation. This is a relatively small cost usually amounting to a few cents per square foot per year. The exact amount can be calculated from the number of hours of operation of pumps, fans, and controls. This information is available only by measurement in the field or by use of detailed computer models with hourly calculation of system performance. Most solar space-heat collection systems operate between 2000 and 2500 h/yr in sunny climates. When this number of hours is multiplied by the cumulative horsepower requirements of all collection subsystem pumps and fans, the cost of operating energy

TABLE 4.5 Fraction of interest in mortgage payment in specified year

Years left on mortgage	*Annual mortgage interest rate, percent*										
	7	*7½*	*8*	*8½*	*9*	*9½*	*10*	*10½*	*11*	*11½*	*12*
20	0.742	0.765	0.785	0.804	0.822	0.837	0.851	0.864	0.876	0.887	0.896
19	0.723	0.747	0.768	0.788	0.806	0.822	0.836	0.850	0.862	0.874	0.884
18	0.704	0.728	0.750	0.770	0.788	0.805	0.820	0.834	0.847	0.859	0.870
17	0.683	0.708	0.730	0.750	0.769	0.786	0.802	0.817	0.830	0.843	0.854
16	0.661	0.686	0.708	0.729	0.748	0.766	0.782	0.798	0.812	0.825	0.837
15	0.638	0.662	0.685	0.706	0.725	0.744	0.761	0.776	0.791	0.805	0.817
14	0.612	0.637	0.660	0.681	0.701	0.719	0.737	0.753	0.768	0.782	0.795
13	0.585	0.609	0.632	0.654	0.674	0.693	0.710	0.727	0.742	0.757	0.771
12	0.556	0.580	0.603	0.624	0.644	0.663	0.681	0.698	0.714	0.729	0.743
11	0.525	0.549	0.571	0.592	0.612	0.631	0.650	0.667	0.683	0.698	0.713
10	0.492	0.515	0.537	0.558	0.578	0.596	0.614	0.632	0.648	0.663	0.678
9	0.456	0.478	0.500	0.520	0.540	0.558	0.576	0.593	0.609	0.625	0.639
8	0.418	0.439	0.460	0.479	0.498	0.516	0.533	0.550	0.566	0.581	0.596
7	0.377	0.397	0.417	0.435	0.453	0.470	0.487	0.503	0.518	0.533	0.548
6	0.334	0.352	0.370	0.387	0.404	0.420	0.436	0.451	0.465	0.480	0.493
5	0.287	0.303	0.319	0.335	0.350	0.365	0.379	0.393	0.407	0.420	0.433
4	0.237	0.251	0.265	0.278	0.292	0.304	0.317	0.329	0.341	0.353	0.364
3	0.184	0.195	0.206	0.217	0.228	0.238	0.249	0.259	0.269	0.279	0.288
2	0.127	0.135	0.143	0.151	0.158	0.166	0.174	0.181	0.188	0.196	0.203
1	0.065	0.070	0.074	0.078	0.083	0.087	0.091	0.095	0.099	0.103	0.107

TABLE 4.6 Mean service life for components in liquid-based solar heating systems*

Component	*Service life, years*
Collectors	
Metal and glass	∞
Wood	3
Elastomer seals carefully specified	30+
Pumps	
Heavy-duty capacitor start	20
Sealed	5
Controls	
Solid-state triac	?
Mechanical relay, 50×10^6 cycles used 50% full load rating	300
Connectors	
Copper bellows	30
Premium elastomer	20
Storage tanks	
Glass-lined†	11
Stone-lined†	9
Galvanized, protected by carbonate deposit	30

*From Ref. 3.

†United States average.

results. To this is added the power consumption of the controller multiplied by its number of hours of operation per year, typically 8760. The operating energy cost in Eq. (4.19) is *levelized*, as described in Chap. 3, by the ratio of capital recovery factors as shown. For income-producing commercial facilities, a portion of the operating energy is deductible as indicated.

Property taxes are based on the assessed value of the solar system, which is related to its resale value. The assessed value is only a fraction of the initial cost $C_{s,tot}$ and is determined by local procedures for calculating the tax basis for property taxes. In many states, solar systems are subject to much-reduced property tax schedules, and these tax incentives must be investigated by the designer prior to making the cost study.

Two real discount rates are used in Eq. (4.19): the first i' applies to the cost of money, and the second i'' applies to the cost of energy.

Equation (4.19) can be used to calculate cash flows in *current* dollars, that is, dollars including the effect of inflation if all discount rates are expressed not in real terms but in current dollar terms. In other words, use i instead of i'. Current-dollar analyses may be required by the federal government in the future.

Maintenance and insurance costs are related to the type of system used and component mean service life. Maintenance costs for passive systems are usually relatively low, occurring only when difficulties in the movable insulation or its control system occur. The balance of the passive system has no moving parts and if properly constructed originally should require no repairs or replacements (barring glass breakage).

The mean service life of a number of components used in active systems is shown in Table 4.6. These estimates of mean service life can be used in Eq. (4.19) to estimate the year k for capital replacement costs R_k. Maintenance costs occur on a more frequent and routine basis and are associated with such things as replacement of filters in an air system, cleaning the collector glazing, or replacing the

TABLE 4.7 U.S. Department of Energy differential inflation rates for fuels and electricity*

Coal	5%
Fuel oil	8%
Gas (natural or LPG)	10%

Electricity

Region	*Electricity, %*
New England:	6.9
Connecticut	
Maine	
Massachusetts	
New Hampshire	
Rhode Island	
Vermont	
Middle Atlantic:	5.9
New Jersey	
New York	
Pennsylvania	
Delaware	
South Atlantic:	5.8
District of Columbia	
Florida	
Georgia	
Maryland	
North Carolina	
South Carolina	
Virginia	
West Virginia	
East South Central:	5.6
Alabama	
Kentucky	
Mississippi	
Tennessee	
Pacific:	7.3
California	
Oregon	
Washington	
East North Central:	5.6
Illinois	
Indiana	
Michigan	
Ohio	
Wisconsin	
West North Central:	5.6
Iowa	
Kansas	
Minnesota	
Nebraska	
North Dakota	
South Dakota	
Missouri	
West South Central:	7.5
Arkansas	
Louisiana	
Oklahoma	
Texas	
Mountain:	5.7
Arizona	
Colorado	
Idaho	
Montana	
Nevada	
New Mexico	
Utah	
Wyoming	

*From Ref. 24.

antifreeze solution in a liquid system. By proper design the maintenance cost M can be reduced to a relatively low level. The designer can estimate M in dollars per year based upon the expected maintenance procedures required for the system. All components present in active systems are similar to those in nonsolar HVAC systems, and the same maintenance schedules for them apply to solar systems. Solar collectors should require very little maintenance other than periodic cleaning in some environments. Insurance charges are determined by an insurance underwriter who insures the solar system. These rates vary depending upon the owner's rate structure.

Fuel inflation rates have been estimated by the U.S. Department of Energy. The DOE recommended values are shown in Table 4.7. Four types of fuels are included—coal, fuel oil, natural gas, and electricity. The electricity inflation rate varies by region because different types of fuels are used in different parts of the country. *The figures shown in Table 4.7 are differential inflation rates*, i.e., rates above the general inflation rate of the economy. The consumer price index averaged over several years should be added to the numbers in Table 4.7 to determine the total inflation rate j_e.

The effective tax rate shown in the list following Eq. (4.19) is based on the assumption that federal taxes can be deducted from state taxes. If state taxes can also be deducted from federal taxes, a slightly different expression applies. In that case, the effective income-tax rate T_{inc} is given by Eq. (4.20) below.

$$T_{inc} = \frac{\text{state rate} + \text{federal rate} - 2 \text{ state rate} \times \text{federal rate}}{1 - \text{state rate} \times \text{federal rate}} \tag{4.20}$$

The costs shown in Eq. (4.19) represent those associated with a specific system size. The optimization of a solar system requires costs for several system sizes. Equation (4.19) is evaluated for each size. The following example illustrates how the solar cost equation is applied.

Example. Tabulate all cash flows from Eq. (4.19) for the following set of economic parameters for a non-income-producing owner. Work the problem in constant dollars.

Initial extra system cost, $C_{s,tot} = \$9000$
Investment tax credit ITC (40 percent federal credit) = \$3600
Assessed value $C_{s,assess} = \$4000$
General inflation rate $j = 7$ percent
[and $i' = (0.09 - 0.07)/1.07 = 0.0187$ if i_m is equal to the owner's discount rate]
Market interest rate, $i_m = 9$ percent for 20 yr
Property tax rate = ½ percent
Energy inflation rate $j_e = 17$ percent per year
[and $i'' = (0.09 - 0.17)/1.17 = -0.068$]
Repairs, one at $t = 10$ yr = \$250 = R_{10}, $R_{k \neq 10} = 0$
Maintenance and insurance, $M + I = \$75$ per year
Tax bracket, $T_{inc} = 30$ percent

Operating energy, C_e (in year 1) = \$20 per year
Salvage value, $C_{s,salv}$ = 0

Solution. Each term in Eq. (4.19) is listed in order in the table below. The economic factors are
$\text{CRF}(i', 20) = 0.0604$
$\text{CRF}(i'', 20) = 0.0220$
$\text{PWF}(i', 10) = 0.830$

Annual extra cost (no down payment)	$(C_{s,tot} - \text{ITC})\text{CRF}(i',20)$	326.16
Salvage	$C_{s,salv} = 0$	0
Replacements	$R_{10}\text{PWF}(i',10)\text{CRF}(i',20)$	12.53
Operating energy	$C_e \dfrac{\text{CRF}(i',20)}{\text{CRF}(i'',20)}$	54.90
Property tax	$T_{prop}\ C_{s,assess}$	20.00
Maintenance and insurance	$M + I$	75.00
Property-tax deduction	$-\ T_{inc}\ T_{prop}\ C_{s,assess}$	−6.00
Income-tax interest deduction	$-\ T_{inc}\ i_m \left[\sum \dfrac{P_k}{(1+i)_k}\right]\text{CRF}(i,N)$ (Table 4.5)	−62.75
Depreciation	N/A	N/A
	Total cost	\$419.84 per year

The total cost per year is expressed in *constant discounted* dollars.

Performance Prediction

Optimum system sizing requires not only an economic calculation as described in the preceding section but also a performance prediction which evaluates the denominator of Eq. (4.18). The prediction of system performance over a lengthy period such as 15 or 20 yr is impossible. The approach typically taken is to predict performance which *would have occurred* over a long period of time (20 to 30 yr) in the past. This method has the advantage that the solar and weather data are all known. They are assumed to apply for a similar period in the future. The standard method for predicting performance is by means of a mathematical model of the system. Various levels of complexity for system models are available and several dozens of these have been derived for both active and passive systems (41).

The most complicated models presumably give the most physical insight but not necessarily better accuracy. The SERI Computer Code Center can provide manuals for active models—TRNSYS or DOE-2—and for the passive model—DEROB. (The passive model PASOLE is available from Los Alamos Laboratory.) The complex models are based on hour-by-hour calculations of solar-system delivery and hourly heating loads. Next in level of complexity is the class of models used for DD calculations. These include the F-chart (42) for active systems and the SLR method (43) for passive systems. They require monthly, not hourly, data and can

TABLE 4.8 F-chart *K* factors

Correction factor	Air or liquid system	Correction factor	Validity range for factor
K_{flow}	A*	$\{[2\ ft^3/(min \cdot ft_c^2)]/\text{actual flow}\}^{-0.28}$	$1\text{–}4\ ft^3/(min \cdot ft_c^2)$
	L	Small effect included in F_R and F_{hx} only	
K_{stor}	A†	$[(0.82\ ft^3/ft_c^2)/\text{actual volume}]^{0.30}$	$0.4\text{–}3.3\ ft^3/ft_c^2$
	L	$[(1.85\ gal/ft_c^2)/\text{actual volume}]^{0.25}$	$0.9\text{–}7.4\ gal/ft_c^2$
K_{DHW}‡	L	$\dfrac{(1.18T_{w,o} + 3.86T_{w,i} - 2.32\ \bar{T}_a - 66.2)}{(212 - \bar{T}_a)}$	
K_{ldhx}**	L	$0.39 + 0.65 \exp[-0.139UA/E_L(\dot{m}c_p)_{air}]$	$0.5 < \dfrac{E_L(\dot{m}c_p)_{air}}{UA} < 50$
	A	N/A	

*User must also include the effect of flow rate in F_R, i.e., in $F_R(\overline{\tau\alpha})$ and $F_R U_c$. Refer to manufacturer data for this.

†For air systems using latent-heat storage, see J. J. Jurinak and S. I. Abdel-Khalik, *Energy*, vol. 4, p. 503 (1979) for the expression for K_{stor}.

‡Only applies for liquid storage DHW systems (air collectors can be used, however); $T_{w,o}$ = hot-water supply temperature, $T_{w,i}$ = cold-water supply temperature to water heater, both °F. Applies only to a specific water use schedule; predictions of performance for other schedules will have reduced accuracy—see Ref. 42.

**UA = unit building heat load, Btu/(h · °F); E_L = load-heat-exchanger effectiveness; $(\dot{m}c_p)_{air}$ = load-heat-exchanger air capacitance rate, Btu/(h · °F) = density × 60 × cubic feet per minute × 0.24.

give solar-performance predictions with only a programmable calculator, not a computer. Through DD, hourly computer models are normally not required unless solar systems of a nonstandard configuration are used. The hourly model may have a function toward the end of DD in verifying F-chart or SLR calculations used to find the final, optimum system configuration. The decision whether an hourly check on the monthly calculations is required can be based on the design budget for the building and the amount of time necessary to prepare input for complex hourly models as well as the computer time to run those models and interpret the output.

The F-Chart Method. The F-chart is a common calculation procedure used in the United States for calculating useful energy delivery of active solar heating systems. Both space and water heating with either air or liquid systems are included. The F-chart applies only to specific system designs of the type shown in Fig. 2.16 for liquid systems and in Fig. 2.19 for air systems. The method, in essence, consists of several empirical equations expressing monthly solar heating fraction f_s as a function of dimensionless groups which relate system properties and weather data for a month to the monthly heating requirement. The several dimensionless parameters are grouped into two dimensionless groups called the solar parameter P_s and the loss parameter P_L.

The solar parameter given in Eq. (4.21) is the ratio of monthly solar energy absorbed by the collector divided by the monthly heating load.

$$P_s = \frac{K_{ldhx} F_{hx} (F_R \overline{\tau\alpha}) \bar{I}_c N}{L} \tag{4.21}$$

where F_{hx} = heat exchanger penalty factor [see Fig. 2.13b or Eq. (2.16)]
$F_R\tau\alpha$ = average collector optical efficiency = 0.95 × collector-efficiency-curve intercept $F_R(\tau\alpha)_n$
$\bar{I}_c$ = monthly average insolation on collector surface (see Chap. 2 or Table A.4)
N = number of days in a month
L = monthly heating load, *net of any passive system delivery* as calculated by the P-chart or SLR method (see page 266), Btu/mo
K_{ldhx} = load-heat-exchanger correction factor for liquid systems

The second parameter used in the F-chart correlation is related to the long-term energy losses from the collector divided by the monthly heating load, as shown in Eq. (4.22).

$$P_L = (K_{stor}\, K_{flow}\, K_{DHW}) \frac{F_{hx}(F_R U_c)(T_r - \bar{T}_a)\, \Delta t}{L} \qquad (4.22)$$

where $F_R U_c$ = magnitude of collector-efficiency-curve slope (can be modified to include piping and duct losses, see page 265)
$\bar{T}_a$ = monthly average ambient temperature, °F*
Δt = number of hours per month = 24 N
T_r = reference temperature = 212°F
K_{stor} = storage volume correction factor
K_{flow} = collector flow rate correction factor
K_{DHW} = conversion factor for parameter P_L when *only* a water heating system is to be studied

The monthly solar fraction depends only on these two parameters. For liquid systems, the monthly solar fraction f_s is given by Eq. (4.23) if $P_s > P_L/12$ (if not, $f_s = 0$).

$$f_s = 1.029P_s - 0.065P_L - 0.245P_s^2 + 0.0018P_L^2 + 0.0215P_s^3 \qquad (4.23)$$

For air-based systems, the monthly solar heating fraction is given by Eq. (4.24) if $P_s > 0.07P_L$ (if not, $f_s = 0$).

$$f_s = 1.040P_s - 0.065P_L - 0.159P_s^2 + 0.00187P_L^2 + 0.0095P_s^3 \qquad (4.24)$$

The flow-rate, storage, load-heat-exchanger, and domestic-hot-water correction factors for use in Eqs. (4.21) and (4.22) are given for both liquid- and air-based systems in Table 4.8 and Fig. 4.41.

The following example illustrates the use of the F-chart method for a liquid system. For any F-chart calculation, use this step-by-step calculation process, in order: collector insolation, collector properties, monthly heat loads, monthly

*Note that $T_a \neq 65°$ − (degree heating days/N), contrary to statements of many U.S. government contractors and in many government reports. The equality is only valid for those months when $T_a < 65$°F every day, i.e., only 3 or 4 mo of the year at the most. The errors propagated through the solar industry by assuming the equality to be true are too many to count. See Table A.4 for $\bar{T}_a$.

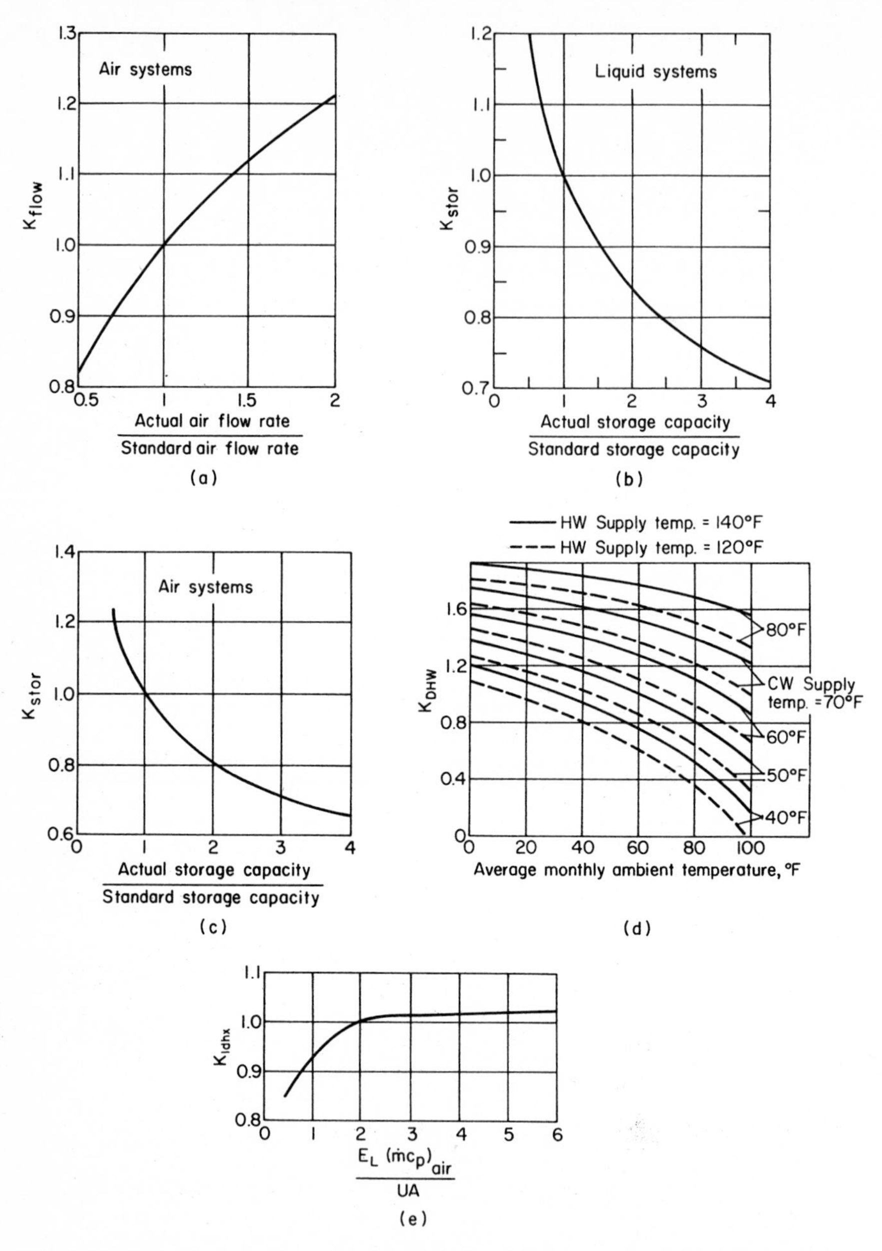

FIG. 4.41 F-chart correction factors: (*a*) K_{flow} [standard valve is 2 standard ft³/(min·ft²)]; (*b*), (*c*) K_{stor} (standard liquid value is 1.85 gal/ft²; standard rock value is 0.82 ft³/ft²); (*d*) K_{DHW}; (*e*) K_{ldhx} (standard value of abscissa is 2.0). *(From Ref. 33.)*

TABLE 4.9 Climatic and solar data for Bismarck, North Dakota

Month	*Average ambient temperature, °F*	*Heating, Fahrenheit degree-days*	*Horizontal solar radiation, langleys/day*
January	8.2	1761	157
February	13.5	1442	250
March	25.1	1237	356
April	43.0	660	447
May	54.4	339	550
June	63.8	122	590
July	70.8	18	617
August	69.2	35	516
September	57.5	252	390
October	46.8	564	272
November	28.9	1083	161
December	15.6	1531	124

ambient temperatures, and finally monthly values of P_s and P_L. Once the parameter values are known, it is an easy matter to calculate the monthly solar fraction and finally the monthly energy delivery. The total of all monthly energy deliveries is the total annual useful energy produced by the solar system. This is the denominator of Eq. (4.18). The annual useful delivered energy divided by the total annual load is the annual solar load fraction, denoted by $\bar{f}_s$.

Example. Calculate the annual heating energy delivery of a solar space-heating system using a double-glazed flat-plate collector in Bismarck, North Dakota. The building and solar system specifications are given below.

Building Specifications

Location: 47°N latitude
Space-heating load: 15,000 Btu/(°F·day)

Solar System Specifications

Collector loss coefficient: $F_R U_C = 0.80$ Btu/(hr·ft²·°F)
Collector optical efficiency (average): $F_R(\overline{\tau\alpha}) = 0.70$
Collector tilt: $\beta = L + 15° = 62°$
Collector area: $A_c = 600$ ft²
Collector fluid flow rate: $\dot{m}_c/A_c = 11.4$ lb/(h·ft$_c^2$) (water)
Collector fluid heat capacity–specific gravity product: $c_{pc} = 0.9$ Btu/(lb·°F) (antifreeze)
Storage capacity: 1.85 gal/ft$_c^2$ (water)
Storage fluid flow rate: $\dot{m}_s/A_c = 20$ lb/(h · ft²) (water)
Storage fluid heat capacity: $c_{ps} = 1$ Btu/(lb·°F) (water)
Heat-exchanger effectiveness: 0.75

Climatic Data

Climatic data from the NWS are tabulated in Table 4.9.

TABLE 4.10 F-chart summary for example

Month	*Collector-plane radiation, Btu/(day·ft²)*	*Monthly energy demand L, million British thermal units*	P_L[a]	P_s[a]	f_s	*Monthly delivery, million British thermal units*
January	1506	26.41	2.68	0.72	0.46	12.15
February	1784	21.63	2.88	0.95	0.60	12.98
March	1795	18.55	3.50	1.22	0.73	13.54
April	1616	9.90	5.75	2.00	0.94	9.31
May	1606	5.08	10.78	>3.00	1.00	5.08
June	1571	1.83	>20.00	>3.00	1.00	1.83
July	1710	0.27	>20.00	>3.00	1.00	0.27
August	1712	0.52	>20.00	>3.00	1.00	0.52
September	1721	3.78	13.76	>3.00	1.00	3.78
October	1722	8.46	6.79	2.58	1.00	8.46
November	1379	16.24	3.79	1.04	0.61	9.91
December	1270	22.96	2.98	0.70	0.43	9.87
Annual		135.66			0.65	87.70

[a]$P_s > 3.0$ or $P_L > 20.0$ implies $P_s = 3.0$ and $P_L = 20$, i.e., $f_s = 1.0$. The annual solar fraction $\bar{f}_s$ is (87.70/135.66), or 6' percent.

Solution. The calculations involved in the steps are shown in Table 4.10. The heat-exchanger penalty factor $F_{hx} = 0.974$ [from Eq. (2.16)].

When using the F-chart, it is important to use a consistent area basis for calculating the efficiency curve information and the solar and loss parameters. The early National Bureau of Standards (NBS) test procedure based the collector efficiency on net glazing area. However, the more recent and more widely used test procedure developed by ASHRAE (93.77) uses the gross area basis. The gross area is the area of the glazing plus the area of opaque weatherstripping, seals, and supports. Therefore, when ASHRAE test data are used, the solar and loss parameters must be based on gross area. The efficiency-curve basis and the F-chart basis must be consistent for proper results.

○ **Use of F-Chart Results to Calculate Performance of Associated Heat-Pump Backup Systems.** Heat pumps are often used as auxiliary heat systems for solar systems. Of course, on very cold days, heat pumps themselves require a backup—resistance heat. The cost of heat produced by a heat pump is the cost of electricity divided by the COP whereas the cost of resistance heat is the same as the cost of electricity. Therefore, to find the total backup energy cost, the relative fractions of backup heat which are heat-pumped and resistance-produced must be known. For a nonsolar system, it is easy to find this quantity by requesting the heat pump manufacturer to calculate it using common in-house computer models.

Heat pumps serving as solar auxiliaries operate during a different set of hours, at a different set of COP values, and hence at a different ratio of resistance to heat-pumped space heating. It has been shown that this relatively complex situation has a simple solution (50). To find the percent of total heat load provided from (1) the atmosphere, (2) the heat-pump compressor, and (3) the strip heaters for each month:

1. Find the atmospheric energy fraction $f_{atm,no\ solar}$ and the compressor work fraction $f_{c,no\ solar}$ for each month *for a stand-alone heat-pump system*, i.e., for the nonsolar heat pump case.

2. Find the monthly solar fraction f_s from the F-chart.

3. Find the monthly atmospheric energy fraction $f_{atm,solar}$ with the solar system from

$$f_{atm,solar} = f_{atm,no\ solar} \times (1 - f_s)$$

4. Likewise, find the monthly amount of heat provided by heat-pump compressor work for the solar case $f_{c,solar}$

$$f_{c,solar} = f_{c,no\ solar} \times (1 - f_s)$$

5. Find the fraction of heat load provided by strip heat with a solar system

$$f_{aux,solar} = (1 - f_{atm,no\ solar} - f_{c,no\ solar}) \times (1 - f_s)$$

6. The total of the four monthly heat fractions—solar, atmospheric, compressor, and auxiliary strip heat—must add up to 1.

The fraction of heat energy which must be purchased, f_{purch}, is the total of compressor power and auxiliary strip heat for each month, or $f_{purch} = (1 - f_{atm,no\ solar}) \times (1 - f_s)$.

When both space and water heating are done by the same solar system, the F-chart predicts only the fraction of the combined load provided by solar heat. If the water heating load is small, the combined load fraction can be used in the preceding heat-pump method (50).

Use of Collectors Connected in Series—Effective Values of F_RU_c and $F_R(\overline{\tau\alpha})$. In the F-chart approach a "collector," effectively with a single inlet and outlet, is used. This "collector," really an array, may be composed of many individual parallel-connected collectors. In this case the array values of F_RU_c and $F_R(\overline{\tau\alpha})$ are the same as the individual collector values if the flow per unit area is the same for all collector panels. However, if collectors are connected in series within the array, the single-panel values of F_RU_c and $F_R(\overline{\tau\alpha})$ cannot be used in the F-chart method. This is a result of progressively higher and higher inlet temperatures at each successive collector panel in a series string and the resulting lower efficiency.

It is possible to define *effective values* of F_RU_c and $F_R(\overline{\tau\alpha})$ for series panels based on the properties of a single panel (55). The correction is made by multiplying F_RU_c and $F_R\overline{\tau\alpha}$ for a single collector by the quantity $\{[1 - (1 - Z)^N] \times [NZ]^{-1}\}$ where N is the number of collectors in series and $Z = F_RU_c/(\dot{m}c_p/A_c)_c$; $\dot{m}/A_c$ is the mass flow rate per unit collector area for the test conditions under which F_RU_c and $F_R(\overline{\tau\alpha})$ were measured, that is, for a *single* collector.

Accounting for Duct and Pipe Losses in the F-Chart Procedure. As shown by Beckman (53), it is possible to modify the parameters F_RU_c and $F_R(\overline{\tau\alpha})$ used in the F-chart procedure to include duct or pipe losses. The modified values, given in

Eqs. (4.25) below, can be used directly in Eqs. (4.21) and (4.22) above. The effective value $[F_R(\overline{\tau\alpha})]'$ is

$$[F_R(\overline{\tau\alpha})]' = F_R(\overline{\tau\alpha})\left[1 + \frac{A_o}{(\dot{m}c_p)_c}\right]^{-1} \tag{4.25a}$$

and the effective value $[F_R U_c]'$ is

$$[F_R U_c]' = F_R U_c\left[1 - \frac{U_p A_i}{(\dot{m}c_p)_c} + \frac{U_p(A_i + A_o)}{A_c F_R U_c}\right] \times \left[1 + \frac{A_o}{(\dot{m}c_p)_c}\right]^{-1} \tag{4.25b}$$

where A_i and A_o are the heat-loss areas for collector inlet and outlet conduits and U_p is the overall thermal conductance between the working fluid and the environment via the pipe or duct and its insulation. Other symbols are as above.

SLR Method for Predicting Passive System Heating Performance. The monthly solar load ratio (SLR) is an empirical method of estimating monthly solar and auxiliary energy requirements for passive solar systems (57). The monthly solar load ratio is a dimensionless correlation parameter defined as follows:

$$\text{SLR} \equiv \frac{\begin{array}{c}\text{solar radiation absorbed by the}\\ \text{passive system per month}\end{array}}{\begin{array}{c}\text{monthly building load (including solar wall}\\ \text{steady-state losses as if it were unilluminated)}\end{array}} \tag{4.26}$$

The SLR is similar to P_s of the F-chart.

The numerator is the product of the solar net aperture area and the monthly solar energy transmitted through 1 ft^2 of "south" glazing times the wall absorptance (for direct gain the absorptance is taken as 1.0; otherwise see appendix Table A.12). The denominator is equal to the building-loss coefficient (including the steady-state conduction through the south solar collection wall) times the monthly heating degree-days to the appropriate base accounting for internal gains.

The SLR can be expressed as follows:

$$\text{SLR} = \frac{\begin{array}{c}\text{collector}\\ \text{wall area}\end{array} \times \text{absorptance} \times \begin{array}{c}\text{monthly solar energy transmitted}\\ \text{through the glazing}\end{array}}{\text{modified building-loss coefficient} \times \text{monthly degree-days}} \tag{4.27}$$

Use the following steps to find passive system performance:

Step 1. Determine the building heat-loss coefficient $24UA$, Btu/(°F·day), and compute a modified building-loss coefficient $(UA)_{mod}$ by adding to $24UA$ the term [(solar wall area A_c) $\times$ U_w], where U_w is taken from Table 4.11. The value of U_w is related to the steady-state conduction coefficient of the combined wall, glazing, and insulation, averaged over the day. UA does not include the passive aperture.

Step 2. Determine SLR for each month of the year. Solar radiation values generally available are measured on a horizontal surface area $\overline{H}_h$, whereas the values required by the SLR are the solar radiation *transmitted through*

TABLE 4.11 U_w values for SLR calculation*

Wall type	*Plain double-glazed*	*With R-9 night insulation added from 5 p.m. to 8 a.m.*
Water wall	5.0	0.7 Btu/(day·°F·ft²)
18-in Trombe wall	3.6	0.5 Btu/(day·°F·ft²)
Direct gain	10.6	2.4 Btu/(day·°F·ft²)

*From Ref. 57.

the vertical south-facing surface. For angles in degrees, this value is given by the following equation (from Ref. 43):

$$\bar{I}_{\text{VERT}} = [0.2260 - 0.002512(L - \delta_s) + 0.0003075(L - \delta_s)^2] \times \overline{H}_h \quad (4.28)$$

for vertical walls with double glazing where L is the latitude and δ_s is the declination.

If the building does not face due south, then this equation cannot be used. It will be necessary to make correction for building orientation. It is felt that a correction factor based on the ASHRAE clear-day tables (6) would probably be a reasonable estimate. The ASHRAE tables give clear-day values for southwest, southeast, and due south orientations, as a function of latitude. A straight proportional correction factor based on these tables is recommended. Note that a separate correction factor will be required for each month.

Step 3. Determine the monthly solar savings fraction f_s and auxiliary heat Q_{aux} for each month of the year based on the values of SLR computed in Step 2 using the equations below

$$g = a_1(\text{SLR}) \qquad \text{SLR} < R \qquad (4.29a)$$
$$g = a_2 - a_3 e^{-a_4(\text{SLR})} \qquad \text{SLR} > R;\ g \leq 1.0$$
$$f_s = 1 - [(UA)_{mod}/24UA] \times (1 - g) \qquad (4.29b)$$

The annual useful energy $Q_u = \sum_{i=1}^{12} f_s \times (24UA)(\text{degree-days/mo})$. The values of the parameters in the f_s function are selected to give a minimum-least-square error in the annual solar heating fraction; but monthly f_s values may be less precise. The values of the least-squares coefficients a_1 to a_4 are given in Table 4.12.

Variations from the Assumed Reference Systems. The monthly SLR equations above are for a very specific passive system as defined below:

Thermal storage = 45 Btu/°F per square foot of glazing
Trombe wall has vents with backdraft dampers; vent area of top or bottom row is fixed at 3 percent of total wall area for all solar fractions

TABLE 4.12 SLR equation coefficients*

Wall type	*R*	a_1	a_2	a_3	a_4
Water wall	1.3	0.4025	0.9872	1.5053	0.9054
Night-insulated water wall	1.2	0.4846	0.9799	1.8495	1.2795
Masonry wall	0.6	0.3698	1.0408	1.0797	0.4607
Night-insulated masonry wall	1.0	0.4556	0.9769	1.2159	0.8469
Direct gain	0.5	0.5213	1.0133	1.0642	0.6927
Night-insulated direct gain	0.7	0.542	0.9866	1.1479	0.9097

*From Ref. 57.

Double glazing (normal transmittance = 0.747)
Temperature range in building = 65 to 75°F
Building mass is negligible except for direct gain which uses mass on floor and nonsouth walls for storage
Night insulation (when used) = R-9; 5:00 p.m. to 8:00 a.m.
Wall-to-room conductance = 1.0 Btu/(h·°F·ft²)
Trombe wall properties: k = 1.0 Btu/(ft·hr·°F)
ρc = 30 Btu/(ft³·°F)
Foreground reflectance = 0.3

If the optimum vent area from Fig. 4.35 is used instead of the assumed 3 percent area, the monthly solar fraction is increased by 7 solar fraction percentage points over that predicted by the SLR method in the solar fraction range from 0.30 to 0.70. That is, $f_{s,\text{opt vents}} = f_{s,3\%\text{ vents}} + 0.07$, from Ref. 57.

If the performance of a system which is different than a reference system is required, then it is necessary to make a correction. The most reliable way of doing this is to refer to results of hour-by-hour calculations which are made for a specific system, varying only the parameter of interest. There are no correction factors analogous to the four F-chart K factors to use for this purpose; however, Klein (67) has generalized the SLR method so that major parameters can be varied in monthly passive calculations.

Effect of a Reflector. A performance improvement can be achieved by a reflector that increases the total flux onto the solar collection wall. The ratio of the total monthly transmitted solar energy with the reflector to that without the reflector (for a specular reflector with $\rho = 0.8$) is given by

$$R_{\text{refl}} = 1.0083 - 1.787\Delta + 19.16\Delta^2 - 40.31\Delta^3 + 24.66\Delta^4 \tag{4.30}$$

where $\Delta \equiv (L - \delta_s)/100$. If a reflector is used, SLR from Eq. (4.26) is multiplied by R_{refl}.

If the reflectance is other than 0.8, the ratio R_{refl} can be computed from the above equation if it is assumed that the difference between unity and the calculated enhancement is proportional to the reflectance.

Discussion of Loads. Monthly heating degree-day values are used in the correlation procedure because they are the only indicators of heating load that are readily available in most localities.

It is possible to distinguish between two solar heat contributions from the solar wall: (1) the energy *saved*, and (2) the energy *supplied*. In this method the energy saved is used to define the solar heating fraction even though it gives a lower value of annual fraction $\bar{f}_s$. The actual solar energy supplied by the solar collection wall will be greater than that estimated by taking the difference between the annual degree-day load and the auxiliary energy. The extra solar heat is the amount used to maintain the building above 65°F during a significant portion of the year. Since *the auxiliary energy required is the most important number to be estimated,* it was felt that this approach is appropriate. In reality, the solar-heated building will generally be warmer than the nonsolar-heated building, if it is assumed that the thermostat is set at 65°F in both cases.

Combined Passive Heating Systems. If two generic types of passive heating systems are combined, the calculation of the solar system performance can still be done by the SLR method. A single value of SLR is calculated for the total aperture area. The solar fraction is then calculated for each type of system, say a water-wall system and a direct-gain system, from Eq. (4.29). The calculated solar fractions for the two systems are then prorated on the basis of the effective aperture area of each, and added. If the system used is 30 percent direct-gain and 70 percent water-wall, for example, the solar fraction calculated for the direct-gain system would be multiplied by 0.3 and added to 0.7 times that calculated for the water-wall system. The sum of the two prorated solar fractions is the monthly solar fraction for the combined system.

System Size Selection

Various criteria can be used for selecting the system size which serves as the basis of the DD documents. One way is to make an arbitrary decision that a given percent of the average annual heat load is to be carried by the solar system. This decision may be based on past system considerations in a specific location and known fuel prices which lead to this solar fraction near the economic optimum.

Another method of selecting the system size is to purchase the largest system which can be acquired for available funds and to accept whatever solar fraction is delivered by this system. If the amount of funding is limited, the system which is purchased may be smaller or larger than the economic optimum. Its use still results in life-cycle savings vs. a 100 percent nonsolar heating system if solar heating is feasible at all. These two methods of system selection are relatively straightforward and may apply to certain situations.

The third selection method chooses that system which results in the minimum cost of providing heat to the load over the life cycle of the solar system, as described in Chap. 3. In the absence of a budgetary constraint or an arbitrary solar fraction decision, this is the proper sizing method.

One of the responsibilities of the solar designer during the DD phase is to establish more accurate costs for the solar system. The DD cost study requires more effort than that for a nonsolar system since a *range* of component sizes must be examined during the optimization process. Several sources for cost data as a function of size exist. For example, Ref. 45 contains many charts and tables of HVAC equipment costs, including labor, vs. equipment size. A standard source of costing used by HVAC designers is the Means cost index published by R. S. Means Co. (44). This manual contains cost as a function of equipment size, as does Ref. 45, but also includes additional information regarding installation cost and geographical variations in labor and equipment costs. The Means index is updated annually.

A third method of estimating a range of costs uses quotes from manufacturers. At least two quotes, each at a limit of the range of expected size, are needed.

Cost Power Laws. Figure 4.42 shows how the extreme-case cost approach is used to establish a *power law* relating storage-tank cost to storage-tank size. The slope of the curve on logarithmic graph paper is the exponent of the power-law relationship between cost and size. The example shows that the lower-limit cost for a 700-gal liquid storage tank is \$1300 and the high-limit cost is \$4300 for a 7000-gal tank. The slope of the cost curve is 0.52. Therefore, the equation relating cost to size for this particular storage tank example is

$$\text{Cost} = \$1300 \times \left(\frac{\text{volume}}{700\text{ gal}}\right)^{0.52} \tag{4.31}$$

Similar relationships can be established for heat exchangers, pumps, and other components of the system. The power-law expressions are used during the optimization process. The simplification has often been made (42) that the cost of all solar systems is directly proportional to size—that is, that the exponent in the power-law relationship is exactly unity. However, this assumption is incorrect and will lead to solar system sizes which are incorrect. If sufficient data or time are not available to establish power-law cost functions, a typical average value of the exponent can be assumed to be 0.9. Of course, the *exponent on the collector cost is 1.0,* since economies of scale are rarely available for solar-collector purchases. However, pumps, storage tanks, heat exchangers, expansion tanks, and most other components have nonlinear cost curves.

Component Nominal Sizes. The rules presented in Table 3.2 can be used to relate the size of the major components of active solar systems to collector area. These rules are based on nominal values of component cost and typical performance characteristics. However, the F-chart permits the designer to vary some component sizes from the nominal values. For example, the designer may wish to investigate the economic consequences of quadrupling the size of storage. This is

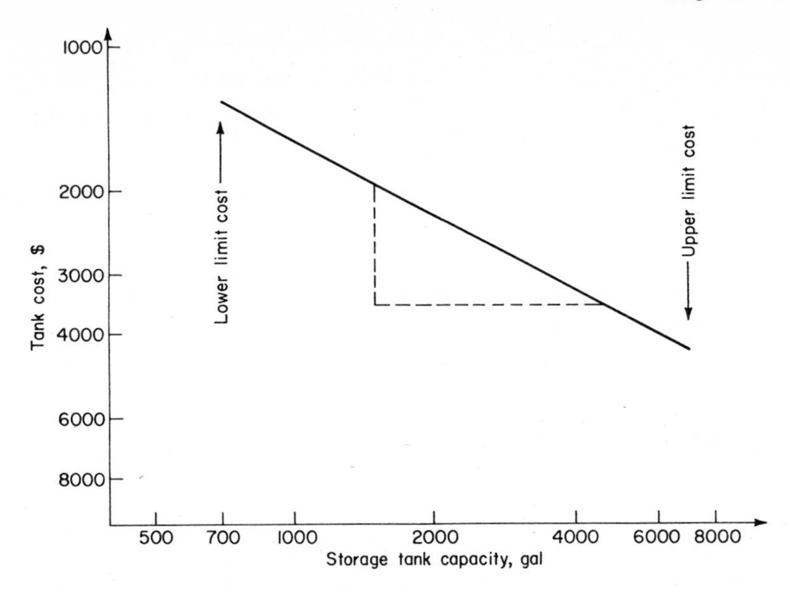

FIG. 4.42 Example of log-log power-law plot estimating a range of solar-system component costs from only two values of the cost at the extremes of the range of interest.

done in the F-chart methodology by varying the correction factor K_{stor}. The associated increase in cost of storage could be calculated from an equation corresponding to Eq. (4.31) derived above. Similar analysis can be carried out regarding collector fluid flow rate, heat-exchanger size, load-heat-exchanger size, storage insulation level, and most first-order design parameters of the solar system. The method is to vary the size of the component, calculate the energy delivery, calculate the effect on system cost, and finally, calculate the total heating cost of this system. The example on page 272 illustrates how such a suboptimization is carried out. A simplified optimization example using this idea for *nominal* sizes only is presented below.

Example. Find the optimum size of a solar system for which the annual F-chart performance is known:

Collector area, ft²	*Yearly delivery Q_u, million British thermal units*
1000	330
1500	436
2000	521
2500	601
3000	661
4000	776
5000	840
6000	898

The annual energy demand L is 1000×10^6 Btu/yr. The following economic conditions apply:

Cost of fuel: \$8 per million British thermal units inflating at 8 percent per year differentially

Annual cost of solar system [from Eq. (4.19) and power-law studies]: $1.4\ A_c + 6.9\ A_c^{1/2}$

Real cost of money, i': 0.02 for 20 yr (10 percent market rate)

Solution. The total cost of heating, C_{tot}, is

$$C_{tot} = \underbrace{(1.4\ A_c + 6.9\ A_c^{1/2})}_{\text{(Solar)}} + \underbrace{(L - Q_u)\overline{C}_{bu}}_{\text{(Backup)}}$$

where $\overline{C}_{bu}$ is the levelized cost,

$$\overline{C}_{bu} = (\$8 \text{ per million British thermal units}) \times \frac{\text{CRF}(0.02, 20)}{\text{CRF}(-0.0556, 20)}$$
$$= \$18.82 \text{ per million British thermal units}$$

The total cost is calculated for various system sizes in Table 4.13. The least-cost configuration in the table uses 4000 ft² of collector.

Fine-Tuning Component Sizes. Rules presented in Table 3.2 are used to find the nominal sizes of ancillary components of active systems. In the design of large systems, more careful sizing studies may be warranted. These studies are suboptimizations of a specific component. To properly optimize every component, the sizes of all should be varied simultaneously; however, most designers will use nominal values for sizing and possibly study only one or two components in further detail. The following example illustrates the method of finding the best collector flow rate.

Example. The designer of a 1000-ft² liquid-based system decides that fine-tuning of the collector flow rate is needed. Use the F-chart to find the best flow rate considering three marginal costs—pumping power, pump cost, and the value of reduction of annual solar system energy delivery resulting from reduced collector flow. For this collector $U_c = 1.0$ Btu/(h·ft²·°F),

TABLE 4.13 Life-cycle heating cost vs. system size

Collector area, ft²	*Solar cost, dollars per year*	*Auxiliary cost, dollars per year*	*C_{tot}, dollars per year*	
1000	1,618	12,609	14,227	
1500	2,367	10,614	12,981	
2000	3,109	9,015	12,124	
2500	3,845	7,509	11,354	
3000	4,578	6,380	10,958	
4000	6,036	4,216	10,252	←Minimum
5000	7,488	3,011	10,499	
6000	8,934	1,920	10,854	
0	0	18,820	18,820	

$(\overline{\tau\alpha}) = 0.75$, $F' = 0.95$ (assuming it is independent of flow rate), and tilt $= 55°$. The system does not use a collector heat exchanger and water is the working fluid. As designed, the pump efficiency is 0.4; the collector-loop pressure drop is 20 ft at a collector flow $\dot{m}$ of 10 lb/(h·ft$_c^2$) and increases with the square of the flow rate; the pump operates 2500 h/yr. The pump base cost is \$200 for $\dot{m} = 10$ lb/(h·ft$_c^2$) flow rate. Pump cost varies with the 0.6 power of flow rate.

The capital recovery factor is 0.100, electric power costs \$0.035 per kilowatt hour, and backup fuel costs \$10 per million British thermal units; the annual heat load is 210.56×10^6 Btu.

Solution. Three costs are involved. The annual pump cost is

$$c_p = 200(\dot{m}/10)^{0.6} \times \text{CRF}$$
$$= 5.02377\dot{m}^{0.6}$$

Annual pumping energy use is given by the product of pump horsepower hp_p and operating hours.

$$hp_p = \Delta_p\,(\text{ft}) \times (\text{gal/min})/3960\eta_p$$
$$= [20\text{ ft} \times (\dot{m}/10)^2][(\dot{m}/500) \times 1000\text{ ft}^2]/(3960 \times 0.4)$$
$$= 0.0002525\dot{m}^3$$

For 2500 h/yr the power cost C_e is

$$C_e = (0.0002525\dot{m}^3) \times (0.746\text{ kW/hp}) \times (\$0.035 \text{ per kilowatt hour}) \times (2500\text{ h/yr}) = 0.016482\dot{m}^3$$

The cost of reduced solar performance C_s is related to the reduced solar fraction $\bar{f}_s$ because of reduced values of the heat-removal factor F_R (the only F-chart factor related to liquid mass flow $\dot{m}$). The highest $\bar{f}_s$ is for an infinite flow rate ($F_{R,\max} = F' = 0.95$) and will serve as the basis for measuring reduced performance. In equation form,

$$C_s = (\bar{f}_{s,\infty} - \bar{f}_{s,\dot{m}}) \times L \times (\$10 \text{ per million British thermal units})$$

where $\bar{f}_s$ values are calculated by the F-chart method for various F_R values. Table 4.14 summarizes the calculations on an annual basis. F-chart results are not given in detail. The minimum-cost flow rate is 9 lb/(h·ft$_c^2$) or 0.018 gal/(min·ft$_c^2$) (18 gal/min total array flow), where ft$_c^2$ represents collector area in square feet. This flow rate value is in the range given for liquid systems in Chap. 3.

Flatness of the Cost Curves. The preceding examples illustrate a common feature of solar system economic analyses. The *total cost,* if plotted vs. solar collector area (or any other system parameter), *shows a relatively flat region near the minimum cost point.* In other words, the minimum life-cycle cost is rather insensitive to collector area within a range of system sizes. In the example, variation of collector area from 3000 to 5000 ft^2, a 67 percent increase, had only a \$450 per year cost effect out of a total cost of approximately \$10,000.

TABLE 4.14 Results of flow study for example

Flow rate, lb/(h·ft²c)	F_R	C_s*	C_e	C_p	*Total cost*
2	0.75623	$175.50	$ 0.13	$ 7.61	$183.24
5	0.86520	72.30	2.06	13.20	87.56
6	0.87861	60.40	3.56	14.72	78.68
7	0.88836	51.90	5.65	16.15	73.70
8	0.89576	45.50	8.44	17.49	71.43
9	0.90158	40.50	12.02	18.77	71.29
10	0.90627	36.50	16.48	20.00	72.98
11	0.91013	33.20	21.94	21.18	76.32
12	0.91337	30.40	28.48	22.31	81.19
13	0.91612	28.10	36.21	23.41	87.72
14	0.91848	26.10	45.23	24.47	95.80
15	0.92054	24.40	55.63	25.51	105.54
∞	0.95000	0			

*$f_{s,\infty}$ = 0.749.

This insensitivity of life-cycle total heating cost to system size near the minimum point indicates that the owner should invest in a solar system at the small end of the interval. The life-cycle heating costs will be roughly the same, but the initial investment size will be significantly smaller. The acquisition of initial capital is one of the problems in solar system purchase in many cases, and if that investment can be reduced without a major consequence on life-cycle cost, it should be done. In the example above, purchase of a system at 3000 ft² would result in an annual life-cycle savings of $8900. The 4000-ft² system corresponds to a yearly saving of $8600, a minute difference, well within the accuracy bands of costs and performance predictions.

Sensitivity Studies. Life-cycle calculations depend on very many parameters. A number of these parameters will not be known precisely, and the sensitivity of the design solution should be determined relative to a range of the imprecisely known parameters. The major member of this group is the projected fuel inflation rate j_e. The effect of inflation is to cause fuel savings to be progressively more valuable in the future as conventional energy costs escalate relative to inflation-proof solar costs. This very important parameter is one of the least precisely known, although guidelines have been given previously for several types of fuels. A careful economic analyst studies a range of fuel inflation rates. If the viability of solar heating is questionable at the lower limit of the range assumed, the owner should be informed of this fact.

Other parameters to which the optimum system size is particularly sensitive include collector and support capital and installation cost and the discount rate i. The collector cost as well as the entire solar system cost can be determined relatively accurately at DD, and so a sensitivity analysis is not required with respect to these variables. The cost of money is also known in most cases, and a sensitivity study is probably not necessary. However, if a project is being planned for the distant future during a period of varying interest rates, it may be useful to examine the interest rate effect on solar life-cycle economics.

○ **Alternative Economic Indices.** The minimum-life-cycle-cost approach described in detail above is the proper method for determining the size of a given solar system. However, if the solar system is but one investment option of the owner, another method is required to rank these options relative to the return on investment. The internal rate of return method (22) can be used for this purpose. This method is not described in this book, since it is a common feature of most engineering economics texts. This is the proper index for ranking several alternative investment options.

The payback period is a common but very poor approach used to rank investment opportunities. The payback period is a faulty economic index since it ignores cash flows beyond the payout period, does not consider the time value of money, and penalizes projects with long lifetimes. The payback period should not be used in solar optimization studies, since it is an insensitive index and will give an improper optimum system size. The payback period may be quoted as an auxiliary economic statistic but it is not the key one.

The calculation of solar economics using Eq. (4.15) is rather laborious, and a simplified method has been used by some designers to select the optimum solar system. Instead of total life-cycle heating cost or average solar energy cost C_{se}, the index which has been proposed is simply the solar cost index (SCI), i.e., the number of British thermal units delivered per year divided by the initial system cost. The SCI is related to the reciprocal of the average cost of solar energy C_{se} defined in Eq. (4.18). If the initial cost is related to the annualized cost by a constant, then the SCI gives the same size results as the average cost of solar energy based on life-cycle economics. In many cases, the annual cost and initial cost are proportional, as seen in Eq. (4.19). The only nonproportional effect in this equation occurs in the replacement term and is a relatively small effect.

The use of the SCI will always result in an optimum collector area different from the proper optimum area, since *the average cost is not the same as the marginal cost.* The marginal cost is the cost of the last unit of solar energy added to the system, whereas the average cost is the average of all units of solar energy and their cost. The minimization process described in the previous section is equivalent to equating the marginal cost of fuel and the marginal cost of solar. If the SCI average cost is used instead, then a nonoptimum system is identified. However, the results from the two methods of analysis differ little, and given the flat optimum of the total cost curve, the SCI method can probably be justified in some cases, although it is not strictly correct from an economic viewpoint. The ease of calculation, however, is its most attractive feature.

OUTPUT OF THE DD PROCESS

The output of the design development process is a set of design development drawings and calculations which represent the solution to all problems identified during the design development process. All questions have been identified and answered by the end of this design phase. An *outline performance specification* is also included as part of the deliverables from the DD effort. The outline specification and DD solar drawings are based on the economically optimum system identified in the preceding economic analysis. If economic optimization is not the selection criterion but one of the other two criteria noted on page 269 is used, the DD drawings are based on that system size. The depth of detail of the economic analysis depends upon the nature of the project and the needs of the owner. In some cases, SD-level economics can be used during DD for smaller jobs to identify the optimum-cost solar system.

The DD documents are the basis of construction documents and working drawings. During the CD phase, little additional refinement of the design should be done. The CD activity is devoted to clarifying the DD outputs and reducing them to legally binding construction documents. In this section, some examples of DD output are shown.

Another product of the DD effort is a detailed cost estimate of the solar system. During the identification of the optimum solar system, the estimate of cost for components uses power-law interpolation for only a limited number of system sizes. After the economically optimum system is identified this way, it is recommended that a formal quote be acquired to reestablish the correctness of the DD cost estimate of the optimum size. The DD cost statement itself must also be adjusted for inflation which is expected to occur between the date of the estimate and the date of bidding following the CD phase.

Example System

The example DD system diagram shown in Fig. 4.43 includes the nominal size of all components in the system, including collector area, pump flow rates, storage-tank and expansion-tank size, as well as other components of the system. The proper arrangement of all piping subsystems is shown, and notes are provided to clarify the function of the system as required. This DD diagram is not a piping diagram although most fittings are shown. The inclusion of every pipe fitting and component in a DD-level drawing is not necessary and is not commensurate with the level of work carried out during this phase.

Specifications at the DD level are also not in the detail required for construction documents. The sizing of all components must be called out, of course, but detailed sections on legal matters, performance penalties, etc., need not be included during DD. The level of the DD specification is dictated by the requirement that an accurate cost estimate be made. In some cases, the owner will use DD documents to request an independent cost study by a cost-estimating firm. Therefore, all components with an associated cost must be included in the DD level outline performance specification.

An example of an outline specification for a passive heating system is given in Fig. 4.44. The specification should be numbered in accordance with the CSI convention. Solar systems are normally included in Division 13 for special construction or Division 15 under mechanical construction. If a separate cost estimate is desired by the owner for the solar system, it is most easily accomplished if the solar specification is included in Division 13, where it can be considered separately from other mechanical systems in Division 15.

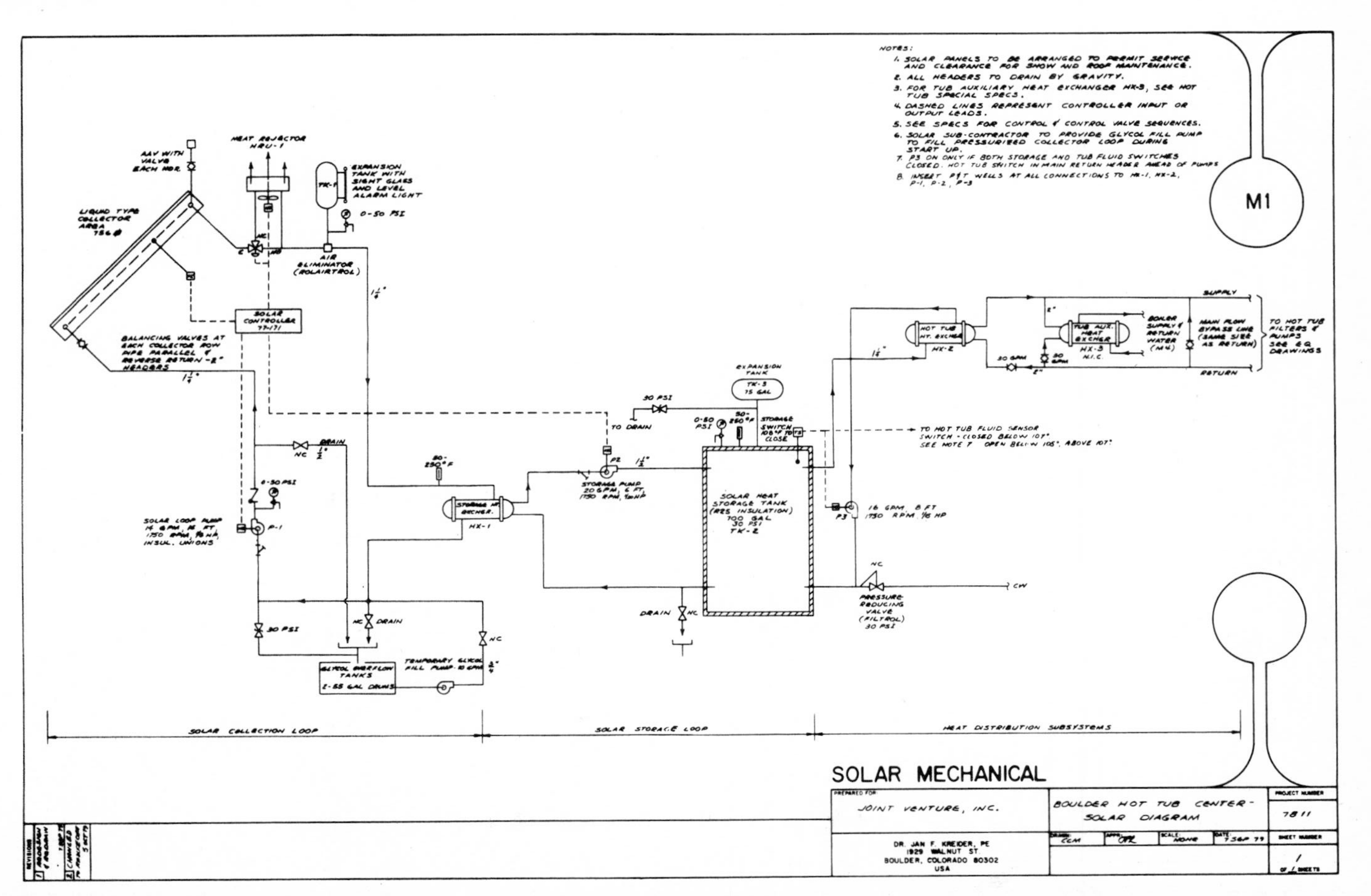

FIG. 4.43 DD-level solar process-heating system one-line diagram showing arrangement of components and nominal system sizes with explanatory notes. *(Courtesy of JFK Associates, Boulder, Colo.)*

1. Scope of Work:

   A. Provide a complete solar thermal wall system consisting of the 12-in concrete wall, custom glazing, insulating curtain, controls, and all miscellaneous items described herein for the Gunnison County Airport.

   B. This scope of the work is limited to the function of the thermal-wall heating system only. For architectural, structural, electrical, and other work, refer to the Working Drawings and Specifications of Associated Architects of Crested Butte, P.C.

2. Concrete:

   A. All concrete shall be in accordance with ACI with 28-day strength of 3000 lb/in^2. Deformed reinforcing steel shall be in accordance with ASTM and shall be detailed, fabricated, and placed as per ACI *Manual of Standard Practice*, latest edition.

   B. The concrete thermal wall shall be 12 in thick and shall have vent openings as per the drawings.

   C. The concrete forming shall be made to the full height of the wall (as per drawings), and all vertical construction joints shall fall behind mullions.

   D. The concrete shall be medium-dark and earth-tone in color to be approved by the architect. The south-exposed surface of the concrete wall shall be finished according to the drawings.

3. Structural Steel and Miscellaneous Metals:

   A. All steel angles, plates, channels, and connectors shall be in accordance with ASTM. All steel shall be detailed and fabricated as per the latest provisions of the AISC *Manual of Steel Construction*.

   B. Miscellaneous metals shall include ¼ × ¼ in W.W. screen mounted over supply openings at the base of the thermal wall and sheet-metal closure at mouth of heat-recovery duct as per drawings.

4. Thermal and Moisture Protection:

   Provide 2-in foam perimeter insulation to below and under the entire foundation. (Refer to drawings for location.)

5. Custom Glazing Sections:

   A. Provide custom glazing sections by LOF or approved equal.

   B. All glazing frames and mullions finishes shall be approved by the architect.

   C. Glass for the thermal wall shall be clear, tempered glass with round edges and double-glazed thickness per drawings.

   D. Caulk and seal glazing sections as per manufacturer's recommendations.

   E. Operable vents at top of glazing section per drawings.

6. Insulating Curtain:

   A. Provide insulating curtains by Thermal Technology Corporation, P.O. Box 130, Aspen, Colorado.

FIG. 4.44 Example DD-level outline passive heating specification. *(Courtesy of JFK Associates, Boulder, Colo.)*

B. Curtain tube shall be standard finish.

C. Operating controls and motor to be provided as per manufacturer's recommendations. The controls shall have a manual override and seasonal changeover.

D. An outer curtain fabric shall be provided and approved by the architect.

E. Access to the curtain tube shall be provided. Allow for curtain tube length plus 2 in in order that the tube can be removed for service.

F. Provide second upper limit switch to be closed when shade is up, 1 A, 24 V ac. Switch coordinates MD-1 damper position with modulation position and operates indicator lamp.

7. Instrumentation:

Minimal instrumentation shall be supplied to monitor the performance of a typical passive wall element regarding operation modes and air and wall temperatures.

A. Supply a digital-readout temperature sensor with selector switch to measure the following temperatures:

1. Inlet vent air
2. Outlet duct air
3. Wall temperature ½ in from south surface
4. Wall temperature 3 in from south surface
5. Wall temperature 6 in from south surface
6. Wall temperature 9 in from south surface
7. Wall temperature ½ in from north surface
(Electrotherm, Denver, Colorado, or equal)

B. Sensors to be placed carefully in the wall during concrete pouring and to be in good contact with concrete. Install in wall section No. 11.

C. Insulated curtain on the same passive wall section (11) to have light panel to show shade up position by red light and down position by blue light. Wire to shade automatic controller per Thermal Technology Corp. instructions.

D. Motorized damper MD-1 signal to activate green light to indicate passive wall airflow for heating zone 11.

E. Install all above instruments on lockable panel in box at location specified by the architect.

8. Submittals:

Submit catalog data and shop drawings for the glazing sections, insulating curtain, and controls prior to installation. Provide an operator's manual for the insulation system.

9. Acceptance Test:

In the presence of the engineer, conduct an acceptance test of the insulation control system and the heating-system control system. Test will include a functional test of both systems in all modes as specified in the control drawing. Adjust as required for proper operation in all modes.

FIG. 4.44 *(Continued)*

DESIGN DEVELOPMENT PHASE CHECKLIST

The checklist below is designed to assist the solar designer by listing many details of each solar component which must be considered during DD. Preceding these lists of component details is a summary outline of the DD phase. Refer to preceding chapter for details on each entry.

DD Phase Solar Activities

- Reevaluate loads by appropriately detailed method using final DD information ☐
- Reevaluate solar economic performance using detailed cost estimate and DD-level economics and SLR, F-chart, or hourly computer simulation ☐
- Resize solar system if necessary on basis of above reevaluation ☐
- Optimize component sizes if justified by size ☐
- Specify all components ☐
- Prepare DD solar drawings and details ☐

Solar Collectors—Active

Materials

- Steel or copper absorber ☐
- Tempered glass covers ☐
- Simple method of glazing replacements is preferred ☐
- Nonfoam insulation adjacent to absorber (see Ref. 69, pp. 78, 79) ☐
- No wood present ☐
- Insulation without organic binders ☐
- Moisture seals ☐
- Dust seals ☐
- No thermal contact, absorber to housing ☐
- Provision for expansion ☐
- Proper mounting system ☐
- Conformance with specification ☐
- Allowable weight ☐
- Internal moisture controlled ☐
- Acceptable glazing seal system ☐
- No dissimilar metals in contact ☐
- Proper riser-to-header joint ☐
- High-temperature solder ☐
- Use longest collector size possible ☐
- Check compatibility with air pollutant species ☐
- Internal manifolding preferred ☐

Performance

- Acceptable thermal efficiency ☐
- ASHRAE test data to be provided by manufacturer ☐
- Proper pressure drop—high enough ☐
  —low enough ☐
- Acceptable incidence-angle modifier ☐
- Proper warranty ☐
- Fluid cross flow with parallel-flow risers ☐
- Flow rate, approx.: 0.025–0.040 gal/(min · ft_c^2), liquid ☐
  2–4 ft^3/(min · ft^2) air ☐
- Annual performance prediction—F-chart for DD system ☐

Mounting

- Design for maximum wind (lift and drag) + snow load ☐
- Design for later collector repair or replacement ☐
- Design to raise collector above roof snow level plus collector slough-off ☐
- Consider added roof load from permanent snow buildup to north of collector ☐
- Design for seismic moments ☐
- If mount metal differs from collector housing, design galvanic break between ☐
- Consider back-surface facade for collector ☐
- Account for array expansion/contraction ☐
- Account for tolerances in designing collector array supports (collectors with no side fasteners permit closer packing) ☐
- Provide snow slide deflector for pedestrian or auto traffic area ☐
- Estimate collector array installed cost ☐

Solar Collectors—Passive Apertures

- Specify size ☐
- Use tempered glazing, inner and outer ☐
- Specify high-transmittance glazing ☐
- Design glazing seals and supports to accommodate expected expansion and contraction ☐
- Calculate glazing wind load ☐

Design movable insulation —tracks ☐
—rollers ☐
—control switch points ☐
—seals ☐
—access for adjustment and repair ☐
—manual controls ☐

Consider full range of possible solar angles to assess internal glare problems in DG systems ☐

Consider full range of possible solar angles to assess external reflected glare problem ☐

Consider transparent infrared mirror to control glazing heat loss; base decision on LC costs ☐

Consider triple glazing to control heat loss; base decision on LC costs ☐

Size overhangs, louvers, or other solar control devices to match annual heat-load phasing ☐

Provide operable windows to exhaust excess heat gain ☐

Analyze LC economics of reflectors to enhance solar gain ☐

Select fabrics, carpets, and other materials inside DG apertures to be insensitive to prolonged ultraviolet exposure ☐

Piping

Use reverse return for collectors and other parallel loops where required ☐

Size headers for proper flow balance (90%/10% rule) ☐

Specify high-temperature solder ☐

No hose connectors to be used ☐

No galvanized pipe above 140°F ☐

Show insulation on all fittings, pipe runs, and joints ☐

Provide clearance for insulation at wall or roof penetrations ☐

Incorporate proper pipe slope for gravity drain ☐

Incorporate proper pipe supports ☐

Account for piping expansion/contraction; consider double-offset collector connections or special coupling, e.g., Victaulic ☐

Pipe sized for velocity below erosion and noise limits but above entrainment velocity (2 ft/s) ☐
Specify dielectric couplings and sacrificial anodes to protect pipe loop materials ☐
Specify high-temperature insulation within 20 ft of collectors ☐
Specify insulation level ☐
Specify insulation jacket based on exposure and usage ☐
Specify pipe loop pressure to avoid boiling and pump cavitation ☐
Provide permanent prime for non-self-priming pumps ☐
Estimate piping installed cost ☐

Valves

Drains

Collector loop (2) ☐
Heat exchanger ☐
Storage tank (use large-diameter pipe for faster drain) ☐
Distribution loop ☐

Isolation Valves

Pumps ☐
Heat exchangers ☐
Storage ☐
Collector ☐
Control valves ☐

Specify check valves or motorized valves in all fluid loops which are able to thermosiphon in reverse ☐
Specify valve body, seat, and packing to be materials compatible with
—fluid ☐
—fluid pressure ☐
—fluid temperature ☐
Mixing valves in DHW systems to fail to minimum input temperature setting ☐
Include balance valves wherever equal flows–parallel circuits are required ☐
Specify precision of balance valve settings (if flows must be known, insert flowmeters in loops where sufficient inlet and outlet straight pipe lengths occur) ☐

- Provide relief valves in any branch which can be isolated with valves while heat is added to that branch ☐
- Provide relief valves in each closed fluid loop ☐
- Size relief valves to provide maximum expected flow rate at peak overpressure conditions ☐
- Show relief valve discharge to safely located drain on drawings ☐
- Minimize use of three-way valves ☐
- Estimate the installed cost of all valves and the LC operating cost of solenoid valves ☐

Dampers

- Specify very low leakage dampers in critical locations in air systems ☐
- Specify damper seal material and adhesive compatible with high-temperature solar-heated air ☐
- Note proper damper linkage adjustment on drawings ☐
- Specify dampers with high-quality sealed bearings ☐

Fluids (for Liquid-Based Systems)

- Confirm specific heat, density, and pour point data ☐
- Specify corrosion inhibitors ☐
- Specify operating temperature range ☐
- Specify freeze protection (specific gravity if aqueous) required ☐
- Specify pH acceptable range (if aqueous) ☐
- Consider effects of low surface tension of some organic fluids on pump and pipe fitting selection ☐
- Design fluid fill and makeup systems; do not use automatic water makeup ☐
- Calculate volume of special fluids needed for each loop involved ☐
- Account for local water hardness by
  - —water softening ☐
  - —using distilled water ☐
- Estimate installed cost of all special fluids and frequency of replacement ☐

Estimate installed cost of fluid fill and makeup systems ☐
Estimate installed cost of all pipe fittings ☐

Ducting

No use of duct tape on duct at outlet of collector; use silicone or equal ☐
Design collector ducts for proper collector flow balance (90%/10% rule) ☐
Show insulation on all ducts and seal gaps ☐
Require sealing all duct joints ☐
Specify maximum allowable duct air leakage ☐
Specify high-temperature insulation within 20 ft of collectors ☐
Size for nominal velocity 800–1200 ft/min ☐
Calculate pressure drops in all duct assemblies for fan sizing in all modes involved ☐
Outdoor intake and exhaust grilles to be at a distance from each other to avoid exhaust recycle ☐
Estimate installed cost of all ducting ☐

Pumps

Size to handle all loop component pressure drops including ☐
Collector ☐
Air separator ☐
Control valves ☐
Isolation valves ☐
Check valves ☐
Heat exchanger ☐
Drain/fill system ☐
Filter ☐
Screen ☐
Pipe runs ☐
Pipe fittings ☐
Sensor wells ☐
Storage ☐

Specify flow rates in accordance with text rules ☐
If system size warrants, optimize pump size, pipe size trade-off ☐
Acquire curves for all pumps ☐
Select pump casing, rotor, and seal materials compatible with fluid pumped ☐

Fans

- Size to handle all loop component pressure drops in all modes ☐
- Collector ☐
- Storage ☐
- Grilles ☐
- Filters ☐
- Dampers ☐
- DHW heating coils ☐
- Duct runs ☐
- Duct fittings ☐
- Backdraft dampers ☐

- Specify flow rates ☐
- Acquire curves for all fans ☐
- Determine if flows in all modes balance or if second fan is required ☐
- If motor is in hot airstream, specify high-temperature type ☐
- Estimate installed costs of all fans and LC operating cost ☐

Storage (Liquid)

- Determine proper size based on LC cost ☐
- Specify tank material ☐
- Specify tank insulation level based on LC cost and environment of tank ☐
- Specify tank pipe fitting size, location, purpose ☐
- Specify tank location—inside/outside ☐
- Specify tank manhole size and location ☐
- Specify tank insulation—no polystyrene ☐
- Design tank fluid inlet diffusers, if to be used (see Ref. 70) ☐
- Use fluid cross flow for both input and output loops ☐
- Use maximum stratification possible without significant extra cost added (consider smaller tanks in series for large systems) ☐
- Specify sensor locations ☐
- Specify tank interior surface coating or lining and anode, if any ☐
- Design expansion tank for storage ☐
- Design tank supports for proper stress, thermal break, and load capacity ☐
- If buried, specify waterproof outer membrane ☐

- If buried, specify waterproof insulation ☐
- If buried, specify hold-down straps if high water table ☐
- If buried, specify anodic protection ☐
- Design bypass for summer to avoid full tank heating when not required ☐
- Provide storage room ventilation for summer ☐
- Specify storage loop operating and maximum pressures ☐
- Provide Filtrol to maintain system pressure and fluid inventory ☐
- Estimate installed storage cost ☐

Storage (Rockbed)

- Determine proper size based on LC cost ☐
- Determine storage location—inside/outside ☐

Specify Rock

- Size range ☐
- Cleanliness ☐
- Type ☐
- Amount ☐

Specify Storage Container

- Material (consider possible wood warpage) ☐
- Airflow path length ☐
- Volume ☐
- Inlet plenum volume ☐
- Outlet plenum volume ☐
- Insulation level and type on basis of LC cost ☐
- Inlet duct location and size for proper flow balance ☐
- Outlet duct locations and size for proper flow balance ☐
- Cover design and materials ☐
- Cover method of sealing and attachment ☐
- Cover R value ☐
- Method of duct-to-container seal ☐
- Fireproof lining ☐
- To use vertical airflow only ☐

- Design sensor location device and support ☐
- Design support for rocks with storage container
  - —screen ☐
  - —support ☐
  - —spacing ☐
- Design rock-bed thermal breaks ☐
- Design summer bypass mode ☐

- Provide for fan, filter, or damper mounts to storage where practical ☐
- Calculate rock-bed Δp (must be above approx. 0.12 in of water, gauge, to avoid channeling) ☐
- Estimate installed storage cost ☐

Storage (Passive)

- Determine required storage capacity [Btu/(°F·ft_c^2)]
  - —by computer simulation ☐
  - —by rules in text ☐
- Select material; determine heat capacity and density ☐
- Determine thickness [required capacity/(ρc_p)] if TSW, or volume if DG ☐
- Specify dark exposed-surface colors subject to aesthetic criteria ☐
- Specify required exposed surface area
  - —aperture area for TSW or WW ☐
  - —3X aperture area for DG ☐
  - (or use computer simulation)
- For TSW, specify wall finish; consider selective surface for heat-loss control ☐
- If floor slab storage for DG is used, specify matte floor finish to control glare ☐
- Specify unobstructed heat-flow path from inner TSW surface to space ☐
- Design TSW support ☐
- Design TSW thermal break with floor slab ☐
- Specify TSW vent dimensions and locations ☐
- Design upper TSW vent backdraft damper and select material ☐
- Specify TSW night insulation, if any ☐
- Specify TSW vent louver inlet screen ☐
- Provide fire dampers in TSW cavities, which may act as chimneys in case of fire ☐
- Estimate installed storage cost increment above that for nonsolar building ☐

Heat Exchangers

- Select effectiveness based on LC cost ☐
- Determine if double-walled heat exchanger required ☐
  - —text recommendation based on usage

Specify Heat Exchanger

- Type ☐
- Materials ☐
- Heat rate ☐

Flow rates or flow velocities ☐
Operating pressure ☐
Operating temperature range ☐
Effectiveness ☐
Allowable pressure drops ☐
Use standard sizes wherever possible ☐

Calculate and specify insulation level based on LC cost ☐
Check for pipe rattle owing to high velocity (S&T exchangers) ☐
If baseboard space heating, calculate length of baseboard considering reduced solar system delivery temperatures ☐
Design humidifier for design peak conditions ☐
Specify submerged coil size and geometry to enhance free-convection transfer in storage tank ☐
Estimate installed costs of heat exchangers ☐

Controls

Function (Identify All Modes)

Collection ☐
Storage ☐
Heat delivery ☐
Ambient reset for delivery loop ☐
Setbacks ☐
Heat rejection ☐
Power failure ☐
Auxiliary heat ☐
Level alarm ☐
Off ☐

Specify Conditions for Each Mode to Be Selected

Temperature ☐
Seasonal ☐
Solar ☐
Other ☐

Specify position of each actuator for each mode ☐
Prepare control diagram ☐
Select sensor type—electrical preferred ☐

Carefully Specify Sensor Locations

Collector ☐
Storage ☐
Ambient ☐
Space ☐

High/low limit switches ☐
Other ☐

Provide trickle flow if collector sensor not upon absorber plate ☐
Specify adequate hysteresis to avoid cycling ☐
For air systems specify collector turnoff if $T_{c,o} < 85°F$ ☐
Design control to protect solar system if power failure ☐
Design control for safe start-up after power failure ☐
Design to avoid cold collector fluid transients at start-up ☐
Provide time delay for solar recovery from night setback ☐
Specify control unit location ($T > 55°F$) ☐
Design seasonal changeover in SH systems to avoid storage heating when no SH load exists ☐
Require uniform hysteresis over entire operating temperature range ☐

Passive Systems

Specify movable insulation mode shift criteria ☐
If active air distribution is used, specify criteria ☐
Solar sensor for movable insulation to be unobstructed ☐
If TSW temperature is measured internally, use waterproof sensors ☐

Estimate total installed cost of control system and LC operating cost including all actuators ☐

Heat Rejection

Determine peak rate, Btu/h; design for collector outlet fluid temperature above 220°F ☐

Select System Type

Hot-water design ☐
Cooling tower ☐
Collector stagnation ☐
Night collector operation ☐
Liquid-to-air exchangers ☐

Design control strategy and sensor location and type ☐

For Automatic Subsystems for Cooling Tower Specify

Water makeup ☐
Water chemistry ☐
Blowdown ☐

Freeze Protection

Base design on lowest probable temperature for 20–30 year period, not 97½% or 99% design temperature ☐

Emphasize simplicity and fail-safe characteristics in freeze protection subsystem design ☐

Select Subsystem Based on LC Cost and Risk

Antifreeze ☐
Hot fluid circulation ☐
Drain down ☐
Drain back ☐
Electric heat ☐

Provide freeze protection for noncollector loop components in unheated spaces ☐

Estimate the installed cost of the freeze-protection subsystem ☐

Expansion Tanks

Size for maximum possible collector loop temperature excursion (exclusive of stagnation) ☐

Specify pressure limit corresponding to collector loop design pressure ☐

Specify acceptance volume and allowable pressure swing ☐

Locate tank fitting at loop pump inlet to provide adequate pump inlet NPSH ☐

Use air separator at tank inlet ☐

Include pressure switch at tank to control fluid makeup pumps ☐

Estimate the installed cost of expansion tanks ☐

Other Components

Locate air vents at high points and points of low fluid velocity—liquid systems (e.g., storage tank) ☐

Provide vacuum breakers on all loops routinely drained ☐

Size auxiliary systems to have 100% backup capacity at design conditions ☐

Provide Filters in Air or Liquid System to Protect

- Rock storage ☐
- Collector airways ☐
- Collector waterways ☐
- Heat exchangers ☐

- Estimate installed cost of all miscellaneous components ☐

Instrumentation

- Measure temperature at all points of heat extraction, addition, or storage ☐
- Measure pressure across each pump, fan, or filter ☐
- Measure pressure at at least one point in each liquid loop ☐
- For system debugging, direct flow measurement in each loop of a system in each mode is very useful ☐
- Use inexpensive photovoltaic pyranometer to measure solar flux for simple system checkout purposes ☐
- Sight glasses on liquid tanks are useful level indicators ☐
- To measure pump or fan hours of operation, specify hour meters at each device ☐
- Estimate installed cost of instrumentation ☐

DD Deliverables

- Drawings—system diagrams, pipe or duct layouts, collector mounting, details showing arrangement of all components listed above ☐
- Outline specifications for all components listed above in checklist ☐

Calculations

- Loads ☐
- Solar performance and expected accuracy thereof ☐
- Sensitivity studies as necessary ☐
- LC economics—LC savings, payback, ROI, etc. ☐

Cost Estimate

- Total of above installed costs ☐
- Compare to project solar budget ☐

- Apply for possible government subsidy for project ☐
- Secure DD approval and authorization to proceed ☐

5 CONSTRUCTION DOCUMENTS AND CONTRACT ADMINISTRATION

I'll tell you how the sun rose—
A ribbon at a time
The steeples swam in amethyst
The news, like squirrels, ran

EMILY DICKINSON

Construction documents consisting of working drawings and specifications set forth in detail the requirements for the construction of a project. The questions resolved during design development are implemented in detail on the working drawings. During the construction documentation phase, the development of any new technical information is minimized. The system developed during DD is that which is shown on the construction documents except for small details.

The refinement of the DD system is focused on resolving questions regarding the interface of the solar system with other building systems. These systems include the structure, the conventional mechanical system, the electrical system, and other subsystems present in a given building. The resolution of interface problems may introduce small changes into the solar design, but major changes should not occur because interface questions have been evaluated continuously since SD.

A statement of probable construction costs may be refined during construction documentation as well. Major changes in cost should not occur. After the completion of CDs, a supplementary detailed cost estimate may be prepared by an experienced cost estimator or construction manager. This is the final check on project cost for the owner prior to the bidding phase.

Although the bidding phase is not a part of CD or CA, it is described briefly in this chapter since it occurs between these two phases. The solar system bidding phase is not very different from that for any other bid package on a building. A few suggestions are made regarding the special requirements of solar bidding because of contractor unfamiliarity.

Contract administration is the responsibility of the architect or construction manager. During the CA phase, the building undergoes construction. This process is terminated upon the acceptance of the building by the owner which follows the inspection of each system by the responsible designer or someone designated by the designer. The CA section of this chapter describes many of the lessons which have been learned regarding the installation of various components of solar systems, the start-up of the system, and the inspection of the system. Some additional material is presented regarding the structure of an acceptance test and typical types of troubleshooting which are occasionally required during the start-up of the system.

SOLAR SYSTEM/BUILDING INTERFACE

The key question resolved during the CD phase is the proper interface of a solar system with the balance of the building systems. This integration is both an architectural and an engineering effort, responding to the aesthetics of the installation as well as the function. During SD and DD, the building–solar system interface enters the design at several points which have cost impacts and which affect the selection of the cost-optimal system. However, the layout of collector arrays, piping, and supports, for example, are not detailed during DD. This type of information is contained on the working drawings prepared during CD.

In this section, a number of questions which arise during the building interface analysis are addressed. Since the collector is the unique feature of a solar system from the HVAC engineer viewpoint, most of these concerns arise from integration of a collector with the building. The installation of fluid piping and other equipment in the mechanical space is not fundamentally different for solar or nonsolar heating systems. It is well beyond the scope of this book to present a complete set of solar-architectural details. The best collection of these is given for active systems in the *Active Solar Energy System Design Practice Manual* prepared by the Department of Energy (Report No. Solar/0802-79/01).

Integration of the Solar Collector Arrays with the Building

Solar collector arrays are mounted on either the tilted roof or the flat roof of a building if the array is to be integrated with the building. Passive apertures, of course, are usually part of south-facing walls. Locating any physical component on the roof of a building can cause problems. The roof is one of the most critical elements in a building, and the imposition of a collector above the roof with requirements for piping connections and structure adds to the difficulty of designing this already critical component. The design of collector supports must respond to several criteria, the primary of which is the support of the collector physical load. Also, the location of other mechanical equipment and stacks is to be accounted for. In addition to the dead load of a solar collector, wind and snow load must also be considered as described in Chap. 4 (for more details see Refs. 33 and 60).

Collector supports must be designed with consideration of aesthetics. The appearance of a collector on a roof depends, for example, on the location of the array on the roof. If the collector is too high on a sloped roof, the building will not look balanced and the collector manifold will be exposed to the wind, reducing performance. If the collector is too low on the roof, similar appearance problems occur and there is a possibility of icing between the eaves and the bottom of the collector. If this icing persists through the winter, water beneath it may wick up along the roof beneath the shingles. Also, collectors mounted on a sloped roof but at a greater tilt angle than the roof will collect snow behind them. This snow will not thaw for extended periods and an ice dam may result. A better approach is to raise the collectors above the roof completely.

Collector support design must also consider future activities such as the routine maintenance including inspection of pipe fittings and cleaning of covers. The collector array, therefore, must have good access. In addition, replacement of glazing or an entire collector may be required in the future. The structure must be designed so that a solar collector can be removed and readily replaced. The need for collector removal also affects piping insulation design. These requirements must be considered carefully since relatively minor features of the design can have major effects on the maintainability of a collector array.

Supports and Attachments. All collector supports require a penetration through the roof membrane. The penetration may be only a local pitch pocket supporting a vertical element of a collector structure; or the penetration may be large enough to accommodate an entire collector array as is the case for collectors integrated with sloped roofs. In both extremes, proper flashing must be used to protect the roofing felt and roof deck from moisture for a very long period of time.

Figure 5.1 shows various sections of a flashing system for a flat-plate collector integrally mounted with the roof. The sections show that a continuous waterproof membrane is used from the collector cap strip around the collector surface and onto the roof deck. If the flashing is made from more than one piece of metal, as is the case when counterflashing is used, each section must be sealed. Lead flashing is considered to be one of the most durable.

Intermediate between a fully flashed collector and one mounted above the roof is the curb mounting method shown in Fig. 5.2. This approach is frequently used on horizontal concrete roofs. The curb is poured in place as part of the roof slab and is later used to support the collector above the basic roof surface. Any curb used for this purpose must have breaks at the appropriate locations to permit water drainage. The location of curb breaks is determined by the slope of the roof, location of drains, and position of curbs. A similar drainage problem must be addressed in the design of fully integrated collectors as shown in Fig. 5.1 where the drainability of the roof must not be altered by the addition of the collectors.

The minimum amount of roof penetration is by vertical supports. The vertical support will have high load-bearing capacity and relatively small area; therefore, the roof penetration will be smaller than for the designs shown in Figs. 5.1 and 5.2. Figure 5.3 shows an example of an upright structural member placed in a pitch pocket, both of which are attached to the roof. The pitch pocket container, on its exterior circumference, should have a lip to accept flashing, which is then continued down onto the roof surface. The pitch pocket is filled with pitch or roofing

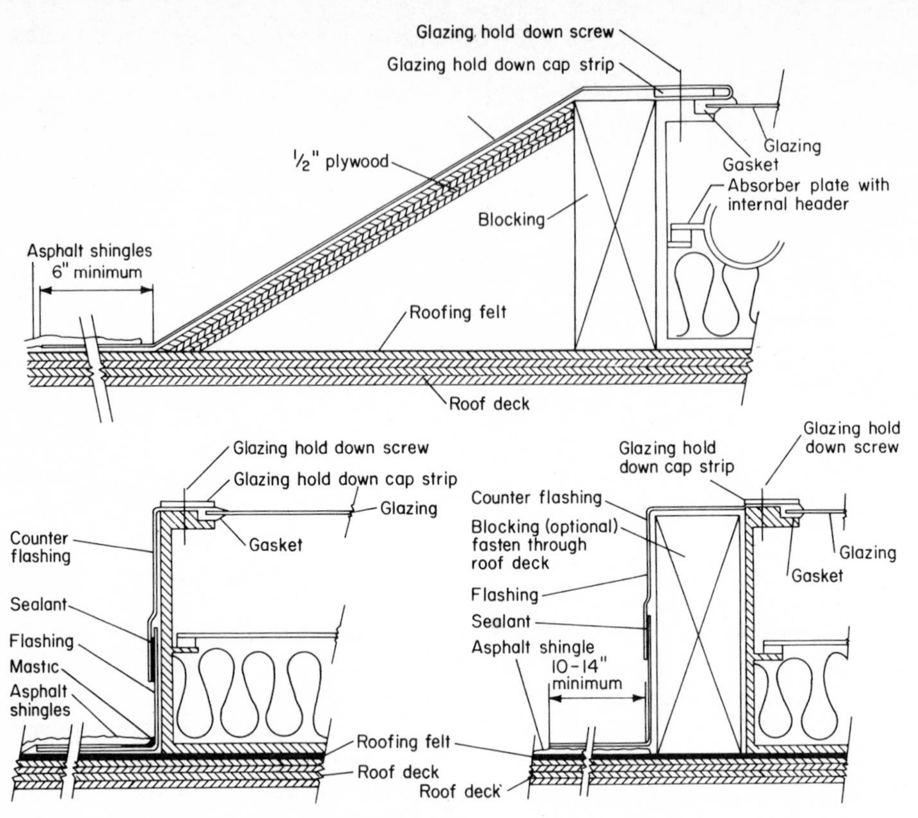

FIG. 5.1 Sections of a roof-integral collector mounting.

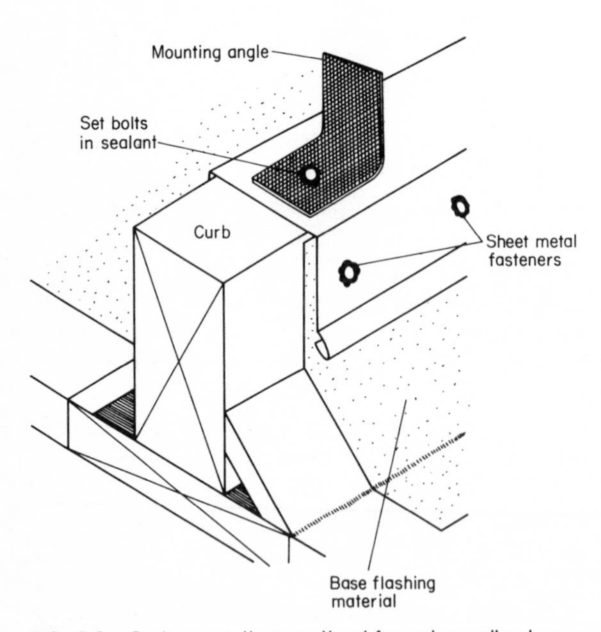

FIG. 5.2 Curb-mounting method for solar collectors.

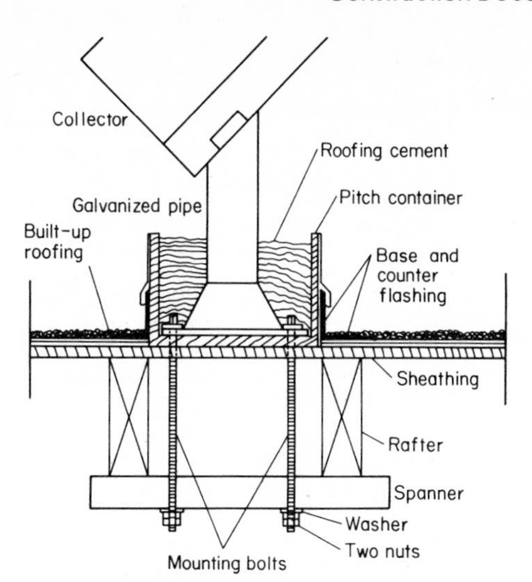

FIG. 5.3 Pitch pocket cross section for mounting collector arrays above a flat, residential wood roof.

cement to afford a weatherproof seal. Pitch pockets, although they require minimal penetration through a roof membrane, are relatively costly and do require frequent inspections since they do not have permanent waterproofing qualities. Only one structural element should be used in each pitch pocket.

The National Association of Roofing Contractors has a set of standard roof penetration detail drawings. Some require neither a pitch pocket nor any rubber diaphragm and can be adapted for solar usage.

Collector-Array Structures

The previous section has briefly described the types of penetration which accommodate collector supports. A structural assembly is used above the roof itself to support each collector, its piping, and its valves. The design of the collector structure is the responsibility of the structural engineer and is based on the expected dynamic loads, principally wind, snow, ice, and hail as well as the dead loads. The HUD Minimum Property Standards (33) require design of collector structure to accommodate 100 mi/h wind speed. This converts to a 25 to 30 lb/ft^2 loading at the surface. Of course, if local wind speeds can be above 100 mi/h, for example in mountain valleys, the local design wind speed must be used for collector structure design.

The materials used for collector supports include aluminum, steel, and wood. The determination of which material to use depends upon price and expected durability. In some cases, stainless steel has been used although the price is very high. Wood may be the most economical material in some locations. However, some wood assemblies have badly warped because of improper treatment; the designer must specify the type of wood pressure treatment to be applied to maintain structural integrity. Codes may restrict the use of wood on roofs.

A collector mounted as shown in Fig. 5.3 causes a point load, although the collector itself is a distributed load. The support for such a point load must be attached to a main structural member. This can be easily accomplished if the collector rows are parallel to the building structure. However, if the building is off the north-south axis, collector support spans may need to vary in length to accommodate the angular discrepancy between the collector support and the building. Typical lengths of structural supports are 10 to 15 ft. Of course, any structural span may be used, but economics usually places an upper limit at approximately 15 ft. The question of collector-to-building alignment vs. collector north-south alignment is answered during DD on the basis of economic considerations.

An additional feature of the support design relates to the requirement that collector inlet and outlet headers have a specific slope to ensure proper drainage. Roofs also slope and may slope in a direction different from that for the collector manifold. The designer must show sufficient information on the drawings to indicate to the plumbers that the pipe slope is independent of the roof slope. Since no roof is absolutely smooth, the support design should accommodate shims for final adjustments.

Figure 5.4 is an example of a flat-plate collector support structure used on a flat-roof building. The steel truss contains triangular sections in both plan and elevation views providing a rigid framework for the collector support. One of the difficulties the support designer may encounter is that the solar collector has not been selected at the time of design, since bidding has not yet occurred. Solar collectors can be purchased in many different sizes. If the manufacturer is not known, the size of the structure cannot be known. The problem can be circumvented either by preselecting or prepurchasing a collector using a prebid process for the solar collector (described later) or by designing a structure which can accommodate a range of collector sizes. For example, the collector face length may be allowed to vary between 6 and 7 ft. Design of a structure to accommodate every possible collector is not economically feasible, and a subgroup of available collectors must be specified for the bidding list if the prebid/prepurchase approach is not used.

Ground-Mounted Collector Arrays

The preceding discussion applies for the most part to new buildings into which the collector can be readily integrated. However, the structural capability or orientation of a building to be retrofitted may be such that integration of the collector on the building is not possible. In that case, the collectors are located nearby on the ground. Such a collector array should be located as near the building to be served as possible, given the constraints of shading and site limitations.

Since the ground-mounted collector array may be considered as a separate structure in some zoning districts, the verification of the status of solar units with respect to zoning must be established as the first step in the design. In some cases a building permit may even be required.

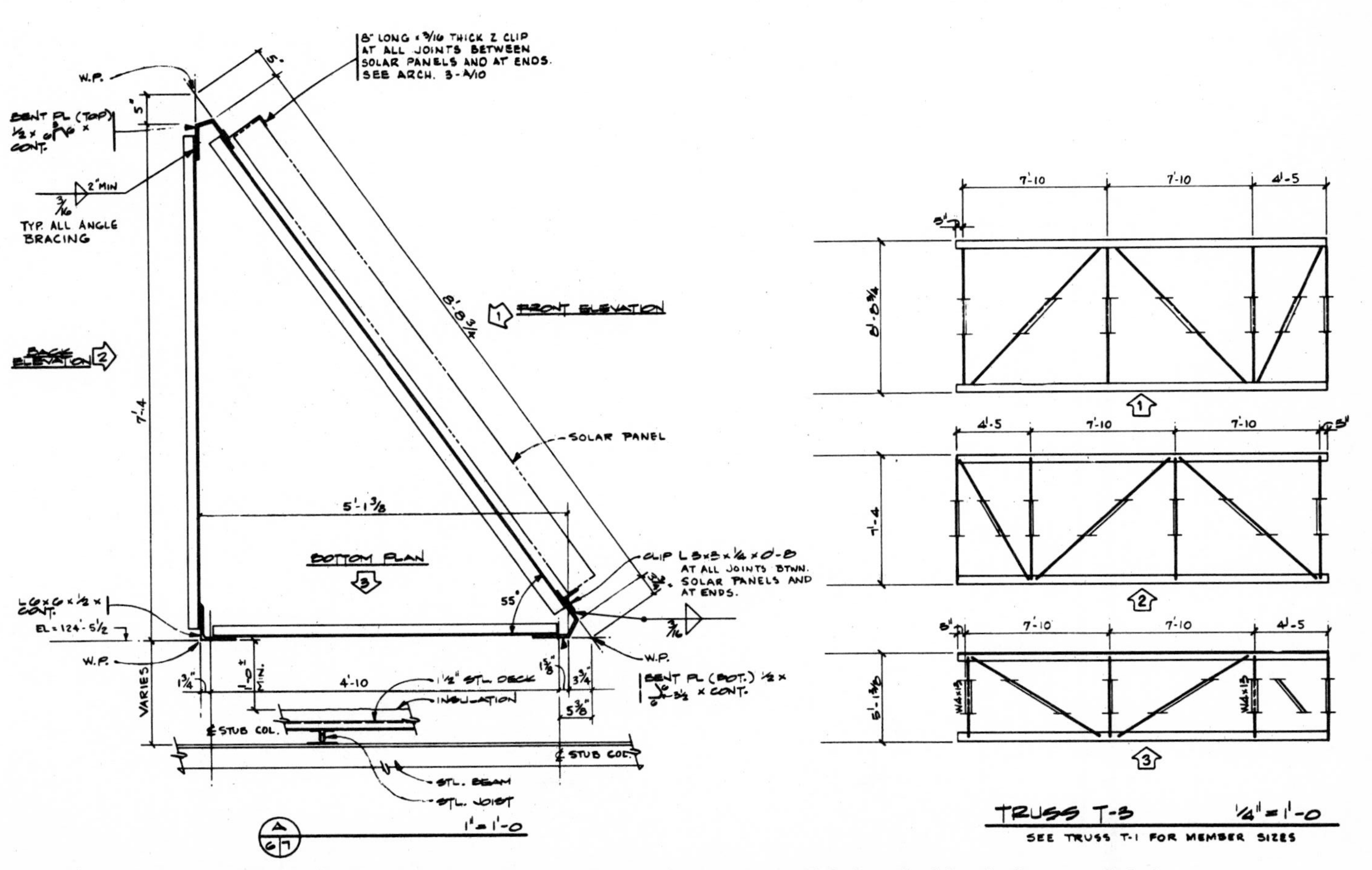

FIG. 5.4 Example structure used to support collector on flat roof. *(Courtesy of C. Lee Architects, Denver, Colo.)*

Ground-mounted collectors are typically supported by concrete piers which are extended below the frost line. The piers contain an embedded threaded rod to which the collector structure is mounted. The threaded rod will need to be sufficiently large to carry expected lift forces from the wind, a typical size being ½ to ¾ in diameter. The collector support assembly may be fabricated from wood, angle iron, pipe, or in some cases aluminum sections, although they are frequently more expensive than the equivalent steel section.

As in the case of roof-mounted arrays, corrosion must be considered in the selection of metal supports once a given collector is selected. Types of wood to be used for ground-mounted arrays are rather limited and cedar, redwood, or pressure-treated wood appear to be the best. The wood must be impervious to attack by moisture over long periods as well as by insects which may be present.

Since the ground-mounted array will not drain onto a roof which has its own drain system, a drain system must be supplied. A simple gutter to avoid soil erosion problems is all that is required at the base of each row of collectors. An alternative approach is to place pebbles below the collectors, thereby preventing soil erosion.

Ground-mounted collector arrays use buried piping. There are specific code requirements for buried piping regarding the number of pipe fittings, for example, per unit length. Buried piping must be very well insulated and completely sealed to avoid excessive heat losses and moisture damage. The pipe must also be positioned below the frost line to avoid any physical stresses. Careful design of buried piping is required since repairs are relatively costly.

A variation of a ground-mounted array is an open array structure located over a parking lot. This idea allows the same land area to serve two purposes. Previous comments apply to design of this structure. In addition, a method of avoiding icicle formation with its attendant potential for injury to passers-beneath is essential.

Design Considerations for Solar-System Mechanical Spaces

All components of active systems other than the collector are most often contained in the mechanical space of a building. Included are the storage tank, heat exchangers, pumps, expansion tanks, controls, and the backup system. This space also accommodates the other mechanical components of the building including air conditioners, water heaters, plumbing fixtures, and other process devices. The effect of a solar system on the mechanical space, of course, is to increase the area requirement. Since building area is very valuable, the added space required for mechanical components should be an absolute minimum.

One of the key considerations in designing a mechanical space, solar or not, is to provide adequate access to all components which may require servicing. In the solar system, this specifically applies to heat exchangers which may need to be disassembled for cleaning and to storage tanks which may require repair or draining. Heat exchangers need a length equal to the shell length in order that the tube bundle can be removed in its entirety for cleaning. Pumps also require servicing and adequate space must be provided for their disassembly.

The layout of the mechanical room should have a logical flow path from one component to the next. For example, the collector pump, collector heat exchanger, storage pump, and storage tank should be mounted adjacent to each other and near a drain. The storage tank, because of its large weight, will require an additional support pad in the mechanical room floor. The layout of the space must also account for the possibility of a storage tank leak. This may dictate that other components be placed above the floor by 4 to 8 in, depending upon the amount of water stored, so that a spill could not cause any damage. The requirement for a large drain near the storage tank is clear. To enhance stratification, the storage tank should be mounted vertically if possible.

Routine servicing of pumps, fans, filters, and collector fluid loops is also a determinant in the equipment layout in a mechanical space. The filters on both air and liquid systems must be changed or cleaned periodically. Depending upon the type of bearings used, pumps and fans also require periodic lubrication. It is best if this can be done without excessive contortions on the part of the owner or maintenance personnel.

The mechanical space may overheat during summer if the active storage tank is kept fully charged. It is recommended that the mechanical space be vented in summer to avoid imposing high temperature stress on electrical components such as pump motors, solar system controllers, and other control devices located in this space.

Instrumentation used to monitor performance of active solar systems should be located in a position where it is easy to read. It should be possible to glance at all temperature and pressure gauges from one or two locations in the mechanical space to very quickly determine if the system is functioning properly. The position of instruments should be called out by a note on the drawings.

Other Details

Passive systems are intimately coupled to the living environment in most designs. The transport of heat from the north surface of a passive wall depends upon the mechanisms of free convection and radiation, neither of which embody large driving forces. Therefore the thermal "connection" between the storage elements of passive heating systems and the space to be heated should be unobstructed.

Finishes. The paints and finishes to be used on visible components of active and passive solar systems should be specified by the designer. In passive systems, the finish on the inner and outer surface of the wall has an effect on both the appearance and the performance of the system. The outer TSW surface dictates the amount of solar radiation absorbed, and the inner surface, the amount of heat released.

Paints to be used for external components, principally collectors, must be selected with the concurrence of the architect, and the colors should coordinate with the building. Since solar collectors are subject to a rather severe environment with large excursions of temperature and high ultraviolet radiation levels, the selection of paint is not to be left to the painter. Proper priming procedures for collectors, for example, are important if paint is to adhere for extended periods. Etching primers are specified for this purpose.

Duct and Pipe Runs. Piping and ducting runs through buildings require space in addition to that for just the fluid conduit. The requirement that all solar piping and ducting be insulated will increase the space needed. The amount of insulation is determined during the DD phase, but the specific space for that insulation must be allocated during the CD phase. Particular problem areas are pipes and ducts running at right angles to structural members or through relatively narrow spaces in interior walls. Insulation must not be sacrificed in these areas, although its installation may prove difficult. The performance effects of inadequate insulation have been described in previous chapters.

Collector Access. Access to solar collector arrays should be provided on a long-term basis for both routine maintenance and replacement purposes in the future. If the collector array is mounted on a flat roof, access is relatively simple, and a walkway should be provided. Gravel roofs are more delicate than they appear, and leaks are difficult to locate. The standard type of walkway for gravel roofs should therefore be used along the front of the collector array if this is to be the location from which maintenance and replacement is done.

Some designers have incorporated walkways on sloping roofs. This seems ill-advised since walkways are expensive but will be used infrequently. It appears more cost-effective to simply use a ladder for access to collectors on tilted surfaces.

The design of either flat- or tilt-mounted collector arrays should consider the necessity for future roof repair. Collector spacing above the roof and flashing used with roof-integrated collectors are the primary factors.

WORKING DRAWINGS AND SPECIFICATIONS

Working drawings document the integration of all building subsystems into a single building system. From the point of view of the solar designer, the working drawings will include information on piping and ducting in active systems; collector mounting; pipe, pump, and heat-exchanger schedules; control system; operational modes; and many other details of the solar system design. Given the relative unfamiliarity of the building trades with solar system installation, *the solar designer should provide extra detail in the working drawings.* This will require an additional fee for the CD effort but is highly recommended since the information, if not on the drawings, will need to be provided in person during the CA phase. This is typically more expensive than doing the drawings properly in the first place.

Do not leave problems for contractors to solve. It is not the responsibility of contractors to design a solar system, and their solutions to design problems will frequently not meet with the approval of the designer. It is better to consider all problems during the preparation of working drawings for this reason. In this section, several suggestions are made regarding the preparation of drawings and the formulation of solar-system specifications. The preparation of working drawings is one of the standard efforts in any building design process and in itself is not treated in this book. The emphasis here is on mechanical drawings, but solar systems are reflected in structural and electrical drawings as well.

Piping

One of the key areas requiring attention of the designer is the location and function of piping. The requirement that all piping drain must be specified on the drawings. Experienced plumbers will carry out this task automatically, but the presence of experienced personnel cannot be guaranteed on every project. Specific locations of piping should be shown with regard to entry into the building. If the balancing techniques described in Chap. 4 are used for liquid collector arrays, the header will be larger than a plumber may consider necessary. Therefore, sizes must be clearly labeled on the drawing.

Any special installation requirements for pumps, heat exchangers, and other components must be clearly noted on the drawings. For example, some pumps require a perfectly level mounting in order that bearing wear not be accelerated. Proper piping symbols should be used for all fittings and the notation of function of valves whether normally open NO or normally closed NC should be shown on the drawings. Most plumbers are unfamiliar with the operation of solar systems and the addition of this NO/NC information will assure that the proper valves are open during start-up and debugging of the system.

Preparation of working drawings has an aesthetic element whether the drawings are concerned with the installation of a collector array on a roof or the location of piping and physical components in a mechanical room. Collector aesthetics are particularly important and should be addressed in the working drawings. The appearance of most flat-plate collectors is not particularly attractive and the positioning of such parts as manifolds out of sight may serve to improve their aesthetics.

Another example of a combined functional and aesthetic arrangement applies in the mechanical space. If components are positioned in a logical flow sequence as suggested earlier, the function of the system can readily be understood by building owners or visitors to the building. Piping will be labeled with arrows and fluid content. The selection of colors, lettering sizes, and lettering locations can add to the aesthetics of this space. These details must be called out in the drawings and specifications if the desired effect is to be achieved, however.

Tolerances

The question of tolerances is addressed during the production of working drawings. Realistic tolerances should be specified. Solar collectors are not manufactured by watchmakers, and the variation from collector to collector may be at the upper limit of the tolerance band shown on the shop drawing. If collector support structures, for example, are designed so that each collector occupies the minimum width shown on the shop drawings, it is quite likely that the last collector mounted will hang over the end of the structure. Full manufacturer's data are needed to establish tolerances and dimensions for pumps, tanks, movable insulation on passive systems, and most components. The layout of equipment in the mechanical space also requires that the size of each component be known.

Instrument Panel

The functional mode of active or passive solar heating systems can be indicated by lights on an instrumentation panel operated by the controller. The panel may also include a temperature readout with a switch used to select various important temperatures throughout the solar system. The design of the instrument panel and its location is a part of the preparation of working drawings. Many building owners

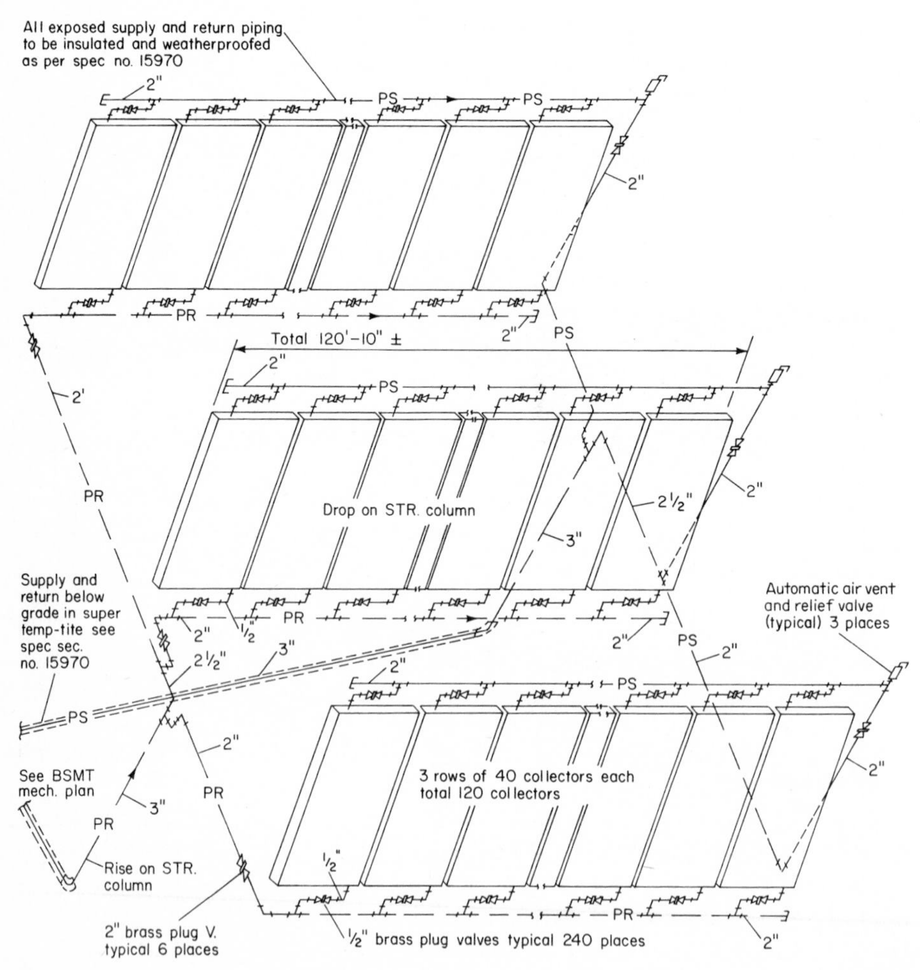

FIG. 5.5 Example solar piping diagram, CD level. *(Courtesy of P. Inman, P. E. and JFK Associates, solar consultant)*

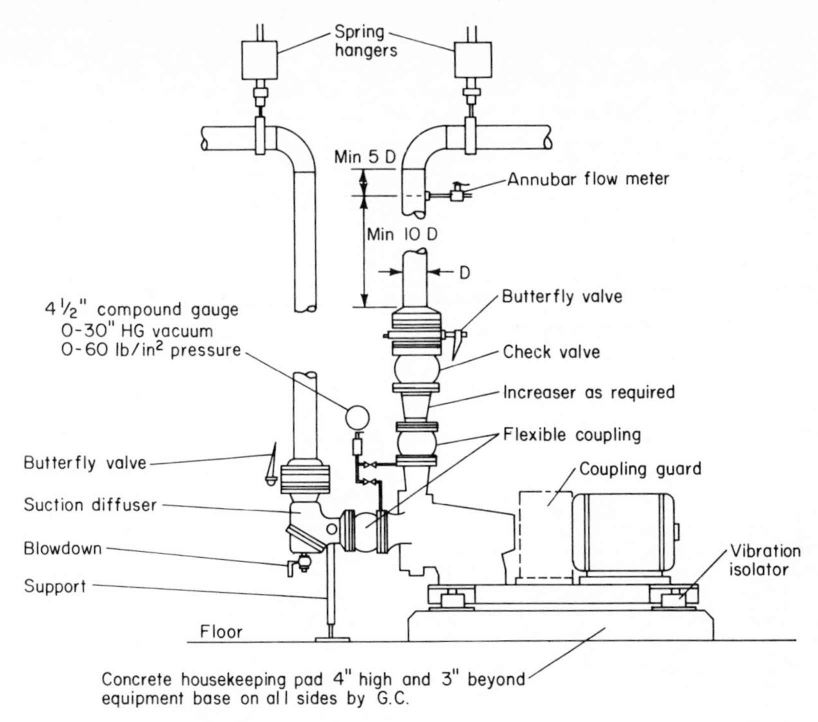

FIG. 5.6 Example pump-mounting detail, CD level. *(Courtesy of James Associates, Indianapolis, Ind.)*

may wish to monitor the solar system more closely than the nonsolar system, and an attractive instrument panel can be included for this purpose.

Examples

Working drawings are relatively complicated and examples cannot be easily shown in the format of a book. However, Figs. 5.5 and 5.6 show selected parts of working drawings for liquid-based solar systems. Figure 5.5 shows an example plumbing diagram for an active collector array showing each pipe fitting and all significant pipe sizes. The sizes have been determined during DD but are reduced to CDs in this phase. Figure 5.6 shows the details of a liquid-pump arrangement to be used in an active system. Details similar to those shown in these two figures are needed for every aspect of any solar heating system.

Figure 5.7 is an example working drawing for a passive heating system. Wall sections are shown, control strategies are called out, and arrangement of subcomponents is indicated. Dimensions of the thermal wall and its ventilation slots as well as the movable insulation track and its location are shown. Such a drawing is frequently included in the architectural section of CDs, whereas the drawings for active systems are often included in the mechanical drawing set.

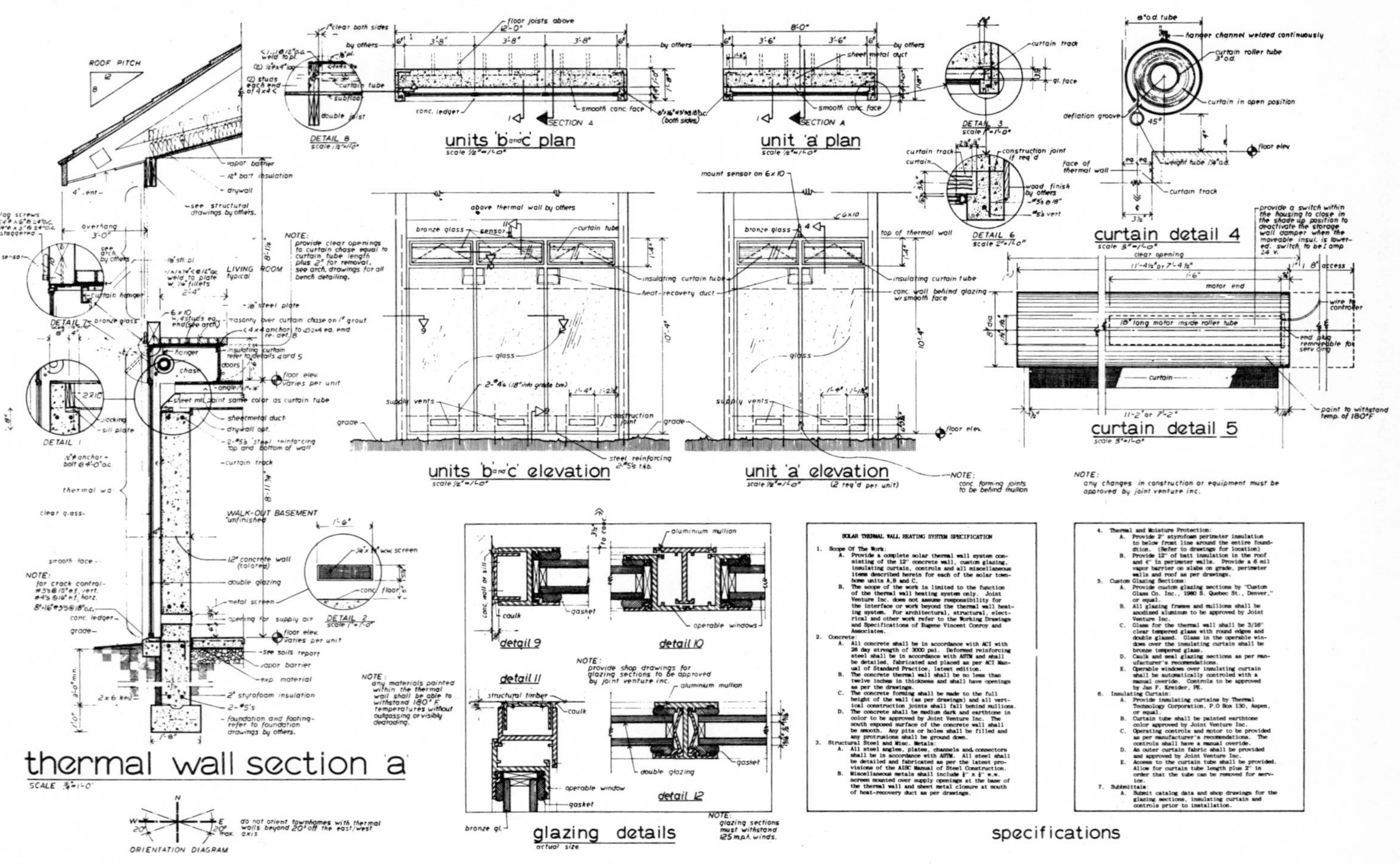

SOLAR THERMAL WALL HEATING SYSTEM SPECIFICATION

1. Scope Of The Work:
   A. Provide a complete solar thermal wall system consisting of the 12" concrete wall, custom glazing, insulating curtain, controls and all miscellaneous items described herein for each of the solar townhome units A,B and C.
   B. The scope of the work is limited to the function of the thermal wall heating system only. Joint Venture Inc. does not assume responsibility for the interface or work beyond the thermal wall heating system. For architectural, structural, electrical and other work refer to the Working Drawings and Specifications of Eugene Vincent Conroy and Associates.
2. Concrete:
   A. All concrete shall be in accordance with ACI with 28 day strength of 3000 psi. Deformed reinforcing steel shall be in accordance with ASTM and shall be detailed, fabricated and placed as per ACI Manual of Standard Practice, latest edition.
   B. The concrete thermal wall shall be no less than twelve inches in thickness and shall have openings as per the drawings.
   C. The concrete forming shall be made to the full height of the wall (as per drawings) and all vertical construction joints shall fall behind mullions.
   D. The concrete shall be medium dark and earthtone in color to be approved by Joint Venture Inc. The south exposed surface of the concrete wall shall be smooth. Any pits or holes shall be filled and any protrusions shall be ground down.
3. Structural Steel and Misc. Metals:
   A. All steel angles, plates, channels and connectors shall be in accordance with ASTM. All steel shall be detailed and fabricated as per the latest provisions of the AISC Manual of Steel Construction.
   B. Miscellaneous metals shall include ½" x ½" w.w. screen mounted over supply openings at the base of the thermal wall and sheet metal closure at mouth of heat-recovery duct as per drawings.
4. Thermal and Moisture Protection:
   A. Provide 2" styrofoam perimeter insulation to below frost line around the entire foundation. (Refer to drawings for location)
   B. Provide 12" of batt insulation in the roof and 4" in perimeter walls. Provide a 6 mil vapor barrier on slabs on grade, perimeter walls and roof as per drawings.
5. Custom Glazing Sections:
   A. Provide custom glazing sections by "Custom Glass Co. Inc., 1980 S. Quebec St., Denver," or equal.
   B. All glazing frames and mullions shall be anodized aluminum to be approved by Joint Venture Inc.
   C. Glass for the thermal wall shall be 3/16" clear tempered glass with round edges and double glazed. Glass in the operable windows over the insulating curtain shall be bronze tempered glass.
   D. Caulk and seal glazing sections as per manufacturer's recomendations.
   E. Operable windows over insulating curtain shall be automatically controled with a manual overide. Controls to be approved by Jan F. Kreider, PE.
6. Insulating Curtain:
   A. Provide insulating curtains by Thermal Technology Corporation, P.O Box 130, Aspen, or equal.
   B. Curtain tube shall be painted earthtone color approved by Joint Venture Inc.
   C. Operating controls and motor to be provided as per manufacturer's recomendations. The controls shall have a manual overide.
   D. An outer curtain fabric shall be provided and approved by Joint Venture Inc.
   E. Access to the curtain tube shall be provided. Allow for curtain tube length plus 2" in order that the tube can be removed for service.
7. Submittals:
   A. Submit catalog data and shop drawings for the glazing sections, insulating curtain and controls prior to installation.

FIG. 5.7 Example working drawing for passive system. *(Courtesy of Phil Tabb Architects, Boulder, Colo.)*

Specifications

Solar specifications are quantitative descriptions of the function of components in active and passive systems. Specifications include other information regarding legal requirements and any details of submittals which must be provided by prospective bidders. Included in this section are a few of the items in solar specifications and a suggested method for selecting one solar contractor from many. Since one collector may produce more energy per square foot than another, the selection of the solar contractor *cannot solely be based on the lowest bid* as is the case in many nonsolar applications.

Example. The contents of a specification of an active system can best be illustrated by the review of one. An example for a liquid system is shown in Fig. 5.8. It was included in Division 13 of the CSI format. This was done for this particular project since a separate bid was requested by the owner for the solar system (a bid alternate could have achieved the same purpose). If separate cost information is not required, the solar system can be included with other mechanical systems in Division 15.

The example includes a large initial section on the solar collector, since this is the principal component of the system. The technical and other requirements as well as warranty to be provided are described, an approved list of manufacturers is given, and a minimum collector efficiency curve is furnished. Criteria for bid selection are given in detail and are based upon an F-chart or G-chart type of performance calculation. This method of bid selection will be described in detail below. Bidders are to calculate the performance of their systems given specific monthly loads and energy requirements as shown.

The remaining sections of the specification include details on solar piping, its identification and insulation. Solar system pumps are described by pump number, flow rate, and size. Heat exchangers are specified with respect to the required heat rate and the flow rates and operating temperatures. Storage tanks and collector fluids are also called out.

As described in Chaps. 3 and 4, the function of a solar system controller is the key to successful solar heat collection. The controller function should be spelled out in detail and the location of sensors clarified. A table of controller operational modes is provided to ensure that the controller subcontractor includes every required mode in the design.

Other CD Contents. Other information may be included in the specification in addition to that shown in the example. Such additional requirements may include the furnishing of spare parts and spare glazings for the collector. Spare sensors may be furnished by the controller manufacturer, and the specification may stipulate that these be installed in the solar collector manufacturer's plant.

In air systems, the type of dampers to be used will be specified. These must be capable of proper adjustment as described in Chap. 4. The rocks which make up storage beds must be called out in the specifications. Their size, type, and cleanliness should be explicitly stated.

DIVISION 13

Special Construction

Section 13A *Solar Collectors*

13A-01 *Solar Collectors:* Nominal 3- by 7-ft panels of 756 ft^2 (36 panels) gross area are required for hot tub and domestic water preheating in the Boulder Hot Tub Center. The collectors are to be liquid-cooled and are to function as described below. The collectors will be roof-mounted on steel supports.

13A-02 *Technical Requirements:*

a. Flat-plate collectors will consist of—from top to bottom—one tempered, white-glass cover forming an insulating air space, a steel or copper absorber plate, rear insulation at the R-13 level or greater, edge insulation at the R-5 level or greater, metal housing and dust/moisture seals. Tubular collectors will not be allowed. Pipe fittings shall be threaded or soldered. No hose connections are permitted.

b. The collector assembly must be weatherproof and dustproof and avoid internal condensation by use of a regenerated desiccant.

c. The net area of each collector must be 85 percent or more of its gross area.

d. Pressure drop at a flow rate of 0.03 gal/(min $\cdot$ ft^2) (40 percent glycol at 110°F) should be less than 30 in water and more than 6 in water, with the collector mounted horizontally.

e. The absorber must be capable of operating at 50 lb/in^2 gauge internal pressure at 250°F.

f. Collector assemblies, filled with water, must weigh less than 11 lb/ft_c^2.

g. Collector mounting shall be able to withstand design wind and snow loads which apply to Boulder, Colorado.

h. Measured steady-state thermal efficiency shall exceed the line shown in Curve 1. Vendor must submit test data plotted on a copy of Curve 1 plus tabulation of data, type of test procedure used, and name and location of testing agency.

13A-03 *Other Requirements:*

a. The exterior housing of the collector modules must be finished in accordance with the architect's specifications.

b. Packaging for shipment shall ensure no breakage or other damage to collector assemblies. All pipe fittings shall be capped.

JOB NO. 7710 SPECIAL CONSTRUCTION S-66

FIG. 5.8 Example of active system performance specification. *(Courtesy of JFK Associates, Boulder, Colo.)*

c. Collectors must be capable of operation at temperatures from −15 to 250°F with fluid flowing. Collectors shall be capable of withstanding, without damage, 400°F or stagnation (no flow with bright sun) whichever is the greater for 3 mo.

d. The service life shall be 15 yr.

13A-04 *Documentation:* The vendor shall submit documentation supporting each of the points above and such others as he deems appropriate. The collector manufacturer shall verify that he has been in business for 2 yr, has at least 50 solar installations in the United States, and has produced at least 50,000 ft^2 of collector.

13A-05 *Warranty:*

Include a 1-yr warranty from the installer against failure of the solar system, any component, or assembly where such failure is caused by a defect in materials, manufacture or installation. The warranty shall cover the full costs of parts and labor required to remedy the defect, including, if necessary, replacement at the site, and shall run from the date of system acceptance by the engineer. The warranty shall also include provision for field inspection at no charge to the consumer, to verify failure, establish probable cause, and determine corrective action required. NOTE: The installer's warranty may include pass-through warranties provided by the supplier or the manufacturer.

Include a five (5) year collector manufacturer's warranty which shall include:

(1) A five (5) year warranty against *defects in materials* or manufacture. The warranty shall cover the full costs of parts and labor required to remedy the defect including, if necessary, replacement at the site, and shall run from the date of installation completion.

(2) A five (5) year warranty against *corrosion* of the absorber plate and other coolant passages. This corrosion warranty shall cover, for the first year, full costs of parts and labor required to remedy the defect including, if necessary, replacement at the site. For the remainder of the warranty period, the corrosion warranty shall cover the cost of all parts, delivered to the site, necessary to remedy the corrosion defect, including the cost of furnishing new collectors, if necessary. The corrosion warranty shall run from the date of installation completion.

(3) All warranties may specify reasonable installation and maintenance procedures for the warranty to be effective. However, no warranty shall be voided or in any way reduced by conditions that may occur in normal operation of the system.

JOB NO. 7710 SPECIAL CONSTRUCTION S-67

FIG. 5.8 *(Continued)*

13A-06 *Manufacturers:* The following manufacturers are acceptable:_____. Other manufacturers must be approved by the solar architect-engineer.

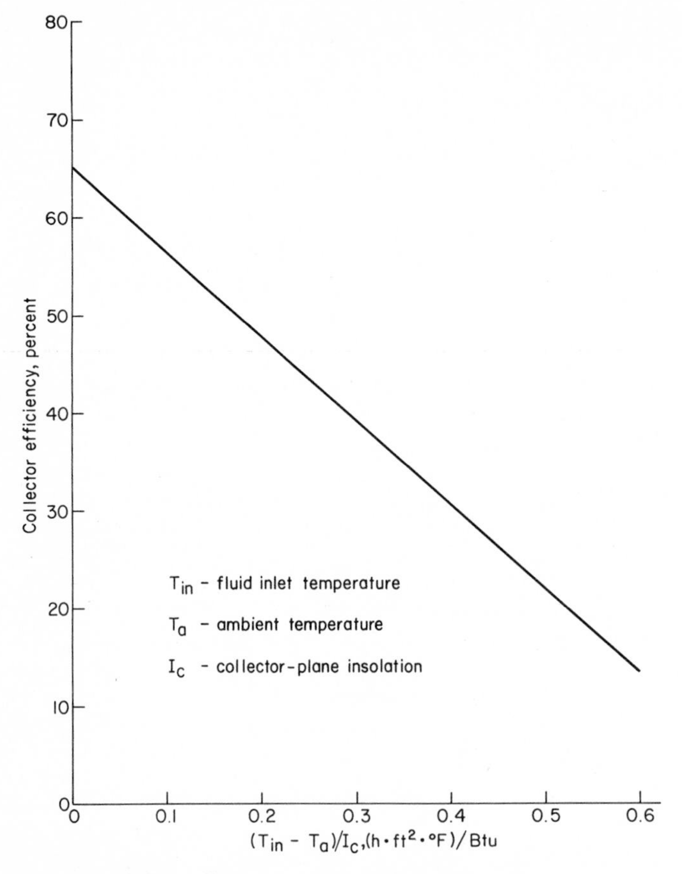

Curve 1 Solar collector minimum efficiency curve.

13A-07 *Cost Data:* The vendor shall supply a total cost figure for 756 ft_c^2 of gross collector area, F.O.B., Boulder, Colorado.

He shall also supply a unit price per collector module which will apply for amounts different from 756 gross ft^2.

Bids will be evaluated on the bases of conformance, experience, financial responsibility; and based upon thermal performance and price as described in Section 13A-08 below.

JOB NO. 7710 SPECIAL CONSTRUCTION S-68

FIG. 5.8 *(Continued)*

13A-08 *Criteria for Bid Selection:* The successful bidder shall be selected based upon *maximum solar cost index* as described below.

The incentive for use of solar heating systems is the savings in conventional fuel costs which are incurred during the life cycle of the system. Although life-cycle costs include factors other than solar collector cost, for purposes of this specification the collector cost and its energy delivery as evaluated below will be used to select the low-cost bidder who shall be awarded the solar collector contract.

a. *Performance Prediction:* The purpose of performance prediction is to determine the expected energy delivery during an average year. The predicted performance will be used along with the cost as described in the next section to select the successful bidder.

The method of performance prediction to be used is known as liquid-based F-chart. In order to make a performance comparison among several collectors, the following items shall be specified for use in the F-chart calculations and shall be the same for all bidders except for manufacturer's data.

Climatic Data and Energy Requirements—See table.
  Monthly averaged ambient temperature—°F
  Monthly averaged collector insolation—$BTU/(day \cdot ft^2)$

System Data
  Collector area—756 gross square feet
  Site latitude—40°N
  Collector tilt—45°
  Monthly load (uniform)—40×10^6 Btu/mo

Manufacturer's Data
  Efficiency curve slope—$F_R U_C$
  Efficiency curve intercept—$F_R(\tau\alpha)$

As a result of the F-chart calculation, each bidder shall tabulate three numbers—the first two shown as "manufacturer's data" immediately above and the third the annual energy delivery in Btu per year for the proposed solar collector array.

The successful bidder shall have the *largest solar cost index* (SCI) defined below. The largest SCI indicates minimum energy delivery cost. The solar cost index is defined as the annual energy delivery as predicted by the F-chart divided by the total cost of the solar system. In equation form

$$SCI = \frac{\text{energy delivery (Btu/yr)}}{\text{total solar system cost}}$$

JOB NO. 7710 SPECIAL CONSTRUCTION S-69

FIG. 5.8 *(Continued)*

Climatic data

Month	*Collector plane insolation, Btu/(day·ft²)*	*Average ambient temperature, °F*
January	1405	32
February	1519	34
March	1813	37
April	1654	48
May	1417	57
June	1491	66
July	1531	73
August	1493	72
September	1721	63
October	1641	54
November	1454	41
December	1369	36

Section 13B *Piping*

13B-01 *Piping:*

1. Provide identification at intervals not exceeding 20 ft and within 12 in of each valve.
   a. Provide stenciled identification of pipe contents and flow direction arrow on each color band by use of legend used on drawings or by approved conventional nomenclature. Apply identification at position most legible from the floor.
   b. Insulate all solar piping with 1 in of formed pipe insulation with minimum R value of 3.0 and with factory-applied fire retardant vapor barrier jacket. PVC jacket if exposed to weather. No gaps are permitted.
2. Collector loop piping shall be of type M copper with dielectric couplings between dissimilar metals. 95/5 solder shall be used for all collector piping above the roofline.
3. Piping shall be arranged so that no air traps or air locks are formed. All piping shall drain by gravity.
4. Collector loop check valve shall be mounted so that gravity will hold the check in the closed position with collector pump off.

Section 13C *Pumps*

13C-01 *Pumps:*

1. Solar collector pump P1 (1 × 1¼ in NPT) shall flow 28 gal/min minimum at 35 ft of head.

JOB NO. 7710 SPECIAL CONSTRUCTION S-70

FIG. 5.8 *(Continued)*

2. Solar preheat exchanger pump P2 (1 × 1¼ in NPT—brass or stainless steel) shall flow 30 gal/min minimum at 10 ft of head.

Section 13D *Heat Exchangers*

13D-01 *Heat Exchangers:*

1. Collector heat exchangers HX-1 and HX-2 shall be shell-and-tube, counterflow with the following performance:
   Shell flow = 28 gal/min, 40 percent glycol
   Shell EWT = 120°F
   Tube flow = 30 gal/min, water
   Tube EWT = 104°F
   Heat rate = 126,000 Btu/h
   Operating pressure = 50 lb/in^2 gauge

   Insulate at the R-6 level.

2. A weatherproof unit heater used as heat rejecter (HRU-1) in series with the collector shall be capable of rejecting 100,000 Btu/h at EWT of 200°F and 28 gal/min (40 percent glycol) and EAT of 100°F.

Section 13E *Tanks*

13E-01 *Tanks:*

1. Collector loop expansion tank TK1 shall be of pressurized diaphragm type with permanent gas charge, have an operating pressure of 30 lb/in^2 gauge, and acceptance volume of 10 gal. It shall be equipped with an Airtrol fitting.

2. Water-heating preheat tank shall be 700 gal, 150 lb/in^2 gauge operating pressure. It shall be glass-lined steel and equipped with magnesium anode and two top and bottom 1½ in NPT female fittings. Insulate at the R-25 level with foam or continuous, taped fiber glass batt.

Section 13F *Collector Fluids*

13F-01 *Collector Fluids:* Collector working fluid shall be a 40 percent aqueous solution of inhibited propylene glycol manufactured by ______ ; pH to be 7.2 to 8.5.

Section 13G *Solar Water-Heating System Controller Specifications*

13G-01 *Controller:*

1. Work Included

   a. The automatic temperature control system shall be of electric components. The system shall be complete in every respect, put in operation, and adjusted under operating conditions.

JOB NO. 7710 SPECIAL CONSTRUCTION S-71

FIG. 5.8 *(Continued)*

b. The system shall be furnished by ______. Controller to be made up of interconnected models ______ with special overheat feature to activate heat rejector blower and heat rejector valve in collector loop at 200 ± 5°F collector temperature by signal to 110 V ac relay. Deactivate standard boil and freeze protect features. For collector controller, insert fixed 35,000-Ω resistor (1 W) in place of storage sensor.

c. The system shall include all control devices, valves, fittings, wire, conduit, etc., as specified and required, and shall be connected so as to perform all functions and operate according to the specified sequences. Coordinate with suppliers of all mechanical equipments and the electrical contractor to provide a complete system without duplication or omission. Low-voltage wiring shall be installed in conduit as per code.

d. All automatic control valves furnished by the temperature control contractor shall be installed under his supervision by the mechanical contractor. All other components shall be installed by competent mechanics who are full-time employees of the control manufacturer.

e. The temperature control system shall completely interface with controls provided with and required for operation of the rooftop heat rejector unit.

13G-02 *Sensors:* Solar collector sensor shall be RTD or thermistor and shall be bonded to the absorber plate center with a thermally conducting material and a bolt. The absorber plate shall be cleaned thoroughly in the area of sensor prior to installation to assure a complete and durable thermal bond.

Preheat tank sensor and solar circulating loop sensor shall be RTD or thermistor and shall be located, respectively, in the lower one-quarter of the storage tank (or at the bottom drain fitting as close to the tank as possible) and in the solar fluid main line. Sensors shall be firmly mounted and shall be of the averaging type.

All sensors shall be of temperature range to fit the range of control, shall be spring-loaded or crimped tip for good transfer and shall have an accuracy of ±1°F from −20 to 400°F.

13G-03 *Sequence of Operation (see controller table)*

13G-04 *Start-up:* After completion of the installation, adjust all thermostats, control valves, motors, and other equipment. Place in complete operating condition prior to requesting the approval of the architect/engineer.

JOB NO. 7710 SPECIAL CONSTRUCTION S-72

FIG. 5.8 *(Continued)*

13G-05 *Solar Water-Heating System Start-up:* Flush solar system with low-sudsing detergent and empty strainer as required to remove refuse. Leak test at 30 lb/in^2 gauge for 24 h with fluid circulating. Fill collector loop with 40 percent propylene glycol; remove all air. Pressure at topmost header of collector to be at least 5 lb/in^2 gauge to avoid vapor lock. When filling collector with fluid, do so at night, during overcast, or early in the morning to avoid steam production in an overheated, stagnated collector.

After system start-up, conduct a full mode test to check all operation modes given in the controller table.

Solar controller operational modes

Sun condition	*Tub fluid temperature, T_t*	*Valve V1*	*Valve V2*	*Boiler water in exchanger HX3*	*Pump P1*	*Pump P2*
No sun	Below 105°F	Open	Closed	On	—*	On
	105–107°F	Modulating	Closed	Off	—*	On
	Above 107°F	Bypass	Closed	Off	—*	On
Low sun (T_1 less than 108°F)	Below 105°F	Open	Closed	On	—*	On
	105–107°F	Modulating	Closed	Off	—*	On
	Above 107°F	Bypass	Closed	Off	—*	On
High sun (T_1 above 108°F)	Below 105°F	Open	Open	On	—*	On
	105–107°F	Modulating	Open	Off	—*	On
	Above 107°F	Bypass	Open	Off	—*	On

NOTES: All above set points are adjustable. DHW Pump P3 on if temperature T_2 is 2°F or more above preheat tank temp T_1; off for T_2 above 140°F.

*Pump P1 comes on at a fixed collector plate temperature of 90°F.

JOB NO. 7710 SPECIAL CONSTRUCTION S-73

FIG. 5.8 *(Continued)*

The specification may require that the collector manufacturer furnish collectors with an opaque cover over the glazing. This is suggested if the collectors are to be exposed to bright sunlight for a long period without operation. This cover should be attached without an adhesive over the surface of the glass, since later removal of the adhesive may be nearly impossible.

Prebid Conference. The solar specifications may call for a prebid conference. For large systems, this is particularly advisable since the investment involved is large but vendor familiarity may be small. At the prebid conference questions regarding the design of the system and its expected function as well as legal requirements are addressed. At a minimum, the building owner's representative, a representative of the architect, and the solar consultant should be present at the prebid conference so that answers can be given immediately. The bidding process usually has tight time constraints. Proceedings of the prebid conference should be furnished to all prospective bidders whether present at the conference or not.

Solar Component Prepurchasing. Solar collectors, storage tanks, and other principal components of the system can be prepurchased by the owner and furnished to the contractor. This is a standard procedure which can be used to reduce the effect of inflation on the project budget. The owner then undertakes the responsibility of inspecting the components and certifying that they meet the specifications.

Prepurchasing of solar collectors has several advantages. If the collector manufacturer is known during the CD process rather than identified at its end, the specific collector model will be known. The collector structure can then be designed for the specific collector to be used rather than for several possible collectors. Piping arrangements, manifold sizes, and other features of the active system can be designed in accordance with the recommendations of the specific collector manufacturer. Prepurchasing of storage tanks and heat exchangers has similar benefits in that the design can be solidified during CDs and addenda need not be issued during the CA phase to accommodate the details of particular solar components.

One of the difficulties with the prepurchase of components is the method of selection. If a solar contractor is selected on the basis of the life-cycle economics of the entire system as described below, the effect of all features of the solar system can be considered. However, if only a collector is purchased it is difficult to include various secondary effects in the selection process. For example, two collectors may have identical thermal performance and identical costs. However, one collector may use internal manifolding and the other not. Internally manifolded collectors require a smaller number of pipe fittings and insulation; hence, the labor costs are lower. The preparation of the accepted list of bidders should consider the amount of labor required for installation, removal, and repair.

A Method of Solar System Contractor Selection. The selection of most HVAC subcontractors is based upon the minimum bid required to provide a system meeting specifications. In the case of a solar system, this will frequently result in the purchase of the least-cost-effective system, since the cheapest system is probably the least efficient one.

The selection of a solar system during the DD phase is based on the concept of cost-effectiveness, not cost. It is impractical for the owner to require that a contractor go through a full life-cycle cost analysis because of limits on both time and expertise.

A simplified method which is related closely to life-cycle costs can be used, however. The SCI method is easy to use:

$$\text{SCI} = \frac{\text{delivered solar heat (Btu/yr)}}{\text{initial installed cost}} \tag{5.1}$$

Since the SCI includes only initial installed cost, it does not show maintenance, insurance, and depreciation effects. However, the principal cost of the solar system is its initial cost and the SCI selection approach is adequate. The calculation of energy delivery in the numerator of Eq. (5.1) must be carried out in exactly the same manner for all manufacturers (see preceding specification example). Since most collector manufacturers are familiar with and use the F-chart, the specification writer can require that this method be used. However, to assure consistent results, the *inputs to the F-chart must be identical for each proposer* except for collector properties. These uniform inputs include monthly heating loads, collector plane insolation for the site in question, monthly average ambient temperature, collector and storage fluid-flow rates, collector tilt, and site latitude. The only variations from one manufacturer to another are in the value of the collector loss coefficient $F_R U_c$ (affects F_{hx}, also) and its optical efficiency $F_R \tau\alpha$. These data must be supplied from formal efficiency tests carried out by an independent testing agency using a standard procedure.

The denominator in the SCI equation is the initial installed cost. This is exactly equivalent to the bid amount proposed by the bidder. The bid selection is not based, however, on this dollar amount but rather on the maximum value of the SCI. The contractor with the maximum SCI value will be providing to the owner the maximum number of British thermal units per year from the solar system for the bid.

The SCI method is best used if a solar contractor is responsible for furnishing and installing the entire system. The SCI idea can also be used to select just a collector, since collector properties are all that are required for the F-chart analysis. However, the cost of the entire system would not be included, and the cost-optimal system is not necessarily associated with highest-SCI-based collector cost alone.

Acceptance Tests. Solar system specifications may also require a formal acceptance test. Although not shown in the example collector specification given above, specifications for large buildings should include a calculation of projected day-long output of the collector array. This projection is made with standard calculation methods given in solar textbooks such as Ref. 1. The solar-system subcontractor at the termination of system installation will be required to run a day-long test. *The results of this day-long test must correlate closely with the day-long calculations submitted as a part of the bid document.* Of course, the calculations provided with the bid are for a specific day; but the performance test of the entire system will not occur on an identical day. However, standard extrapolation methods in the efficiency equation and incidence-angle-modifier equation can be used. The acceptance test is the only basis of verifying that the contractor has installed a system which performs in the manner described in the bid response.

If the formal acceptance test result does not correspond to the calculated result, the specification must include some method of assessing a penalty to the contractor. A reasonable tolerance range for the result should be included. The day-long performance prediction as well as day-long measured performance should be reviewed by the solar consultant.

Parts List. Solar-system specifications may optionally include a parts list. A parts list serves as a conveniently condensed specification in that all components in the solar system are listed. A parts list is recommended for possible contractors who are unfamiliar with solar system installations and may neglect to include some components either in the bid or in the work. The parts list will include all components present in the working drawings specifications. Therefore, the parts list is not an additional list of parts but rather a summary of parts already listed.

For an active system, the parts list might include the following components:

Solar collectors
Collector pump
Collector heat exchanger
Gate valves
Bleed valves
Pressure-relief valve
Heat-transfer fluid
Piping connectors
Insulation for storage
Pipe tees
Pipe dope
Storage-tank pump
Storage-tank drains
Water inhibitors
Storage relief valve
Pipe labels and stencils
Mode selector
Delivery-fan coil unit
Domestic-water controller
Water-preheat pumps
P/T valve
Heat-rejection unit
Wiring for controller and sensors
Collector mounting brackets
Control system
Collector-loop check valve
Air separator
Drains
Temperature and pressure gauges
Piping
Insulation for piping
Insulation for heat exchangers
Elbows
Thermometer wells
Storage-tank expansion tank
Storage-tank pipe fittings
Storage and collector sensors
Paint
Collector- and storage-loop filters
Two- or three-way valves
Water-preheat tank
Water-preheat heat exchanger
Delivery-loop expansion tank
Energy-delivery pump
Heat-rejection-unit control valve
Miscellaneous sealing compounds
Insulation
Masonry
Thermal cement
Roofing material
Sheet metal

A parts list for a passive heating system would include the following components.

Thermal-storage-wall material
Glazing seals
Movable insulation
Movable insulation track
Backdraft dampers
Glazing overhang
Control panel
Movable insulation housing, etc.
Glazing units
Glazing cap strips and screws
Movable-insulation controller
Storage-wall paint or dye
Hybrid air mover and dampers
Instrumentation—thermistors, selector switch
Miscellaneous seals and fasteners

The parts list or *bill of materials* will include each component listed above and its part number or size taken from the specification. The parts list can serve later as a checklist as well as a cost-estimating list.

NOTES ON THE BIDDING PROCESS

At the conclusion of CD, an invitation to bid is released to all bidders who are qualified and have expressed an interest in the solar construction effort. A package of materials including drawings, specifications, and general provisions are provided to each bidder. A fee is frequently charged for the bid package on large projects. The response to the solar system bid invitation will be a dollar amount (plus a unit collector cost if extra collectors may later be added or subtracted), and a value of the SCI. The SCI is the basis for successful bidder selection, but the total cost must also fall within the constraints of the overall construction budget.

The bid submission must include all information requested by the specifications. In the example specification (Fig. 5.8), information regarding the amount of experience of the firm as well as independent test information should be submitted. A warranty should be supplied as delineated in the specifications, including pass-through warranties from equipment manufacturers as well as the contractor's installation warranty. A management chart for the project may be requested if it is a relatively large project and responsible personnel must be identified.

A construction schedule may be requested in the specifications and penalties for overruns also delineated.

The execution and legal provisions of the bidding phase are similar to those for other HVAC systems and are not described here. The structure of a prebid conference which would be included as part of the bidding phase has been described earlier.

Trades

During the bid preparation effort, the potential conflict between trades should be clarified by the contractor. Many trades are involved in the installation of solar systems, including plumbers, steam fitters, sheet-metal workers, steel workers, carpenters, crane operators, control installers, glaziers, insulators, roofers, electricians, mechanics, and laborers.

Conflicts between these trades can occur since the delineation of each solar craft area has not been universally established. For example, in the installation of a collector, a crane operator, a plumber, and a flashing laborer may all be present.

One method of avoiding conflict of trades which may occur during the CA process is to use prefabricated modules. In a liquid system, one preassembled module might include the collector and storage pumps, the controller, the collector-to-storage heat exchanger, expansion tanks, check valves, and control valves. This unit can then be connected directly to the collector piping and the storage tank, thereby completing system installation by only one trade.

Precedence in some areas may dictate which trade carries out which activity. As the familiarity of contractors with solar system installation increases, the potential for conflict among trades will be reduced.

TABLE 5.1 Example cost submittal from solar contractor

Quantity	*Part and description*	*Price*
30	Brand X solar collectors, model Z, 3 × 7 ft nominal dimension	$12,840
6	Brand X solar collectors, model Z1 (with heat sensor)	2,658
	Steel racks, to mount 36 collectors. In accordance with note 1 on blueprint. (Racks will be provided and installed by ________, approximate cost, $2,000)	Not in contract price
1	Pump, brand Y (P1) circulator. 18 gal/min minimum @ 16 ft of head	160
1	Pump, brand Y (P2) brass, 111 circulator. 33 gal/min @ 6 ft of head	150
1	Pump, brand Y (P3) brass, 111 circulator. 16 gal/min Min. @ 8 ft of head	150
1	HX-1 heat exchanger, shell-and-tube type, counterflow with the following performance: Shell flow = 14 gal/min (40% glycol) Shell EWT = 130°F Tube flow = 20 gal/min water Tube EWT = 104°F Heat rate = 126,000 Btu/h Insulated to R-6	815
	Brand Z, model T1	
1	HX-2 heat exchanger, shell-and-tube type, counterflow with the following performance: Shell flow = 16 gal/min water Shell EWT = 120° F Tube flow = 30 gal/min water Tube EWT = 105°F Heat rate = 80,000 Btu/h	
	Brand Z, model T2	1,900
1	HRU-1 heat rejector, to reject 100,000 Btu/h, brand Z, model U	760
1	Controller model V	85
1	Temperature control switch, to operate heat rejecter	35
12	Temperature pressure wells	70
2	Temperature-pressure well adapters	12
4	Thermometers	50
4	Pressure gauge	50
1	Expansion tank, TK3, pressurized diaphragm type, operating pressure 30 lb/in^2 gauge with air eliminator valve and Airtrol™ fitting, 75 gal	300

TABLE 5.1 (*Continued*)

Quantity	*Part and description*	*Price*
1	Expansion tank, TK1, pressurized diaphragm type, operating pressure 30 lb/in² gauge and acceptance volume 10 gal; equipped with an Airtrol™ fitting	100
1	Storage tank TK2, 700 gal, 30 lb/in² gauge, steel, unlined with manhole and 4 fittings (2-in NPT); R-25 foam insulation and saddles per drawing	1,300
	Hardware, miscellaneous	3,320
	Pipe and insulation	
	Fittings	
	Pipe hangers	
	Labor	3,600
	Propylene glycol—40% solution	130
2	Temperature switch, brand A, model Q, 0–250°F range with closed differential switch, thermal, including thermal wells and electric relays; repeatability to ½°F and local readout	300
	Subtotal	$28,785
	Sales tax (exempt)	0
	Total	$28,785
	Tax benefits on the system are:	
	10% solar credit	2,876
	10% investment tax credit	2,876
	State tax credit	288
	Net cost	$22,745
		($30.09 per square foot of collector)

Price Lists

Both the architect and solar consultant should review the solar-system bids. The bid with the highest SCI would be the first to be reviewed; if prepared properly, it would be the approved bid and other bids need not be reviewed. If the initial total cost of the system appears higher than estimated by the designer, the contractor should furnish a detailed breakdown of prices. The specifications may even call for a submittal of such a price list. Table 5.1 is an example cost breakdown for a liquid system for a process heat application in a building. Each component is listed, the quantity is shown, and the total price is given. The designer can very rapidly identify points of cost overrun by reviewing this list.

Table 5.2 shows the range of costs of large solar systems installed in the late 1970s. These system costs apply for the specific locations shown but indicate that the average cost per square foot of collector is approximately $30 to $35 (in 1980 dollars). If a bid at $40 per square foot were to be submitted (in comparable dollars), the cost might be acceptable. However, if a bid were to come in for $75 per square

TABLE 5.2 Example system costs—solar heating systems*

Project	*Location*	*Solar cost, dollars*	*Solar collector area, ft²*	*Solar cost/area, dollars per square foot*
Denver RTD Platte facility (1977)	Denver, CO	1,200,000	39,000	$30.77
Boulder RTD facility (1979)	Boulder, CO	860,000	30,000	28.67
Aurora RTD facility (1980)	Aurora, CO	1,400,000	45,300	30.91
Student housing (1976)	Boulder, CO	22,500	850	26.47
University of Colorado Credit Union (1978)	Boulder, CO	83,000	2,250	36.89
Seattle First National Bank (1979)	Spokane, WA	~100,000	2,500	~40.00
Gunnison Airport (1980)	Gunnison, CO	45,000	1,000 (passive)	45.00

*JFK Associates, solar consultant.

foot for a standard space-heating system, the solar consultant and architect should carefully determine why the cost is beyond the range of most systems installed in the United States.

One reason for inflated bids on solar contracts is the so-called fear factor. This is an arbitrary factor indirectly applied by contractors who are unfamiliar with the installation of solar systems. The installation of the first few systems may not be profit-making ventures, and the increased cost for these systems may reflect the uncertainty of contractors regarding their expertise in this area. As solar expertise develops in a given locality, bids will become more competitive and should drop to the levels indicated above. Typical high-cost items which may be overpriced are the solar collector, the pumps (which may have been oversized by the contractor), labor hours, and storage tank. Current guidelines for installation labor are not consistent and are not presented in this book for that reason. Means cost data can be one source of these data (44).

CONTRACT ADMINISTRATION

Contract administration is the final phase of design and construction of a building. During this phase, the various components of a building are assembled. The activities of the designer will include the review of shop drawings, issuing of field reports and change notices, and advising the installers about special considerations which apply to solar systems.

This section does not include a step-by-step description of the CA process but rather is a listing of various items which must be carefully observed during assembly of the solar system to the building. The installation of collectors, storage, piping, working fluids, controls, and so forth for active systems and for the corresponding components of passive systems are described. Emphasis is placed upon problem areas which have occurred in the past. Installation steps which are identical to nonsolar HVAC system installation procedures are not described. Only those specific to the installation of a solar system are emphasized.

This section also includes recommendations on the start-up and inspection phases of the project and the performance of an acceptance test. Various types of acceptance tests can be run, depending upon the size of the building and the amount of time available for this purpose. Finally, a method of checking year-long solar performance is presented. In the absence of detailed instrumentation, this method can be used to determine whether the system is operating properly during the course of the year and whether its performance agrees with that projected by computer studies during SD and DD. Instrumentation to monitor performance is also described.

Solar System Installation

The installation of active systems differs markedly from that for passive systems. Since passive systems are more integrated with buildings than active systems and since the number of mechanical components is smaller, the installation effort is less for passive than for active systems. However, inspection and advice to contractors during CA are equally important for both.

The installation of a solar system will require approval by a building inspector. Many building inspectors have limited familiarity with solar systems. Hence, it is advisable to get the local inspection authority involved in the project as early as possible so that any decisions requiring rework will be eliminated. If the building inspector is involved in the entire process, the system installation can go quite smoothly.

The contractor must also have good communications established with the designer. Although the construction documents for a solar system will be more detailed than those for nonsolar systems if the recommendations of the previous section are followed, there still will be questions regarding solar features. These are best and most quickly resolved by conversations with the solar-system designer. If a construction manager (CM) is used for the process, the designer and CM should also have good communications established. The role of a CM has not been described in detail in this book; the CM is normally involved with the design from very early stages in the process, provides cost estimates, and should be familiar with the solar-system operation and design.

Because of the relative unfamiliarity of contractors with solar-system installation, the A/E project manager should allocate additional time for site visits and additional consultation during CA.

Installation Manuals. One major method of assisting the system installer is to acquire all instruction manuals and installation manuals provided by component manufacturers. Most solar-collector manufacturers, for example, provide detailed instructions for the assembly of a collector to supports. Likewise, controller manufacturers provide step-by-step procedures for control-subsystem installation. These manuals must be reviewed, however, before they are supplied to an installer, since some of the steps may not apply to the specific system or location. If the contractor or CM does not have sufficient expertise in these areas, the solar consultant or solar-system designer should review the manuals.

Shop Drawings. The review of shop drawings is a critical activity during the CA phase. Shop drawings of storage tanks, collectors, expansion tanks, dampers, passive-system movable insulation, and heat exchangers should be reviewed carefully. The most critical shop drawings for active systems are those for controls, storage, and collectors.

The proper size and location of piping connections must be verified when reviewing collector shop drawings. If collector tolerances have changed between the working drawings and the shop drawings, the size of collector arrays may be affected. The location of piping is particularly critical since collector array manifolds may be prefabricated in the plumbing shop. They must be able to connect directly to the collector. Further, the finished collector model number must be checked against the bid submittal to ensure that the proper type of glazing, insulation, and absorber surface have been provided. Most collector manufacturers offer several sizes of collectors. Hence, the proper overall dimensions of the collectors must be verified.

The review of storage-tank shop drawings should address the location and size of piping connections, the type of insulation installed on the tank, if factory-installed insulation is used, and the presence of a manhole with an appropriate gasket. The mounting provisions for the tank should be checked, including the type of supports used, their location, and method of insulation. The pressure rating of the tank must obviously be checked, and so must the tank material and lining.

Shop drawings of the movable insulation in passive systems are to be reviewed. The type of insulation material, its color, and its insulating value should be checked. The method of operation and the wiring diagram for the control system, if an automatic control is used for the movable insulation, should also be examined. Each controller mode should be checked against the specifications.

The review of shop drawings is not fundamentally different for solar and non-solar systems. The key steps in the process are to compare shop drawings both to working drawings and to specifications to ensure that every requirement of the construction documents has been met. The review of shop drawings is the responsibility of the architect/engineer or solar consultant.

Scheduling. Some components of solar systems may be in short supply in a specific location. Then, the long-lead-time items should be ordered by the contractor well in advance of the planned installation date. The most common long-lead-time components for solar systems are the solar collectors, the insulated storage tank, the controls, and the movable insulation for passive systems. Most other components are off-the-shelf items—heat exchangers, pumps, expansion tanks, piping, insulation, backup heating systems, and terminal load devices. In some cases, these components may also be in short supply, but this is unusual.

The installation and assembly of a solar system must be carried out with proper time sequencing. For example, the storage tank, if located in a mechanical space in a basement of a residence, must be in place before the subflooring is installed, since storage tanks are relatively large. Likewise, the leak testing of collector piping in a liquid system or ducts in an air system should take place before pipe or duct insulation. The contractor or CM should prepare either a critical path method (CPM) diagram or a sequence list for the building to ensure that the proper sequence of installation events takes place. Recommendations about these steps are given in the remainder of this section. Figure 5.9 is an example of a CPM diagram for a residential solar heating system. The solar activities are presented in this figure above the center line, and the major standard building activities are presented below the line.

The activities earliest in the system installation sequence include the installation of the storage tank and the storage tank sensor as well as the installation of collector supports on the roof. In subsequent steps, the collector is installed, the piping is checked, and a fundamental check of all modes is completed. Once leak testing has been undertaken, the insulation is installed, and wiring is completed. The final step, of course, is an inspection which is treated later. Information which can be used in

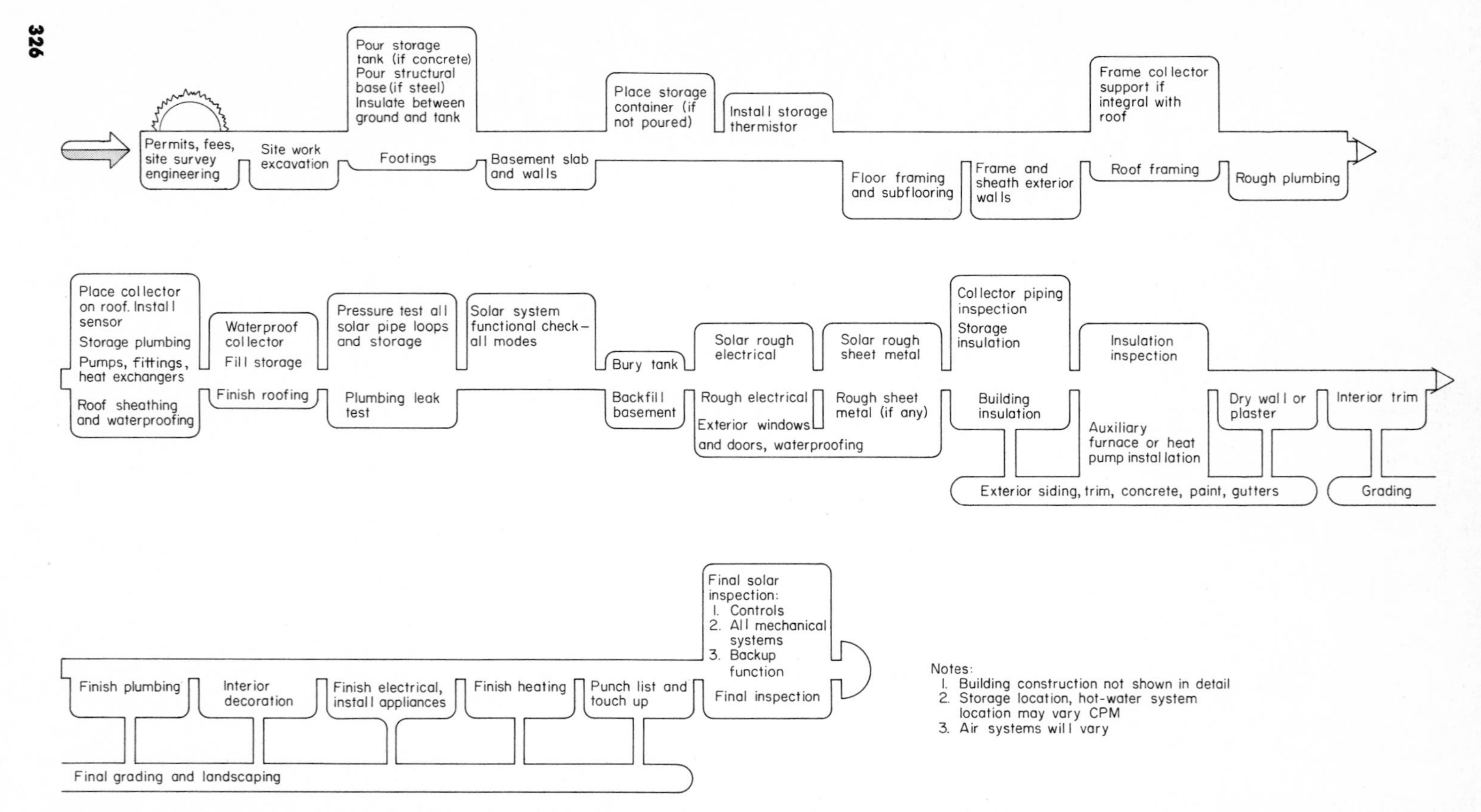

FIG. 5.9 Critical-path diagram for installation of liquid-based solar space-heating system in a residence. *(Courtesy of JFK Associates, Boulder, Colo.)*

preparing a CPM diagram for other types of systems is contained in Ref. 47, Sec. 9.

Active-Collector Installation. If a large system is to be installed, it may be worthwhile to present a half-day or 1-day "school" for the collector installers. This meeting should cover proper methods of piping, connection of ducts, leak testing, installation of insulation, and attachment of the collector to the support. If the collector is integrated with the roof, additional requirements regarding flashing and sealing should be described. The laborers should then be shown how collectors are installed. Installation of the collectors in the first array should be monitored closely.

Installation of most collectors will occur on sunny days. Since no fluid is flowing through the collectors, they will be operating at their stagnation temperature, which can be 300 to 400°F. Installation of collectors, which weigh several hundred pounds, on a pitched roof is hard work, particularly if the weather is hot. The result may be a relatively sloppy installation, particularly if the collectors must be hauled onto the roof by hand. An alternative method for installation under these conditions should be considered. One method of reducing the temperature of the collector is to use a collector cover. This should be called out in the specifications for the collector as noted earlier.

Preassembly. Another approach is to preassemble collector subarrays on the ground. These arrays are inspected, leak-tested, checked out, and drained before they are lifted to the roof. The assembly can be done in the shade if the job is being performed in summer, and quality control should therefore be improved.

Whether collectors are installed individually or in arrays, a hoist or crane will normally be required to raise them to a roof. The contractor must make arrangements for this machine to be present at the expected time of installation. Cranes are relatively expensive on an hourly basis, but in most cases turn out to be cost-effective when compared with other, slower methods.

Collectors mounted on a sloping roof have little space available for piping, particularly if one pipe must cross another. The requirement that headers slope is also a difficulty for roof installations since a sloping header may appear from the ground to be misinstalled. The entire array may even appear to be installed at a slant on the roof. The designer should provide detailed instructions and piping layouts for collectors to be installed on tilted roofs.

If screwed rather than soldered fittings are used for piping, the plumber must be instructed not to overtorque the collector fittings. Many collectors do not provide a collector header rotation stop within the assembly. If the header is overtorqued, the connections between collector risers and headers may be damaged. Manufacturer's installation instructions specify the proper torque.

Most collectors should be installed with the flow channels running vertically. This is a prerequisite for proper flow balance in many collectors and should be called out on the drawings. However, an on-site inspection should also be made to ensure that the collectors are being installed in this manner.

As collector rows are assembled, they should be individually leak-checked instead of being leak-checked as an entire system when completed. A row-by-row check helps find the leaks. They are repaired more rapidly, since workers are available at the particular location to make the repair expeditiously. Of course, after a leak test of a liquid system is made, all water must be drained from the collector in winter to avoid freezing overnight.

Roof Mounting. Some collector manufacturers have shipped collectors in a very inconvenient form for unloading from the truck on site. The collectors should be shipped so they can be easily unloaded by crane. Collectors shipped from a manufacturer's warehouse will have been loaded by a forklift, in all probability. However, if this method is not to be used for unloading, the contractor should request that the method of shipment be compatible with the method of unloading, not with the method of loading.

Collectors mounted on a sloped roof require a starter strip to be placed along the bottom of the first row to ensure that the row is level and at the proper position. The starter strip in some installations has been a piece of warped scrap lumber which is essentially useless for this purpose. The starter strip should be a straight piece of finished lumber with sufficient strength to support the first row of collectors in the proper fashion. Its location is important since the configuration of the entire roof array depends upon it.

Collectors located on a sloped roof are installed after the roof felt is installed. If the collectors are integral with the roof, the flashing is installed next and finally the shingles. If the collectors are to be installed above the roof surface, the roof penetrations for support mounts should be installed. Then shingles and flashing can be completed. A chalk outline on the felt will save time when laying out the array. This can be done before the installation of any collector. The location of roof penetrations can also be marked.

The solar collector sensor should be installed in a collector by the manufacturer as described in an earlier part of this chapter. This special collector must be identified by the installer and placed in a location to which wires can be run conveniently. If more than one collector is furnished with a sensor, a similar consideration applies. These collectors must be handled with care and should be installed in an area, preferably in the top row of the array, where shading cannot occur even on the shortest day of the year in early morning or late afternoon. Further, attention should be paid to possible snow-coverage problems, etc.

Many flat-plate collectors have vents and weep holes to eliminate water vapor from the interior of the collector. These must be uncovered and remain uncovered after the collector array is installed. Some collectors have weep holes only along the bottom edge of the box. If these collectors are installed upside down, the function of the holes is eliminated. Therefore, the installer must be aware of the top and bottom.

Filling. The filling of solar collector arrays is described later. If the array is to be filled for leak-testing purposes, however, during the assembly step described here, this should be done at night if uncovered collectors are used. During a sunny day, the injection of city water into a stagnating collector can cause extensive damage from steam formation, pipe and absorber plate buckling, and thermal shock problems. This problem is particularly acute for tubular or concentrating solar collectors, since they will operate at very high temperatures. If the collector is covered as recommended earlier, this stagnation problem and thermal shock problem can be avoided during filling for leak testing. For those collectors which do not use covers, the fill must be done very early in the morning or preferably at night after the collector has cooled.

In summary, the installation of a collector can present problems because of the collector's weight, its possible high temperature, and the requirement that it be installed on a roof which may be inconvenient to work upon. Unfamiliarity on the part of laborers may cause problems unless a training session is provided at the outset. Detailed instructions should be presented at this time with respect to the collector and the structure installation. Problems can be avoided during collector assembly if the arrays are preassembled at the construction site. Pipe locations, model numbers, sizes, and proper pipe angles should all be checked before collectors are raised to a roof for installation.

Installation of Storage. Storage containers can be installed properly without much difficulty. In this section, a discussion of the installation of liquid storage is followed by a similar discussion for rock beds used in air systems. The storage tank should be installed relatively early in the building construction process if it is contained within a building. If the tank is buried outside, the installation will occur still earlier.

After installation of the tank on insulated saddles on a proper foundation, a pressure test should be carried out. This pressure test is best done using ambient-temperature water. The tank can be pressurized to 50 percent above its working pressure for several hours. Then an accurate pressure gauge can show whether any leaks occur. If the storage tank is in a location whose temperature differs from that of the water used to test the tank, temperature changes in the water will affect the pressure reading. For example, if a storage tank is placed in a relatively warm mechanical space and filled with cold city water, the water will warm and a pressure increase will be indicated on the gauge.

The collector-tank sensor should be installed before the pressure test. The tank should not be permanently filled until a source for heat is available in the building. If the building is unheated at the time of the leak test, fluid from the tank must be drained after the test.

Insulation. After the successful pressure test, piping connections can be made. If the tank has not been factory-insulated, insulation is to be applied after the leak check. If foam is used, the temperature must be above 40°F for proper foam curing. Shrinkage will ultimately result in cracks through which heat can be lost. Relatively small gaps in insulation have large potential for heat loss, contrary to the conventional wisdom of many contractors.

The drain valves and isolation valves can next be installed on the tank. Pressure-relief-valve piping can also be run to the drain. The end of the storage pressure relief and drain lines should be unthreaded to discourage a possible connection to these pipes at some future time. Storage water is not potable and the owner should be made aware of this fact by labeling of these pipes. The drain valves specified on most storage tanks are of the boiler-drain type. Buried liquid-storage tanks must be fully waterproofed as described in Chap. 2. The success of buried storage is dependent upon the water-sealing effort. Most buried storage tanks have not worked in the past, since proper waterproofing could not be achieved.

Anodic protection for buried tanks is installed at this time.

Storage tanks placed outside and above ground require high insulation levels to control heat losses during the coldest periods of winter. The pipes connecting outdoor storage tanks to the system must also be well insulated, since an antifreeze solution is not used in outdoor storage.

Storage expansion tanks in a mechanical space need not be insulated if they are installed below the connecting piping. Since hot fluids rise, an expansion tank which is placed above the pipe will lose heat. Expansion tanks should only be connected into fluid loops after the system pressure test has been completed; otherwise the membrane in the tank may be damaged by overpressure during the system leak test.

Rock Storage. Wood or concrete boxes are most often used for pebble-bed storage. These boxes must be tightly sealed with a sealant capable of withstanding the highest temperature to be expected.

The rock used in pebble-bed storage must be clean and should be inspected during the period of fill to ensure proper cleanliness and proper size. The size constraints on rock beds are important, as noted in earlier chapters. For rock beds using vertical airflow, the rock-bed sensor should be installed near the bottom of the bed. Sometimes a second sensor near the top is used to measure the available temperature for the storage-to-load mode. Since the thermistor sensors are very fragile they must be installed with care.

When a large rock bed is filled, the larger rocks roll to the bottom of the pile. Larger rocks create a smaller pressure drop and flow channeling results. Therefore, the installer should distribute sizes uniformly.

Rock boxes are often insulated inside the inner surface of the box with rigid-foam insulation. This insulation can be damaged during filling if rocks are dumped carelessly into the box. After filling of the box, the top of the rock bed should be covered to keep the rocks clean during the balance of the construction process. On many jobs, the permanent lid is not installed on the rock box until some weeks after it has been filled.

Piping Installation. Collector piping has been described briefly above, and the requirements for high-temperature solder have been indicated. On threaded piping, the control of leaks is more difficult. Teflon-base pipe dope can help control them. Piping insulation is basically the same as for nonsolar systems and is not described in detail herein.

In previous chapters, the problems associated with galvanic corrosion between dissimilar metals in liquid piping systems have been noted. Dielectric couplings between dissimilar metals do not stop galvanic corrosion. However, the best solution to galvanic corrosion is to minimize usage of dissimilar metals in the piping loop.

Fittings. Piping layouts should require the minimum number of elbows. Pipe bends have a lower pressure drop than elbows and should be used if possible with copper piping. Piping arrangements follow energy-flow diagrams governed by the logical flow of fluid from one component to the next. Pipe runs must be made with the proper size of pipe as indicated on the drawings, since the sizes of pressure drops in certain parts of the solar system are important, particularly in collector headers. Pumps, check valves, and heat exchangers must be installed in the proper flow direction. Types of valves should not be mixed, and normally open and normally closed positions of three-way valves and two-way solenoid valves must be checked. Connections to storage tanks are particularly important as noted in Chap. 4.

Air vents should be installed at all high points in liquid loops. Their installation at points of low fluid velocity is also recommended—for example, in the heat exchanger shell and storage tank. Vacuum breakers are required on drain-down systems. P/T wells for measuring pressures and temperatures at various points in the system should be installed at the points specified on the drawings.

The slope of piping in the collector loop is important to ensure proper draining of the collector array. This is necessary both in drain-down and nonfreezing systems since collector systems must be drainable under conditions of power outage. The slope of piping should be called out on the CDs and amounts to approximately ⅛ in per foot. This nominal slope must include the effect of pipe sag between supports. Slope is needed in the collector headers, in piping runs, in mechanical space, and within the building walls wherever horizontal runs are present.

Supports. The support of piping in solar systems is the same as in any other application, and some rules of thumb follow. For vertical pipe, copper should be supported every story, threaded pipe every two stories, and plastic pipe every half story. (Of course, plastic pipe is not used in any high-temperature loop of a solar system.) Horizontal piping must be supported at closer intervals than vertical piping. Copper piping should be supported every 8 ft if drawn and every 6 ft if annealed, threaded pipe every 12 ft, and plastic pipe every 1 to 2 ft (46).

Some components can be damaged if high soldering temperatures are used. One sensitive fitting is the water tempering valve in domestic hot-water systems. The sensor should be removed from the tempering valve during soldering and then reinstalled after the valve body has cooled. As described elsewhere, collectors can also be damaged by overtemperature. P/T valves are also damaged by high temperatures.

An additional subtle detail concerning liquid piping must be emphasized. No inverted U-shaped loops should be present anywhere in a liquid system. These loops act as permanent air traps from which air can never be totally removed. The presence of trapped air in a liquid system must be avoided. Air trapped in these loops expands much more for a given temperature rise than does liquid. The calculated acceptance volume of expansion tanks in loops containing air is too small relative to the much greater expansion volume of air. It is important, therefore, that *local air traps be present nowhere in the liquid system*.

The presence of an upright U-loop should also be avoided since these act as dirt traps and can eventually result in blockage. All piping from the inlet of the array to the outlet must slope uniformly upward. No local downward piping should be present at any point in the pipe loop. This also applies for piping running from the mechanical space to the inlet header of the collector and from the outlet header of the collector to the mechanical space. Another source of possible air entrapment is in pipe reducers. Eccentric reducers can be used to solve this problem.

Pipe Joints and Fittings. Flanges or unions should be used near each component in liquid systems, including pumps, heat exchangers, storage tanks, expansion tanks, and heat-rejection units. These permit efficient removal for possible service in the future. The fittings are to be shown on the CD piping diagrams.

Piping expansion joints are not capable of carrying a load. Therefore, they must be supported. A guide should be provided for horizontal expansion joints; otherwise, they may sag. Expansion joints must also be anchored at one end, as described in their installation instructions, to a building component capable of providing proper counterforce to the expansion force.

All valves must be accessible for rapid turnoff if emergency shutdown is required. Gauges, thermometers, and flowmeters also must be easily viewed and positioned so that system status can be identified quickly. Valves in collector loops using toxic fluids should have their handles removed after system start-up. This will reduce the probability that fluid can be drawn from this loop and used for a purpose for which it is not intended. Drain valves in toxic-fluid loops should be so labeled, and the tanks should indicate that this fluid must not be used for drinking. Valves located outside should be positioned with their stems downward so that no penetration of moisture through the stem opening into the insulation will occur. Air vents must be installed in a vertical position if they are of the float type.

City-water pressure-reducing valves should be connected with isolating valves and a normally closed bypass loop around them. Pressure-reducing valves may need service; it is desirable that building water pressure not be lost during these periods. A globe valve placed in parallel with the pressure-reducing valve can be used as a coarse pressure control while the pressure-reducing valve is removed.

Unwanted Thermosiphons. Another subtle problem which can occur in liquid systems is establishment of a thermosiphon in a single vertical element of piping. This occurs in large pipes and can cause control malfunctions. For example, if a pipe rising from the collector-to-storage heat exchanger to a collector array is sufficiently large, hot fluid from the heat exchanger can rise through the inner core of the pipe to the collector inlet header. The collector array itself is cold on a winter night, and cold fluid from it sinks into the mechanical space to replace the rising hot fluid. The thermosiphon thus set up can result in heat losses from the heat exchanger but can also cause the collector sensor to receive a relatively warm signal. If this sensor represents a temperature 10 or 15°F warmer than the sensor at the bottom of storage, the collector pump may even turn on and cold fluid from the collector will circulate through the heat exchanger. The cycling of the collector pump at night is to be avoided.

One method of eliminating the single-pipe thermosiphon is to put a convection-block loop in the main line. This loop will not be an air trap since high-velocity fluid flows through the pipe, and air will not accumulate.

The sensor for P/T valves must be in good contact with the fluid in the tank upon which it is mounted. Therefore, this valve must be screwed directly into the fitting on the tank and not connected indirectly via a length of pipe. The sensors on most P/T valves are relatively short and require intimate contact with the tank to be controlled.

The installation of each solar system differs from that of other solar systems. The preceding notes are applicable to most but do not necessarily compose an exhaustive list. The importance of common sense and experience cannot be overemphasized in the successful installation of a liquid solar system.

Leak Test. After the completion of all liquid loops, a pressure and leak test comes next. This should be done prior to any insulation or installation of drywall or sheathing in the building. During the pressure test, all pressure-sensitive components should be blocked off—for example, the P/T valve, the expansion tank, and pressure gauges. Leak tests run for several hours.

Air can be used for leak testing, but temperature variations during the test may cause an erroneous reading. Liquid pressure tests have the difficulty that the liquid must be drained after the pressure test if the building is not heated in winter. Ambient-temperature water should be used for leak tests in warmer seasons; otherwise pipes will sweat and leaks will be concealed.

Insulation. After completion of the pressure test, the piping can be insulated. By design, sufficient space in walls, mechanical spaces, and elsewhere is available for insulation. If piping is buried, the insulation must be sealed with pitch. Most codes do not permit any joints to be present in buried piping, and a complete seal for all buried piping is essential to avoid parasitic heat losses.

The insulation of unburied indoor and outdoor piping is as important as that of buried piping. All components of liquid loops are insulated, with the following exceptions: tempering valves, tops of air vents, frost sensors, pump motors, and P/T relief valves. Outdoor piping must be insulated with no gaps and covered with a weatherproof lagging. The joints in insulation covers outdoors should face downward to reduce the tendency for moisture to penetrate through the joint. Some insulation covers require painting. A proper paint for the insulation must be selected. Some insulations are sensitive to oil-based paints.

Labels. The final step in piping installation is to label all pipes with fluid contents and flow-direction arrows. The arrows should be spaced within a few feet of each valve or component in the system, and the fluid, if not water, should be noted on the arrow. All valves should be tagged with numbers or functions corresponding to the flow diagram for the system. This is particularly important for valves which must be operated during an emergency shutdown or draining. Likewise, valves on toxic-fluid loops should be carefully labeled. Domestic hot-water systems should not be filled until utilities are available to keep the tanks warm. This also applies for the storage loop and delivery loop for space-heating systems.

Installation of Ducts and Dampers. Ducts are installed in accordance with long-standing practices of the sheet metal and air-conditioning industry. Leakage control and insulation requirements are more stringent for solar than for nonsolar systems.

The control of leakage from ducting in solar systems is important, as emphasized in previous chapters. In the installation step, the leak-control effort reaches its final step. The securing of ducts with rivets or screws is adequate for mechanical purposes but not for leak-control purposes. Each duct joint must be sealed. Tape is one standard method of sealing HVAC ducts, but silicone sealant (or equivalent) is required for solar collector ducts.

The control of leakage through dampers is also important. The damper must be properly adjusted.

Motorized dampers and backdraft dampers should not be confused, nor should backdraft dampers and manual dampers be mixed. Obviously the proper flow direction for backdraft dampers is essential. Motorized dampers are controlled by signals from control systems. The proper control of the two critical dampers as described in Chap. 2 is essential.

The material for fabrication of ducts is specified in the CD, and should be the smallest gauge possible to avoid transient heat-storage effects. Heat used to heat a duct is not useful heat delivered to a space. The requirement of small-gauge metal, of course, is limited by the structural considerations in designing the duct and pressure differences across the duct.

Dampers. Leak testing of dampers is as important as leak testing of piping; however, it is more difficult. One method of checking for leaks is to pass scented air or other tracer gas through the system. This scent can be introduced at the inlet side of the blower and each joint tested to determine whether a leak is present. Another method of testing for leaks is to pass a moistened surface of the hand near joints; if a cooling sensation is felt, there is a leak present.

Insulation. After sealing, duct insulation can be applied if the insulation is not already integral with the duct. Insulation must be protected from the environment if used out of doors. Covers of appropriate material, either metal or plastic, should be used.

If an electronic air purifier is part of the circulation system, it is installed on the cold side of the circulation loop, since its operating temperature must be below approximately 120°F. The location of the purifier is called out in the CD.

Some Notes on Control Installation. It has been emphasized repeatedly in this book that the control system can make or break a solar system. The desi of controls has received attention in detail, but the installation is equally impuortant. In systems used in commercial buildings and residences, the controller may be installed by the electrician, with sensors installed by the solar contractor. Problems have sometimes arisen in interfacing these two crafts. In larger buildings, a controller subcontractor is involved and the installation of the controls can be accomplished somewhat more reliably.

Sensors. The location of sensors in storage and on the collector has been described in detail in Chap. 4. The details are shown on the construction documents and must be adhered to. The collector sensor should be installed by the collector manufacturer to assure proper installation and uninterrupted warranty on the collector. Storage sensors are located in a fitting provided near the bottom of the tank. Proper location should be reconfirmed by inspection of the tank in the field to follow up on review of shop drawings. Storage sensors must have the sensing element located well within the tank.

Heat-rejection-unit sensors, if separate from the control thermistor in the collector, should also be installed by the collector manufacturer. Sensors should not be soldered; they are sensitive to high temperature and must be attached by a screw, bolt, or pipe connection.

The sensors must sense the temperature for which they are designed and not a spurious temperature. Heat leaks from a high-temperature source in the vicinity of a sensor can cause malfunction of the control. Storage sensors located in mechanical spaces are particularly subject to this problem. If a storage sensor is located near a surface which is above the storage-tank temperature, convection and radiation heat transport can cause the control sensor to heat and have an incorrect signal. The method of achieving a good signal for control sensors has been described in detail in Chap. 4; every item in that section must be adhered to as shown on the drawings to ensure proper control function. All sensors must be insulated.

Wiring. The wiring used to connect sensors to the control unit must be impervious to moisture and must be flexible. If moisture-sensitive insulation is used, it will provide an improper sensor reading, since relatively low voltages and high impedances are involved. This problem is very difficult to trace if a system malfunctions. Stranded wire appears to be more satisfactory than single-conductor wire since it is more flexible and does not place a physical load on the sensor. In high-temperature locations, Teflon insulation is most appropriate. This is particularly important in the vicinity of the collector sensor and collector piping.

Sensor wiring should be routed away from sources of electronic interference such as 110-V ac lines in long parallel runs. The sensor wire can be carried in a separate conduit. Alternatively, if the sensor and 110-V wiring are each twisted, a cancellation of the electric fields occurs and interference is reduced. Outside, sensor wiring from collector to mechanical space should be kept above the roof so it is not exposed to any standing water. The wires must be well attached so that they are not whipped about by the wind. In many cases, the sensor wiring is attached to the collector piping with cable ties for this purpose. Wiring can be run within the insulation of collector piping but future troubleshooting is rather difficult. The wiring should also be kept away from foot traffic. Sensor and controller wiring should be of the proper gauge, soldered, and color-coded.

Thermostats. The zone thermostats should be located in accordance with standard practice. They are located on interior walls and not exposed to either sunlight (important in direct-gain systems) or drafts. Likewise, they are to be kept away from radiant heat sources, including fireplaces, stoves, electronic appliances, and reflected sunlight.

Dual-point thermostats used in solar heating systems should be adjusted at the factory. The temperature band should be checked, however. The second contact should close 2 to 3°F below the closure of the first contact. This can be checked by setting the thermostat at a relatively high temperature. An ohmmeter across both contacts checks the thermostat as it is heated slightly with a hair dryer or other low-temperature heat source. The temperature of the thermostat can be monitored using a standard dial thermometer located near the temperature sensor.

Controller Location. The location of the control unit is important. Most controllers use solid-state devices requiring a relatively uniform temperature throughout the year. The location chosen should be ventilated, have easy access, and be away from heat sources such as the solar storage tank or safety valve release points where steam could be present.

The controller should be wired on an unswitched circuit so that an inadvertent operation of the switch would not deactivate the control circuit, with associated serious problems of overheated collectors, etc. The control should be capable of being deactivated, however, by means of an internally grounded shutoff switch. If the control unit is located in the mechanical space where moisture may be present on the floor, a wooden platform should be provided for persons working on the control box.

Before the control system is tested, all wiring should be rechecked. Solid-state circuits are easy to destroy by application of a voltage of the wrong type at the wrong place. Warranties do not cover this. Color-coded wires are useful. Control triac failures are the most common and result from incorrect wiring, shorts, or voltage spikes.

Passive Heating Systems. It is noted at the beginning of this section that the installation of passive systems is less complicated vis-à-vis active systems. The opportunity for error is reduced since the mechanical complexity of the system is less and the number of components involved is smaller.

Backdraft dampers used in TSW systems require careful installation. There is only a very small density difference available to hold backdraft dampers closed; therefore, a flat-seal surface must be provided for these light-gauge plastic valves. Rectangular or round openings can be used as specified on the drawings. The backdraft damper must be free to open.

Concrete Work. Passive systems using either water or masonry storage require larger foundations as shown on the drawings. The inspection of these pours must be made early in the construction process to assure that the proper thickness has been used.

The pouring of large masonry storage walls is a major process. The size of walls used in passive systems is of the proportion of major structural elements of bridges or large buildings. Walls are typically 8 to 10 ft high, 10 to 20 in thick, and as long as the side of a building. The pressure imposed on forms used for pouring walls of this size is enormous. It is possible to pour the wall in sections. However, a monolithic wall is best relative to good heat transfer.

Plywood forms are normally used for passive walls unless a specific wall texture is required, but metal and plastic forms can also be used and reused. Ties holding both sides of the form together are usually made from metal and remain within the wall after the forms are removed. Framing for the forms uses sheeting lumber, nails, wire, and cable ties. Large 2 × 4 or 2 × 6 units (called whalers) are used for additional structure and support near the bottom edge of the wall. The basic wood materials used are 4 × 8 × ¾ plywood and 2 × 4's.

The openings in thermal-storage walls are formed by knockout inserts within the form. The location of window openings and thermal-storage-wall vents can be drawn on the form itself during its assembly. As one form wall is located (perfectly plumb and level), the window openings in that area should be laid out on the form. Beveled window openings ensure proper removal of their forms. One piece of the window form is attached to the standing form panel and the second half is suspended from the first. A chamfer strip around windows and along the top of the wall will eliminate sharp edges on the concrete which would be potential breakage points.

The placement of reinforcing steel occurs next. Horizontal steel is wired to vertical steel on 16-in centers, and the vertical steel is wired to anchors projecting from the wall foundation. Additional steel over window openings or vent areas is installed at this time. If temperature sensors are to be used in the wall, they can be supported by the steel during this part of the wall-construction process. Next, the second surface of the form is installed, and bulkheads are added to close out the ends of the wall. The form is completed by the addition of whalers.

The concrete is injected into the form by pump or crane and bucket. The mixture must be vibrated to prevent air pockets from occurring and to distribute concrete evenly within a form. The pour should be made all at once if possible. Particular attention to areas beneath TSW lower vents is suggested. Following wall curing, the form is removed by breaking off the snap ties just inside the wall surface. Final touch-up and repair of the wall are done at the same time.

The curing of concrete occurs in two separate ways. Concrete is able to support its design load within a matter of weeks. However, the hydration reaction which converts concrete to a masonry unit may take several years. During this time, moisture deep within the wall reacts to form a chemical hydrate. The heat of reaction is in part solar heat (causing the energy delivery of the passive system to be below design for the first 2 or 3 yr). The wall can be painted, however, in a much shorter time. See Ref. 71 for further details.

Glazing and Insulation. Glazing is installed to the south of the thermal-storage wall after it has been poured and the forms removed. (Glazing is difficult to install in winter.) Since glazing inner surfaces are not accessible after installation, it is absolutely essential that they be perfectly clean. Construction dust can collect on the inner surface of the glass, and therefore all thermal-storage-wall openings and vents should be taped since the glass cannot be cleaned easily once installed.

Automatic movable insulation requires adjustment of the limit switches controlling the position of the insulation. If the roll-up type of insulation is used, the limit switches are adjusted at the bottom and top of specified travel. Proper insertion of the shade in its track is also essential to proper function. Adjustment of the insolation sensor is completed during start-up, as described below.

Liquids. After the leak test, the working fluid can be installed. If the collector fluid is to be mixed on site to achieve the specific gravity required for freeze protection, this can be done using the information given in Table 5.3. The pH is also to be checked at this time if the solution is an electrolyte. A toxic fluid which has not been color-coded can be dyed with a vegetable dye at this time. The dye identifies the fluid as a nonpotable liquid in case it leaks in the future. If a collector loop is filled from the roof with glycol, it is important to keep it off the roof. Glycols dissolve some roof materials. (See page 339, "System Cleaning.")

Instruments. Instrumentation installation for system performance monitoring is detailed in several references produced by the National Solar Data Network. This exercise is not a part of the normal system installation effort and is not described herein. However, most solar systems will have instrumentation present at least for simple functional checks. The location of this instrumentation is called out on the working drawings.

Sensors for temperature readout should be isolated from external heat sources by insulation. Pressure gauges should be installed with shutoff valves. Flowmeters require the proper length of straight pipe both upstream and downstream in accordance with manufacturer's requirements (see Fig. 5.6).

TABLE 5.3 Glycol specific gravity vs. freeze point*

	Ethylene glycol		*Propylene glycol*	
Freeze point, °F	*Percent by weight*	*Specific gravity*	*Percent by weight*	*Specific gravity*
20	16	1.028	19	1.021
10	26	1.044	29	1.035
0	33	1.055	36	1.047
−10	40	1.069	43	1.058
−20	45	1.078	48	1.066
−30	49	1.086	52	1.072
−40	53	1.094	55	1.077
−50	56	1.102	57	1.081

*Adapted from Ref. 32 with permission.

If a solar-radiation sensor is to be installed in the vicinity of the collector array, it must not be shaded at any time during the year. Other instrumentation such as level alarms, site gauges, and pump hour meters are installed in the normal manner.

System Start-up

At the completion of construction, the solar system will have been leak-checked and the controls debugged. The start-up of liquid systems is treated first in this section and is followed by a discussion of air systems and passive heating systems.

System Cleaning. Before the working fluid is installed in a liquid system, the system should be flushed with a low-sudsing detergent. Alternatively, solutions of trisodium phosphate (1 lb to 50 gal), sodium carbonate (1 lb to 30 gal) or sodium hydroxide (1 lb to 50 gal) can be used. Commercial products such as Betz Labs 346, Calgon Vantage, Chemed Corp. BC-45, Drewsperse-78, or Nal Clean 8910 work well. The purpose of flushing the system is to remove any metal particles, solder flux, or other material left from system assembly. The refuse entrained in the circulating water is trapped in the strainer and drained periodically during flushing. After the coarse material has been collected by the strainer, the filter may be installed. Increasing pressure drop across the filter indicates that the filter element is removing additional material from the fluid stream. The element may be replaceable or cleanable. The element is cleaned and reinstalled until a steady filter pressure drop occurs. A strainer or filter must be present in the fluid loop from the first operation in order that the collector, pumps, and valves remain clean.

If several collector subarrays are connected in parallel to form the full array, each collector bank can be isolated in turn by shutting off all others to increase the flow velocity through that bank. For example, if four rows of collectors are involved in an array, the shutoff valves for three of the four can be closed and all available fluid pumped at high velocity through the remaining bank. This higher flow rate will entrain any material in relatively low-flow-velocity areas of the array. This procedure is carried out in turn for each row of the entire collector array, ensuring the maximum possible cleaning effect. At the end of the row-by-row cleaning step, of course, each collector row valve will be reopened. After the system has been flushed, the flushing solution should be drained and washed from the fluid loops. The volume drained can be used to estimate the required amount of fluid needed for the final fill.

The flushing of systems which do not use water-based working fluids is somewhat more difficult. Since organic and silicone oils are very expensive, it is not practical to flush the system with these oils and then discard the fluid. The only method which seems appropriate for these fluids is to repeatedly empty the strainer and filter.

System Charging. The first step of start-up for a liquid system is charging the various fluid loops. Prior to filling the collector loop with antifreeze solution, the system should be filled with water and as much air as possible ejected through the bleed valves. Initial charging uses the flow loop shown in Fig. 4.32*b*. The drain valve is opened and the water is drained through this valve (if the drain valve is opened too far, air may enter the system through it). At the same time antifreeze solution is injected above the check valve by the pump. In this manner, the water can be smoothly ejected from the loop. The presence eventually of colored glycol in the drain stream will indicate that the fluid loop has been filled.

After the collector loop has been filled with fluid, the system is pressurized (with a small amount of line pressure applied through one of the fill lines). This design pressure is sufficient to cause 3 to 5 lb/in^2 gauge at the topmost header of the system. This pressurization prevents fluid from boiling at this lowest-pressure, highest-temperature point in the loop. After system pressurization, the collector-loop expansion tank can be connected to the system. Typical cold collector loop pressures are 15 to 30 lb/in^2 gauge.

Liquid systems can also be filled from a fitting at the highest point in the system. Using this method avoids trapped air in the system. The filling rate can be controlled by the rate at which the initial water fill is drained from the system.

As noted in several previous sections, collectors should be filled only when no sun is present—at night, very early in the morning, or before protective opaque covers have been removed. This is particularly important for tubular and concentrating-type collectors.

Initial Operation. During the first few days of operation, any remaining air in system loops will be eliminated via air vents. At the end of the period, system pressure will stabilize and the operating pressure can be checked by the pressure gauge at the expansion tank. If pressure is higher than desired at this point, opening a drain valve will reduce system pressure. If for some reason an airlock should occur in part of the collector loop, the flowmeter installed in the circuit will immediately indicate this. If no flowmeter is used, the pressure rise across the collector pump or storage pump can be converted to a flow rate from the pump curve. Alternatively, an infrared thermograph of the operating collector can locate flow blockages (66).

Balancing. Once operation is stable, the flow balancing subcontractor will balance all flows according to the design. Flow balancing on small systems can be done by the HVAC contractor, who may use either a pressure-drop or a temperature-rise approach for each collector loop. In the pressure-drop approach, a differential pressure gauge (e.g., a Magnehelic™) is installed across the balancing valves. Pressure differences can be converted to flows depending on the temperature and type of fluid in use. The flow in each loop is shown on the drawings which are the basis of the balancing. System balancing is an iterative process. The balancing valve for the longest loop is first opened completely and the flow through that loop checked. If the flow is above design, the valve is modulated. The remaining, progressively shorter loops are then adjusted and the cycle is repeated until the proper balance of flow occurs. If the final flow adjustment gives a total flow above or below the design level, no serious system performance problems will occur if the variation is no more than 10 percent from design. If flow varies by more than ± 10 percent, the balancing of each loop should be based on a proportional ratio of the design flow rates. For example, if two fluid loops have an installed total flow of 100 gal/min but the design flow rates are 40 and 50 gal/min, then the installed 100 gal/min total flow is apportioned on the same ratio in both loops. The final adjustment would result in flows of 44.4 and 55.6 gal/min. Each balancing valve should be labeled with its final setting.

In the absence of precise differential-pressure-measurement equipment, the temperature-rise approach for flow balancing can be used. This method is based on the requirement that each subloop in a collector system have the same temperature rise under bright sun. This is assured in the design by requiring that each collector have the same flow rate per unit area. Therefore, the temperature rise in 1 collector or 100 collectors in parallel will be the same for the same flow rate per square foot. Thermometers can be used to check for equivalent temperature rises in array subloops until flow balance is achieved. Since solar collectors have a slow response time, the temperature-rise approach will require more time in the field than the pressure-difference method. However, both are satisfactory.

Controller Checkout. During system start-up, the operation of the controller should be examined in all modes—storage, collection, energy delivery, and heat rejection. The examination will take some time and should be carried out over a period of a week or two during which weather changes occur. A day with sun followed by a day with no sun is ideal.

A significant heat load should be present to check day/night energy delivery from storage. If a sunless day or two passes, the backup system should begin operation automatically. These weather conditions are not always present at the time of system start-up, so control fine tuning must be delayed until a significant heat load and some weather variation occur.

Air Systems. The start-up of air systems is similar to that for liquid systems except that some steps such as loop flushing need not be done. System balancing is required for air systems after all leaks have been checked and all dampers properly adjusted. The operation of the domestic hot-water-preheat controller is separate from that of the heating system and should be checked on both air- and liquid-based systems.

Passive Systems. Passive solar systems present little difficulty during the start-up phase, the only moving part being the movable insulation. The control for the movable insulation is easily checked by exposing it to sunlight and then covering it to ensure that the insulation positions itself in the appropriate manner for sun and no-sun conditions. This check can be made without any requirement for weather changes and can be done in a matter of minutes.

Backdraft dampers should be inspected. At night backdraft dampers should be held closed by the density difference of air in the wall cavity and the room. Smoke may be used to check for reverse circulation at night.

SYSTEM INSPECTION AND ACCEPTANCE TEST

After system start-up by the contractor, the system must be inspected by the architect/engineer or solar consultant. The inspection should be made after all phases of start-up and system debugging have been completed. The purpose of the inspection is not to find mistakes which must be repaired prior to a second inspection, but rather to examine the system for the final time to ensure that all components are installed and functioning properly.

An acceptance test is not always a requirement of a solar system installation. For large commercial systems an acceptance test will frequently be required, however.

System Inspection

First, a visual inspection of the entire system ensures that all components are correctly installed. After the visual inspection, a functional check is made to determine whether the system operates properly in all modes. In this section, the important items to be checked during an inspection are outlined. A checklist is a convenient way to ensure that no part of the system is overlooked during the inspection. This checklist will be different for each project and is therefore not presented here in detail. However, the major items to be examined are given.

Visual Inspection. During a visual inspection, the inspector should examine all flow loops, following the direction of flow. For example, starting at the collector pump in a liquid collector loop, the inspector examines the pump, piping connecting the pump to the check valve, piping to and from the collector, connections at the heat exchanger, and expansion tanks, as well as any instrumentation contained in the fluid line. If the direction of flow is followed, each component of the system will be examined and none left out. During the visual inspection, the part number and rating of all components should be determined. The horsepower and speed for pump and circulator motors should be checked against the specifications; the ratings on all safety valves should be checked to ensure their conformance with the specifications. Collector part numbers should be checked as well as sizes of filters, expansion tanks, and heat exchangers.

The direction of flow in all components should be checked. This is particularly important in check valves, heat exchangers, pumps, solenoid valves (two-port or three-port), backdraft dampers, storage tanks, and auxiliary heating systems. Check valves and pumps are occasionally installed backward or the proper cross flow in a storage tank is not achieved. Insulation on all piping and ducting should be carefully examined. Any gaps in the insulation should be noted for future repair. In liquid systems leakage through insulation will damage it. Ducts are checked for leaks as described in the previous section on system installation.

Storage tanks should be inspected to ensure that the design operating pressure exists, that all insulation is intact, and that all valves are in the proper position (NO or NC); storage-expansion-tank isolation valves should be open if present. The position of the storage sensor should be verified. In liquid tanks, the sensor is to be centrally located in the bottom one-third to one-quarter of the tank. In pebble-bed storage, the sensor is located within approximately 6 in of the bottom of the pebble bed.

Collector arrays should be inspected for proper pipe connections and slope, piping insulation, and the presence of all required air vents and vacuum breakers. Moisture and dirt seals on collectors should be examined to ensure that no leaks are present and that no seals have been damaged during soldering or positioning of collectors during installation. The glazing on collectors should be clear, without any internal condensation. Internal condensation indicates that moisture is present in the collector and either that the vent system is not functioning or that there is a leak in piping in liquid-based collectors. The source of any moisture in a collector must be identified and repairs made, since collector deterioration will occur if moisture is allowed to remain. Collector sensor wiring should also be checked at this time. The sensor itself cannot be examined since it is inside a collector.

All wiring should be inspected, with particular attention to control wiring. The proper sensors must be connected to the proper terminals on the control as indicated in the instructions. Output signals from the controllers to pumps, solenoid valves, heat-rejection units, and auxiliary heating systems should also be checked. During start-up the function of these components has been verified.

In liquid systems, the specific gravity and pH of the working fluid should be checked. A small sample (1 or 2 fluid oz) of fluid is drawn from the drain valve and a hygrometer is used to verify specific gravity, with proper attention to fluid temperature. Fluid pH can be checked by any number of methods, the most common of which is litmus or other color-indicating papers. The pH can also be checked with kits available from aquarium-supply houses or chemical-supply houses. If continuous monitoring of pH is needed, an electronic pH meter can be used.

The auxiliary furnace, boiler, or heat pump should also be inspected in the usual manner for nonsolar systems. If the jacket of the backup domestic water heater feels warm to the touch, it is likely that the insulation is installed improperly in the jacket. This happens when some manufacturers slide the jacket over a tank which has been wrapped with bat insulation. The jacket may catch the insulation and tear it from the tank.

After the visual inspection of the system, a punchlist is made for contractor repairs. If the repairs noted during the visual inspection will affect the subsequent functional check, these repairs must be made in advance of the second step of the inspection process.

Functional Check. The functional or operational test of a system is as important as the visual inspection described above. The functional test should be carried out on a sunny day so that proper operation of the control and collection system can be verified. Checks on cloudy days cannot be complete. After a solar system has begun operation on a sunny day, a list of performance data should be made and checked against the design. The data to be noted include the operating pressure in each loop—collector, storage, DHW, and energy delivery. In air systems, this measurement need not be made. The second type of information recorded during the operational check is pressure rises and drops across various components. In a liquid system, the pressure rise across the pump should correspond to the design flow rate through the system according to the pump curve. The pump curve must apply for the fluid in use (which may not be water in some cases, although it is the basis of most common pump curves).

Pressure drops through each component should also be checked by inserting pressure gauges in P/T fittings. The largest pressure drops will occur in collector loops across control valves, heat exchangers, filters, and balancing valves. These pressure drops can be checked against the design if the overall system pressure drop differs from design. If system pressure drop is within a few percent of design, a component-by-component check is not necessary.

The third type of data recorded is temperatures during system operation. Collector inlet and outlet temperatures measured, for example, at the heat-exchanger shell inlet and outlet connections are most important. The collector efficiency curve can be used to predict what temperature rise should occur given the inlet temperature and the local solar radiation level. A precise measurement of solar radiation need not be made, and a simple solar-cell device accurate to 10 percent is adequate for this inspection operation. Typical temperature rises across collectors are 10 to 15°F for liquid collectors and 40 to 60°F for air collectors.

The temperature rise in the heat exchangers should correspond to flow rates on shell and tube sides. For example, in the collector exchanger, the storage side flow rate is frequently 50 to 100 percent higher than on the collector side. The temperature rise on the tube side will then be proportionately less than the temperature drop in the lower-flow shell side. The specific heat of both fluids also enters into the relative magnitude of temperature drop and rise in heat exchangers.

Temperatures of storage as well as temperatures in the heated space should be reported. On a sunny day, the solar system may have adequate capacity to heat the building and the backup system may not come on. A manual override of the controller will cause the backup system to operate for a short time so that pressure and temperature measurements may be recorded. The pressure at the inlet of each pump and circulator must be above the NPSH for that unit to ensure that no cavitation occurs.

The record of temperatures, pressures, and flow rates made during the formal inspection should be inserted in the operator's manual described later. These measured conditions during a sunny period represent the performance of the new, properly operating solar system. During subsequent years, inspections will be repeated and comparisons can be made to the original performance baseline established during the formal inspection. Deviations from these measured data in the future will indicate possible sources of system deterioration or malfunction.

The final step of the operational check involves exercising the control system through all of its functional modes to determine whether control mode switches occur according to the design. On a sunny day, the system will normally operate only in one mode; therefore, the system is placed in other modes by using manually operated sensors instead of system sensors. The thermistors are disconnected from the controller and a potentiometer is connected in their place. The apparent temperature sensed by the controller can then be manipulated by adjusting the resistance of the potentiometer. Resistance values for various temperatures can be read from a table such as Table 4.3 for the controller in question.

The controller can be placed in its various operational modes using potentiometers as follows. On a sunny day, the system will be operating in its collector-to-storage mode. To disengage this mode, the resistance of the collector and storage sensor potentiometers can be made approximately equal within 1 or 2°F equivalent. As the collector potentiometer is reduced in its temperature setting, the collector and strage pumps in a liquid system or the dampers in an air system will turn off or reposition themselves, respectively. The sensor resistance at which this happens should be recorded and compared with the resistance chart to determine whether the proper hysteresis is present in the controller. As noted in previous chapters, the turnoff temperature difference in liquid systems is approximately 3°F, in air systems 15 to 20°F.

After the turnoff temperature difference check, the turnon temperature difference can be checked by increasing the temperature equivalent setting of the collector sensor. At the point where the collector and storage pump begin operation, the resistance on the potentiometer should again be checked and converted to a temperature reading. The turnon temperature difference can then be determined and should be 15 to 20°F for liquid systems and 35 to 40°F for air systems.

The heat-rejection unit, if operated as a part of the solar controller, can also be checked. The collector temperature seen by the controller can be increased by changing the collector sensor potentiometer setting. As the resistance is continuously changed, it will eventually correspond to that for a 200 to 220°F collector temperature, which should activate the heat-rejection system, depending upon the precise set point for high-limit heat rejection. The analogous method is used if heat rejection is controlled by storage, not collector, temperature. During the heat-rejection phase in liquid systems, the heat-rejection fan should begin to operate and the solenoid valves diverting flow through the heat-rejection unit should be activated. In most liquid heat-rejection systems, the storage pump will also cease operation at this time. Deactivation of the heat-rejection mode can be checked by progressively reversing the resistance until the minimum point for heat rejection is passed; the system should then operate in the collector-to-storage mode. In domestic hot-water systems, hot water should be dumped automatically if this is the heat-rejection mechanism.

The operation of the auxiliary heating system is also checked at this time. In both air and liquid systems, this control function is actuated by zone thermostats. The backup check is particularly simple in these cases, since turning the thermostat up should cause the backup system to turn on and turning the thermostat down should cause it to turn off. At the point where the lower temperature contact is closed and the upper temperature contact is open on a dual-point thermostat, the circulation pump P3 (see Fig. 4.20) will be operating but the backup system will not be on.

If an ambient reset is used on the boiler, its function can be checked by providing a false ambient temporarily at the ambient sensor and checking the boiler outlet fluid temperature. Usually the outlet fluid temperature specified at two values of ambient temperature, and two checks at two temperatures during inspection, should be adequate to verify the function of the ambient-reset feature of the backup system.

The final check to be made of the control system is the function of the fail-safe system used to deal with power failures. In liquid systems described earlier, the fluid is drained from the collector during the power failure and is not recharged until the following night. In air systems, there may be no particular control feature to accommodate a power failure. Power failures are simulated by disconnecting the power supply to the solar system to check for this operational mode. Of course, complete draining of liquid systems is not essential to pass this test.

During the inspection of the control system and its actuator, the function of each component in each mode should be examined. During mode switches, no relay chatter or control-valve flutter should occur, solenoid valves should undergo the full travel, and dampers should seat properly.

After the visual and operational inspection, a punchlist is prepared for the contractor, giving all repairs to be made. If a second inspection is required, it can be performed during the acceptance test described in the next section, if those items requiring correction do not have a bearing on system performance during the test.

Acceptance Test

Acceptance tests are used to determine that the solar system is functioning properly from a thermal point of view over a time period of 1 or 2 days. The acceptance test is fundamentally different from the control test or inspection described above in that it is used to verify that collector energy delivery and energy flow to storage are at the design levels. Acceptance tests are most often used for larger commercial systems, since additional expense is required to carry out this formal test. If a formal acceptance test is not carried out for smaller systems, the functional check during which detailed pressures and temperatures are recorded as noted above can be used in its place.

The acceptance test will vary according to the type of system used. The test procedure should be a formal procedure agreed to during contract negotiation between the solar system subcontractor and the owner or architect/engineer. The formal test procedure should delineate all data to be recorded and how frequently it is to be recorded. The following information should be included in the formal test report:

Date
Test location
Testing firm
Test engineer
Test procedure

Test equipment serial numbers and calibration dates
Manufacturer and model numbers
System information including collector tilt, collector area
Nominal collector flow rate
Storage size
Heat exchanger size, etc.
Test data:
- Collector-loop flow rate
- Storage-loop flow rate
- Collector inlet bulk temperature
- Collector outlet bulk temperature
- Storage temperature
- Storage inlet bulk temperature, if heat exchanger is used
- Storage outlet bulk temperature, if heat exchanger is used
- Solar radiation on collector plane
- Wind speed
- Ambient temperature
- Sun time at data collection point
- Collector inlet pressure
- Collector outlet pressure
- Storage tank pressure

Collector information should include net collector area, gross collector area, number of covers, type of absorber surface, etc.

During the formal system performance test, the system should be disconnected from the load since heat to load would cause a removal of energy from storage, which is not considered part of the test. A daylong test is normally specified. Performance data listed above should be recorded every 15 min for all hours of system operation from early morning to late afternoon. Information calculated from the raw data above will include the following: 15-min energy collection, collector efficiency, heat delivered to storage, flow rate, and pump energy consumption. Useful energy delivery for each 15-min period will be totaled at the end of the day to determine the total useful energy delivered over the test period.

The daily energy collected is to be compared to the prediction of daylong performance contained in the solar system bid submittal. This is the most important aspect of the system inspection and acceptance effort, since it determines whether the solar system operates as designed. Of course, the acceptance test will not necessarily be carried out on a day with the same weather which was assumed during the daylong calculation included with the system bid submittal. The method of correlating the two sets of numbers is to recalculate (see example below) the projected energy delivery using the same procedure as used during the bid submittal but for solar, wind, and ambient-temperature data collected during the formal test procedure. The daylong calculation and the daylong measured performance should agree within a few percent as specified in the contract. A performance penalty should be included in the contract if the daylong system performance is not within the acceptance band agreed upon. This penalty can be deducted from the 10 to 20 percent retainer withheld by the owner until completion of work. It is suggested that all details of the acceptance test be included as part of the contract to avoid any difficulty in resolving problems which may occur during the acceptance test.

During the night following the daytime acceptance test, an acceptance test of liquid storage can be carried out (rock storage is difficult to test). If no energy is actively extracted from storage, the temperature of storage will drop overnight because of parasitic losses through the insulation and possible evaporation if the tank is an open tank. The temperature drop during this test can be used to check the effective insulation R value of the storage tank with Eq. (4.1). If the calculated resistance measured in this functional test is different from that specified for the tank, the reason should be determined. Presumably during the prior visual inspection of the tank the amount of insulation has been determined. If an overnight temperature drop exceeds that for the observed R value, some other source of heat loss may be in operation, such as conduction through tank supports, thermosiphoning of the collector, and storage loops or other parasitic heat losses. If the tank is well insulated, this test may need to be extended over a period of 2 or more days in order to have a measurable temperature drop.

If no heat is added to or removed from storage during the test, fluid in the tank will tend to stratify and a single tank temperature may not be appropriate for use in Eq. (2.9). In this case, it may be advisable to operate the heat-exchanger pump to mix fluid in the storage tank and maintain a relatively uniform temperature. If this fluid "stirring" is done, however, the amount of energy put into the fluid by the pump must be included in the energy balance before Eq. (2.9) is used.

Example. An acceptance test on a large commercial system (26,378 ft^2 net) gave the data shown in Table 5.4. The system began operation at 0843 and turned off at 1500. Measured fluid flow was 18.93 $lb/(h \cdot ft^2)$ of glycol solution whose specific gravity was 1.038 and specific heat was 0.86 $Btu/(lb \cdot °F)$.

In the bid submittal, the manufacturer presented a linear collector curve with intercept $F_R(\tau\alpha) = 0.691$ and slope $F_RU_c = 0.869$. The specification

TABLE 5.4 Test data*

Time	*I_c, collector-plane solar flux, Btu/(h·ft²)*	*Collector inlet array fluid temperature, °F*	*Ambient temperature T_a, °F*	*Collector fluid temperature rise, °F*
900	223.6	100.8	37	7.8
1000	288.6	103.5	40	8.8
1100	329.5	110.7	45	10.3
1200	340.4	115.6	45	10.6
1300	323.5	119.8	46	10.3
1400	277.8	122.8	48	9.2
1500	202.5	123.4	47	6.9

*For hour preceding noted time.

TABLE 5.5 Measured and calculated performance

Time	Measured q,* Btu/(ft²·day)	Calculated q: $(T_{in} - T_a)/I_c$	Efficiency†	q, Btu/(ft²·day)
900	40.0‡	0.29	0.44	28.0‡
1000	143.3	0.22	0.50	144.1
1100	167.7	0.20	0.52	171.3
1200	172.6	0.21	0.51	173.6
1300	167.7	0.23	0.49	158.5
1400	149.8	0.27	0.46	127.8
1500	112.3	0.38	0.36	72.9
Total	953.4			876.2

*$q_{meas} = [(\dot{m}c_p)/A_c](T_{out} - T_{in})$

†$\eta_c = 0.691 - 0.869\ (T_{in} - T_a/I_c$; incidence angle was less than 40° for the test and angle modifier was negligible.

‡System operated for only 17 min of this hour.

required that performance based on efficiency data be within 10 percent of measured data. Was the acceptance test passed?

Solution. To determine whether the test was passed, calculate the measured energy delivery by multiplying fluid flow rate, density, specific heat, and temperature rise together. Compare this number to the expected energy delivery q based on bid collector properties, that is, $q = F_R(\tau\alpha)I_c - F_RU_c(T_{in} - T_a)$. Table 5.5 summarizes these calculations. The table shows that the system produced 8 percent more heat during the test than predicted in the bid information. Therefore, the acceptance test is passed.

Long-Term System Performance, Approximate Method

The viability of a solar system over a long period depends upon the amount of energy delivered to the heated space. The precise measurement of this quantity and its relationship to the total building load requires a very complex monitoring system costing (in 1980 dollars) approximately $50,000, plus periodic computer and data processing charges which occur.

An approximate method of estimating system performance described in this section requires minimal instrumentation but can indicate whether the system is operating as predicted. The method determines the as-built heating load on the building. This number always differs from the calculated heat load. To determine the real heat load on a building, the building can be operated without any solar energy input for a week or two during a very cold period of the winter. From the measured utility energy consumption and known conversion efficiencies, the amount of energy consumed during the test period can be determined relatively accurately. This amount of measured consumption plus internal gains when divided by the concurrent number of degree-days will give the as-built heat load on the building in British thermal units per heating degree-day.

The heating load measured on the building as noted above when multiplied by the total annual degree-days for the test year will estimate the amount of energy required to heat the building. This number is adjusted by accounting for internal gains. The net number of British thermal units required by the building is then known.

In addition, the amount of backup energy consumed during the test year is also known from utility bills. The solar load fraction for the year is then easily calculated from Eq. (5.2).

$$\bar{f}_{s,exp} = 1.0 - \frac{Q_{aux}}{Q_{tot}} \tag{5.2}$$

where $\bar{f}_{s,exp}$ is the experimentally measured annual solar fraction, Q_{aux} is the amount of auxiliary energy required, and Q_{tot} is the total net heating demand on the building. This annual solar fraction can be compared with that found by computer studies of the building given the same weather data.

In determining the total amount of energy consumed for space and water heating, care must be taken to subtract any other energy end uses in the building such as hot tubs, swimming pools, electrical appliances, etc.

The accuracy of this method is not known, but its use over a 6- to 12-mo period should eliminate the effects of thermal storage in the building, since storage is only a 1- or 2-day-long phenomenon.

Acceptance Tests for Passive Systems

A short-term acceptance test as described above cannot be carried out on a daily basis for a passive system. In fact, no direct method of measuring energy delivery for various types of passive systems is available. The only method of verifying the performance of a passive system uses the long-term performance check described in the preceding section. The procedure would be the same as for an active system except that the passive system insulation would need to be in place during the test. Also, no solar radiation can be permitted to strike the surface of the thermal-storage wall or storage elements of the passive system. This measured load would therefore be slightly different than during the operation of the passive system since a net outward flow of heat would occur through passive apertures during the test period, although relatively high levels of insulation could reduce this heat loss to a low level. A calculated correction could be made to the annual projected energy consumption using the degree-day method. For more precise calculations, the bin method of energy consumption estimation could be used.

OPERATOR'S MANUAL

Solar systems are relatively unfamiliar to HVAC service personnel and building owners. Therefore, the assembly of a detailed operating manual for any solar system should be included in the architect/engineer's or solar consultant's contract.

The manual serves exactly the same function as the operating manual for an automobile, a piece of industrial equipment, or a computer. It describes the function of the system in detail for each operational mode; every component in the system is described by means of its specifications and shop drawings as well as brochures. The manual should also include the name of equipment suppliers as well as a log of maintenance procedures and other items noted in the following section.

The manual should include a table of contents, an index, a description of system and all components, and all special operating instructions. Starting and stopping procedures as well as emergency shutdown procedures should be described in detail. Routine maintenance, servicing, and lubrication schedules should be presented. Parts lists for the system, serial numbers, model numbers, and equipment ratings are to be included along with a copy of shop drawings, solar flow diagrams, and wiring diagrams. Sources and part numbers for spare parts and contacts for system service and repair, preferably all from local sources, should be noted.

The manual should be bound in a looseleaf book with durable paper. Each page should be of uniform size, with foldout diagrams and illustrations provided as required. The information contained in the original copy of the operator's manual should be in such a form that it is reproducible by a dry copier. Moisture-, oil-, and weather-resistant cover materials should be used. The number of manuals to be provided is specified in the contract.

The amount of detail contained in the operating manual varies with system size. For residential systems, a binding of all component specifications, descriptive brochures, piping diagrams, and controller-mode tabulation should suffice. A maintenance schedule and log should be added.

For larger systems, a much more detailed manual should be prepared, since maintenance personnel are hired specifically to maintain complicated systems. These people require more detailed data than the owner of a residential system. An outline of the contents of such a manual is presented in the next few pages.

Controls

The system operating instructions contained in the manual should include start-up procedures and normal and abnormal operating modes accounted for in the design (including a table showing which components are operating in which system mode). A maintenance schedule for lubricating pumps, dampers, solenoid valves, and the like should be given with specific monthly or other time periods noted. A log should also be provided in the same section to encourage reporting of routine as well as nonroutine maintenance activities and problems. The maintenance schedule is prepared by the engineer by reference to component-manufacturer information.

In addition to routine lubrication and servicing, an inspection of the solar system should be carried out once or twice a year. During this inspection, the collector should be examined for condition of glazing, seals, gaskets, absorber plates, collector-sensor contact, and housing surfaces. Any deterioration of collector insulation and seals should be noted and corrected. During inspection of the collector array, the roof penetration areas could also be examined for evidence of leaks. In liquid systems, the condition of the fluid should be checked periodically. In large systems, a specific gravity and pH check should be made monthly. System operating pressure should also be examined on a schedule more frequent than once a month. A computer with alarms for nondesign conditions can be used for large systems.

Solenoid valves in liquid systems and dampers in air systems should be checked for full travel. The two critical dampers present in standard air systems should receive particular attention and the adjustment of these dampers should be verified during each inspection. If possible, the seals on these dampers should also be examined visually through a removable panel near the dampers. Leakage during various modes of operation should also be checked. Pipes and valves should be inspected with particular attention to leaks and deterioration of insulation.

Expansion tanks should be checked to determine whether the pressure charge is still present. A site glass in the expansion tank will immediately indicate the level of fluid in the system. Since continuous inspection of this indicator is not possible, it has been suggested in previous chapters that a level alarm of some sort be used to draw immediate attention to any fluid loss in the system.

Storage tanks should be examined for leaks, insulation deterioration, and—for liquid storage tanks—corrosion. Any evidence of moisture collection or leakage in pebble-bed containers should also be analyzed. Storage tanks for domestic hot-water systems should be flushed periodically. It should not be necessary to flush the main space-heating-system storage tank if there has been proper attention to corrosion prevention. The status of corrosion inhibitors should also be determined during the storage tank inspection.

The function of the control system can quickly be checked by manually overriding the controller to cause all pumps, valves, heat-rejection units, and dampers to assume positions corresponding to each possible operating mode. During mode switches, attention should be paid to any noises which develop arising from pumps, check valves, solenoid valves, or other parts. Such noises may indicate functional deterioration. Thermostat settings should be checked.

Temperature, pressure, and flow data recorded during the final inspection can be used to quickly check the operation of the system from a functional point of view. Of course, the temperature measured on a given day for a well-operating system will be different than on the next day. However, on a sunny day temperature rises in collectors and across heat exchangers should be relatively the same although the absolute values of the temperatures will be different. In addition, pressures and pressure rises and drops across various components should also be similar, irrespective of the precise operating condition. The data recorded during the acceptance test can be quickly used to determine whether the system is operating approximately correctly.

It is advisable to annually make a detailed inspection of the function of the system and record a full set of data similar to that recorded during the inspection period in the same type of weather. The data should check closely if the storage temperature is similar at the beginning of each of these test days. Any malfunction should be repaired. If the system is not equipped with a flowmeter, the pressure rise across the air or liquid mover and the current draw are the only methods of checking for proper flow in the system.

The pressure drops across various components such as liquid and air filters can indicate when replacement or cleaning is required. System operating pressure and relief-valve function should also be checked during the routine inspection. If any debris is caught in an air vent, it will sputter. Collector glazings should be cleaned periodically as required. If temperature rises on a sunny day through each row of a collector array are not the same as originally measured, rebalancing may be necessary. Antifreeze should be replaced periodically in accordance with the instructions of the manufacturer. If the pH is not between 8 and 10, the antifreeze or the inhibitor must be replaced immediately.

The operating manual should also delineate the owner's responsibilities. The collector array should be kept free of all debris including dirt, leaves, soot, etc. If the building is to be unoccupied for an extended period, the procedure for shutting down the system should be noted. Before calling for service in the case of a residential system malfunction, the homeowner should be sure that all electric cords are plugged in, low-voltage wires are connected, fuses or circuit breakers are intact, and all manual valves are open.

Malfunctions can be checked sequentially. For example, if the pump does not operate when it appears that it should be operating, a thermometer should be used to first check that the proper differential exists between the collector and the storage. If the differential exists, the voltage available at the pump and the controller should be checked. If adequate voltage is present, a control check can be made. If the storage sensor is disconnected from the controller, this condition simulates a very high resistance and a very low temperature if the sensor has the characteristics shown in Table 4.3. This should cause the pump or fan to begin operation. Similarly the collector sensor can be shorted, simulating very high collector temperature, and the collector pump or fan should operate. If the controller is found to be functioning properly, the next step is to manually override the controller to activate the pump. If the pump operates, the controller triac or relay is malfunctioning. If all other measures fail, the controller unit is malfunctioning and must be replaced. Similar simple step-by-step checks of this type can be identified for most components in the system, and a system owner can carry out most of these troubleshooting sequences on smaller systems.

Inspection and maintenance for passive systems is relatively simple. The only moving part requiring periodic service is the movable insulation. Any motors involved should be lubricated and their temperatures checked. The duration of travel for movable insulation and the seal between the insulation and its track should also be checked periodically. If the inner surface of passive glazing can be cleaned, periodic cleaning is advised.

The operating manual should also contain information regarding the sources for spare and special parts from local manufacturers. Manufacturers' names, addresses, and phone numbers should be listed at one place in the manual. Warranties passed through by the contractor or those provided by the contractor should also be included.

A section on safety can be included in the operator's manual, including first aid for electric shock and for exposure to high temperatures or to toxic fluids. In addition to these first-aid measures, a list of don'ts could be provided to ensure safe operation of the system: "Don't turn off the controller," "Don't withdraw fluid from the collector loop," "Don't close isolation valves," "Don't fill a hot collector," etc.

Besides the flow diagram included in the operator's manual, a permanent framed copy of it may be mounted on the wall of the mechanical room with all components and operational modes labeled. The diagram is useful both for troubleshooting the system and for explaining the system to visitors. Also posted on the wall should be a list, permanently framed, of emergency telephone numbers. These will include the usual emergency numbers for the location and the numbers for solar-system service and troubleshooting personnel.

Insuring the Solar System*

The prevailing attitude among insurance underwriters seems to be that solar energy is just another type of home heating that will not necessarily present greater risks. In fact, even those agents in a HUD survey who expressed some concern over such problems as wind damage or the need for special repairs to equipment did not include a clause in their policies to offset these risks.

Accordingly, agents are free to cover solar-equipped homes under standard policies without special exclusions or rate adjustments. The systems are being insured as part of the building, with a premium reflecting the increased replacement value of the structure. This procedure is expected to continue unless future claims data indicate that solar energy systems are more hazardous than conventional ones. In truth, unless the owner chooses to inform the agent that the building is solar-heated, the insurance company is generally unaware of this information (unless it is one of the few whose questionnaires have a place to indicate the presence of solar equipment). It is advisable, however, for policyholders to inform their agents of the presence of solar equipment, especially when solar additions have not previously been considered in the insured value of the property.

Some companies have specific standards they expect solar-equipped homes to meet, such as reinforced roofs or safely insulated collectors, but the primary concern of all the companies is that the system is properly constructed and installed. Solar equipment presently must meet the same standards as those set for plumbing, electrical wiring, and flammable materials in conventional systems; insurers expect that the solar industry or government eventually will set uniform standards specific to the equipment.

Agents' Concerns. The willingness of insurance companies to write policies for the solar home indicates an assumption that properly installed solar equipment is considered to present a risk no greater than conventional heating equipment. This assumption is based largely on the fact that, with the exception of collectors, most solar energy hardware has been used successfully in conventional heating systems. Therefore, any system installed by a professional contractor is likely to be insured with little question.

Owners who have installed solar systems themselves, on the other hand, may

*Prepared by U.S. Department of Housing and Urban Development, publication HUD-401, 1979.

run into obstacles in obtaining insurance. An agent approached to write a policy on such a system would probably insist on proof that the system meets all local building codes and standards. Adherence to these standards may be determined by an engineer or insurance company investigator during an inspection, which is part of the routine of securing a policy on any home. If the system proved to be safely constructed, the home would probably be insurable.

While the industry representatives largely agree that properly installed solar systems are reliable and safe, they have identified a number of potential hazards for the solar home. Note that these hazards are unlikely to occur in well-designed and -installed systems. The analysis is simply an effort on the part of the insurance industry to determine the problems which might arise with solar heating equipment.

Potential Hazards. The insurance industry has identified a number of potential hazards for solar-equipped homes, although it agrees that many of these possibilities are remote and as far as insurers know have not occurred. Insurers believe that most of these problems can be avoided if the system is properly installed and well constructed.

Broken Glass. Insurers believe that loss is most likely to occur from the breaking of glass collector plates by hail, high winds, the weight of ice or snow, or vandalism (especially with ground-mounted collectors). The use of thicker glass, screening, strong plastic panes, or tempered glass will minimize such losses. In any case, this type of loss usually involves only minor claims.

Roof Collapse. Stress from the weight of collectors and piping might cause the collapse of an improperly reinforced roof. Insurance agents make it a point to check that precautions against this risk have been taken. This situation is not expected to be a problem in professionally installed systems.

Roof Damage. In a retrofit installation, where collectors are installed directly atop the roof shingles, water can collect under the panels and cause mildew or rotting. In new installations, however, collectors may be mounted directly on felt paper, then properly flashed and sealed prior to shingling, thereby avoiding this problem. Alternatively, the collectors can be supported above the roof surface.

Broken Pipes. In any plumbing system, there is a risk that pipes will burst if liquid freezes in them. In a solar-equipped home this can be avoided by using an automatic or manual drainage system or by adding antifreeze to the collector fluid.

Improperly welded joints or corrosion due to the use of two incompatible metals can cause pipes to leak. Again, the companies stress that they anticipate few losses in systems installed by experienced contractors.

Fire. Some underwriters believe there is little danger of fire with solar energy systems because they have few moving parts and do not normally use combustible fluids. A small risk might be added by the amount of wiring and controls required for two heating systems—the solar energy system and the conventionally fueled backup. Insufficient insulation around the collector could cause excessive heat to build up and dry out wood supports, and the use of flammable insulation (such as polyurethane foam) could create hazardous conditions.

The failure of pressure-relief valves on storage tanks can cause explosions just as in conventional systems. Adequately sized and responsive valves and venting systems must be provided in parts of the systems containing pressurized fluids.

Personal Injury. Most insurers think that the chance of personal injury is remote. The slight risk of accidental poisoning by antifreeze leakage into the drinking water, injury from ruptured pipes, or burns from ground-mounted collectors can be guarded against by proper system design.

If a solar energy system is shut down by an insured peril, coverage under most policies would reimburse the homeowner for the additional fuel costs of the backup system. The problem for the company is determining how much expense would be covered.

Claims. There have not been a sufficient number of loss claims attributable to solar energy systems for underwriters to draw any firm conclusions about them, especially since the few losses that have occurred are regarded by insurers as unique situations. Furthermore, since solar-equipped homes represent a very small portion of any insurer's business, few statistics have been gathered.

Tools for Use in Troubleshooting Solar Systems

Active solar systems require only a few tools for initial troubleshooting and repair. In addition to the standard tools used for repair of any HVAC system, the following tools can be useful in working on solar systems:

Magnetic compass
Torch
Solder
Level
Tubing cutter
Ohmmeter
Ammeter
Sensor substitution potentiometers or rheostats
Thermometers with adapters for P/T wells
Flowmeter in removable straight pipe section
Spare wire
Pyranometer for solar flux measurement
24-V power supply
Calipers
Pressure gauge on hose
Stethoscope

TABLE 5.6 Liquid-based solar water-heating system troubleshooting summary*

Symptom	*Source*	*Cause*	*Action*
Drop in system pressure	Collectors	Leak in absorber	Repair leak. Replace wet insulation. Replace absorber. Replace collector.
(NOTE: After leaks have been fixed, remove all wet insulation and replace with new, dry insulation.)	Piping	Leak from soldered joint	Resweat joint.
		Leak from threaded joint	Tighten, replace Teflon tape, add pipe dope.
		Leak from clamped joint	Tighten hose clamp.
	Valves (gate, globe, etc.)	Leak from valve packing	Tighten packing nut, replace packing.
	Drains	Leak from drain	Close drain tightly, add cap.
	Air vents	Leak from needle valve	Inspect, clean mechanism, repair or replace.
		Leak from coin vent	Inspect, tighten.
	Pressure-relief valve	Preset relief valve has expelled excess pressure	Install relief valve with higher setting. Recharge system to operating pressure. Lower pressure in system.
		Adjustable relief valve has expelled excess pressure	Set for higher temperatures, recharge system. Lower pressure in system.
	Expansion tank (diaphragm)	Bladder exceeded	Inspect for fluid in air chamber, replace.
	Expansion tank (standard)	Inadequate air space	Adjust water level.
	Heat exchanger	Leak from exchanger	Inspect; repair or replace.
Drop in water pressure at both hot and cold outlets	Inlet water pressure	Well pump problems	Check wiring, check pressure tank and controls, check water level in well, check pump motor and impellers, repair or replace.
		Frozen inlet pipe	Thaw frozen section. Keep water running until hard freeze is over.
		Strainer in house supply line clogged	Clean basket.

TABLE 5.6 Liquid-based solar water-heating system troubleshooting summary* (*Continued*)

Symptom	*Source*	*Cause*	*Action*
Drop in water pressure at both hot and cold outlets (*continued*)		Problem with community water supply	Notify local water department.
		Broken or compressed pipe between main and house	Replace pipe.
		Pressure-reducing valve	Check for plugged ports and proper operation. Clean; repair or replace.
Drop in water pressure at hot outlets only; cold outlet pressure is OK	Obstruction in domestic hot-water system	Valve improperly closed	Open manual valves.
			Check controls for automatic valves, check automatic valve operation, repair or replace.
			Look at flow direction arrows on check valves and backflow preventer. Reinstall correctly. Check for jammed elements. Repair or replace.
		Strainers clogged	Clean basket.
		Bent or crushed pipe restricting flow	Repair or replace.
Inadequate or no circulation of coolant	Collector	Passages clogged	Remove collector, backflush passages.
		Airlocked	Install air vents at all high points.
	Pump	Undersized	Check specs, add larger motor and/or impeller, change pump if necessary.
		Airlocked	If present, loosen vent on pump head.
		Impeller bound	If possible, turn impeller manually to free it; repair or replace.
		Installed backward	Check flow direction arrow and reverse.

		Installed incorrectly	Check manufacturer literature for proper operating position.
		Pump isolated	Make sure all shutoff valves are open.
	Air vents	Inadequate number	Install additional vents at high points.
		Improperly placed	Install at correct positions.
		Faulty, causing airlock	Operate needle valve manually, check float control, inspect, clean, repair or replace.
	Shutoff valves	Closed	Open.
	Automatic valves	Closed	Check controls, operation, and springs; repair or replace.
	Piping	Airlocked	Install vents if necessary, check pitch.
		Too small	Increase size, install larger pump or pump motor, or install second pump.
	Check valve	Installed backward	Check flow direction arrow and reverse.
		Jammed	Repair or replace.
Pump functioning improperly (cycles excessively, or always active or inactive); freeze protection modes not operating	Differential controller and control system	Sensors improperly located	Place sensors as specified by manufacturer.
		Sensor improperly secured	Assure tight bond between sensor materials.
		Break in wire	Inspect, repair.
		Wired incorrectly	Check all wiring with drawings, inspect connections.
		Controller settings	Assure proper settings.
		Sensors faulty	Test by heating one sensor, cooling others. Use ohmmeter to determine whether sensor is producing proper resistance at various temperatures.
		Controller faulty	Check fuses. Check relays. Check for loose circuit boards; repair, replace.

TABLE 5.6 Liquid-based solar water-heating system troubleshooting summary* (*Continued*)

Symptom	*Source*	*Cause*	*Action*
Pump functioning improperly *(continued)*	Power wiring	No power to controller	Inspect electrical connections and wiring, repair.
		No power to pump	Inspect electrical connections and wiring, repair.
Decrease in performance of solar portions of system	Collectors	Outgassing	Clean surface; contact supplier.
		Condensation	Inspect and repair glazing seal. Inspect and repair pipe gaskets. Inspect weep holes for clogging and proper location. Replace dessicant.
		Absorber coating degredation	Recoat absorber. Contact installer.
		Dirt on glazing	If rain or dew offers inadequate washing, rinse cover (check with manufacturer for suggested methods).
	Piping	Night losses; reverse thermosiphon	Install check valve on collector supply.
		Insulation seams and joints deteriorated	Reglue, tape, or staple pipe insulation.
Excessive vibration noise in system	Pipe hanger	Vibration from equipment transmitted to framing members	Locate area of noise, disconnect hangers from framing, replace with flexible type or wrap hanger on outside of insulation.
	Pump	Impeller shaft misaligned	Inspect, repair, replace.
		Pump vibrates excessively	Clean and rebalance impeller. Install vibration isolators on both sides of pump (strainers and check valves between insolators increase mass).
Fluid noise	Piping	Entrapped air	Force-purge system, install additional vents.
		Air purge installed backward	Check flow direction arrow and reverse.
		Gravity return line oversized, creates waterfall noise	Replace with smaller-diameter tubing.

	Air vents	Not venting	Check for tight cap, vent manually.
		Air locked	Force-purge, vent pump if it has bleeder.
Water hammer	Piping	Pressure	Install shock suppressor.
		Loose support of pipes	Secure pipe supports to structure.
		Entrapped air	Install additional vents.
	Automatic valves	Open or close too fast	Install valves with slower action. Modify controls to slow valve action.
Squeaking noise	Piping	Pipe expansion in contact with building materials	Cut holes for pipes ¼ in larger than pipe OD. Pack penetration with insulation.
Humming of mechanical equipment motors	Differential controller or control system	Controller is not fully deactivating component, allows slight continuous current to motors	Use differential controls troubleshooting procedure. Check solenoid at motor starter for complete break. Adjust or replace.
Insufficient hot water	Collector loop	Failure in fluid system circulators, controls	Use appropriate troubleshooting procedure.
	Tank	Too small	Install second tank.
		Conventional heater problems	Check backup source, wiring, gas or oil lines, burners, and set point. Check for deposits on heater.
		High storage losses	Insulate tank with additional layer.
	Piping	Reverse thermosiphon	Install check valve on collector supply.
	Mixing valve	Improperly adjusted	Check adjustment temperature indicator, set higher.
		Faulty	Replace.
Hot water in cold-water lines	Mixing valve	Hot water from top of tank expanding into cold feed of mixing valve	Increase distance so that top of tank to bottom of mixing valve is at least 16 in. Install check valve in cold-water line at mixing valve.

*From Rick Schwolsky, with permission.

The potentiometer is used to exercise the controller through its various modes as described earlier. Pyranometers, flowmeters, and thermometers can be used to measure collector energy delivery and to verify that the collector loop heat exchangers and storage tanks are functioning properly. Electronic thermometers with point-contact sensors which can be placed at various points on a pipe quickly are particularly useful. Compasses and levels are used to determine proper collector orientation. If a system seems to be operating for an excessive number of hours per day, temporary installation of a recording hour meter can verify the extent of the problem. A troubleshooting summary is contained in Table 5.6 for solar DHW systems.

SUMMARY

This section completes the description of the solar design and installation process for active and passive systems in this book. Although inspection and acceptance tests are not strictly a part of the design process, they are as important as any step in the design process in assuring that a building owner is delivered a properly functioning solar heating system. Since many architectural and engineering firms are unfamiliar with these matters, the role of the solar consultant in supplementing information of this type will be important in the 1980s.

APPENDIX

TABLE A.1 Conversion Factors

Physical quantity	*Symbol*	*Conversion factor*
Area	A	1 ft^2 = 0.0929 m^2
		1 in^2 = 6.452×10^{-4} m^2
Density	ρ	1 lb_m/ft^3 = 16.018 kg/m^3
		1 $slug/ft^3$ = 515.379 kg/m^3
Heat, energy, or work	Q or W	1 Btu = 1055.1 J
		1 cal = 4.186 J
		1 ft/lb_f = 1.3558 J
		1 hp/h = 2.685×10^6 J
Force	F	1 lb_f = 4.448 N
Heat flow rate	q	1 Btu/h = 0.2931 W
		1 Btu/s = 1055.1 W
Heat flux	q/A	1 $Btu/(h \cdot ft^2)$ = 3.1525 W/m^2
Heat-transfer coefficient	h	1 $Btu/(h \cdot ft^2 \cdot °F)$ = 5.678 $W/(m^2 \cdot K)$
Length	L	1 ft = 0.3048 m
		1 in = 2.54 cm
		1 mi = 1.6093 km
Mass	m	1 lb_m = 0.4536 kg
		1 slug = 14.594 kg
Mass flow rate	$\dot{m}$	1 lb_m/h = 0.000126 kg/s
		1 lb_m/s = 0.4536 kg/s
Power	$\dot{W}$	1 hp = 745.7 W
		1 $ft/(lb \cdot s)$ = 1.3558 W
		1 Btu/s = 1055.1 W
		1 Btu/h = 0.293 W
Pressure	p	1 lb_f/in^2 = 6894.8 Pa (N/m^2)
		1 lb_f/ft^2 = 47.88 Pa (N/m^2)
		1 atm = 101,325 Pa (N/m^2)
Radiation	I	1 langley = 41,860 J/m^2
Specific heat capacity	c	1 $Btu/(lb_m \cdot °F)$ = 4187 $J/(kg \cdot K)$
Internal energy or enthalpy	e or h	1 Btu/lb_m = 2326.0 J/kg
		1 cal/g = 4184 J/kg

TABLE A.1 Conversion Factors (*Continued*)

Physical quantity	*Symbol*	*Conversion factor*
Temperature	T	T(°R) = (9/5)T(K)
		T(°F) = [T(°C)](9/5) + 32
		T(°F) = [T(K) − 273.15](9/5) + 32
Thermal conductivity	k	1 Btu/(h·ft·°F) = 1.731 W/(m·K)
Thermal resistance	R_{th}	1 b/(°F·Btu) = 1.8958 K/W
Velocity	V	1 ft/s = 0.3048 m/s
		1 mi/h = 0.44703 m/s
Viscosity, dynamic	μ	1 lb_m/(ft·s) = 1.488 N/(s·m^2)
		1 cP = 0.00100 N/(s·m^2)
Viscosity, kinematic	ν	1 ft^2/s = 0.09029 m^2/s
		1 ft^2/h = 2.581 × 10^{-5} m^2/s
Volume	V	1 ft^3 = 0.02832 m^3
		1 in^3 = 1.6387 × 10^{-5} m^3
		1 gal (U.S. liq.) = 0.003785 m^3
Volumetric flow rate	$\dot{Q}$	1 ft^3/min = 0.000472 m^3/s

TABLE A.2a The Average Solar Ephemeris*

	Declination		*Equation of time*			*Declination*		*Equation of time*	
Date	*Degrees*	*Minutes*	*Minutes*	*Seconds*	*Date*	*Degrees*	*Minutes*	*Minutes*	*Seconds*
Jan. 1	−23	4	− 3	14	Feb. 1	−17	19	−13	34
5	22	42	5	6	5	16	10	14	2
9	22	13	6	50	9	14	55	14	17
13	21	37	8	27	13	13	37	14	20
17	20	54	9	54	17	12	15	14	10
21	20	5	11	10	21	10	50	13	50
25	19	9	12	14	25	9	23	13	19
29	18	8	13	5					
Mar. 1	− 7	53	−12	38	Apr. 1	+ 4	14	− 4	12
5	6	21	11	48	5	5	46	3	1
9	4	48	10	51	9	7	17	1	52
13	3	14	9	49	13	8	46	− 0	47
17	1	39	8	42	17	10	12	+ 0	13
21	− 0	5	7	32	21	11	35	1	6
25	+ 1	30	6	20	25	12	56	1	53
29	3	4	5	7	29	14	13	2	33
May 1	+14	50	+ 2	50	June 1	+21	57	+ 2	27
5	16	2	3	17	5	22	28	1	49
9	17	9	3	35	9	22	52	1	6
13	18	11	3	44	13	23	10	+ 0	18
17	19	9	3	44	17	23	22	− 0	33
21	20	2	3	34	21	23	27	1	25
25	20	49	3	16	25	23	25	2	17
29	21	30	2	51	29	23	17	3	7
July 1	+23	10	− 3	31	Aug. 1	+18	14	− 6	17
5	22	52	4	16	5	17	12	5	59
9	22	28	4	56	9	16	6	5	33
13	21	57	5	30	13	14	55	4	57
17	21	21	5	57	17	13	41	4	12
21	20	38	6	15	21	12	23	3	19
25	19	50	6	24	25	11	2	2	18
29	18	57	6	23	29	9	39	1	10
Sep. 1	+ 8	35	− 0	15	Oct. 1	− 2	53	+10	1
5	7	7	+ 1	2	5	4	26	11	17
9	5	37	2	22	9	5	58	12	27
13	4	6	3	45	13	7	29	13	30
17	2	34	5	10	17	8	58	14	25
21	+ 1	1	6	35	21	10	25	15	10
25	− 0	32	8	0	25	11	50	15	46
29	2	6	9	22	29	13	12	16	10
Nov. 1	−14	11	+16	21	Dec. 1	−21	41	+11	16
5	15	27	16	23	5	22	16	9	43
9	16	38	16	12	9	22	45	8	1
13	17	45	15	47	13	23	6	6	12
17	18	48	15	10	17	23	20	4	17
21	19	45	14	18	21	23	26	2	19
25	20	36	13	15	25	23	25	+ 0	20
29	21	21	11	59	29	23	17	− 1	39

*Since each year is 365.25 days long, the precise value of declination varies from year to year. *The American Ephemeris and Nautical Almanac* published each year by the U.S. Government Printing Office contains precise values for each day of each year.

TABLE A.2*b* Average Length of a Day for Various Latitudes in the Northern Hemisphere

North latitude, degrees	*Months*											
	Jan.	*Feb.*	*Mar.*	*Apr.*	*May*	*June*	*July*	*Aug.*	*Sep.*	*Oct.*	*Nov.*	*Dec.*
20	10.9	11.3	11.9	12.5	12.9	13.2	13.1	12.7	12.1	11.5	11.0	10.8
25	10.6	11.2	11.8	12.6	13.2	13.5	13.4	12.9	12.1	11.4	10.8	10.5
30	10.3	10.9	11.8	12.7	13.5	13.9	13.8	13.1	12.2	11.3	10.5	10.1
35	9.9	10.7	11.7	12.9	13.8	14.3	14.1	13.3	12.2	11.1	10.1	9.7
40	9.4	10.5	11.7	13.1	14.2	14.8	14.6	13.6	12.3	10.9	9.7	9.2
45	8.9	10.2	11.6	13.3	14.6	15.4	15.1	13.9	12.3	10.7	9.3	8.6
50	8.3	9.8	11.6	13.5	15.2	16.1	15.7	14.3	12.4	10.5	8.7	7.9
55	7.5	9.4	11.5	13.8	15.9	17.1	16.6	14.7	12.4	10.1	8.0	6.9
60	6.3	8.8	11.4	14.2	16.8	18.4	17.7	15.4	12.5	9.7	7.1	5.6

[a]From *Solar Heating and Cooling*, revised 1st ed., by J. F. Kreider and F. Kreith. Copyright 1977 Hemisphere Publishing Corp. Used by permission. "Normal" component is beam radiation only.

[b]1 Btu/(h·ft^2) = 3.152 W/m^2. Ground reflection not included on normal or horizontal surfaces.

TABLE A.3a Solar Position and Insolation Values for 24° North Latitude[a]

Date	Solar time		Solar position		BTUH/sq. ft. total insolation on surface[b]						
	AM	PM	Alt	Azm			South facing surface angle with horiz.				
					Normal	Horiz.	14	24	34	44	90
Jan 21	7	5	4.8	65.6	71	10	17	21	25	28	31
	8	4	16.9	58.3	239	83	110	126	137	145	127
	9	3	27.9	48.8	288	151	188	207	221	228	176
	10	2	37.2	36.1	308	204	246	268	282	287	207
	11	1	43.6	19.6	317	237	283	306	319	324	226
	12		46.0	0.0	320	249	296	319	332	336	232
	Surface daily totals				2766	1622	1984	2174	2300	2360	1766
Feb 21	7	5	9.3	74.6	158	35	44	49	53	56	46
	8	4	22.3	67.2	263	116	135	145	150	151	102
	9	3	34.4	57.6	298	187	213	225	230	228	141
	10	2	45.1	44.2	314	241	273	286	291	287	168
	11	1	53.0	25.0	321	276	310	324	328	323	185
	12		56.0	0.0	324	288	323	337	341	335	191
	Surface daily totals				3036	1998	2276	2396	2436	2424	1476
Mar 21	7	5	13.7	83.3	194	60	63	64	62	59	27
	8	4	27.2	76.8	267	141	150	152	149	142	64
	9	3	40.2	67.9	295	212	226	229	225	214	95
	10	2	52.3	54.8	309	266	285	288	283	270	120
	11	1	61.9	33.4	315	300	322	326	320	305	135
	12		66.0	0.0	317	312	334	339	333	317	140
	Surface daily totals				3078	2270	2428	2456	2412	2298	1022
Apr 21	6	6	4.7	100.6	40	7	5	4	4	3	2
	7	5	18.3	94.9	203	83	77	70	62	51	10
	8	4	32.0	89.0	256	160	157	149	137	122	16
	9	3	45.6	81.9	280	227	227	220	206	186	46
	10	2	59.0	71.8	292	278	282	275	259	237	61
	11	1	71.1	51.6	298	310	316	309	293	269	74
	12		77.6	0.0	299	321	328	321	305	280	79
	Surface daily totals				3036	2454	2458	2374	2228	2016	488
May 21	6	6	8.0	108.4	86	22	15	10	9	9	5
	7	5	21.2	103.2	203	98	85	73	59	44	12
	8	4	34.6	98.5	248	171	159	145	127	106	15
	9	3	48.3	93.6	269	233	224	210	190	165	16
	10	2	62.0	87.7	280	281	275	261	239	211	22
	11	1	75.5	76.9	286	311	307	293	270	240	34
	12		86.0	0.0	288	322	317	304	281	250	37
	Surface daily totals				3032	2556	2447	2286	2072	1800	246
Jun 21	6	6	9.3	111.6	97	29	20	12	12	11	7
	7	5	22.3	106.8	201	103	87	73	58	41	13
	8	4	35.5	102.6	242	173	158	142	122	99	16
	9	3	49.0	98.7	263	234	221	204	182	155	18
	10	2	62.6	95.0	274	280	269	253	229	199	18
	11	1	76.3	90.8	279	309	300	283	259	227	19
	12		89.4	0.0	281	319	310	294	269	236	22
	Surface daily totals				2994	2574	2422	2230	1992	1700	204

TABLE A.3a Solar Position and Insolation Values for 24° North Latitude[a] *(Continued)*

Date	Solar time AM	Solar time PM	Solar position Alt	Solar position Azm	Normal	Horiz.	14	24	34	44	90
					BTUH/sq. ft. total insolation on surfaces[b]		South facing surface angle with horiz.				
Jul 21	6	6	8.2	109.0	81	23	16	11	10	9	6
	7	5	21.4	103.8	195	98	85	73	59	44	13
	8	4	34.8	99.2	239	169	157	143	125	104	16
	9	3	48.4	94.5	261	231	221	207	187	161	18
	10	2	62.1	89.0	272	278	270	256	235	206	21
	11	1	75.7	79.2	278	307	302	287	265	235	32
	12		86.6	0.0	280	317	312	298	275	245	36
	Surface daily totals				2932	2526	2412	2250	2036	1766	246
Aug 21	6	6	5.0	101.3	35	7	5	4	4	4	2
	7	5	18.5	95.6	186	82	76	69	60	50	11
	8	4	32.2	89.7	241	158	154	146	134	118	16
	9	3	45.9	82.9	265	223	222	214	200	181	39
	10	2	59.3	73.0	278	273	275	268	252	230	58
	11	1	71.6	53.2	284	304	309	301	285	261	71
	12		78.3	0.0	286	315	320	313	296	272	75
	Surface daily totals				2864	2408	2402	2316	2168	1958	470
Sep 21	7	5	13.7	83.8	173	57	60	60	59	56	26
	8	4	27.2	76.8	248	136	144	146	143	136	62
	9	3	40.2	67.9	278	205	218	221	217	206	93
	10	2	52.3	54.8	292	258	275	278	273	261	116
	11	1	61.9	33.4	299	291	311	315	309	295	131
	12		66.0	0.0	301	302	323	327	321	306	136
	Surface daily totals				2878	2194	2342	2366	2322	2212	992
Oct 21	7	5	9.1	74.1	138	32	40	45	48	50	42
	8	4	22.0	66.7	247	111	129	139	144	145	99
	9	3	34.1	57.1	284	180	206	217	223	221	138
	10	2	44.7	43.8	301	234	265	277	282	279	165
	11	1	52.5	24.7	309	268	301	315	319	314	182
	12		55.5	0.0	311	279	314	328	332	327	188
	Surface daily totals				2868	1928	2198	2314	2364	2346	1442
Nov 21	7	5	4.9	65.8	67	10	16	20	24	27	29
	8	4	17.0	58.4	232	82	108	123	135	142	124
	9	3	28.0	48.9	282	150	186	205	217	224	172
	10	2	37.3	36.3	303	203	244	265	278	283	204
	11	1	43.8	19.7	312	236	280	302	316	320	222
	12		46.2	0.0	315	247	293	315	328	332	228
	Surface daily totals				2706	1610	1962	2146	2268	2324	1730
Dec 21	7	5	3.2	62.6	30	3	7	9	11	12	14
	8	4	14.9	55.3	225	71	99	116	129	139	130
	9	3	25.5	46.0	281	137	176	198	214	223	184
	10	2	34.3	33.7	304	189	234	258	275	283	217
	11	1	40.4	18.2	314	221	270	295	312	320	236
	12		42.6	0.0	317	232	282	308	325	332	243
	Surface daily totals				2624	1474	1852	2058	2204	2286	1808

[a]From *Solar Heating and Cooling,* revised 1st ed., by J. F. Kreider and F. Kreith. Copyright 1977 Hemisphere Publishing Corp. Used by permission.

TABLE A.3*b* Solar Position and Insolation Values for 32° North Latitude[a]

Date	Solar time AM	Solar time PM	Solar position Alt	Solar position Azm	BTUH/sq. ft. total insolation on surfaces[b] Normal	Horiz.	South facing surface angle with horiz. 22	32	42	52	90
Jan 21	7	5	1.4	65.2	1	0	0	0	0	1	1
	8	4	12.5	56.5	203	56	93	106	116	123	115
	9	3	22.5	46.0	269	118	175	193	206	212	181
	10	2	30.6	33.1	295	167	235	256	269	274	221
	11	1	36.1	17.5	306	198	273	295	308	312	245
	12		38.0	0.0	310	209	285	308	321	324	253
	Surface daily totals				2458	1288	1839	2008	2118	2166	1779
Feb 21	7	5	7.1	73.5	121	22	34	37	40	42	38
	8	4	19.0	64.4	247	95	127	136	140	141	108
	9	3	29.9	53.4	288	161	206	217	222	220	158
	10	2	39.1	39.4	306	212	266	278	283	279	193
	11	1	45.6	21.4	315	244	304	317	321	315	214
	12		48.0	0.0	317	255	316	330	334	328	222
	Surface daily totals				2872	1724	2188	2300	2345	2322	1644
Mar 21	7	5	12.7	81.9	185	54	60	60	59	56	32
	8	4	25.1	73.0	260	129	146	147	144	137	78
	9	3	36.8	62.1	290	194	222	224	220	209	119
	10	2	47.3	47.5	304	245	280	283	278	265	150
	11	1	55.0	26.8	311	277	317	321	315	300	170
	12		58.0	0.0	313	287	329	333	327	312	177
	Surface daily totals				3012	2084	2378	2403	2358	2246	1276
Apr 21	6	6	6.1	99.9	66	14	9	6	6	5	3
	7	5	18.8	92.2	206	86	78	71	62	51	10
	8	4	31.5	84.0	255	158	156	148	136	120	35
	9	3	43.9	74.2	278	220	225	217	203	183	68
	10	2	55.7	60.3	290	267	279	272	256	234	95
	11	1	65.4	37.5	295	297	313	306	290	265	112
	12		69.6	0.0	297	307	325	318	301	276	118
	Surface daily totals				3076	2390	2444	2356	2206	1994	764
May 21	6	6	10.4	107.2	119	36	21	13	13	12	7
	7	5	22.8	100.1	211	107	88	75	60	44	13
	8	4	35.4	92.9	250	175	159	145	127	105	15
	9	3	48.1	84.7	269	233	223	209	188	163	33
	10	2	60.6	73.3	280	277	273	259	237	208	56
	11	1	72.0	51.9	285	305	305	290	268	237	72
	12		78.0	0.0	286	315	315	301	278	247	77
	Surface daily totals				3112	2582	2454	2284	2064	1788	469
Jun 21	6	6	12.2	110.2	131	45	26	16	15	14	9
	7	5	24.3	103.4	210	115	91	76	59	41	14
	8	4	36.9	96.8	245	180	159	143	122	99	16
	9	3	49.6	89.4	264	236	221	204	181	153	19
	10	2	62.2	79.7	274	279	268	251	227	197	41
	11	1	74.2	60.9	279	306	299	282	257	224	56
	12		81.5	0.0	280	315	309	292	267	234	60
	Surface daily totals				3084	2634	2436	2234	1990	1690	370

[b]1 Btu/(h·ft^2) = 3.152 W/m^2. Ground reflection not included on normal or horizontal surfaces.

TABLE A.3*b* Solar Position and Insolation Values for 32° North Latitude[a] (*Continued*)

Date	Solar time		Solar position		BTUH/sq. ft. total insolation on surfaces[b]						
	AM	PM	Alt	Azm	Normal	Horiz.	South facing surface angle with horiz. 22	32	42	52	90
Jul 21	6	6	10.7	107.7	113	37	22	14	13	12	8
	7	5	23.1	100.6	203	107	87	75	60	44	14
	8	4	35.7	93.6	241	174	158	143	125	104	16
	9	3	48.4	85.5	261	231	220	205	185	159	31
	10	2	60.9	74.3	271	274	269	254	232	204	54
	11	1	72.4	53.3	277	302	300	285	262	232	69
	12		78.6	0.0	279	311	310	296	273	242	74
	Surface daily totals				3012	2558	2422	2250	2030	1754	458
Aug 21	6	6	6.5	100.5	59	14	9	7	6	6	4
	7	5	19.1	92.8	190	85	77	69	60	50	12
	8	4	31.8	84.7	240	156	152	144	132	116	33
	9	3	44.3	75.0	263	216	220	212	197	178	65
	10	2	56.1	61.3	276	262	272	264	249	226	91
	11	1	66.0	38.4	282	292	305	298	281	257	107
	12		70.3	0.0	284	302	317	309	292	268	113
	Surface daily totals				2902	2352	2388	2296	2144	1934	736
Sep 21	7	5	12.7	81.9	163	51	56	56	55	52	30
	8	4	25.1	73.0	240	124	140	141	138	131	75
	9	3	36.8	62.1	272	188	213	215	211	201	114
	10	2	47.3	47.5	287	237	270	273	268	255	145
	11	1	55.0	26.8	294	268	306	309	303	289	164
	12		58.0	0.0	296	278	318	321	315	300	171
	Surface daily totals				2808	2014	2288	2308	2264	2154	1226
Oct 21	7	5	6.8	73.1	99	19	29	32	34	36	32
	8	4	18.7	64.0	229	90	120	128	133	134	104
	9	3	29.5	53.0	273	155	198	208	213	212	153
	10	2	38.7	39.1	293	204	257	269	273	270	188
	11	1	45.1	21.1	302	236	294	307	311	306	209
	12		47.5	0.0	304	247	306	320	324	318	217
	Surface daily totals				2696	1654	2100	2208	2252	2232	1588
Nov 21	7	5	1.5	65.4	2	0	0	0	1	1	1
	8	4	12.7	56.6	196	55	91	104	113	119	111
	9	3	22.6	46.1	263	118	173	190	202	208	176
	10	2	30.8	33.2	289	166	233	252	265	270	217
	11	1	36.2	17.6	301	197	270	291	303	307	241
	12		38.2	0.0	304	207	282	304	316	320	249
	Surface daily totals				2406	1280	1816	1980	2084	2130	1742
Dec 21	8	4	10.3	53.8	176	41	77	90	101	108	107
	9	3	19.8	43.6	257	102	161	180	195	204	183
	10	2	27.6	31.2	288	150	221	244	259	267	226
	11	1	32.7	16.4	301	180	258	282	298	305	251
	12		34.6	0.0	304	190	271	295	311	318	259
	Surface daily totals				2348	1136	1704	1888	2016	2086	1794

[a]From *Solar Heating and Cooling*, revised 1st ed., by J. F. Kreider and F. Kreith. Copyright 1977 Hemisphere Publishing Corp. Used by permission.

TABLE A.3*c* **Solar Position and Insolation Values for 40° North Latitude**[a]

Date	Solar time AM	Solar time PM	Solar position Alt	Solar position Azm	BTUH/sq. ft. total insolation on surface[b] Normal	Horiz.	South facing surface angle with horiz. 30	40	50	60	90
Jan 21	8	4	8.1	55.3	142	28	65	74	81	85	84
	9	3	16.8	44.0	239	83	155	171	182	187	171
	10	2	23.8	30.9	274	127	218	237	249	254	223
	11	1	28.4	16.0	289	154	257	277	290	293	253
	12		30.0	0.0	294	164	270	291	303	306	263
	Surface daily totals				2182	948	1660	1810	1906	1944	1726
Feb 21	7	5	4.8	72.7	69	10	19	21	23	24	22
	8	4	15.4	62.2	224	73	114	122	126	127	107
	9	3	25.0	50.2	274	132	195	205	209	208	167
	10	2	32.8	35.9	295	178	256	267	271	267	210
	11	1	38.1	18.9	305	206	293	306	310	304	236
	12		40.0	0.0	308	216	306	319	323	317	245
	Surface daily totals				2640	1414	2060	2162	2202	2176	1730
Mar 21	7	5	11.4	80.2	171	46	55	55	54	51	35
	8	4	22.5	69.6	250	114	140	141	138	131	89
	9	3	32.8	57.3	282	173	215	217	213	202	138
	10	2	41.6	41.9	297	218	273	276	271	258	176
	11	1	47.7	22.6	305	247	310	313	307	293	200
	12		50.0	0.0	307	257	322	326	320	305	208
	Surface daily totals				2916	1852	2308	2330	2284	2174	1484
Apr 21	6	6	7.4	98.9	89	20	11	8	7	7	4
	7	5	18.9	89.5	206	87	77	70	61	50	12
	8	4	30.3	79.3	252	152	153	145	133	117	53
	9	3	41.3	67.2	274	207	221	213	199	179	93
	10	2	51.2	51.4	286	250	275	267	252	229	126
	11	1	58.7	29.2	292	277	308	301	285	260	147
	12		61.6	0.0	293	287	320	313	296	271	154
	Surface daily totals				3092	2274	2412	2320	2168	1956	1022
May 21	5	7	1.9	114.7	1	0	0	0	0	0	0
	6	6	12.7	105.6	144	49	25	15	14	13	9
	7	5	24.0	96.6	216	214	89	76	60	44	13
	8	4	35.4	87.2	250	175	158	144	125	104	25
	9	3	46.8	76.0	267	227	221	206	186	160	60
	10	2	57.5	60.9	277	267	270	255	233	205	89
	11	1	66.2	37.1	283	293	301	287	264	234	108
	12		70.0	0.0	284	301	312	297	274	243	114
	Surface daily totals				3160	2552	2442	2264	2040	1760	724
Jun 21	5	7	4.2	117.3	22	4	3	3	2	2	1
	6	6	14.8	108.4	155	60	30	18	17	16	10
	7	5	26.0	99.7	216	123	92	77	59	41	14
	8	4	37.4	90.7	246	182	159	142	121	97	16
	9	3	48.8	80.2	263	233	219	202	179	151	47
	10	2	59.8	65.8	272	272	266	248	224	194	74
	11	1	69.2	41.9	277	296	296	278	253	221	92
	12		73.5	0.0	279	304	306	289	263	230	98
	Surface daily totals				3180	2648	2434	2224	1974	1670	610

[b] 1 Btu/(h·ft²) = 3.152 W/m². Ground reflection not included on normal or horizontal surfaces.

TABLE A.3*c* Solar Position and Insolation Values for 40° North Latitude[a] (*Continued*)

Date	Solar time		Solar position		BTUH/sq. ft. total insolation on surfaces[b]						
	AM	PM	Alt	Azm	Normal	Horiz.	South facing surface angle with horiz. 30	40	50	60	90
Jul 21	5	7	2.3	115.2	2	0	0	0	0	0	0
	6	6	13.1	106.1	138	50	26	17	15	14	9
	7	5	24.3	97.2	208	114	89	75	60	44	14
	8	4	35.8	87.8	241	174	157	142	124	102	24
	9	3	47.2	76.7	259	225	218	203	182	157	58
	10	2	57.9	61.7	269	265	266	251	229	200	86
	11	1	66.7	37.9	275	290	296	281	258	228	104
	12		70.6	0.0	276	298	307	292	269	238	111
	Surface daily totals				3062	2534	2409	2230	2006	1728	702
Aug 21	6	6	7.9	99.5	81	21	12	9	8	7	5
	7	5	19.3	90.9	191	87	76	69	60	49	12
	8	4	30.7	79.9	237	150	150	141	129	113	50
	9	3	41.8	67.9	260	205	216	207	193	173	89
	10	2	51.7	52.1	272	246	267	259	244	221	120
	11	1	59.3	29.7	278	273	300	292	276	252	140
	12		62.3	0.0	280	282	311	303	287	262	147
	Surface daily totals				2916	2244	2354	2258	2104	1894	978
Sep 21	7	5	11.4	80.2	149	43	51	51	49	47	32
	8	4	22.5	69.6	230	109	133	134	131	124	84
	9	3	32.8	57.3	263	167	206	208	203	193	132
	10	2	41.6	41.9	280	211	262	265	260	247	168
	11	1	47.7	22.6	287	239	298	301	295	281	192
	12		50.0	0.0	290	249	310	313	307	292	200
	Surface daily totals				2708	1788	2210	2228	2182	2074	1416
Oct 21	7	5	4.5	72.3	48	7	14	15	17	17	16
	8	4	15.0	61.9	204	68	106	113	117	118	100
	9	3	24.5	49.8	257	126	185	195	200	198	160
	10	2	32.4	35.6	280	170	245	257	261	257	203
	11	1	37.6	18.7	291	199	283	295	299	294	229
	12		39.5	0.0	294	208	295	308	312	306	238
	Surface daily totals				2454	1348	1962	2060	2098	2074	1654
Nov 21	8	4	8.2	55.4	136	28	63	72	78	82	81
	9	3	17.0	44.1	232	82	152	167	178	183	167
	10	2	24.0	31.0	268	126	215	233	245	249	219
	11	1	28.6	16.1	283	153	254	273	285	288	248
	12		30.2	0.0	288	163	267	287	298	301	258
	Surface daily totals				2128	942	1636	1778	1870	1908	1686
Dec 21	8	4	5.5	53.0	89	14	39	45	50	54	56
	9	3	14.0	41.9	217	65	135	152	164	171	163
	10	2	20.,	29.4	261	107	200	221	235	242	221
	11	1	25.0	15.2	280	134	239	262	276	283	252
	12		26.6	0.0	285	143	253	275	290	296	263
	Surface daily totals				1978	782	1480	1634	1740	1796	1646

[a]From *Solar Heating and Cooling*, revised 1st ed., by J. F. Kreider and F. Kreith. Copyright 1977 Hemisphere Publishing Corp. Used by permission.

TABLE A.3*d* Solar Position and Insolation Values for 48° North Latitude[a]

Date	Solar time		Solar position		BTUH/sq. ft. total insolation on surfaces[b]						
							South facing surface angle with horiz.				
	AM	PM	Alt	Azm	Normal	Horiz.	38	48	58	68	90
Jan 21	8	4	3.5	54.6	37	4	17	19	21	22	22
	9	3	11.0	42.6	185	46	120	132	140	145	139
	10	2	16.9	29.4	239	83	190	206	216	220	206
	11	1	20.7	15.1	261	107	231	249	260	263	243
	12		22.0	0.0	267	115	245	264	275	278	255
	Surface daily totals				1710	596	1360	1478	1550	1578	1478
Feb 21	7	5	2.4	72.2	12	1	3	4	4	4	4
	8	4	11.6	60.5	188	49	95	102	105	106	96
	9	3	19.7	47.7	251	100	178	187	191	190	167
	10	2	26.2	33.3	278	139	240	251	255	251	217
	11	1	30.5	17.2	290	165	278	290	294	288	247
	12		32.0	0.0	293	173	291	304	307	301	258
	Surface daily totals				2330	1080	1880	1972	2024	1978	1720
Mar 21	7	5	10.0	78.7	153	37	49	49	47	45	35
	8	4	19.5	66.8	236	96	131	132	129	122	96
	9	3	28.2	53.4	270	147	205	207	203	193	152
	10	2	35.4	37.8	287	187	263	266	261	248	195
	11	1	40.3	19.8	295	212	300	303	297	283	223
	12		42.0	0.0	298	220	312	315	309	294	232
	Surface daily totals				2780	1578	2208	2228	2182	2074	1632
Apr 21	6	6	8.6	97.8	108	27	13	9	8	7	5
	7	5	18.6	86.7	205	85	76	69	59	48	21
	8	4	28.5	74.9	247	142	149	141	129	113	69
	9	3	37.8	61.2	268	191	216	208	194	174	115
	10	2	45.8	44.6	280	228	268	260	245	223	152
	11	1	51.5	24.0	286	252	301	294	278	254	177
	12		53.6	0.0	288	260	313	305	289	264	185
	Surface daily totals				3076	2106	2358	2266	2114	1902	1262
May 21	5	7	5.2	114.3	41	9	4	4	4	3	2
	6	6	14.7	103.7	162	61	27	16	15	13	10
	7	5	24.6	93.0	219	118	89	75	60	43	13
	8	4	34.7	81.6	248	171	156	142	123	101	45
	9	3	44.3	68.3	264	217	217	202	182	156	86
	10	2	53.0	51.3	274	252	265	251	229	200	120
	11	1	59.5	28.6	279	274	296	281	258	228	141
	12		62.0	0.0	280	281	306	292	269	238	149
	Surface daily totals				3254	2482	2418	2234	2010	1728	982
Jun 21	5	7	7.9	116.5	77	21	9	9	8	7	5
	6	6	17.2	106.2	172	74	33	19	18	16	12
	7	5	27.0	95.8	220	129	93	77	59	39	15
	8	4	37.1	84.6	246	181	157	140	119	95	35
	9	3	46.9	71.6	261	225	216	198	175	147	74
	10	2	55.8	54.8	269	259	262	244	220	189	105
	11	1	62.7	31.2	274	280	291	273	248	216	126
	12		65.5	0.0	275	287	301	283	258	225	133
	Surface daily totals				3312	2626	2420	2204	1950	1644	874

[b] $1 \text{ Btu/(h·ft}^2) = 3.152 \text{ W/m}^2$. Ground reflection not included on normal or horizontal surfaces.

TABLE A.3*d* Solar Position and Insolation Values for 48° North Latitude[a] (*Continued*)

Date	Solar time		Solar position		BTUH/sq. ft. total insolation on surfaces[b]						
	AM	PM	Alt	Azm			South facing surface angle with horiz.				
					Normal	Horiz.	38	48	58	68	90
Jul 21	5	7	5.7	114.7	43	10	5	5	4	4	3
	6	6	15.2	104.1	156	62	28	18	16	15	11
	7	5	25.1	93.5	211	118	89	75	59	42	14
	8	4	35.1	82.1	240	171	154	140	121	99	43
	9	3	44.8	68.8	256	215	214	199	178	153	83
	10	2	53.5	51.9	266	250	261	246	224	195	116
	11	1	60.1	29.0	271	272	291	276	253	223	137
	12		62.6	0.0	272	279	301	286	263	232	144
	Surface daily totals				3158	2474	2386	2200	1974	1694	956
Aug 21	6	6	9.1	98.3	99	28	14	10	9	8	6
	7	5	19.1	87.2	190	85	75	67	58	47	20
	8	4	29.0	75.4	232	141	145	137	125	109	65
	9	3	38.4	61.8	254	189	210	201	187	168	110
	10	2	46.4	45.1	266	225	260	252	237	214	146
	11	1	52.2	24.3	272	248	293	285	268	244	169
	12		54.3	0.0	274	256	304	296	279	255	177
	Surface daily totals				2898	2086	2300	2200	2046	1836	1208
Sep 21	7	5	10.0	78.7	131	35	44	44	43	40	31
	8	4	19.5	66.8	215	92	124	124	121	115	90
	9	3	28.2	53.4	251	142	196	197	193	183	143
	10	2	35.4	37.8	269	181	251	254	248	236	185
	11	1	40.3	19.8	278	205	287	289	284	269	212
	12		42.0	0.0	280	213	299	302	296	281	221
	Surface daily totals				2568	1522	2102	2118	2070	1966	1546
Oct 21	7	5	2.0	71.9	4	0	1	1	1	1	1
	8	4	11.2	60.2	165	44	86	91	95	95	87
	9	3	19.3	47.4	233	94	167	176	180	178	157
	10	2	25.7	33.1	262	133	228	239	242	239	207
	11	1	30.0	17.1	274	157	266	277	281	276	237
	12		31.5	0.0	278	166	279	291	294	288	247
	Surface daily totals				2154	1022	1774	1860	1890	1866	1626
Nov 21	8	4	3.6	54.7	36	5	17	19	21	22	22
	9	3	11.2	42.7	179	46	117	129	137	141	135
	10	2	17.1	29.5	233	83	186	202	212	215	201
	11	1	20.9	15.1	255	107	227	245	255	258	238
	12		22.2	0.0	261	115	241	259	270	272	250
	Surface daily totals				1668	596	1336	1448	1518	1544	1442
Dec 21	9	3	8.0	40.9	140	27	87	98	105	110	109
	10	2	13.6	28.2	214	63	164	180	192	197	190
	11	1	17.3	14.4	242	86	207	226	239	244	231
	12		18.6	0.0	250	94	222	241	254	260	244
	Surface daily totals				1444	446	1136	1250	1326	1364	1304

[a]From *Solar Heating and Cooling*, revised 1st ed., by J. F. Kreider and F. Kreith. Copyright 1977 Hemisphere Publishing Corp. Used by permission.

TABLE A.3*e* Solar Position and Insolation Values for 56° North Latitude[a]

Date	Solar time		Solar position		BTUH/sq. ft. total insolation on surfaces[b]						
	AM	PM	Alt	Azm	Normal	Horiz.	South facing surface angle with horiz.				
							46	56	66	76	90
Jan 21	9	3	5.0	41.8	78	11	50	55	59	60	60
	10	2	9.9	28.5	170	39	135	146	154	156	153
	11	1	12.9	14.5	207	58	183	197	206	208	201
	12		14.0	0.0	217	65	198	214	222	225	217
	Surface daily totals				1126	282	934	1010	1058	1074	1044
Feb 21	8	4	7.6	59.4	129	25	65	69	72	72	69
	9	3	14.2	45.9	214	65	151	159	162	161	151
	10	2	19.4	31.5	250	98	215	225	228	224	208
	11	1	22.8	16.1	266	119	254	265	268	263	243
	12		24.0	0.0	270	126	268	279	282	276	255
	Surface daily totals				1986	740	1640	1716	1742	1716	1598
Mar 21	7	5	8.3	77.5	128	28	40	40	39	37	32
	8	4	16.2	64.4	215	75	119	120	117	111	97
	9	3	23.3	50.3	253	118	192	193	189	180	154
	10	2	29.0	34.9	272	151	249	251	246	234	205
	11	1	32.7	17.9	282	172	285	288	282	268	236
	12		34.0	0.0	284	179	297	300	294	280	246
	Surface daily totals				2586	1268	2066	2084	2040	1938	1700
Apr 21	5	7	1.4	108.8	0	0	0	0	0	0	0
	6	6	9.6	96.5	122	32	14	9	8	7	6
	7	5	18.0	84.1	201	81	74	66	57	46	29
	8	4	26.1	70.9	239	129	143	135	123	108	82
	9	3	33.6	56.3	260	169	208	200	186	167	133
	10	2	39.9	39.7	272	201	259	251	236	214	174
	11	1	44.1	20.7	278	220	292	284	268	245	200
	12		45.6	0.0	280	227	303	295	279	255	209
	Surface daily totals				3024	1892	2282	2186	2038	1830	1458
May 21	4	8	1.2	125.5	0	0	0	0	0	0	0
	5	7	8.5	113.4	93	25	10	9	8	7	6
	6	6	16.5	101.5	175	71	28	17	15	13	11
	7	5	24.8	89.3	219	119	88	74	58	41	16
	8	4	33.1	76.3	244	163	153	138	119	98	63
	9	3	40.9	61.6	259	201	212	197	176	151	109
	10	2	47.6	44.2	268	231	259	244	222	194	146
	11	1	52.3	23.4	273	249	288	274	251	222	170
	12		54.0	0.0	275	255	299	284	261	231	178
	Surface daily totals				3340	2374	2374	2188	1962	1682	1218
Jun 21	4	8	4.2	127.2	21	4	2	2	2	2	1
	5	7	11.4	115.3	122	40	14	13	11	10	8
	6	6	19.3	103.6	185	86	34	19	17	15	12
	7	5	27.6	91.7	222	132	92	76	57	38	15
	8	4	35.9	78.8	243	175	154	137	116	92	55
	9	3	43.8	64.1	257	212	211	193	170	143	98
	10	2	50.7	46.4	265	240	255	238	214	184	133
	11	1	55.6	24.9	269	258	284	267	242	210	156
	12		57.5	0.0	271	264	294	276	251	219	164
	Surface daily totals				3438	2526	2388	2166	1910	1606	1120

[b] 1 Btu/(h·ft²) = 3.152 W/m². Ground reflection not included on normal or horizontal surfaces.

TABLE A.3e Solar Position and Insolation Values for 56° North Latitude[a] (*Continued*)

Date	Solar time		Solar position		BTUH/sq. ft. total insolation on surfaces[b]						
	AM	PM	Alt	Azm	Normal	Horiz.	South facing surface angle with horiz.				
							46	56	66	76	90
Jul 21	4	8	1.7	125.8	0	0	0	0	0	0	0
	5	7	9.0	113.7	91	27	11	10	9	8	6
	6	6	17.0	101.9	169	72	30	18	16	14	12
	7	5	25.3	89.7	212	119	88	74	58	41	15
	8	4	33.6	76.7	237	163	151	136	117	96	61
	9	3	41.4	62.0	252	201	208	193	173	147	106
	10	2	48.2	44.6	261	230	254	239	217	189	142
	11	1	52.9	23.7	265	248	283	268	245	216	165
	12		54.6	0.0	267	254	293	278	255	225	173
	Surface daily totals				3240	2372	2342	2152	1926	1646	1186
Aug 21	5	7	2.0	109.2	1	0	0	0	0	0	0
	6	6	10.2	97.0	112	34	16	11	10	9	7
	7	5	18.5	84.5	187	82	73	65	56	45	28
	8	4	26.7	71.3	225	128	140	131	119	104	78
	9	3	34.3	56.7	246	168	202	193	179	160	126
	10	2	40.5	40.0	258	199	251	242	227	206	166
	11	1	44.8	20.9	264	218	282	274	258	235	191
	12		46.3	0.0	266	225	293	285	269	245	200
	Surface daily totals				2850	1884	2218	2118	1966	1760	1392
Sep 21	7	5	8.3	77.5	107	25	36	36	34	32	28
	8	4	16.2	64.4	194	72	111	111	108	102	89
	9	3	23.3	50.3	233	114	181	182	178	168	147
	10	2	29.0	34.9	253	146	236	237	232	221	193
	11	1	32.7	17.9	263	166	271	273	267	254	223
	12		34.0	0.0	266	173	283	285	279	265	233
	Surface daily totals				2368	1220	1950	1962	1918	1820	1594
Oct 21	8	4	7.1	59.1	104	20	53	57	59	59	57
	9	3	13.8	45.7	193	60	138	145	148	147	138
	10	2	19.0	31.3	231	92	201	210	213	210	195
	11	1	22.3	16.0	248	112	240	250	253	248	230
	12		23.5	0.0	253	119	253	263	266	261	241
	Surface daily totals				1804	688	1516	1586	1612	1588	1480
Nov 21	9	3	5.2	41.9	76	12	49	54	57	59	58
	10	2	10.0	28.5	165	39	132	143	149	152	148
	11	1	13.1	14.5	201	58	179	193	201	203	196
	12		14.2	0.0	211	65	194	209	217	219	211
	Surface daily totals				1094	284	914	986	1032	1046	1016
Dec 21	9	3	1.9	40.5	5	0	3	4	4	4	4
	10	2	6.6	27.5	113	19	86	95	101	104	103
	11	1	9.5	13.9	166	37	141	154	163	167	164
	12		10.6	0.0	180	43	159	173	182	186	182
	Surface daily totals				748	156	620	678	716	734	722

[a]From *Solar Heating and Cooling,* revised 1st ed., by J. F. Kreider and F. Kreith. Copyright 1977 Hemisphere Publishing Corp. Used by permission.

TABLE A.3*f* Solar Position and Insolation Values for 64° North Latitude[a]

Date	Solar time AM	Solar time PM	Solar position Alt	Solar position Azm	BTUH/sq. ft. total insolation on surfaces[b] Normal	Horiz.	South facing surface angle with horiz. 54	64	74	84	90
Jan 21	10	2	2.8	28.1	22	2	17	19	20	20	20
	11	1	5.2	14.1	81	12	72	77	80	81	81
	12		6.0	0.0	100	16	91	98	102	103	103
	Surface daily totals				306	45	268	290	302	306	304
Feb 21	8	4	3.4	58.7	35	4	17	19	19	19	19
	9	3	8.6	44.8	147	31	103	108	111	110	107
	10	2	12.6	30.3	199	55	170	178	181	178	173
	11	1	15.1	15.3	222	71	212	220	223	219	213
	12		16.0	0.0	228	77	225	235	237	232	226
	Surface daily totals				1432	400	1230	1286	1302	1282	1252
Mar 21	7	5	6.5	76.5	95	18	30	29	29	27	25
	8	4	20.7	62.6	185	54	101	102	99	94	89
	9	3	18.1	48.1	227	87	171	172	169	160	153
	10	2	22.3	32.7	249	112	227	229	224	213	203
	11	1	25.1	16.6	260	129	262	265	259	246	235
	12		26.0	0.0	263	134	274	277	271	258	246
	Surface daily totals				2296	932	1856	1870	1830	1736	1656
Apr 21	5	7	4.0	108.5	27	5	2	2	2	1	1
	6	6	10.4	95.1	133	37	15	9	8	7	6
	7	5	17.0	81.6	194	76	70	63	54	43	37
	8	4	23.3	67.5	228	112	136	128	116	102	91
	9	3	29.0	52.3	248	144	197	189	176	158	145
	10	2	33.5	36.0	260	169	246	239	224	203	188
	11	1	36.5	18.4	266	184	278	270	255	233	216
	12		97.6	0.0	268	190	289	281	266	243	225
	Surface daily totals				2982	1644	2176	2082	1936	1736	1594
May 21	4	8	5.8	125.1	51	11	5	4	4	3	3
	5	7	11.6	112.1	132	42	13	11	10	9	8
	6	6	17.9	99.1	185	79	29	16	14	12	11
	7	5	24.5	85.7	218	117	86	72	56	39	28
	8	4	30.9	71.5	239	152	148	133	115	94	80
	9	3	36.8	56.1	252	182	204	190	170	145	128
	10	2	41.6	38.9	261	205	249	235	213	186	167
	11	1	44.9	20.1	265	219	278	264	242	213	193
	12		46.0	0.0	267	224	288	274	251	222	201
	Surface daily totals				3470	2236	2312	2124	1898	1624	1436
Jun 21	3	9	4.2	139.4	21	4	2	2	2	2	1
	4	8	9.0	126.4	93	27	10	9	8	7	6
	5	7	14.7	113.6	154	60	16	15	13	11	10
	6	6	21.0	100.8	194	96	34	19	17	14	13
	7	5	27.5	87.5	221	132	91	74	55	36	23
	8	4	34.0	73.3	239	166	150	133	112	88	73
	9	3	39.9	57.8	251	195	204	187	164	137	119
	10	2	44.9	40.4	258	217	247	230	206	177	157
	11	1	48.3	20.9	262	231	275	258	233	202	181
	12		49.5	0.0	263	235	284	267	242	211	189
	Surface daily totals				3650	2488	2342	2118	1862	1558	1356

[b] 1 Btu/(h·ft²) = 3.152 W/m². Ground reflection not included on normal or horizontal surfaces.

TABLE A.3*f* Solar Position and Insolation Values for 64° North Latitude[a] (*Continued*)

Date	Solar time		Solar position		BTUH/sq. ft. total insolation on surfaces[b]						
	AM	PM	Alt	Azm	Normal	Horiz.	South facing surface angle with horiz.				
							54	64	74	84	90
Jul 21	4	8	6.4	125.3	53	13	6	5	5	4	4
	5	7	12.1	112.4	128	44	14	13	11	10	9
	6	6	18.4	99.4	179	81	30	17	16	13	12
	7	5	25.0	86.0	211	118	86	72	56	38	28
	8	4	31.4	71.8	231	152	146	131	113	91	77
	9	3	37.3	56.3	245	182	201	186	166	141	124
	10	2	42.2	39.2	253	204	245	230	208	181	162
	11	1	45.4	20.2	257	218	273	258	236	207	187
	12		46.6	0.0	259	223	282	267	245	216	195
	Surface daily totals				3372	2248	2280	2090	1864	1588	1400
Aug 21	5	7	4.6	108.8	29	6	3	3	2	2	2
	6	6	11.0	95.5	123	39	16	11	10	8	7
	7	5	17.6	81.9	181	77	69	61	52	42	35
	8	4	23.9	67.8	214	113	132	123	112	97	87
	9	3	29.6	52.6	234	144	190	182	169	150	138
	10	2	34.2	36.2	246	168	237	229	215	194	179
	11	1	37.2	18.5	252	183	268	260	244	222	205
	12		38.3	0.0	254	188	278	270	255	232	215
	Surface daily totals				2808	1646	2108	1008	1860	1662	1522
Sep 21	7	5	6.5	76.5	77	16	25	25	24	23	21
	8	4	12.7	72.6	163	51	92	92	90	85	81
	9	3	18.1	48.1	206	83	159	159	156	147	141
	10	2	22.3	32.7	229	108	212	213	209	198	189
	11	1	25.1	16.6	240	124	246	248	243	230	220
	12		26.0	0.0	244	129	258	260	254	241	230
	Surface daily totals				2074	892	1726	1736	1696	1608	1532
Oct 21	8	4	3.0	58.5	17	2	9	9	10	10	10
	9	3	8.1	44.6	122	26	86	91	93	92	90
	10	2	12.1	30.2	176	50	152	159	161	159	155
	11	1	14.6	15.2	201	65	193	201	203	200	195
	12		15.5	0.0	208	71	207	215	217	213	208
	Surface daily totals				1238	358	1088	1136	1152	1134	1106
Nov 21	10	2	3.0	28.1	23	3	18	20	21	21	21
	11	1	5.4	14.2	79	12	70	76	79	80	79
	12		6.2	0.0	97	17	89	96	100	101	100
	Surface daily totals				302	46	266	286	298	302	300
Dec 21	11	1	1.8	13.7	4	0	3	4	4	4	4
	12		2.6	0.0	16	2	14	15	16	17	17
	Surface daily totals				24	2	20	22	24	24	24

[a]From *Solar Heating and Cooling*, revised 1st ed., by J. F. Kreider and F. Kreith. Copyright 1977 Hemisphere Publishing Corp. Used by permission.

[b]1 Btu/(h·ft^2) = 3.152 W/m^2. Ground reflection not included on normal or horizontal surfaces.

TABLE A.4 Monthly Solar and Climatic Data for Locations in the Conterminous United States*

STA NO: 3133	HORIZ INSOL:	953.	1273.	1764.	2246.	2577.	2734.	2653.	2382.	2022.	1516.	1041.	853.	1835.
YUCCA FLATS	TILT = LAT:	1635.	1851.	2147.	2265.	2259.	2247.	2247.	2266.	2306.	2126.	1722.	1534.	0
NV	TILT = LAT+15:	1748.	1897.	2078.	2036.	1908.	1841.	1867.	1986.	2174.	2151.	1824.	1657.	0
LAT = 36.95	TILT = 90:	1541.	1499.	1363.	983.	680.	552.	607.	850.	1281.	1629.	1572.	1499.	0
	KT:	.62	.64	.68	.71	.72	.73	.73	.72	.72	.70	.64	.61	0
	AMB TEMP:	0	0	0	0	0	0	0	0	0	0	0	0	0
	HTG DEG DAYS:	0	0	0	0	0	0	0	0	0	0	0	0	0
STA NO: 3812	HORIZ INSOL:	722.	971.	1306.	1668.	1804.	1854.	1776.	1627.	1361.	1147.	849.	658.	1312.
ASHEVILLE	TILT = LAT:	1086.	1285.	1492.	1642.	1594.	1564.	1532.	1527.	1460.	1471.	1266.	1032.	0
NC	TILT = LAT+15:	1139.	1297.	1429.	1478.	1368.	1314.	1301.	1348.	1365.	1468.	1323.	1094.	0
LAT = 35.43	TILT = 90:	961.	983.	905.	718.	541.	479.	498.	608.	787.	1065.	1099.	950.	0
	KT:	.45	.47	.49	.52	.50	.50	.49	.49	.48	.51	.49	.44	0
	AMB TEMP:	37.9	39.4	45.9	55.9	63.7	70.8	73.5	72.8	66.7	56.8	46.3	38.7	55.7
	HTG DEG DAYS:	840	717	592	279	100	14	0	0	50	269	561	815	4237
STA NO: 3813	HORIZ INSOL:	769.	1020.	1363.	1736.	1895.	1919.	1785.	1718.	1439.	1247.	940.	729.	1379.
MACON	TILT = LAT:	1087.	1295.	1526.	1702.	1671.	1630.	1550.	1613.	1526.	1552.	1335.	1077.	0
GA	TILT = LAT+15:	1139.	1307.	1464.	1535.	1436.	1371.	1321.	1427.	1430.	1551.	1396.	1143.	0
LAT = 32.70	TILT = 90:	932.	954.	881.	688.	509.	444.	459.	586.	777.	1084.	1129.	967.	0
	KT:	.44	.46	.50	.53	.52	.52	.49	.51	.49	.53	.50	.45	0
	AMB TEMP:	47.8	50.4	56.5	65.8	73.5	79.6	81.4	80.9	75.8	65.7	55.2	48.3	65.1
	HTG DEG DAYS:	543	423	298	66	6	0	0	0	0	82	304	518	2240
STA NO: 3820	HORIZ INSOL:	751.	1015.	1338.	1728.	1865.	1904.	1803.	1667.	1410.	1220.	917.	721.	1362.
AUGUSTA	TILT = LAT:	1075.	1303.	1504.	1696.	1652.	1614.	1563.	1565.	1498.	1528.	1317.	1084.	0
GA	TILT = LAT+15:	1125.	1316.	1442.	1529.	1419.	1357.	1329.	1384.	1403.	1526.	1377.	1151.	0
LAT = 33.37	TILT = 90:	927.	970.	879.	698.	517.	453.	471.	585.	774.	1076.	1121.	981.	0
	KT:	.43	.47	.49	.53	.52	.51	.49	.49	.49	.52	.50	.45	0
	AMB TEMP:	45.8	48.3	54.6	63.8	71.7	78.2	80.4	79.6	74.2	64.1	53.7	46.4	63.4
	HTG DEG DAYS:	601	475	346	90	10	0	0	0	0	104	344	577	2547
STA NO: 3822	HORIZ INSOL:	795.	1044.	1399.	1761.	1852.	1844.	1783.	1621.	1364.	1217.	941.	754.	1365.
SAVANNAH	TILT = LAT:	1117.	1318.	1562.	1725.	1645.	1572.	1551.	1521.	1435.	1496.	1319.	1105.	0
GA	TILT = LAT+15:	1170.	1331.	1500.	1557.	1415.	1327.	1322.	1347.	1344.	1493.	1378.	1173.	0
LAT = 32.13	TILT = 90:	954.	965.	893.	684.	495.	431.	451.	554.	723.	1031.	1106.	988.	0
	KT:	.44	.47	.51	.54	.51	.50	.49	.48	.46	.51	.50	.45	0
	AMB TEMP:	49.9	52.1	58.0	66.1	73.3	79.1	81.1	80.6	76.2	67.1	57.1	50.4	65.9
	HTG DEG DAYS:	483	379	256	63	0	0	0	0	0	60	253	458	1952

TABLE A.4 Monthly Solar and Climatic Data for Locations in the Conterminous United States *(Continued)*

STA NO: 3860	HORIZ INSOL:	526.	757.	1067.	1448.	1710.	1844.	1769.	1580.	1306.	1004.	638.	467.	1176.
HUNTINGTON	TILT = LAT:	787.	1002.	1219.	1425.	1505.	1542.	1515.	1486.	1424.	1322.	958.	726.	0
WV	TILT = LAT+15:	815.	1002.	1160.	1279.	1289.	1292.	1282.	1307.	1329.	1315.	990.	760.	0
LAT = 38.37	TILT = 90:	691.	775.	770.	680.	569.	521.	540.	641.	813.	989.	832.	660.	0
	KT:	.36	.39	.42	.46	.48	.49	.48	.48	.48	.48	.41	.35	0
	AMB TEMP:	34.3	36.1	44.3	55.7	64.5	72.4	75.3	73.9	67.7	57.1	45.5	36.0	55.2
	HTG DEG DAYS:	952	809	649	293	115	11	0	0	46	265	585	899	4624
STA NO: 3870	HORIZ INSOL:	730.	982.	1329.	1697.	1839.	1918.	1830.	1699.	1406.	1180.	680.	670.	1347.
GREENVILLE-SPARTANBU	TILT = LAT:	1083.	1288.	1513.	1671.	1625.	1618.	1579.	1597.	1508.	1507.	1305.	1038.	0
SC	TILT = LAT+15:	1136.	1301.	1450.	1504.	1394.	1358.	1340.	1409.	1411.	1505.	1366.	1101.	0
LAT = 34.90	TILT = 90:	953.	978.	910.	718.	538.	477.	497.	619.	804.	1035.	1130.	951.	0
	KT:	.44	.47	.50	.53	.51	.51	.50	.51	.49	.52	.50	.44	0
	AMB TEMP:	42.3	44.4	50.9	61.0	69.1	75.9	78.3	77.5	71.7	61.7	51.0	42.9	60.6
	HTG DEG DAYS:	704	577	450	144	29	0	0	0	9	145	420	685	3163
STA NO: 3927	HORIZ INSOL:	805.	1069.	1409.	1616.	1890.	2153.	2155.	1983.	1621.	1293.	938.	766.	1475.
FORT WORTH	TILT = LAT:	1157.	1373.	1585.	1580.	1676.	1816.	1856.	1865.	1739.	1622.	1336.	1151.	0
TX	TILT = LAT+15:	1215.	1390.	1522.	1426.	1440.	1518.	1567.	1646.	1633.	1624.	1398.	1225.	0
LAT = 32.83	TILT = 90:	1001.	1020.	919.	650.	512.	456.	491.	649.	884.	1139.	1133.	1044.	0
	KT:	.46	.49	.51	.50	.53	.58	.59	.59	.56	.55	.51	.47	0
	AMB TEMP:	44.8	48.7	55.0	65.2	72.5	80.6	84.8	84.9	77.7	67.6	55.8	47.9	65.5
	HTG DEG DAYS:	626	456	335	88	0	0	0	0	0	60	287	530	2382
STA NO: 3928	HORIZ INSOL:	784.	1058.	1405.	1783.	2036.	2264.	2239.	2032.	1616.	1250.	871.	690.	1502.
WICHITA	TILT = LAT:	1302.	1497.	1663.	1774.	1790.	1680.	1907.	1924.	1798.	1699.	1401.	1196.	0
KS	TILT = LAT+15:	1379.	1522.	1598.	1593.	1526.	1559.	1599.	1690.	1686.	1705.	1472.	1279.	0
LAT = 37.65	TILT = 90:	1207.	1198.	1055.	812.	622.	556.	597.	773.	1014.	1287.	1261.	1147.	0
	KT:	.52	.54	.55	.55	.57	.60	.61	.61	.58	.59	.54	.51	0
	AMB TEMP.	31.3	36.3	43.6	56.6	66.1	75.8	80.7	79.7	70.6	59.6	44.8	34.5	56.6
	HTG DEG DAYS:	1045	804	671	275	90	7	0	0	32	211	606	946	4587
STA NO: 3937	HORIZ INSOL:	728.	1010.	1313.	1570.	1849.	1970.	1788.	1657.	1485.	1381.	917.	706.	1365.
LAKE CHARLES	TILT = LAT:	954.	1228.	1434.	1528.	1649.	1684.	1563.	1556.	1558.	1682.	1220.	962.	0
LA	TILT = LAT+15:	990.	1236.	1375.	1382.	1422.	1419.	1335.	1381.	1462.	1686.	1269.	1011.	0
LAT = 30.12	TILT = 90:	775.	866.	788.	589.	463.	408.	424.	530.	748.	1135.	987.	822.	0
	KT:	.38	.43	.46	.47	.51	.53	.49	.49	.50	.56	.46	.40	0
	AMB TEMP:	52.3	55.1	60.3	68.9	75.2	80.7	82.4	82.2	78.4	70.0	60.2	54.3	68.3
	HTG DEG DAYS:	415	306	200	26	0	0	0	0	0	36	177	338	1498

STA NO: 3940 JACKSON MS LAT = 32.32													
HORIZ INSOL:	754.	1026.	1369.	1708.	1941.	2024.	1909.	1780.	1509.	1271.	902.	709.	1409.
TILT = LAT:	1049.	1296.	1528.	1672.	1721.	1716.	1655.	1672.	1604.	1578.	1257.	1028.	0
TILT = LAT+15:	1096.	1308.	1466.	1509.	1478.	1440.	1406.	1479.	1504.	1578.	1311.	1087.	0
TILT = 90:	890.	950.	876.	671.	509.	443.	465.	594.	808.	1097.	1050.	912.	0
KT:	.42	.46	.50	.52	.54	.54	.52	.53	.51	.54	.48	.43	0
AMB TEMP:	47.1	49.8	56.1	65.7	72.7	79.4	81.7	81.2	76.0	65.8	55.3	48.9	65.0
HTG DEG DAYS:	569	442	313	74	6	0	0	0	0	91	301	504	2300

STA NO: 3945 COLUMBIA MO LAT = 38.82													
HORIZ INSOL:	612.	875.	1179.	1526.	1880.	2090.	2116.	1878.	1450.	1101.	703.	523.	1328.
TILT = LAT:	981.	1215.	1376.	1510.	1652.	1736.	1802.	1778.	1609.	1492.	1105.	865.	0
TILT = LAT+15:	1028.	1226.	1314.	1354.	1410.	1444.	1513.	1561.	1505.	1490.	1150.	913.	0
TILT = 90:	892.	967.	882.	723.	613.	561.	604.	751.	928.	1135.	982.	808.	0
KT:	.43	.46	.47	.49	.53	.56	.58	.57	.53	.53	.46	.40	0
AMB TEMP:	29.3	33.6	41.7	55.0	64.4	73.0	77.3	76.0	68.3	58.0	43.9	32.8	54.4
HTG DEG DAYS:	1107	879	730	314	117	11	0	5	42	247	633	998	5083

STA NO: 3947 KANSAS CITY MO LAT = 39.30													
HORIZ INSOL:	648.	895.	1203.	1575.	1873.	2080.	2102.	1862.	1452.	1092.	737.	562.	1340.
TILT = LAT:	1080.	1265.	1417.	1564.	1645.	1726.	1789.	1764.	1619.	1493.	1196.	975.	0
TILT = LAT+15:	1136.	1278.	1355.	1403.	1403.	1436.	1502.	1548.	1514.	1492.	1250.	1035.	0
TILT = 90:	998.	1017.	917.	755.	620.	568.	612.	756.	942.	1144.	1078.	929.	0
KT:	.46	.47	.48	.50	.52	.55	.58	.57	.54	.53	.49	.44	0
AMB TEMP:	27.1	32.3	40.7	54.2	64.1	73.0	77.5	76.5	68.0	57.6	42.3	31.3	53.7
HTG DEG DAYS:	1175	916	753	336	127	15	0	0	50	259	681	1045	5357

STA NO: 4725 BINGHAMTON NY LAT = 42.22													
HORIZ INSOL:	386.	576.	861.	1242.	1496.	1681.	1659.	1425.	1131.	779.	414.	297.	996.
TILT = LAT:	591.	768.	986.	1221.	1309.	1396.	1411.	1339.	1249.	1044.	604.	445.	0
TILT = LAT+15:	606.	760.	932.	1091.	1120.	1168.	1191.	1174.	1160.	1031.	613.	456.	0
TILT = 90:	524.	605.	653.	639.	568.	548.	573.	642.	763.	805.	517.	394.	0
KT:	.31	.33	.36	.41	.42	.45	.46	.44	.43	.41	.31	.27	0
AMB TEMP:	22.0	22.8	31.3	44.7	55.1	64.8	69.1	67.3	60.2	50.3	38.2	25.4	46.0
HTG DEG DAYS:	1333	1182	1045	609	320	75	21	40	172	456	804	1228	7285

STA NO:12832 APALACHICOLA FL LAT = 29.73													
HORIZ INSOL:	853.	1126.	1474.	1879.	2091.	1998.	1814.	1688.	1535.	1371.	1040.	818.	1474.
TILT = LAT:	1150.	1387.	1624.	1836.	1858.	1709.	1586.	1585.	1611.	1660.	1409.	1145.	0
TILT = LAT+15:	1205.	1403.	1562.	1659.	1595.	1439.	1355.	1407.	1514.	1663.	1476.	1215.	0
TILT = 90:	957.	986.	886.	668.	475.	403.	420.	530.	765.	1112.	1156.	1000.	0
KT:	.44	.48	.52	.57	.58	.54	.50	.50	.51	.55	.52	.45	0
AMB TEMP:	53.7	55.8	60.7	68.3	74.9	80.0	81.4	81.5	78.6	70.8	61.1	55.2	68.5
HTG DEG DAYS:	368	290	175	30	0	0	0	0	0	22	158	318	1361

TABLE A.4 Monthly Solar and Climatic Data for Locations in the Conterminous United States *(Continued)*

Station	Parameter													
STA NO:12834 DAYTONA BEACH FL LAT = 29.18	HORIZ INSOL:	958.	1213.	1548.	1884.	1968.	1826.	1784.	1682.	1478.	1251.	1036.	870.	1458.
	TILT = LAT:	1308.	1500.	1706.	1839.	1755.	1573.	1564.	1580.	1542.	1485.	1385.	1219.	0
	TILT = LAT+15:	1380.	1522.	1643.	1663.	1511.	1332.	1338.	1403.	1449.	1482.	1449.	1298.	0
	TILT = 90:	1101.	1067.	921.	657.	457.	392.	411.	520.	726.	976.	1126.	1068.	0
	KT:	.49	.51	.54	.57	.55	.50	.49	.49	.49	.50	.51	.48	0
	AMB TEMP:	58.4	59.6	63.9	69.7	75.0	79.4	81.0	81.1	79.5	73.3	65.1	59.6	70.5
	HTG DEG DAYS:	241	210	120	17	0	0	0	0	0	5	97	212	902
STA NO:12839 MIAMI FL LAT = 25.80	HORIZ INSOL:	1057.	1314.	1603.	1859.	1844.	1708.	1763.	1630.	1456.	1303.	1119.	1019.	1473.
	TILT = LAT:	1369.	1568.	1730.	1807.	1660.	1495.	1563.	1534.	1498.	1495.	1421.	1361.	0
	TILT = LAT+15:	1445.	1594.	1670.	1639.	1438.	1275.	1342.	1369.	1410.	1493.	1487.	1455.	0
	TILT = 90:	1111.	1063.	869.	581.	399.	354.	369.	461.	653.	930.	1109.	1163.	0
	KT:	.50	.52	.54	.55	.51	.47	.49	.48	.47	.49	.51	.51	0
	AMB TEMP:	67.2	67.8	71.3	75.0	78.0	81.0	82.3	82.9	81.7	77.8	72.2	68.3	75.5
	HTG DEG DAYS:	53	67	17	0	0	0	0	0	0	0	13	56	205
STA NO:12841 ORLANDO FL LAT = 28.55	HORIZ INSOL:	999.	1244.	1582.	1898.	1989.	1831.	1801.	1673.	1497.	1304.	1096.	926.	1487.
	TILT = LAT:	1356.	1529.	1739.	1851.	1775.	1581.	1581.	1572.	1559.	1544.	1463.	1294.	0
	TILT = LAT+15:	1432.	1553.	1676.	1675.	1529.	1339.	1353.	1397.	1465.	1543.	1534.	1382.	0
	TILT = 90:	1138.	1079.	927.	647.	448.	384.	404.	509.	722.	1008.	1188.	1135.	0
	KT:	.50	.52	.55	.57	.55	.50	.50	.49	.49	.52	.53	.50	0
	AMB TEMP:	60.3	61.5	65.9	71.3	76.4	80.2	81.4	81.8	80.1	74.3	66.6	61.5	71.8
	HTG DEG DAYS:	197	184	94	13	0	0	0	0	0	0	75	170	733
STA NO:12842 TAMPA FL LAT = 27.97	HORIZ INSOL:	1011.	1259.	1594.	1908.	1998.	1847.	1753.	1653.	1492.	1346.	1108.	935.	1492.
	TILT = LAT:	1357.	1538.	1745.	1860.	1785.	1597.	1543.	1553.	1550.	1590.	1464.	1291.	0
	TILT = LAT+15:	1433.	1563.	1682.	1684.	1538.	1353.	1323.	1382.	1457.	1590.	1535.	1378.	0
	TILT = 90:	1131.	1077.	918.	637.	438.	377.	394.	496.	709.	1030.	1180.	1124.	0
	KT:	.50	.52	.55	.57	.56	.50	.48	.48	.49	.53	.53	.49	0
	AMB TEMP:	60.4	61.8	66.0	72.0	77.2	81.0	81.9	82.2	80.8	74.7	66.8	61.6	72.2
	HTG DEG DAYS:	203	176	90	9	0	0	0	0	0	0	71	169	718
STA NO:12844 WEST PALM BEACH FL LAT = 26.68	HORIZ INSOL:	1000.	1233.	1556.	1814.	1845.	1706.	1779.	1663.	1419.	1224.	1060.	958.	1438.
	TILT = LAT:	1304.	1475.	1685.	1764.	1657.	1488.	1571.	1564.	1462.	1407.	1356.	1290.	0
	TILT = LAT+15:	1374.	1496.	1625.	1599.	1435.	1269.	1348.	1394.	1375.	1402.	1416.	1376.	0
	TILT = 90:	1063.	1008.	863.	590.	411.	362.	379.	479.	653.	864.	1064.	1105.	0
	KT:	.48	.50	.53	.54	.51	.47	.49	.49	.46	.47	.49	.49	0
	AMB TEMP:	65.5	66.1	69.8	73.9	77.5	80.5	81.9	82.3	81.5	77.2	71.0	66.8	74.5
	HTG DEG DAYS:	83	91	25	0	0	0	0	0	0	0	22	78	299

STA NO:12907	HORIZ INSOL:	959.	1196.	1516.	1727.	1952.	2073.	2131.	2009.	1705.	1406.	1041.	890.	1551.
LAREDO	TILT = LAT:	1263.	1440.	1647.	1679.	1747.	1782.	1862.	1888.	1782.	1663.	1348.	1203.	0
TX	TILT = LAT+15:	1328.	1459.	1586.	1522.	1507.	1500.	1580.	1673.	1679.	1667.	1408.	1278.	0
LAT = 27.53	TILT = 90:	1036.	995.	859.	586.	429.	371.	394.	541.	796.	1074.	1069.	1030.	0
	KT:	.47	.49	.52	.52	.54	.57	.59	.59	.56	.55	.49	.46	0
	AMB TEMP:	56.5	60.9	67.6	76.3	81.3	86.0	87.9	87.7	82.9	75.5	65.2	58.6	73.9
	HTG DEG DAYS:	299	177	87	0	0	0	0	0	0	8	74	231	876

STA NO:12916	HORIZ INSOL:	835.	1112.	1415.	1780.	1968.	2004.	1814.	1717.	1514.	1335.	973.	779.	1437.
NEW ORLEANS	TILT = LAT:	1126.	1372.	1555.	1738.	1751.	1712.	1585.	1612.	1588.	1615.	1307.	1086.	0
LA	TILT = LAT+15:	1179.	1388.	1494.	1571.	1507.	1441.	1354.	1430.	1492.	1617.	1364.	1150.	0
LAT = 29.98	TILT = 90:	937.	978.	853.	647.	471.	406.	424.	540.	760.	1084.	1065.	944.	0
	KT:	.44	.48	.50	.54	.55	.54	.50	.50	.51	.54	.49	.44	0
	AMB TEMP:	52.9	55.6	60.7	68.6	75.1	80.4	81.9	81.9	78.2	69.8	60.1	54.8	68.3
	HTG DEG DAYS:	403	299	188	29	0	0	0	0	0	40	179	327	1465

STA NO:12917	HORIZ INSOL:	800.	1071.	1353.	1610.	1871.	2011.	1846.	1736.	1527.	1321.	953.	754.	1404.
PORT ARTHUR	TILT = LAT:	1067.	1312.	1480.	1566.	1668.	1718.	1613.	1630.	1603.	1596.	1274.	1042.	0
TX	TILT = LAT+15:	1114.	1324.	1421.	1417.	1438.	1446.	1376.	1446.	1506.	1596.	1328.	1100.	0
LAT = 29.95	TILT = 90:	881.	930.	811.	598.	463.	406.	425.	543.	766.	1069.	1035.	899.	0
	KT:	.42	.46	.48	.49	.52	.54	.51	.51	.51	.53	.48	.42	0
	AMB TEMP:	52.0	55.1	60.1	68.9	75.0	80.8	83.0	83.1	78.9	69.9	60.2	54.2	68.5
	HTG DEG DAYS:	420	302	202	33	0	0	0	0	0	35	184	342	1518

STA NO:12919	HORIZ INSOL:	913.	1135.	1458.	1737.	1927.	2115.	2213.	2027.	1694.	1439.	1055.	862.	1548.
BROWNSVILLE	TILT = LAT:	1153.	1330.	1563.	1687.	1732.	1827.	1938.	1905.	1757.	1672.	1329.	1118.	0
TX	TILT = LAT+15:	1206.	1343.	1505.	1531.	1497.	1537.	1643.	1691.	1657.	1676.	1386.	1182.	0
LAT = 25.90	TILT = 90:	913.	888.	788.	559.	403.	348.	364.	509.	753.	1050.	1029.	927.	0
	KT:	.43	.45	.49	.52	.54	.58	.61	.59	.55	.55	.48	.43	0
	AMB TEMP:	60.3	63.4	67.7	74.9	79.3	82.8	84.4	84.4	81.6	75.7	68.1	62.8	73.8
	HTG DEG DAYS:	225	151	89	0	0	0	0	0	0	5	35	145	650

STA NO:12921	HORIZ INSOL:	895.	1154.	1450.	1612.	1895.	2069.	2121.	1947.	1638.	1350.	1009.	847.	1499.
SAN ANTONIO	TILT = LAT:	1214.	1423.	1592.	1568.	1690.	1767.	1843.	1829.	1725.	1626.	1352.	1190.	0
TX	TILT = LAT+15:	1276.	1441.	1531.	1420.	1457.	1485.	1562.	1619.	1622.	1628.	1414.	1265.	0
LAT = 29.53	TILT = 90:	1016.	1012.	865.	592.	458.	400.	428.	573.	812.	1084.	1101.	1043.	0
	KT:	.46	.49	.51	.49	.53	.56	.58	.57	.54	.54	.50	.47	0
	AMB TEMP:	50.7	54.5	60.8	69.6	76.0	82.2	84.7	84.7	79.3	70.5	59.7	53.2	68.8
	HTG DEG DAYS:	451	310	194	31	0	0	0	0	0	32	179	373	1570

TABLE A.4 Monthly Solar and Climatic Data for Locations in the Conterminous United States *(Continued)*

Station														
STA NO:12924	HORIZ INSOL:	898.	1147.	1430.	1642.	1866.	2094.	2186.	1991.	1687.	1416.	1043.	845.	1521.
CORPUS CHRISTI	TILT = LAT:	1173.	1379.	1549.	1596.	1672.	1798.	1906.	1870.	1764.	1679.	1357.	1136.	0
TX	TILT = LAT+15:	1229.	1394.	1489.	1446.	1445.	1511.	1615.	1657.	1661.	1683.	1418.	1203.	0
LAT = 27.77	TILT = 90:	954.	951.	812.	570.	429.	374.	397.	543.	793.	1089.	1080.	967.	0
	KT:	.44	.47	.49	.49	.52	.57	.60	.58	.55	.55	.49	.44	0
	AMB TEMP:	56.3	59.6	64.9	72.8	77.9	82.4	84.8	85.1	81.0	73.9	64.9	59.1	71.9
	HTG DEG DAYS:	304	199	120	0	0	0	0	0	0	7	81	219	930
STA NO:12928	HORIZ INSOL:	912.	1161.	1435.	1663.	1864.	2036.	2111.	1922.	1625.	1390.	1034.	849.	1500.
KINGSVILLE	TILT = LAT:	1189.	1393.	1552.	1615.	1671.	1752.	1845.	1805.	1693.	1639.	1338.	1137.	0
TX	TILT = LAT+15:	1247.	1409.	1493.	1465.	1444.	1476.	1567.	1602.	1594.	1641.	1397.	1204.	0
LAT 27.52	TILT = 90:	966.	958.	810.	571.	425.	371.	394.	529.	760.	1057.	1059.	965.	0
	KT:	.45	.47	.49	.50	.52	.56	.58	.56	.53	.54	.49	.44	0
	AMB TEMP:	0	0	0	0	0	0	0	0	0	0	0	0	0
	HTG DEG DAYS	0	0	0	0	0	0	0	0	0	0	0	0	0
STA NO:12960	HORIZ INSOL:	772.	1034.	1297.	152[illegible].	1775.	1898.	1828.	1586	1471.	1276.	924.	730.	1351.
HOUSTON	TILT = LAT:	1023.	1261.	1414.	1479.	1585.	1627.	1597.	1583	1541.	1533.	1228.	1000.	0
TX	TILT = LAT+15:	1066.	1270.	1355.	1339.	1369.	1373	1363.	1405	1446.	1532.	1278.	1054.	0
LAT = 29.98	TILT = 90:	839.	890.	774.	573.	454.	404.	424.	534.	738.	1024.	993.	858.	0
	KT:	.40	.44	.46	.46	.49	.51	.50	.50	.49	.52	.46	.41	0
	AMB TEMP:	52.1	55.3	60.8	69.4	75.8	81.1	83.3	83.4	79.2	70.9	61.1	54.6	68.9
	HTG DEG DAYS:	416	294	189	23	0	0	0	0	0	24	155	333	1434
STA NO:13721	HORIZ INSOL:	608.	862.	1181.	1538.	1763.	1893.	1817.	1627.	1357.	1021.	707.	537.	1243.
PATUXENT RIVER	TILT = LAT:	955.	1178.	1371.	1520.	1551.	1582.	1555.	1531.	1487.	1347.	1093.	878.	0
MD	TILT = LAT+15:	998.	1187.	1309.	1364.	1327.	1323.	1315.	1347.	1388.	1340.	1138.	926.	0
LAT = 38.28	TILT = 90:	860.	928.	871.	719.	579.	526.	547.	656.	848.	1008.	965.	817.	0
	KT:	.42	.45	.47	.49	.49	.51	.50	.49	.49	.49	.45	.40	0
	AMB TEMP:	0	0	0	0	0	0	0	0	0	0	0	0	0
	HTG DEG DAYS:	0	0	0	0	0	0	0	0	0	0	0	0	0
STA NO:13722	HORIZ INSOL:	694.	943.	1276.	1644.	1808.	1864.	1776.	1611.	1377.	1105.	812.	636.	1295.
RALEIGH-DURHAM	TILT = LAT:	1046.	1250.	1459.	1620.	1596.	1570.	1530.	1513.	1484.	1418.	1212.	1002.	0
NC	TILT = LAT+15:	1096.	1261.	1397.	1457.	1369.	1318.	1298.	1335.	1387.	1413.	1265.	1062.	0
LAT = 35.87	TILT = 90:	926.	959.	891.	717.	549.	486.	504.	610.	807.	1029.	1053.	923.	0
	KT:	.43	.46	.48	.51	.50	.50	.49	.48	.49	.50	.48	.43	0
	AMB TEMP:	40.5	42.2	49.2	59.5	67.4	74.4	77.5	76.5	70.6	60.2	50.0	41.2	59.1
	HTG DEG DAYS:	760	638	502	180	48	0	0	0	12	186	450	738	3514

STA NO:13723	HORIZ INSOL:	715.	970.	1313.	1683.	1868.	1953.	1864.	1697.	1418.	1141.	839.	659.	1343.
GREENSBORO	TILT = LAT:	1096.	1300.	1511.	1661.	1648.	1640.	1602.	1595.	1535.	1478.	1272.	1058.	0
NC	TILT = LAT+15:	1151.	1313.	1448.	1494.	1411.	1373.	1357.	1406.	1435.	1476.	1330.	1125.	0
LAT = 36.08	TILT = 90:	979.	1004.	928.	737.	563.	499.	519.	639.	838.	1081.	1114.	985.	0
	KT:	.45	.47	.50	.53	.52	.52	.51	.51	.50	.52	.50	.45	0
	AMB TEMP:	38.7	40.6	47.8	58.6	67.1	74.4	77.2	76.0	69.7	59.2	48.3	39.6	58.1
	HTG DEG DAYS:	815	683	544	203	59	0	0	0	24	209	501	787	3825

STA NO:13737	HORIZ INSOL:	678.	932.	1281.	1677.	1888.	2000.	1853.	1680.	1396.	1083.	811.	624.	1325.
NORFOLK	TILT = LAT:	1051.	1259.	1482.	1659.	1663.	1674.	1590.	1581.	1517.	1409.	1249.	1016.	0
VA	TILT = LAT+15:	1102.	1271.	1419.	1490.	1422.	1399.	1346.	1392.	1418.	1404.	1306.	1079.	0
LAT = 36.90	TILT = 90:	942.	980.	923.	751.	581.	517.	531.	649.	842.	1038.	1102.	949.	0
	KT:	.44	.46	.49	.53	.53	.53	.51	.51	.50	.50	.49	.44	0
	AMB TEMP:	40.5	41.4	48.1	57.8	66.7	74.5	78.3	76.9	71.8	61.7	51.6	42.3	59.3
	HTG DEG DAYS:	760	661	532	226	53	0	0	0	9	141	402	704	3488

STA NO:13739	HORIZ INSOL:	555.	795.	1108.	1434.	1660.	1811.	1758.	1575.	1281.	959.	619.	470.	1169.
PHILADELPHIA	TILT = LAT:	898.	1104.	1297.	1417.	1458.	1510.	1501.	1483.	1412.	1288.	970.	783.	0
PA	TILT = LAT+15:	938.	1110.	1236.	1270.	1247.	1263.	1268.	1302.	1316.	1280.	1005.	824.	0
LAT = 39.88	TILT = 90:	817.	882.	843.	699.	580.	539.	560.	664.	829.	981.	860.	732.	0
	KT:	.40	.43	.45	.46	.46	.48	.48	.48	.48	.47	.42	.38	0
	AMB TEMP:	32.3	33.9	41.9	52.9	63.2	72.3	76.8	74.8	68.1	57.4	46.2	35.2	54.6
	HTG DEG DAYS:	1014	871	716	367	122	0	0	0	38	249	564	924	4865

STA NO:13740	HORIZ INSOL:	632.	877.	1210.	1565.	1762.	1872.	1774.	1601.	1348.	1033.	733.	567.	1248.
RICHMOND	TILT = LAT:	977.	1183.	1398.	1546.	1552.	1569.	1523.	1504.	1466.	1345.	1118.	915.	0
VA	TILT = LAT+15:	1022.	1192.	1336.	1388.	1329.	1315.	1290.	1325.	1369.	1338.	1163.	967.	0
LAT = 37.50	TILT = 90:	874.	922.	877.	716.	566.	512.	528.	634.	823.	995.	980.	849.	0
	KT:	.42	.44	.47	.49	.49	.50	.49	.48	.49	.48	.46	.41	0
	AMB TEMP:	37.5	39.4	46.9	57.8	66.5	74.2	77.9	76.3	70.0	59.3	49.0	39.0	57.8
	HTG DEG DAYS:	853	717	569	226	64	0	0	0	21	203	480	806	3939

STA NO:13741	HORIZ INSOL:	661.	899.	1236.	1581.	1764.	1882.	1796.	1620.	1358.	1080.	765.	591.	1270.
ROANOKE	TILT = LAT:	1029.	1216.	1429.	1561.	1554.	1578.	1542.	1523.	1477.	1416.	1173.	960.	0
VA	TILT = LAT+15:	1079.	1226.	1367.	1402.	1331.	1322.	1305.	1341.	1379.	1411.	1224.	1017.	0
LAT = 37.32	TILT = 90:	925.	948.	895.	719.	564.	510.	528.	637.	826.	1050.	1032.	895.	0
	KT:	.44	.45	.48	.50	.49	.50	.49	.49	.49	.50	.47	.43	0
	AMB TEMP:	36.4	38.1	45.3	55.9	64.4	71.7	75.2	74.1	68.0	57.8	46.7	37.4	55.9
	HTG DEG DAYS:	887	753	611	283	101	0	0	0	32	235	549	856	4307

TABLE A.4 Monthly Solar and Climatic Data for Locations in the Conterminous United States *(Continued)*

Station														
STA NO:13754	HORIZ INSOL:	757.	1025.	1387.	1794.	1925.	1939.	1830.	1634.	1427.	1170.	907.	718.	1376.
CHERRY POINT	TILT = LAT:	1135.	1357.	1598.	1770.	1700.	1634.	1578.	1534.	1534.	1491.	1354.	1135.	0
NC	TILT = LAT+15:	1193.	1373.	1524.	1593.	1456.	1371.	1340.	1355.	1435.	1489.	1419.	1209.	0
LAT = 34.90	TILT = 90:	1004.	1037.	957.	754.	552.	478.	497.	601.	817.	1072.	1178.	1051.	0
	KT:	.46	.49	.52	.56	.54	.52	.50	.49	.50	.52	.52	.47	0
	AMB TEMP:	0	0	0	0	0	0	0	0	0	0	0	0	0
	HTG DEG DAYS:	0	0	0	0	0	0	0	0	0	0	0	0	0
STA NO:13781	HORIZ INSOL:	571.	827.	1149.	1480.	1710.	1883.	1823.	1615.	1318.	984.	645.	489.	1206.
WILMINGTON	TILT = LAT:	925.	1156.	1349.	1465.	1502.	1568.	1556.	1521.	1455.	1324.	1016.	818.	0
DE	TILT = LAT+15:	968.	1164.	1238.	1313.	1284.	1309.	1313.	1337.	1357.	1317.	1054.	862.	0
LAT = 39.67	TILT = 90:	843.	925.	876.	717.	589.	546.	569.	675.	851.	1008.	903.	767.	0
	KT:	.41	.44	.46	.47	.48	.50	.50	.49	.49	.48	.43	.39	0
	AMB TEMP:	32.0	33.6	41.6	52.3	62.4	71.4	75.8	74.1	67.9	57.2	45.7	34.7	54.0
	HTG DEG DAYS:	1023	879	725	381	128	0	0	0	32	254	579	939	4940
STA NO:13865	HORIZ INSOL:	744.	1012.	1328.	1662.	1860.	1963.	1823.	1739.	1454.	1258.	897.	699.	1370.
MERIDIAN	TILT = LAT:	1033.	1276.	1477.	1625.	1650.	1667.	1583.	1633.	1540.	1559.	1249.	1011.	0
MS	TILT = LAT+15:	1079.	1286.	1416.	1466.	1420.	1401.	1348.	1445.	1444.	1558.	1302.	1068.	0
LAT = 32.33	TILT = 90:	875.	933.	846.	656.	500.	440.	458.	584.	777.	1083.	1043.	895.	0
	KT:	.42	.45	.48	.51	.52	.53	.50	.51	.50	.53	.48	.42	0
	AMB TEMP:	46.9	49.8	56.1	65.4	72.4	79.2	81.2	80.7	75.3	64.8	54.2	47.9	64.5
	HTG DEG DAYS:	575	443	312	79	7	0	0	0	0	111	331	530	2388
STA NO:13866	HORIZ INSOL:	498.	707.	1010.	1356.	1639.	1776.	1682.	1514.	1272.	972.	613.	440.	1123.
CHARLESTON	TILT = LAT:	730.	917.	1142.	1329.	1443.	1488.	1443.	1422.	1383.	1271.	909.	668.	0
WV	TILT = LAT+15:	753.	913.	1085.	1192.	1238.	1248.	1224.	1252.	1290.	1262.	938.	695.	0
LAT = 38.37	TILT = 90:	634.	702.	719.	638.	553.	511.	524.	619.	789.	947.	785.	598.	0
	KT:	.34	.37	.40	.43	.46	.47	.46	.46	.45	.46	.39	.33	0
	AMB TEMP:	34.5	36.5	44.5	55.9	64.5	72.0	75.0	73.6	67.5	57.0	45.4	36.2	55.2
	HTG DEG DAYS:	946	798	642	287	113	10	0	0	46	267	588	893	4590
STA NO:13874	HORIZ INSOL:	718.	969.	1304.	1686.	1854.	1914.	1812.	1709.	1422.	1200.	883.	674.	1345.
ATLANTA	TILT = LAT:	1022.	1238.	1464.	1654.	1642.	1621.	1569:	1605.	1515.	1506.	1266.	1004.	0
GA	TILT = LAT+15:	1068.	1248.	1402.	1491.	1410.	1362.	1334.	1418.	1419.	1503.	1322.	1061.	0
LAT = 33.65	TILT = 90:	879.	920.	859.	689.	520.	457.	476.	600.	787.	1064.	1077.	902.	0
	KT:	.42	.45	.48	.52	.52	.51	.50	.51	.49	.52	.49	.43	0
	AMB TEMP:	42.4	45.0	51.1	61.1	69.1	75.6	78.0	77.5	72.3	62.4	51.4	43.5	60.8
	HTG DEG DAYS:	701	560	443	144	27	0	0	0	8	137	408	667	3095

STA NO:13876	HORIZ INSOL:	707.	967.	1296.	1674.	1857.	1919.	1810.	1724.	1455.	1211.	858.	661.	1345.
BIRMINGHAM	TILT = LAT:	1000.	1234.	1453.	1641.	1645.	1625.	1567.	1619.	1552.	1520.	1219.	976.	0
AL	TILT = LAT+15:	1043.	1243.	1392.	1479.	1413.	1366.	1333.	1431.	1454.	1518.	1271.	1031.	0
LAT = 33.57	TILT = 90:	856.	915.	851.	683.	519.	456.	474.	602.	804.	1073.	1031.	873.	0
	KT:	.41	.45	.48	.51	.52	.52	.50	.51	.50	.52	.47	.42	0
	AMB TEMP:	44.2	46.9	53.3	63.2	70.5	77.4	79.9	79.2	73.9	63.3	52.1	45.2	62.4
	HTG DEG DAYS:	654	517	389	116	20	0	0	0	6	137	391	614	2844

STA NO:13880	HORIZ INSOL:	744.	995.	1339.	1732.	1860.	1844.	1799.	1585.	1394.	1193.	934.	721.	1345.
CHARLESTON	TILT = LAT:	1049.	1262.	1498.	1698.	1649.	1568.	1561.	1487.	1476.	1478.	1332.	1068.	0
SC	TILT = LAT+15:	1097.	1272.	1436.	1532.	1418.	1322.	1329.	1317.	1382.	1475.	1393.	1133.	0
LAT = 32.90	TILT = 90:	897.	930.	868.	690.	509.	442.	464.	557.	755.	1030.	1129.	960.	0
	KT:	.42	.45	.49	.53	.52	.50	.49	.47	.48	.51	.50	.44	0
	AMB TEMP:	48.6	50.5	56.5	64.6	72.1	77.9	80.2	79.6	75.2	66.1	56.3	49.3	64.7
	HTG DEG DAYS:	521	419	300	69	5	0	0	0	0	74	271	487	2146

STA NO:13881	HORIZ INSOL:	719.	971.	1318.	1695.	1856.	1921.	1831.	1695.	1416.	1173.	866.	672.	1344.
CHARLOTTE	TILT = LAT:	1073.	1279.	1503.	1670.	1639.	1619.	1578.	1593.	1523.	1505.	1289.	1054.	0
NC	TILT = LAT+15:	1125.	1291.	1441.	1503.	1405.	1358.	1339.	1405.	1425.	1503.	1348.	1119.	0
LAT = 35.22	TILT = 90:	946.	975.	909.	724.	546.	482.	502.	624.	817.	1088.	1119.	971.	0
	KT:	.44	.46	.50	.53	.52	.51	.50	.51	.50	.52	.50	.45	0
	AMB TEMP:	42.1	44.0	50.6	60.8	68.8	75.9	78.5	77.7	72.0	61.7	51.0	42.5	60.5
	HTG DEG DAYS:	710	588	461	145	34	0	0	0	10	152	420	698	3218

STA NO:13882	HORIZ INSOL:	631.	859.	1176.	1550.	1732.	1831.	1735.	1630.	1336.	1108.	773.	580.	1245.
CHATTANOOGA	TILT = LAT:	900.	1096.	1319.	1519.	1532.	1548.	1499.	1530.	1426.	1402.	1112.	863.	0
TN	TILT = LAT+15:	935.	1099.	1260.	1368.	1317.	1302.	1275.	1351.	1333.	1396.	1155.	906.	0
LAT = 35.03	TILT = 90:	772.	819.	791.	665.	523.	471.	487.	602.	763.	1005.	946.	771.	0
	KT:	.38	.41	.44	.48	.48	.49	.48	.49	.47	.49	.44	.38	0
	AMB TEMP:	40.2	42.9	49.8	60.5	68.5	76.0	78.8	78.0	71.9	60.8	48.9	41.2	59.8
	HTG DEG DAYS:	769	625	483	165	51	0	0	0	9	182	483	738	3505

STA NO:13883	HORIZ INSOL:	762.	1021.	1355.	1747.	1895.	1947.	1842.	1703.	1439.	1211.	921.	722.	1380.
COLUMBIA	TILT = LAT:	1113.	1325.	1533.	1717.	1676.	1646.	1592.	1599.	1538.	1530.	1346.	1107.	0
SC	TILT = LAT+15:	1167.	1339.	1470.	1547.	1438.	1381.	1352.	1413.	1440.	1529.	1409.	1177.	0
LAT = 33.95	TILT = 90:	971.	996.	906.	717.	531.	464.	484.	604.	804.	1087.	1157.	1012.	0
	KT:	.45	.47	.50	.54	.53	.52	.50	.51	.50	.53	.51	.46	0
	AMB TEMP:	45.4	47.6	54.2	64.1	72.1	78.8	81.2	80.2	74.5	64.2	53.8	46.0	63.5
	HTG DEG DAYS:	608	493	360	83	12	0	0	0	0	112	341	589	2598

TABLE A.4 Monthly Solar and Climatic Data for Locations in the Conterminous United States *(Continued)*

STA NO:13889	HORIZ INSOL:	900.	1164.	1522.	1856.	1956.	1885.	1802.	1694.	1442.	1223.	996.	818.	1438.
JACKSONVILLE	TILT = LAT:	1250.	1460.	1692.	1815.	1740.	1614.	1573.	1591.	1512.	1471.	1360.	1168.	0
FL	TILT = LAT+15:	1316.	1480.	1629.	1640.	1496.	1362.	1343.	1411.	1419.	1467.	1423.	1242.	0
LAT = 30.50	TILT = 90:	1063.	1055.	938.	678.	479.	411.	430.	544.	734.	987.	1122.	1033.	0
	KT:	.48	.50	.54	.56	.54	.51	.50	.50	.48	.50	.50	.47	0
	AMB TEMP:	54.6	56.3	61.2	68.1	74.3	79.2	81.0	81.0	78.2	70.5	61.2	55.4	68.4
	HTG DEG DAYS:	348	282	176	24	0	0	0	0	0	19	161	317	1327
STA NO:13891	HORIZ INSOL:	621.	863.	1191.	1599.	1803.	1902.	1804.	1666.	1383.	1121.	759.	569.	1273.
KNOXVILLE	TILT = LAT:	903.	1120.	1348.	1573.	1592.	1600.	1554.	1566.	1491.	1440.	1110.	864.	0
TN	TILT = LAT+15:	939.	1125.	1288.	1415.	1365.	1342.	1318.	1381.	1394.	1436.	1154.	909.	0
LAT = 35.82	TILT = 90:	783.	848.	820.	698.	547.	489.	507.	626.	810.	1047.	953.	780.	0
	KT:	.39	.42	.45	.50	.50	.51	.49	.50	.49	.51	.45	.39	0
	AMB TEMP:	40.6	42.8	49.9	60.3	68.4	75.5	78.2	77.3	71.6	60.9	49.2	41.5	59.7
	HTG DEG DAYS:	756	630	484	173	47	0	0	0	10	175	474	729	3478
STA NO:13893	HORIZ INSOL:	683.	945.	1278.	1639.	1885.	2045.	1972.	1824.	1471.	1205.	817.	629.	1366.
MEMPHIS	TILT = LAT:	999.	1233.	1450.	1611.	1665.	1718.	1696.	1717.	1587.	1548.	1193.	959.	0
TN	TILT = LAT+15:	1043.	1242.	1388.	1450.	1427.	1437.	1434.	1513.	1486.	1547.	1243.	1014.	0
LAT = 35.05	TILT = 90:	870.	934.	873.	699.	548.	489.	516.	656.	848.	1120.	1024.	871.	0
	KT:	.42	.45	.48	.51	.52	.55	.54	.54	.52	.54	.47	.42	0
	AMB TEMP:	40.5	43.8	51.0	62.5	70.9	78.6	81.6	80.4	73.6	63.0	50.9	42.7	61.6
	HTG DEG DAYS:	760	594	457	131	22	0	0	0	7	142	423	691	3227
STA NO:13894	HORIZ INSOL:	828.	1100.	1408.	1722.	1872.	1869.	1715.	1642.	1449.	1299.	955.	759.	1385.
MOBILE	TILT = LAT:	1134.	1369.	1555.	1681.	1666.	1599.	1500.	1541.	1522.	1580.	1298.	1070.	0
AL	TILT = LAT+15:	1188.	1385.	1493.	1519.	1436.	1350.	1283.	1367.	1427.	1580.	1355.	1133.	0
LAT = 30.68	TILT = 90:	953.	986.	864.	644.	474.	413.	426.	536.	741.	1070.	1067.	937.	0
	KT:	.44	.48	.50	.52	.52	.50	.47	.48	.49	.53	.49	.43	0
	AMB TEMP:	51.2	54.0	59.4	67.9	74.8	80.3	81.6	81.5	77.5	68.9	58.5	52.9	67.4
	HTG DEG DAYS:	451	337	221	40	0	0	0	0	0	39	211	385	1684
STA NO:13895	HORIZ INSOL:	752.	1013.	1341.	1729.	1897.	1972.	1841.	1746.	1468.	1262.	915.	719.	1388.
MONTGOMERY	TILT = LAT:	1046.	1276.	1492.	1693.	1683.	1674.	1598.	1639.	1556.	1564.	1280.	1047.	0
AL	TILT = LAT+15:	1092.	1287.	1431.	1527.	1447.	1407.	1360.	1450.	1459.	1563.	1336.	1108.	0
LAT = 32.30	TILT = 90:	887.	933.	855.	677.	503.	440.	459.	585.	784.	1086.	1072.	931.	0
	KT:	.42	.45	.49	.53	.53	.53	.51	.52	.50	.53	.49	.43	0
	AMB TEMP:	47.5	50.6	56.5	65.2	72.4	78.9	81.0	80.7	76.0	65.8	55.0	48.5	64.8
	HTG DEG DAYS:	556	419	299	76	8	0	0	0	0	93	306	512	2269

STA NO:13897	HORIZ INSOL:	580.	824.	1130.	1544.	1825.	1963.	1891.	1737.	1398.	1114.	711.	521.	1270.
NASHVILLE	TILT = LAT:	832.	1063.	1273.	1517.	1610.	1648.	1625.	1634.	1511.	1437.	1031.	774.	0
TN	TILT = LAT+15:	862.	1065.	1214.	1364.	1380.	1379.	1375.	1440.	1413.	1433.	1068.	809.	0
LAT = 36.12	TILT = 90:	716.	803.	776.	682.	556.	500.	523.	652.	826.	1048.	880.	689.	0
	KT:	.37	.40	.43	.48	.51	.52	.52	.52	.50	.51	.42	.36	0
	AMB TEMP:	38.3	41.0	48.7	60.1	68.5	76.6	79.6	78.5	72.0	60.9	48.4	40.4	59.4
	HTG DEG DAYS:	828	672	524	176	45	0	0	0	10	180	498	763	3696

STA NO:13923	HORIZ INSOL:	794.	1037.	1366.	1610.	1852.	2114.	2077.	1932.	1580.	1268.	919.	744.	1441.
SHERMAN	TILT = LAT:	1164.	1346.	1543.	1577.	1640.	1780.	1788.	1818.	1701.	1608.	1333.	1141.	0
TX	TILT = LAT+15:	1224.	1361.	1481.	1421.	1408.	1488.	1511.	1603.	1596.	1610.	1395.	1215.	0
LAT = 33.72	TILT = 90:	1020.	1010.	909.	664.	521.	471.	502.	656.	882.	1144.	1141.	1044.	0
	KT:	.46	.48	.50	.50	.51	.57	.57	.57	.55	.55	.51	.47	0
	AMB TEMP:	41.7	45.9	52.3	63.7	71.2	79.4	83.6	83.7	76.0	65.8	53.4	44.8	63.5
	HTG DEG DAYS:	722	535	411	114	13	0	0	0	0	90	353	626	2864

STA NO:13957	HORIZ INSOL:	762.	1038.	1342.	1613.	1886.	2065.	2014.	1877.	1554.	1303.	929.	731.	1426.
SHREVEPORT	TILT = LAT:	1069.	1318.	1496.	1576.	1673.	1748.	1741.	1764.	1657.	1628.	1308.	1073.	0
LA	TILT = LAT+15:	1118.	1331.	1434.	1422.	1438.	1464.	1475.	1559.	1555.	1631.	1367.	1137.	0
LAT = 32.47	TILT = 90:	911.	970.	860.	642.	505.	447.	475.	618.	836.	1138.	1100.	959.	0
	KT:	.43	.47	.49	.49	.52	.56	.55	.56	.53	.55	.50	.44	0
	AMB TEMP:	47.2	50.5	56.8	66.4	73.4	80.2	83.2	83.2	77.4	67.5	56.2	49.2	65.9
	HTG DEG DAYS:	552	416	291	65	5	0	0	0	0	70	278	490	2167

STA NO:13958	HORIZ INSOL:	865.	1125.	1429.	1605.	1834.	2072.	2106.	1931.	1606.	1333.	987.	825.	1476.
AUSTIN	TILT = LAT:	1184.	1397.	1576.	1563.	1634.	1765.	1826.	1814.	1696.	1620.	1339.	1175.	0
TX	TILT = LAT+15:	1243.	1414.	1515.	1413.	1410.	1482.	1547.	1605.	1594.	1621.	1399.	1250.	0
LAT = 30.30	TILT = 90:	997.	1003.	870.	603.	465.	412.	442.	586.	814.	1093.	1100.	1038.	0
	KT:	.46	.48	.50	.49	.51	.56	.58	.57	.54	.54	.50	.47	0
	AMB TEMP:	49.7	53.3	59.5	68.6	75.2	81.6	84.6	84.7	78.9	70.1	59.1	52.3	68.1
	HTG DEG DAYS:	483	344	223	44	0	0	0	0	0	39	205	399	1737

STA NO:13959	HORIZ INSOL:	833.	1096.	1428.	1612.	1774.	2112.	2130.	1958.	1601.	1301.	957.	803.	1467.
WACO	TILT = LAT:	1168.	1385.	1591.	1573.	1578.	1790.	1841.	1841.	1702.	1605.	1329.	1178.	0
TX	TILT = LAT+15:	1226.	1402.	1529.	1420.	1361.	1499.	1557.	1626.	1599.	1606.	1389.	1254.	0
LAT = 31.62	TILT = 90:	997.	1012.	901.	627.	478.	434.	466.	618.	842.	1105.	1109.	1057.	0
	KT:	.46	.48	.51	.49	.49	.57	.58	.58	.54	.54	.50	.47	0
	AMB TEMP:	47.0	50.9	57.2	67.3	74.5	81.9	85.6	85.7	78.9	69.1	57.5	49.8	67.1
	HTG DEG DAYS:	558	401	280	56	0	0	0	0	0	51	241	471	2058

TABLE A.4 Monthly Solar and Climatic Data for Locations in the Conterminous United States *(Continued)*

STA NO:13960	HORIZ INSOL:	822.	1071.	1422.	1627.	1889.	2135.	2122.	1950.	1587.	1276.	936.	780.	1468.
DALLAS	TILT = LAT:	1187.	1376.	1601.	1591.	1674.	1801.	1829.	1834.	1699.	1598.	1334.	1180.	0
TX	TILT = LAT+15:	1248.	1393.	1538.	1435.	1438.	1506.	1545.	1619.	1595.	1599.	1396.	1257.	0
LAT = 32.85	TILT = 90:	1031.	1023.	929.	654.	512.	456.	489.	642.	864.	1121.	1131.	1073.	0
	KT:	.47	.49	.52	.50	.52	.57	.58	.58	.54	.54	.51	.48	0
	AMB TEMP:	45.4	49.4	55.8	66.4	73.8	81.6	85.7	85.8	78.2	68.0	55.9	48.2	66.2
	HTG DEG DAYS:	608	437	314	71	0	0	0	0	0	55	284	521	2290
STA NO:13962	HORIZ INSOL:	924.	1183.	1576.	1843.	2037.	2209.	2139.	1956.	1598.	1315.	1008.	863.	1554.
ABILENE	TILT = LAT:	1358.	1537.	1790.	1810.	1804.	1862.	1845.	1840.	1707.	1645.	1445.	1323.	0
TX	TILT = LAT+15:	1436.	1562.	1724.	1632.	1546.	1554.	1558.	1624.	1603.	1648.	1517.	1417.	0
LAT = 32.43	TILT = 90:	1194.	1149.	1033.	717.	521.	451.	482.	635.	860.	1150.	1230.	1216.	0
	KT:	.52	.53	.57	.56	.57	.59	.59	.58	.55	.56	.54	.52	0
	AMB TEMP:	43.7	47.9	54.5	65.2	72.4	80.3	83.9	83.6	76.1	66.1	54.1	46.4	64.5
	HTG DEG DAYS:	660	479	354	104	11	0	0	0	0	89	336	577	2610
STA NO:13963	HORIZ INSOL:	731.	1003.	1313.	1611.	1929.	2107.	2032.	1861.	1518.	1228.	847.	674.	1404.
LITTLE ROCK	TILT = LAT:	1081.	1317.	1490.	1581.	1704.	1769.	1747.	1751.	1639.	1576.	1238.	1039.	0
AR	TILT = LAT+15:	1133.	1331.	1428.	1423.	1460.	1477.	1476.	1544.	1536.	1576.	1292.	1102.	0
LAT = 34.73	TILT = 90:	948.	1000.	893.	683.	550.	488.	516.	660.	869.	1136.	1064.	950.	0
	KT:	.44	.47	.49	.50	.54	.56	.56	.55	.53	.54	.48	.44	0
	AMB TEMP:	39.5	42.9	50.3	61.7	69.8	78.1	81.4	80.6	73.3	62.4	50.3	41.6	61.0
	HTG DEG DAYS:	791	619	470	139	21	0	0	0	5	143	441	725	3354
STA NO:13964	HORIZ INSOL:	744.	999.	1312.	1616.	1912.	2089.	2065.	1877.	1502.	1201.	851.	682.	1404.
FORT SMITH	TILT = LAT:	1125.	1326.	1498.	1588.	1688.	1752.	1772.	1768.	1626.	1550.	1267.	1077.	0
AR	TILT = LAT+15:	1182.	1341.	1435.	1424.	1445.	1463.	1495.	1558.	1524.	1549.	1324.	1145.	0
LAT = 35.33	TILT = 90:	999.	1017.	907.	696.	557.	498.	531.	676.	874.	1126.	1099.	996.	0
	KT:	.46	.48	.49	.50	.53	.56	.57	.56	.53	.54	.49	.46	0
	AMB TEMP:	39.0	43.3	50.3	62.2	70.1	78.0	82.2	81.4	74.0	63.2	50.4	41.5	61.3
	HTG DEG DAYS:	806	608	471	132	17	0	0	0	0	135	438	729	3336
STA NO:13966	HORIZ INSOL:	862.	1123.	1472.	1763.	2017.	2221.	2166.	1969.	1602.	1291.	957.	799.	1520.
WICHITA FALLS	TILT = LAT:	1301.	1487.	1683.	1734.	1782.	1864.	1861.	1854.	1729.	1650.	1412.	1260.	0
TX	TILT = LAT+15:	1375.	1510.	1618.	1562.	1525.	1553.	1569.	1634.	1623.	1653.	1482.	1348.	0
LAT = 33.97	TILT = 90:	1159.	1132.	999.	723.	548.	480.	513.	671.	901.	1181.	1221.	1171.	0
	KT:	.51	.52	.54	.54	.56	.60	.59	.59	.56	.56	.53	.51	0
	AMB TEMP:	41.5	45.9	52.5	64.3	72.3	81.3	85.8	85.5	77.0	66.0	52.9	44.2	64.1
	HTG DEG DAYS:	729	535	409	112	13	0	0	0	0	92	369	645	2904

STA NO:13967
OKLAHOMA CITY
OK
LAT = 35.40

HORIZ INSOL:	801.	1055.	1400.	1725.	1918.	2144.	2128.	1950.	1554.	1233.	901.	725.	1461.
TILT = LAT:	1239.	1419.	1614.	1702.	1693.	1795.	1824.	1839.	1690.	1601.	1363.	1171.	0
TILT = LAT+15:	1308.	1439.	1550.	1531.	1449.	1497.	1537.	1619.	1585.	1602.	1429.	1249.	0
TILT = 90:	1115.	1097.	982.	739.	560.	503.	538.	698.	909.	1167.	1194.	1094.	0
KT:	.49	.51	.53	.54	.53	.57	.58	.58	.55	.55	.52	.49	0
AMB TEMP:	36.8	41.3	48.2	60.4	68.3	76.8	81.5	81.1	73.0	62.4	49.2	40.0	59.9
HTG DEG DAYS:	874	664	532	180	36	0	0	0	12	148	474	775	3695

STA NO:13968
TULSA
OK
LAT = 36.20

HORIZ INSOL:	732.	978.	1306.	1603.	1822.	2021.	2030.	1965.	1473.	1164.	827.	659.	1373.
TILT = LAT:	1133.	1316.	1503.	1578.	1608.	1693.	1740.	1759.	1602.	1516.	1254.	1064.	0
TILT = LAT+15:	1191.	1331.	1440.	1419.	1377.	1415.	1468.	1548.	1500.	1515.	1310.	1132.	0
TILT = 90:	1017.	1020.	925.	707.	557.	507.	543.	691.	877.	1113.	1097.	992.	0
KT:	.46	.48	.50	.50	.51	.54	.56	.56	.52	.53	.49	.46	0
AMB TEMP:	36.6	41.2	48.3	60.8	68.8	77.3	82.1	81.4	73.3	62.9	49.4	39.8	60.2
HTG DEG DAYS:	880	666	528	176	28	0	0	0	10	143	468	781	3680

STA NO:13970
BATON ROUGE
LA
LAT = 30.53

HORIZ INSOL:	785.	1054.	1379.	1681.	1871.	1926.	1746.	1677.	1464.	1301.	920.	737.	1379.
TILT = LAT:	1057.	1300.	1519.	1640.	1666.	1647.	1526.	1574.	1537.	1580.	1237.	1027.	0
TILT = LAT+15:	1104.	1312.	1458.	1482.	1436.	1388.	1305.	1396.	1443.	1580.	1288.	1084.	0
TILT = 90:	878.	929.	841.	629.	472.	413.	427.	541.	746.	1068.	1008.	891.	0
KT:	.42	.46	.49	.51	.52	.52	.48	.49	.49	.53	.47	.42	0
AMB TEMP:	51.0	53.9	59.7	68.4	74.8	80.3	82.0	81.6	77.5	68.5	58.6	52.9	67.4
HTG DEG DAYS:	451	335	208	33	0	0	0	0	0	54	208	381	1670

STA NO:13985
DODGE CITY
KS
LAT = 37.77

HORIZ INSOL:	827.	1122.	1476.	1886.	2090.	2358.	2295.	2055.	1687.	1301.	894.	732.	1560.
TILT = LAT:	1401.	1614.	1763.	1885.	1837.	1953.	1953.	1948.	1889.	1788.	1454.	1299.	0
TILT = LAT+15:	1488.	1646.	1696.	1693.	1564.	1616.	1636.	1710.	1773.	1798.	1530.	1394.	0
TILT = 90:	1309.	1303.	1124.	859.	634.	566.	606.	783.	1068.	1363.	1315.	1257.	0
KT:	.55	.57	.58	.60	.58	.63	.63	.62	.61	.61	.56	.54	0
AMB TEMP:	30.8	35.2	41.2	54.0	64.0	73.7	79.2	78.1	68.9	57.9	42.6	33.4	54.9
HTG DEG DAYS:	1060	834	738	344	115	21	0	0	41	247	666	980	5046

STA NO:13994
ST. LOUIS
MO
LAT = 38.75

HORIZ INSOL:	627.	886.	1205.	1564.	1871.	2092.	2049.	1817.	1459.	1100.	718.	531.	1327.
TILT = LAT:	1013.	1232.	1410.	1550.	1645.	1739.	1747.	1717.	1619.	1488.	1134.	881.	0
TILT = LAT+15:	1062.	1244.	1348.	1390.	1404.	1447.	1469.	1508.	1514.	1487.	1182.	930.	0
TILT = 90:	924.	981.	904.	739.	610.	560.	593.	729.	933.	1131.	1010.	825.	0
KT:	.44	.46	.48	.50	.52	.56	.56	.55	.53	.53	.47	.41	0
AMB TEMP:	31.3	35.1	43.3	56.5	65.8	74.9	78.6	77.2	69.6	59.1	45.0	34.6	55.9
HTG DEG DAYS:	1045	837	682	272	103	10	0	0	35	224	600	942	4750

TABLE A.4 Monthly Solar and Climatic Data for Locations in the Conterminous United States *(Continued)*

STA NO:13995	HORIZ INSOL:	684.	926.	1235.	1604.	1882.	2075.	2063.	1873.	1481.	1144.	775.	603.	1362.
SPRINGFIELD	TILT = LAT:	1074.	1258.	1426.	1585.	1657.	1732.	1763.	1769.	1625.	1514.	1191.	983.	0
MO	TILT = LAT+15:	1128.	1270.	1364.	1424.	1417.	1444.	1485.	1555.	1521.	1513.	1243.	1042.	0
LAT = 37.23	TILT = 90:	969.	983.	892.	728.	585.	530.	566.	715.	908.	1128.	1049.	918.	0
	KT:	.45	.47	.48	.50	.52	.55	.56	.56	.53	.53	.48	.43	0
	AMB TEMP:	32.9	37.0	44.0	56.5	65.1	73.6	77.8	77.1	69.3	59.0	45.5	36.0	56.1
	HTG DEG DAYS:	995	784	660	275	94	10	0	6	35	227	585	899	4570
STA NO:13996	HORIZ INSOL:	681.	941.	1257.	1642.	1915.	2126.	2128.	1910.	1516.	1147.	772.	584.	1385.
TOPEKA	TILT = LAT:	1143.	1340.	1487.	1633.	1683.	1764.	1811.	1910.	1696.	1576.	1259.	1016.	0
KS	TILT = LAT+15:	1205.	1357.	1424.	1465.	1435.	1466.	1520.	1588.	1587.	1578.	1318.	1081.	0
LAT = 39.07	TILT = 90:	1060.	1080.	961.	781.	625.	570.	611.	767.	983.	1210.	1138.	971.	0
	KT:	.48	.50	.50	.52	.54	.57	.58	.58	.56	.56	.51	.45	0
	AMB TEMP:	28.0	33.4	41.2	54.5	64.5	73.5	78.2	77.2	68.2	57.6	42.9	31.8	54.3
	HTG DEG DAYS:	1147	885	745	329	118	13	0	0	55	259	663	1029	5243
STA NO:14601	HORIZ INSOL:	455.	725.	1094.	1440.	1729.	1857.	1859.	1611.	1255.	839.	471.	379.	1143.
BANGOR	TILT = LAT:	856.	1131.	1370.	1452.	1511.	1526.	1571.	1533.	1449.	1225.	817.	749.	0
ME	TILT = LAT+15:	897.	1142.	1308.	1296.	1283.	1267.	1316.	1340.	1349.	1218.	844.	792.	0
LAT = 44.80	TILT = 90:	817.	962.	966.	794.	676.	623.	662.	768.	930.	991.	751.	735.	0
	KT:	.41	.45	.48	.48	.49	.49	.51	.51	.50	.47	.39	.39	0
	AMB TEMP:	0	0	0	0	0	0	0	0	0	0	0	0	0
	HTG DEG DAYS:	0	0	0	0	0	0	0	0	0	0	0	0	0
STA NO:14607	HORIZ INSOL:	419.	724.	1133.	1414.	1578.	1757.	1762.	1501.	1103.	688.	366.	311.	1063.
CARIBOU	TILT = LAT:	861.	1214.	1485.	1439.	1374.	1438.	1485.	1428.	1275.	1004.	629.	642.	0
ME	TILT = LAT+15:	905.	1231.	1421.	1282.	1166.	1193.	1242.	1246.	1183.	991.	644.	677.	0
LAT = 46.87	TILT = 90:	839.	1064.	1086.	820.	657.	630.	669.	753.	843.	819.	575.	633.	0
	KT:	.42	.49	.52	.48	.45	.47	.49	.48	.45	.41	.33	.36	0
	AMB TEMP:	10.7	12.9	23.6	36.7	49.7	59.6	64.9	62.3	54.1	43.8	31.4	16.1	38.8
	HTG DEG DAYS:	1683	1459	1283	849	474	170	84	122	327	657	1003	1516	9632
STA NO:14732	HORIZ INSOL:	548.	795.	1118.	1457.	1690.	1802.	1784.	1583.	1280.	951.	593.	457.	1171.
NEW YORK CITY (LA GU	TILT = LAT:	914.	1130.	1324.	1445.	1483.	1498.	1520.	1493.	1421.	1299.	946.	784.	0
NY	TILT = LAT+15:	956.	1137.	1263.	1294.	1256.	1252.	1282.	1310.	1325.	1292.	980.	826.	0
LAT = 40.77	TILT = 90:	842.	914.	875.	726.	600.	550.	579.	682.	849.	1003.	844.	740.	0
	KT:	.41	.44	.46	.47	.47	.48	.49	.49	.48	.48	.41	.38	0
	AMB TEMP:	32.1	33.1	40.6	51.7	61.8	71.5	76.7	74.9	68.1	58.1	47.3	35.6	54.3
	HTG DEG DAYS:	1020	893	756	399	145	0	0	0	30	224	531	911	4909

STA NO:14733	HORIZ INSOL:	349.	546.	889.	1315.	1596.	1804.	1776.	1513.	1152.	784.	403.	283.	1034.
BUFFALO	TILT = LAT:	526.	729.	1034.	1303.	1397.	1491.	1507.	1428.	1285.	1071.	599.	429.	0
NY	TILT = LAT+15:	536.	720.	978.	1164.	1191.	1242.	1267.	1251.	1194.	1058.	608.	439.	0
LAT = 42.93	TILT = 90:	462.	576.	695.	689.	606.	583.	611.	691.	795.	836.	518.	382.	0
	KT:	.29	.32	.38	.43	.45	.48	.49	.47	.45	.42	.31	.26	0
	AMB TEMP:	23.7	24.4	32.1	44.9	55.1	65.7	70.1	68.4	61.6	51.5	39.8	27.9	47.1
	HTG DEG DAYS:	1280	1137	1020	603	321	58	12	33	138	419	756	1150	6927

STA NO:14734	HORIZ INSOL:	552.	793.	1109.	1449.	1687.	1795.	1760.	1565.	1273.	951.	596.	454.	1165.
NEWARK	TILT = LAT:	920.	1125.	1311.	1437.	1480.	1493.	1500.	1475.	1411.	1298.	950.	775.	0
NJ	TILT = LAT+15:	963.	1132.	1250.	1286.	1264.	1248.	1266.	1294.	1315.	1290.	984.	816.	0
LAT = 40.70	TILT = 90:	848.	909.	864.	721.	599.	548.	573.	674.	842.	1000.	848.	731.	0
	KT:	.42	.44	.45	.47	.47	.48	.48	.48	.48	.48	.42	.38	0
	AMB TEMP:	31.4	32.6	40.6	51.7	61.9	71.4	76.4	74.6	67.8	57.5	46.2	34.5	53.9
	HTG DEG DAYS:	1042	907	756	399	143	0	0	0	34	243	564	946	5034

STA NO:14735	HORIZ INSOL:	457.	688.	986.	1335.	1570.	1730.	1725.	1499.	1170.	817.	457.	356.	1066.
ALBANY	TILT = LAT:	775.	990.	1171.	1325.	1373.	1433.	1464.	1414.	1306.	1124.	713.	604.	0
NY	TILT = LAT+15:	806.	992.	1112.	1183.	1172.	1196.	1233.	1238.	1214.	1113.	730.	630.	0
LAT = 42.75	TILT = 90:	716.	810.	792.	697.	596.	565.	596.	682.	807.	879.	629.	564.	0
	KT:	.38	.40	.42	.44	.44	.46	.47	.46	.45	.43	.35	.33	0
	AMB TEMP:	21.5	23.5	33.4	46.9	57.7	67.5	72.0	69.6	61.9	51.4	39.6	25.9	47.6
	HTG DEG DAYS:	1349	1162	980	543	253	39	9	22	135	422	762	1212	6888

STA NO:14737	HORIZ INSOL:	528.	764.	1078.	1410.	1637.	1777.	1765.	1546.	1238.	926.	568.	430.	1139.
ALLENTOWN	TILT = LAT:	863.	1070.	1267.	1395.	1436.	1479.	1504.	1456.	1367.	1254.	888.	716.	0
PA	TILT = LAT+15:	901.	1075.	1207.	1249.	1228.	1237.	1269.	1278.	1273.	1246.	918.	751.	0
LAT = 40.65	TILT = 90:	789.	860.	833.	700.	585.	544.	573.	666.	814.	963.	786.	667.	0
	KT:	.40	.42	.44	.45	.46	.47	.48	.47	.46	.47	.40	.36	0
	AMB TEMP:	27.8	29.4	38.1	49.9	60.1	69.5	74.1	71.7	64.7	54.1	42.3	30.7	51.0
	HTG DEG DAYS:	1153	997	834	453	190	21	0	6	85	344	681	1063	5827

STA NO:14740	HORIZ INSOL:	478.	715.	979.	1315.	1568.	1686.	1649.	1422.	1154.	853.	497.	385.	1058.
HARTFORD	TILT = LAT:	795.	1016.	1147.	1299.	1374.	1400.	1403.	1335.	1276.	1164.	777.	650.	0
CT	TILT = LAT+15:	827.	1019.	1089.	1161.	1174.	1172.	1185.	1171.	1186.	1154.	798.	679.	0
LAT = 41.93	TILT = 90:	730.	826.	765.	673.	585.	545.	566.	637.	776.	903.	686.	607.	0
	KT:	.38	.41	.41	.43	.44	.45	.45	.44	.44	.44	.36	.34	0
	AMB TEMP:	24.8	26.8	35.6	47.7	58.3	67.8	72.7	70.4	62.8	52.6	41.3	28.2	49.1
	HTG DEG DAYS:	1246	1070	911	519	226	24	0	12	106	384	711	1141	6350

TABLE A.4 Monthly Solar and Climatic Data for Locations in the Conterminous United States *(Continued)*

STA NO:14742	HORIZ INSOL:	385.	607.	940.	1296.	1574.	1729.	1721.	1475.	1122.	741.	375.	283.	1021.
BURLINGTON	TILT = LAT:	660.	880.	1133.	1291.	1374.	1425.	1456.	1394.	1266.	1033.	575.	470.	0
VT	TILT = LAT+15:	684.	879.	1075.	1151.	1170.	1187.	1223.	1219.	1175.	1020.	584.	485.	0
LAT = 44.47	TILT = 90:	611.	727.	785.	704.	621.	589.	620.	699.	804.	820.	504.	434.	0
	KT:	.34	.38	.41	.43	.45	.46	.47	.46	.44	.41	.30	.29	0
	AMB TEMP:	16.8	18.6	29.1	43.0	54.8	65.2	69.8	67.4	59.3	48.8	37.0	22.6	44.4
	HTG DEG DAYS:	1494	1299	1113	660	331	63	20	49	191	502	840	1314	7876
STA NO:14745	HORIZ INSOL:	460.	686.	974.	1317.	1582.	1705.	1675.	1455.	1140.	817.	463.	362.	1053.
CONCORD	TILT = LAT:	799.	999.	1160.	1307.	1384.	1410.	1421.	1371.	1273.	1136.	739.	636.	0
NH	TILT = LAT+15:	834.	1002.	1102.	1167.	1180.	1178.	1197.	1201.	1182.	1125.	759.	665.	0
LAT = 43.20	TILT = 90:	745.	823.	790.	695.	606.	566.	590.	670.	792.	895.	659.	601.	0
	KT:	.39	.41	.42	.43	.45	.45	.46	.45	.44	.44	.36	.34	0
	AMB TEMP:	20.6	22.6	32.3	44.2	55.1	64.7	69.7	67.2	59.5	49.3	38.0	24.8	45.6
	HTG DEG DAYS:	1376	1187	1014	624	315	58	16	45	182	487	810	1246	7360
STA NO:14751	HORIZ INSOL:	536.	771.	1083.	1411.	1652.	1805.	1764.	1551.	1267.	934.	579.	447.	1150.
HARRISBURG	TILT = LAT:	866.	1072.	1267.	1394.	1451.	1503.	1505.	1460.	1397.	1256.	896.	741.	0
PA	TILT = LAT+15:	903.	1076.	1207.	1248.	1240.	1257.	1270.	1282.	1302.	1247.	925.	778.	0
LAT = 40.22	TILT = 90:	788.	856.	827.	694.	583.	542.	566.	661.	826.	959.	790.	690.	0
	KT:	.40	.42	.44	.45	.46	.48	.48	.47	.47	.47	.40	.37	0
	AMB TEMP:	30.1	32.3	41.0	52.8	63.1	72.0	76.1	73.9	67.0	55.8	43.8	32.6	53.4
	HTG DEG DAYS:	1082	916	744	370	128	0	0	0	51	293	636	1004	5224
STA NO:14764	HORIZ INSOL:	450.	682.	970.	1304.	1567.	1712.	1659.	1461.	1158.	822.	459.	363.	1051.
PORTLAND	TILT = LAT:	794.	1004.	1162.	1295.	1370.	1414.	1407.	1378.	1302.	1159.	746.	655.	0
ME	TILT = LAT+15:	829.	1008.	1104.	1156.	1167.	1180.	1185.	1206.	1210.	1149.	767.	687.	0
LAT = 43.65	TILT = 90:	744.	833.	797.	695.	608.	573.	592.	680.	816.	919.	670.	625.	0
	KT:	.39	.41	.42	.43	.44	.46	.46	.46	.45	.45	.36	.35	0
	AMB TEMP:	21.5	22.9	31.8	42.7	52.7	62.2	68.0	66.4	58.7	49.1	38.6	25.7	45.0
	HTG DEG DAYS:	1349	1179	1029	669	381	106	27	55	200	493	792	1218	7498
STA NO:14765	HORIZ INSOL:	506.	739.	1032.	1374.	1655.	1775.	1695.	1499.	1209.	907.	538.	419.	1112.
PROVIDENCE	TILT = LAT:	855.	1055.	1220.	1361.	1450.	1473.	1443.	1411.	1343.	1252.	858.	726.	0
RI	TILT = LAT+15:	893.	1060.	1160.	1218.	1237.	1230.	1217.	1238.	1250.	1243.	886.	763.	0
LAT = 41.73	TILT = 90:	790.	858.	814.	701.	606.	559.	574.	665.	815.	975.	766.	686.	0
	KT:	.40	.42	.43	.45	.47	.47	.47	.46	.46	.47	.39	.37	0
	AMB TEMP:	28.4	29.4	36.9	47.3	56.9	66.4	72.1	70.4	63.4	53.7	43.3	31.5	50.0
	HTG DEG DAYS:	1135	997	871	531	259	36	0	10	93	350	651	1039	5972

STA NO:14768	HORIZ INSOL:	364.	560.	903.	1339.	1606.	1817.	1781.	1519.	1160.	782.	404.	281.	1043.
ROCHESTER	TILT = LAT:	566.	757.	1058.	1331.	1405.	1501.	1510.	1435.	1297.	1072.	605.	428.	0
NY	TILT = LAT+15:	580.	749.	1002.	1188.	1198.	1250.	1270.	1256.	1206.	1059.	615.	438.	0
LAT = 43.12	TILT = 90:	505.	603.	714.	706.	612.	588.	615.	696.	806.	838.	525.	382.	0
	KT:	.30	.33	.39	.44	.45	.48	.49	.47	.45	.42	.31	.26	0
	AMB TEMP:	24.0	24.8	33.0	46.1	56.5	66.9	71.2	69.3	62.3	52.3	40.5	28.3	47.9
	HTG DEG DAYS:	1271	1126	992	567	285	46	9	26	126	398	735	1138	6719
STA NO:14771	HORIZ INSOL:	385.	571.	890.	1324.	1578.	1778.	1758.	1504.	1165.	777.	399.	285.	1034.
SYRACUSE	TILT = LAT:	615.	779.	1039.	1314.	1380.	1469.	1491.	1419.	1305.	1064.	594.	439.	0
NY	TILT = LAT+15:	633.	772.	984.	1174.	1177.	1225.	1254.	1243.	1213.	1051.	602.	449.	0
LAT = 43.12	TILT = 90:	555.	623.	701.	697.	604.	580.	609.	690.	811.	831.	513.	393.	0
	KT:	.32	.34	.38	.44	.45	.47	.48	.47	.45	.42	.31	.27	0
	AMB TEMP:	23.6	24.6	33.2	46.5	56.8	66.9	71.5	69.7	62.8	52.5	41.0	28.1	48.1
	HTG DEG DAYS:	1283	1131	986	555	272	46	11	18	120	392	720	1144	6678
STA NO:14777	HORIZ INSOL:	455.	689.	991.	1339.	1591.	1760.	1746.	1513.	1199.	897.	490.	368.	1086.
WILKES-BARRE-SCRANTO	TILT = LAT:	723.	952.	1156.	1322.	1394.	1463.	1486.	1425.	1326.	1223.	742.	589.	0
PA	TILT = LAT+15:	748.	952.	1098.	1183.	1192.	1223.	1253.	1250.	1233.	1214.	760.	611.	0
LAT = 41.33	TILT = 90:	651.	762.	764.	676.	582.	551.	579.	665.	798.	946.	647.	538.	0
	KT:	.35	.39	.41	.43	.45	.47	.48	.47	.45	.46	.35	.32	0
	AMB TEMP:	26.0	27.3	36.0	48.5	58.9	67.9	72.2	70.0	62.9	52.6	40.8	29.1	49.4
	HTG DEG DAYS:	1209	1056	899	495	219	28	7	18	116	391	726	1113	6277
STA NO:14780	HORIZ INSOL:	560.	797.	1109.	1456.	1672.	1775.	1703.	1533.	1261.	956.	621.	475.	1160.
LAKEHURST	TILT = LAT:	913.	1113.	1300.	1441.	1468.	1480.	1455.	1442.	1388.	1287.	980.	799.	0
NJ	TILT = LAT+15:	954.	1119.	1239.	1291.	1255.	1239.	1230.	1267.	1293.	1279.	1016.	842.	0
LAT = 40.03	TILT = 90:	834.	891.	847.	712.	585.	535.	552.	651.	817.	983.	871.	750.	0
	KT:	.41	.43	.45	.47	.47	.47	.47	.47	.47	.47	.42	.39	0
	AMB TEMP:	0	0	0	0	0	0	0	0	0	0	0	0	0
	HTG DEG DAYS:	0	0	0	0	0	0	0	0	0	0	0	0	0
STA NO:14819	HORIZ INSOL:	507.	760.	1107.	1459.	1789.	2007.	1944.	1719.	1354.	969.	566.	402.	1215.
CHICAGO	TILT = LAT:	859.	1095.	1327.	1453.	1568.	1658.	1650.	1631.	1531.	1362.	923.	685.	0
IL	TILT = LAT+15:	897.	1102.	1266.	1300.	1335.	1377.	1385.	1429.	1428.	1357.	958.	718.	0
LAT = 41.78	TILT = 90:	794.	898.	891.	746.	643.	602.	629.	755.	933.	1070.	831.	644.	0
	KT:	.40	.43	.46	.48	.50	.53	.53	.53	.52	.50	.41	.35	0
	AMB TEMP:	24.3	27.4	36.8	49.9	60.0	70.5	74.7	73.7	65.9	55.4	40.4	28.5	50.6
	HTG DEG DAYS:	1262	1053	874	453	208	26	0	8	57	316	738	1132	6127

TABLE A.4 Monthly Solar and Climatic Data for Locations in the Conterminous United States *(Continued)*

STA NO:14820	HORIZ INSOL:	388.	601.	922.	1350.	1681.	1843.	1828.	1583.	1240.	867.	466.	318.	1091.
CLEVELAND	TILT = LAT:	576.	797.	1061.	1334.	1474.	1529.	1554.	1494.	1378.	1174.	694.	474.	0
OH	TILT = LAT+15:	589.	790.	1005.	1193.	1258.	1275.	1309.	1310.	1283.	1164.	709.	486.	0
LAT = 41.40	TILT = 90:	502.	624.	698.	683.	608.	567.	598.	693.	832.	905.	600.	419.	0
	KT:	.30	.34	.38	.44	.47	.49	.50	.49	.47	.44	.33	.28	0
	AMB TEMP:	26.9	27.9	36.1	48.3	58.3	67.9	71.4	70.0	63.9	53.8	41.6	30.3	49.7
	HTG DEG DAYS:	1181	1039	896	501	244	40	9	17	95	354	702	1076	6154
STA NO:14821	HORIZ INSOL:	459.	677.	980.	1353.	1647.	1813.	1755.	1641.	1282.	945.	538.	387.	1123.
COLUMBUS	TILT = LAT:	693.	900.	1122.	1332.	1446.	1510.	1498.	1548.	1414.	1268.	805.	596.	0
OH	TILT = LAT+15:	714.	897.	1065.	1193.	1237.	1263.	1265.	1359.	1318.	1260.	826.	617.	0
LAT = 40.00	TILT = 90:	610.	703.	724.	662.	578.	541.	561.	691.	832.	967.	698.	536.	0
	KT:	.34	.37	.40	.43	.46	.48	.48	.50	.48	.47	.37	.31	0
	AMB TEMP:	28.4	30.3	39.2	51.2	61.1	70.4	73.6	71.9	65.2	54.2	41.7	30.7	51.5
	HTG DEG DAYS:	1135	972	800	418	176	13	0	8	76	342	699	1063	5702
STA NO:14822	HORIZ INSOL:	417.	680.	1000.	1399.	1716.	1866.	1835.	1576.	1253.	876.	478.	344.	1120.
DETROIT	TILT = LAT:	670.	966.	1186.	1392.	1503.	1543.	1558.	1489.	1409.	1218.	749.	562.	0
MI	TILT = LAT+15:	692.	967.	1127.	1244.	1280.	1285.	1309.	1305.	1313.	1209.	768.	583.	0
LAT = 42.42	TILT = 90:	605.	785.	798.	726.	633.	587.	616.	708.	867.	955.	662.	518.	0
	KT:	.34	.39	.42	.46	.48	.50	.50	.49	.48	.46	.36	.31	0
	AMB TEMP:	25.5	26.9	35.4	48.1	58.4	69.1	73.3	71.9	64.5	54.3	41.1	29.6	49.9
	HTG DEG DAYS:	1225	1067	918	507	238	26	0	11	80	342	717	1097	6228
STA NO:14826	HORIZ INSOL:	383.	636.	957.	1339.	1658.	1813.	1797.	1555.	1196.	829.	429.	309.	1075.
FLINT	TILT = LAT:	606.	897.	1132.	1330.	1451.	1498.	1524.	1470.	1342.	1150.	657.	492.	0
MI	TILT = LAT+15:	623.	896.	1075.	1188.	1236.	1248.	1281.	1287.	1248.	1140.	670.	508.	0
LAT = 42.97	TILT = 90:	544.	728.	767.	703.	625.	585.	616.	709.	832.	904.	575.	448.	0
	KT:	.32	.38	.41	.44	.47	.48	.49	.48	.46	.44	.33	.29	0
	AMB TEMP:	22.3	23.8	32.6	45.9	55.8	65.8	69.7	68.2	61.0	51.2	38.3	26.8	46.8
	HTG DEG DAYS:	1324	1154	1004	573	306	65	14	36	147	433	801	1184	7041
STA NO:14827	HORIZ INSOL:	455.	698.	982.	1361.	1672.	1842.	1787.	1594.	1274.	924.	516.	370.	1123.
FORT WAYNE	TILT = LAT:	713.	960.	1139.	1344.	1466.	1529.	1522.	1504.	1416.	1261.	789.	583.	0
IN	TILT = LAT+15:	737.	960.	1081.	1203.	1252.	1276.	1283.	1320.	1320.	1253.	810.	605.	0
LAT = 41.00	TILT = 90:	636.	766.	748.	682.	599.	560.	583.	691.	849.	973.	691.	530.	0
	KT:	.35	.39	.40	.44	.47	.49	.49	.49	.48	.47	.36	.31	0
	AMB TEMP:	25.3	27.6	36.5	49.3	59.6	69.5	73.0	71.3	64.5	53.6	40.2	28.6	49.9
	HTG DEG DAYS:	1231	1047	884	471	216	23	0	12	90	363	744	1128	6209

STA NO:14837	HORIZ INSOL:	515.	804.	1136.	1398.	1743.	1948.	1934.	1708.	1299.	911.	504.	389.	1191.
MADISON	TILT = LAT:	936.	1227.	1396.	1395.	1526.	1605.	1638.	1624.	1481.	1302.	832.	704.	0
WI	TILT = LAT+15:	983.	1241.	1334.	1247.	1298.	1333.	1374.	1421.	1380.	1297.	860.	741.	0
LAT = 43.13	TILT = 90:	887.	1031.	961.	738.	652.	615.	652.	778.	924.	1037.	753.	674.	0
	KT:	.43	.48	.49	.46	.49	.52	.53	.53	.50	.49	.39	.37	0
	AMB TEMP:	16.8	20.3	30.2	45.3	56.0	65.8	70.1	68.7	59.7	49.9	34.7	21.9	44.9
	HTG DEG DAYS:	1494	1252	1079	591	297	72	14	39	173	474	909	1336	7730

STA NO:14839	HORIZ INSOL:	479.	737.	1089.	1443.	1768.	1977.	1962.	1719.	1310.	908.	525.	378.	1191.
MILWAUKEE	TILT = LAT:	838.	1088.	1323.	1443.	1548.	1629.	1662.	1634.	1492.	1291.	873.	669.	0
WI	TILT = LAT+15:	876.	1096.	1262.	1289.	1317.	1352.	1393.	1430.	1391.	1285.	903.	702.	0
LAT = 42.95	TILT = 90:	784.	902.	905.	759.	657.	617.	655.	779.	928.	1025.	792.	635.	0
	KT:	.40	.43	.46	.47	.50	.53	.54	.53	.51	.48	.40	.35	0
	AMB TEMP:	19.4	22.5	31.4	44.7	54.2	64.5	69.9	69.2	61.1	51.0	36.5	24.2	45.7
	HTG DEG DAYS:	1414	1190	1042	609	348	90	15	36	140	440	855	1265	7444

STA NO:14847	HORIZ INSOL:	325.	603.	1029.	1383.	1683.	1811.	1835.	1523.	1049.	673.	332.	253.	1042.
SAULT STE. MARIE	TILT = LAT:	572.	935.	1308.	1401.	1473.	1483.	1547.	1449.	1196.	962.	528.	449.	0
MI	TILT = LAT+15:	591.	938.	1248.	1248.	1248.	1229.	1293.	1264.	1108.	949.	536.	466.	0
LAT = 46.47	TILT = 90:	533.	796.	944.	793.	689.	638.	684.	757.	783.	778.	470.	424.	0
	KT:	.32	.40	.47	.47	.48	.48	.51	.48	.43	.40	.30	.29	0
	AMB TEMP:	14.2	15.2	24.0	38.2	49.0	58.7	63.8	63.2	55.3	46.2	32.8	20.1	40.0
	HTG DEG DAYS:	1575	1394	1271	804	496	200	96	125	291	583	966	1392	9193

STA NO:14848	HORIZ INSOL:	416.	660.	993.	1387.	1723.	1922.	1852.	1666.	1291.	909.	497.	340.	1138.
SOUTH BEND	TILT = LAT:	645.	909.	1164.	1376.	1510.	1590.	1574.	1578.	1449.	1255.	769.	533.	0
IN	TILT = LAT+15:	663.	907.	1105.	1231.	1287.	1324.	1324.	1382.	1350.	1247.	790.	551.	0
LAT = 41.70	TILT = 90:	574.	727.	774.	707.	623.	586.	608.	732.	880.	978.	677.	483.	0
	KT:	.33	.37	.41	.45	.48	.51	.51	.51	.49	.47	.36	.30	0
	AMB TEMP:	24.0	26.3	35.3	48.1	58.4	68.6	72.3	71.0	63.8	53.4	39.6	28.2	49.1
	HTG DEG DAYS:	1271	1084	921	507	245	35	6	24	98	368	762	1141	6462

STA NO:14850	HORIZ INSOL:	311.	568.	1001.	1405.	1729.	1912.	1910.	1609.	1165.	754.	377.	257.	1083.
TRAVERSE CITY	TILT = LAT:	483.	810.	1228.	1413.	1511.	1571.	1614.	1531.	1327.	1065.	587.	409.	0
MI	TILT = LAT+15:	492.	806.	1169.	1261.	1283.	1302.	1351.	1338.	1233.	1053.	597.	419.	0
LAT = 44.73	TILT = 90:	429.	665.	860.	772.	675.	634.	674.	766.	849.	850.	518.	371.	0
	KT:	.28	.35	.44	.47	.49	.51	.53	.51	.46	.42	.31	.26	0
	AMB TEMP:	20.8	20.7	28.7	42.7	52.8	63.7	68.7	67.5	59.4	49.8	36.9	25.9	44.8
	HTG DEG DAYS:	1370	1240	1125	669	367	104	33	66	178	471	843	1212	7698

TABLE A.4 Monthly Solar and Climatic Data for Locations in the Conterminous United States *(Continued)*

Station														
STA NO:14852	HORIZ INSOL:	385.	587.	890.	1278.	1586.	1759.	1734.	1506.	1194.	851.	457.	315.	1045.
YOUNGSTOWN	TILT = LAT:	566.	768.	1015.	1257.	1390.	1462.	1476.	1418.	1318.	1144.	670.	465.	0
OH	TILT = LAT+15:	577.	760.	960.	1124.	1189.	1222.	1246.	1244.	1226.	1133.	683.	476.	0
LAT = 41.27	TILT = 90:	491.	598.	664.	644.	580.	550.	576.	661.	793.	878.	576.	409.	0
	KT:	.30	.33	.37	.41	.45	.47	.48	.46	.45	.44	.33	.27	0
	AMB TEMP:	25.7	26.7	35.3	47.7	57.6	67.0	70.7	69.2	62.7	52.6	40.3	28.8	48.7
	HTG DEG DAYS:	1218	1072	921	519	258	42	9	22	118	384	741	1122	6426
STA NO:14858	HORIZ INSOL:	244.	484.	933.	1366.	1660.	1838.	1838.	1521.	1010.	671.	291.	192.	1004.
HOUGHTON	TILT = LAT:	379.	704.	1174.	1387.	1447.	1501.	1547.	1451.	1152.	978.	447.	297.	0
MI	TILT = LAT+15:	383.	698.	1116.	1234.	1226.	1243.	1293.	1265.	1066.	965.	450.	301.	0
LAT = 47.17	TILT = 90:	337.	587.	850.	796.	691.	656.	697.	769.	762.	799.	393.	266.	0
	KT:	.25	.33	.43	.47	.47	.49	.51	.49	.42	.40	.27	.23	0
	AMB TEMP:	0	0	0	0	0	0	0	0	0	0	0	0	0
	HTG DEG DAYS:	0	0	0	0	0	0	0	0	0	0	0	0	0
STA NO:14860	HORIZ INSOL:	346.	577.	920.	1359.	1646.	1847.	1833.	1455.	1201.	827.	416.	278.	1059.
ERIE	TILT = LAT:	499.	767.	1068.	1347.	1442.	1529.	1557.	1369.	1338.	1124.	606.	397.	0
PA	TILT = LAT+15:	506.	759.	1011.	1204.	1230.	1274.	1309.	1200.	1245.	1112.	614.	403.	0
LAT = 42.08	TILT = 90:	429.	603.	710.	699.	608.	578.	610.	653.	817.	871.	518.	344.	0
	KT:	.28	.33	.39	.44	.46	.49	.50	.45	.46	.43	.31	.25	0
	AMB TEMP:	25.1	25.2	32.9	44.8	54.6	64.6	68.7	67.5	61.4	51.6	40.1	29.1	47.1
	HTG DEG DAYS:	1237	1114	995	606	336	80	24	43	141	415	747	1113	6851
STA NO:14895	HORIZ INSOL:	428.	650.	964.	1357.	1668.	1839.	1787.	1596.	1272.	908.	505.	353.	1111.
AKRON-CANTON	TILT = LAT:	650.	872.	1113.	1340.	1463.	1528.	1522.	1505.	1413.	1230.	762.	543.	0
OH	TILT = LAT+15:	669.	868.	1056.	1199.	1249.	1275.	1283.	1321.	1317.	1221.	781.	560.	0
LAT = 40.92	TILT = 90:	574.	687.	729.	679.	597.	559.	582.	690.	846.	947.	663.	487.	0
	KT:	.33	.36	.40	.44	.47	.49	.49	.49	.48	.46	.35	.30	0
	AMB TEMP:	26.3	27.7	36.2	48.5	58.7	68.3	71.7	70.3	63.7	53.3	40.7	29.4	49.6
	HTG DEG DAYS:	1200	1044	893	495	231	33	9	16	101	369	729	1104	6224
STA NO:14898	HORIZ INSOL:	451.	725.	1104.	1439.	1719.	1908.	1889.	1622.	1218.	821.	465.	350.	1143.
GREEN BAY	TILT = LAT:	832.	1118.	1378.	1448.	1503.	1568.	1597.	1542.	1394.	1181.	790.	652.	0
WI	TILT = LAT+15:	871.	1128.	1317.	1293.	1276.	1301.	1337.	1348.	1297.	1172.	815.	685.	0
LAT = 44.48	TILT = 90:	790.	947.	968.	786.	667.	629.	664.	767.	889.	949.	721.	628.	0
	KT:	.40	.45	.48	.48	.49	.51	.52	.51	.48	.46	.38	.35	0
	AMB TEMP:	15.4	18.0	28.6	43.8	54.5	64.5	69.2	67.7	58.9	49.2	34.1	20.9	43.7
	HTG DEG DAYS:	1538	1316	1128	636	338	91	22	54	191	490	927	1367	8098

STA NO:14913 DULUTH MN LAT = 46.83														
	HORIZ INSOL:	389.	673.	1034.	1373.	1643.	1767.	1854.	1547.	1095.	725.	381.	292.	1064.
	TILT = LAT:	769.	1099.	1326.	1392.	1432.	1446.	1562.	1476.	1264.	1074.	665.	581.	0
	TILT = LAT+15:	805.	1110.	1266.	1239.	1214.	1199.	1305.	1287.	1173.	1064.	683.	610.	0
	TILT = 90:	742.	955.	963.	793.	679.	632.	696.	776.	835.	881.	612.	567.	0
	KT:	.39	.45	.48	.47	.47	.47	.51	.49	.45	.43	.35	.34	0
	AMB TEMP:	8.5	12.1	23.5	38.6	49.4	59.0	65.6	64.1	54.4	45.3	28.4	14.4	38.6
	HTG DEG DAYS:	1751	1481	1287	792	484	194	67	104	318	611	1098	1569	9756

STA NO:14914 FARGO ND LAT = 46.90														
	HORIZ INSOL:	415.	706.	1098.	1476.	1835.	1994.	2120.	1825.	1304.	874.	457.	337.	1203.
	TILT = LAT:	850.	1175.	1429.	1509.	1604.	1628.	1786.	1764.	1557.	1374.	873.	730.	0
	TILT = LAT+15:	893.	1189.	1367.	1345.	1356.	1343.	1486.	1538.	1451.	1373.	908.	774.	0
	TILT = 90:	828.	1027.	1043.	860.	748.	691.	774.	918.	1039.	1152.	826.	729.	0
	KT:	.42	.47	.51	.50	.52	.53	.59	.58	.54	.52	.42	.40	0
	AMB TEMP:	5.9	10.7	24.2	42.3	54.6	64.7	70.7	69.2	57.9	47.0	28.6	13.0	40.8
	HTG DEG DAYS:	1832	1520	1265	681	334	97	13	33	234	558	1092	1612	9271

STA NO:14918 INTERNATIONAL FALLS MN LAT = 48.57														
	HORIZ INSOL:	356.	663.	1046.	1444.	1716.	1853.	1921.	1618.	1121.	704.	346.	272.	1088.
	TILT = LAT:	758.	1153.	1391.	1488.	1497.	1509.	1615.	1558.	1331.	1092.	636.	598.	0
	TILT = LAT+15:	795.	1168.	1330.	1324.	1265.	1246.	1345.	1357.	1236.	1083.	655.	631.	0
	TILT = 90:	743.	1025.	1039.	877.	735.	683.	748.	848.	907.	916.	594.	596.	0
	KT:	.40	.47	.50	.50	.49	.50	.53	.52	.47	.44	.34	.36	0
	AMB TEMP:	1.9	7.0	20.6	38.2	50.1	60.4	65.8	63.2	53.0	43.5	24.9	8.7	36.5
	HTG DEG DAYS:	1956	1624	1376	804	462	168	66	112	364	667	1203	1745	10547

STA NO:14920 LA CROSSE WI LAT = 43.87														
	HORIZ INSOL:	481.	765.	1101.	1426.	1713.	1905.	1900.	1666.	1242.	864.	494.	370.	1161.
	TILT = LAT:	883.	1176.	1360.	1430.	1498.	1568.	1608.	1585.	1416.	1240.	836.	682.	0
	TILT = LAT+15:	926.	1188.	1298.	1277.	1273.	1302.	1348.	1386.	1318.	1233.	864.	717.	0
	TILT = 90:	838.	993.	945.	767.	655.	618.	657.	774.	894.	993.	762.	655.	0
	KT:	.42	.46	.48	.47	.48	.51	.52	.52	.49	.47	.39	.36	0
	AMB TEMP:	16.1	20.0	31.1	47.6	59.0	68.5	72.8	71.4	61.8	51.8	35.4	21.8	46.4
	HTG DEG DAYS:	1516	1260	1051	522	224	39	10	17	130	421	888	1339	7417

STA NO:14922 MINNEAPOLIS-ST. PAUL MN LAT = 44.88														
	HORIZ INSOL:	464.	764.	1104.	1442.	1737.	1928.	1970.	1687.	1255.	860.	480.	353.	1170.
	TILT = LAT:	885.	1214.	1387.	1455.	1518.	1582.	1664.	1610.	1450.	1267.	843.	679.	0
	TILT = LAT+15:	930.	1229.	1325.	1298.	1289.	1311.	1391.	1407.	1350.	1261.	873.	716.	0
	TILT = 90:	849.	1040.	980.	796.	680.	640.	693.	805.	933.	1030.	778.	660.	0
	KT:	.42	.48	.49	.48	.49	.51	.54	.53	.50	.48	.40	.37	0
	AMB TEMP:	12.2	16.5	28.3	45.1	57.1	66.9	71.9	70.2	60.0	50.0	32.4	18.6	44.1
	HTG DEG DAYS:	1637	1358	1138	597	271	65	11	21	173	472	978	1438	8159

TABLE A.4 Monthly Solar and Climatic Data for Locations in the Conterminous United States *(Continued)*

STA NO:14923	HORIZ INSOL:	535.	812.	1119.	1459.	1754.	1969.	1939.	1715.	1357.	996.	595.	433.	1224.
MOLINE	TILT = LAT:	912.	1183.	1338.	1452.	1537.	1629.	1647.	1625.	1529.	1398.	974.	751.	0
IL	TILT = LAT+15:	955.	1194.	1276.	1299.	1310.	1355.	1383.	1424.	1427.	1394.	1012.	790.	0
LAT = 41.45	TILT = 90:	846.	971.	894.	740.	628.	590.	622.	746.	926.	1096.	880.	711.	0
	KT:	.42	.46	.46	.47	.49	.52	.53	.53	.51	.51	.43	.38	0
	AMB TEMP:	21.5	25.7	35.7	50.6	61.1	70.8	74.5	72.9	64.6	54.4	39.2	26.6	49.8
	HTG DEG DAYS:	1349	1100	908	436	184	20	0	11	79	344	774	1190	6395
STA NO:14925	HORIZ INSOL:	477.	753.	1082.	1410.	1696.	1902.	1909.	1662.	1250.	870.	494.	370.	1156.
ROCHESTER	TILT = LAT:	874.	1154.	1333.	1413.	1483.	1565.	1615.	1581.	1428.	1253.	838.	685.	0
MN	TILT = LAT+15:	916.	1165.	1271.	1261.	1261.	1300.	1353.	1383.	1329.	1246.	867.	721.	0
LAT = 43.92	TILT = 90:	829.	973.	926.	759.	651.	618.	660.	774.	902.	1005.	765.	660.	0
	KT:	.41	.46	.47	.47	.48	.51	.52	.52	.49	.48	.39	.36	0
	AMB TEMP:	12.9	16.9	27.8	44.5	56.2	66.0	70.1	68.6	59.3	49.6	32.6	18.9	43.6
	HTG DEG DAYS:	1615	1347	1153	615	292	78	21	35	185	485	972	1429	8227
STA NO:14931	HORIZ INSOL:	579.	859.	1165.	1538.	1876.	2121.	2085.	1828.	1416.	1061.	664.	481.	1306.
BURLINGTON	TILT = LAT:	987.	1248.	1391.	1533.	1645.	1753.	1770.	1735.	1596.	1488.	1100.	843.	0
IA	TILT = LAT+15:	1036.	1261.	1329.	1373.	1401.	1454.	1484.	1521.	1491.	1488.	1147.	891.	0
LAT = 40.78	TILT = 90:	918.	1021.	922.	767.	647.	602.	639.	776.	956.	1163.	999.	803.	0
	KT:	.44	.47	.48	.50	.53	.57	.57	.56	.53	.54	.46	.40	0
	AMB TEMP:	22.9	27.3	36.9	51.3	61.8	71.4	75.4	73.9	65.4	55.3	39.8	27.6	50.8
	HTG DEG DAYS:	1305	1056	871	416	172	16	0	8	70	320	756	1159	6149
STA NO:14933	HORIZ INSOL:	581.	861.	1181.	1557.	1868.	2125.	2097.	1828.	1434.	1068.	658.	487.	1312.
DES MOINES	TILT = LAT:	1024.	1278.	1428.	1558.	1637.	1753.	1778.	1738.	1630.	1528.	1121.	891.	0
IA	TILT = LAT+15:	1077.	1293.	1365.	1394.	1393.	1452.	1489.	1522.	1524.	1528.	1171.	945.	0
LAT = 41.53	TILT = 90:	963.	1057.	959.	791.	659.	617.	656.	793.	991.	1208.	1028.	861.	0
	KT:	.45	.49	.49	.51	.52	.57	.57	.56	.54	.55	.47	.42	0
	AMB TEMP:	19.4	24.2	33.9	49.5	60.9	70.5	75.1	73.3	64.3	54.3	37.8	25.0	49.0
	HTG DEG DAYS:	1414	1142	964	465	186	26	0	13	94	350	816	1240	6710
STA NO:14935	HORIZ INSOL:	661.	917.	1265.	1692.	1972.	2242.	2216.	1939.	1509.	1138.	739.	569.	1405.
GRAND ISLAND	TILT = LAT:	1189.	1363.	1537.	1701.	1729.	1848.	1878.	1847.	1719.	1630.	1276.	1075.	0
NE	TILT = LAT+15:	1258.	1383.	1473.	1524.	1470.	1528.	1570.	1618.	1608.	1635.	1339.	1148.	0
LAT = 40.97	TILT = 90:	1129.	1127.	1028.	848.	674.	623.	666.	822.	1035.	1287.	1178.	1052.	0
	KT:	.50	.51	.52	.55	.55	.60	.61	.59	.57	.58	.52	.48	0
	AMB TEMP:	22.3	27.7	35.5	49.9	60.7	70.7	76.3	75.0	64.4	53.7	38.2	27.0	50.1
	HTG DEG DAYS	324	1044	915	461	184	35	6	5	107	362	804	1178	6425

STA NO:14936 HURON SD LAT = 44.38														
HORIZ INSOL:	488.	745.	1114.	1530.	1872.	2101.	2183.	1892.	1418.	988.	577.	405.	1276.
TILT = LAT:	926.	1155.	1391.	1549.	1638.	1722.	1843.	1817.	1663.	1493.	1063.	806.	0
TILT = LAT+15:	973.	1166.	1328.	1383.	1388.	1423.	1536.	1588.	1553.	1494.	1110.	855.	0
TILT = 90:	887.	979.	976.	837.	713.	668.	735.	887.	1065.	1222.	998.	793.	0
KT:	.43	.46	.49	.51	.53	.56	.60	.59	.56	.55	.47	.41	0
AMB TEMP:	12.5	17.9	29.0	45.8	57.0	67.1	73.7	72.1	60.7	49.6	32.4	19.2	44.8
HTG DEG DAYS:	1627	1319	1116	576	273	72	9	13	169	482	978	1420	8054

STA NO:14940 MASON CITY IA LAT = 43.15													
HORIZ INSOL:	554.	836.	1168.	1519.	1395.	2114.	2084.	1833.	1405.	1010.	600.	443.	1288.
TILT = LAT:	1035.	1291.	1444.	1527.	1660.	1737.	1764.	1750.	1622.	1485.	1059.	852.	0
TILT = LAT+15:	1092.	1309.	1381.	1365.	1409.	1437.	1474.	1531.	1514.	1485.	1105.	904.	0
TILT = 90:	990.	1091.	996.	804.	696.	647.	686.	831.	1015.	1196.	982.	832.	0
KT:	.46	.50	.50	.50	.53	.56	.57	.57	.55	.54	.46	.42	0
AMB TEMP:	14.2	18.5	29.0	45.7	57.4	67.2	71.3	69.9	60.2	50.5	33.6	20.1	44.8
HTG DEG DAYS:	1575	1302	1116	579	265	64	13	31	165	457	942	1392	7901

STA NO:14943 SIOUX CITY IA LAT = 42.40													
HORIZ INSOL:	569.	842.	1170.	1578.	1901.	2124.	2122.	1845.	1421.	1038.	643.	469.	1310.
TILT = LAT:	1035.	1273.	1431.	1587.	1666.	1748.	1797.	1759.	1629.	1507.	1125.	886.	0
TILT = LAT+15:	1091.	1289.	1368.	1419.	1415.	1447.	1503.	1539.	1522.	1507.	1176.	941.	0
TILT = 90:	983.	1064.	975.	821.	684.	634.	679.	819.	1006.	1204.	1041.	863.	0
KT:	.46	.49	.49	.52	.54	.57	.58	.57	.55	.55	.48	.43	0
AMB TEMP:	18.0	23.4	33.2	49.4	60.9	70.3	75.3	73.5	63.4	53.1	36.3	23.5	48.4
HTG DEG DAYS:	1457	1165	986	474	189	33	0	10	113	378	861	1287	6953

STA NO:14944 SIOUX FALLS SD LAT = 43.57													
HORIZ INSOL:	533.	802.	1152.	1543.	1894.	2100.	2150.	1845.	1410.	1005.	608.	441.	1290.
TILT = LAT:	1002.	1240.	1430.	1557.	1658.	1725.	1817.	1764.	1635.	1492.	1098.	867.	0
TILT = LAT+15:	1056.	1255.	1367.	1391.	1407.	1426.	1517.	1542.	1527.	1493.	1147.	921.	0
TILT = 90:	960.	1048.	992.	827.	704.	652.	709.	846.	1032.	1209.	1025.	852.	0
KT:	.45	.48	.50	.51	.54	.56	.59	.57	.55	.55	.48	.43	0
AMB TEMP:	14.2	19.4	30.0	46.1	57.7	67.6	73.3	71.8	60.9	50.2	33.1	20.0	45.4
HTG DEG DAYS:	1575	1277	1085	567	259	65	10	18	165	465	957	1395	7838

STA NO:14991 EAU CLAIRE WI LAT = 44.87													
HORIZ INSOL:	452.	746.	1090.	1426.	1681.	1872.	1886.	1621.	1196.	826.	451.	341.	1132.
TILT = LAT:	851.	1177.	1366.	1437.	1468.	1538.	1594.	1543.	1371.	1204.	768.	643.	0
TILT = LAT+15:	892.	1190.	1304.	1282.	1247.	1276.	1334.	1348.	1275.	1196.	792.	675.	0
TILT = 90:	813.	1006.	965.	787.	661.	627.	671.	774.	879.	973.	702.	621.	0
KT:	.41	.47	.48	.48	.48	.50	.52	.51	.48	.46	.37	.35	0
AMB TEMP:	11.7	15.4	27.3	44.5	56.2	66.1	70.5	68.4	58.7	48.7	32.0	18.0	43.1
HTG DEG DAYS:	1652	1389	1169	615	293	65	14	37	202	505	990	1457	8388

TABLE A.4 Monthly Solar and Climatic Data for Locations in the Conterminous United States *(Continued)*

STA NO:22010	HORIZ INSOL:	958.	1206.	1580.	1700.	1827.	2024.	2054.	1936.	1584.	1360.	1060.	903.	1516.
DEL RIO	TILT = LAT:	1314.	1494.	1747.	1655.	1632.	1732.	1789.	1819.	1663.	1636.	1429.	1281.	0
TX	TILT = LAT+15:	1386.	1516.	1684.	1498.	1409.	1457.	1518.	1611.	1563.	1638.	1498.	1367.	0
LAT = 29.37	TILT = 90:	1109.	1065.	947.	613.	450.	398.	424.	568.	781.	1088.	1170.	1132.	0
	KT:	.49	.51	.55	.51	.51	.55	.57	.57	.53	.54	.52	.50	0
	AMB TEMP:	50.8	55.7	62.6	72.0	78.2	84.3	86.7	86.1	80.2	71.2	59.6	52.3	70.0
	HTG DEG DAYS:	449	283	163	16	0	0	0	0	0	34	184	394	1523
STA NO:23023	HORIZ INSOL:	1081.	1383.	1839.	2192.	2430.	2562.	2389.	2210.	1844.	1522.	1176.	1000.	1802.
MIDLAND-ODESSA	TILT = LAT:	1627.	1833.	2118.	2165.	2142.	2142.	2052.	2081.	1989.	1935.	1724.	1565.	0
TX	TILT = LAT+15:	1735.	1876.	2050.	1951.	1822.	1770.	1723.	1833.	1872.	1950.	1824.	1689.	0
LAT = 31.93	TILT = 90:	1455.	1386.	1216.	809.	537.	431.	478.	672.	984.	1362.	1491.	1462.	0
	KT:	.60	.62	.66	.67	.68	.69	.66	.65	.63	.64	.62	.60	0
	AMB TEMP:	43.6	47.8	54.3	64.3	72.3	79.9	82.3	81.8	75.4	65.8	53.3	45.9	63.9
	HTG DEG DAYS:	663	482	349	98	0	0	0	0	0	81	356	592	2621
STA NO:23034	HORIZ INSOL:	962.	1208.	1606.	1851.	2031.	2186.	2123.	1966.	1607.	1337.	1044.	895.	1566.
SAN ANGELO	TILT = LAT:	1386.	1547.	1810.	1813.	1801.	1850.	1836.	1848.	1707.	1649.	1469.	1339.	0
TX	TILT = LAT+15:	1467.	1572.	1744.	1637.	1545.	1547.	1553.	1632.	1604.	1652.	1542.	1434.	0
LAT = 31.37	TILT = 90:	1207.	1140.	1023.	696.	501.	431.	461.	614.	839.	1134.	1237.	1219.	0
	KT:	.52	.53	.57	.56	.56	.59	.58	.58	.54	.55	.54	.52	0
	AMB TEMP:	46.4	50.4	57.1	67.2	74.5	81.6	84.7	84.5	76.8	67.2	55.5	48.3	66.2
	HTG DEG DAYS:	577	413	287	74	0	0	0	0	0	73	298	518	2240
STA NO:23042	HORIZ INSOL:	1031.	1332.	1762.	2168.	2396.	2544.	2412.	2208.	1820.	1468.	1116.	935.	1766.
LUBBOCK	TILT = LAT:	1613.	1813.	2056.	2151.	2108.	2119.	2063.	2083.	1986.	1910.	1694.	1522.	0
TX	TILT = LAT+15:	1721.	1855.	1988.	1937.	1792.	1750.	1729.	1833.	1868.	1924.	1792.	1641.	0
LAT = 33.65	TILT = 90:	1468.	1403.	1221.	852.	578.	473.	518.	717.	1024.	1379.	1491.	1441.	0
	KT:	.60	.62	.65	.67	.67	.68	.66	.66	.63	.64	.62	.59	0
	AMB TEMP:	39.1	42.7	48.9	60.0	68.5	77.1	79.7	78.4	71.0	61.0	48.8	41.3	59.7
	HTG DEG DAYS:	803	624	508	190	29	0	0	0	8	162	486	735	3545
STA NO:23043	HORIZ INSOL:	1047.	1373.	1807.	2218.	2459.	2610.	2441.	2242.	1913.	1527.	1131.	952.	1810.
ROSWELL	TILT = LAT:	1631.	1871.	2111.	2201.	2163.	2171.	2087.	2115.	2094.	1994.	1711.	1545.	0
NM	TILT = LAT+15:	1741.	1917.	2042.	1982.	1837.	1789.	1749.	1860.	1972.	2012.	1810.	1667.	0
LAT = 33.40	TILT = 90:	1483.	1448.	1249.	860.	575.	463.	512.	717.	1071.	1440.	1503.	1462.	0
	KT:	.61	.63	.66	.68	.68	.70	.67	.66	.66	.66	.62	.60	0
	AMB TEMP:	38.1	42.9	49.3	59.7	68.5	77.0	79.2	77.9	70.4	59.6	46.9	39.3	59.1
	HTG DEG DAYS:	834	619	487	185	20	0	0	0	17	195	543	797	3697

STA NO:23044	HORIZ INSOL:	1125.	1480.	1909.	2364.	2601.	2682.	2450.	2284.	1987.	1639.	1244.	1031.	1900.
EL PASO	TILT = LAT:	1703.	1984.	2208.	2341.	2287.	2235.	2102.	2151.	2157.	2110.	1842.	1620.	0
TX	TILT = LAT+15:	1820.	2037.	2140.	2109.	1938.	1839.	1763.	1894.	2034.	2134.	1955.	1751.	0
LAT = 31.80	TILT = 90:	1529.	1510.	1265.	850.	532.	416.	474.	681.	1059.	1494.	1603.	1517.	0
	KT:	.62	.66	.69	.72	.72	.72	.67	.67	.67	.68	.65	.61	0
	AMB TEMP:	43.6	48.4	54.6	63.9	72.2	80.3	82.3	80.5	74.2	64.0	51.6	44.4	63.4
	HTG DEG DAYS:	663	465	328	89	0	0	0	0	0	92	402	639	2678

STA NO:23047	HORIZ INSOL:	960.	1244.	1631.	2019.	2212.	2393.	2280.	2103.	1760.	1404.	1033.	872.	1659.
AMARILLO	TILT = LAT:	1552.	1726.	1916.	2007.	1947.	1993.	1950.	1985.	1938.	1864.	1611.	1474.	0
TX	TILT = LAT+15:	1654.	1763.	1849.	1806.	1659.	1651.	1637.	1747.	1821.	1876.	1701.	1588.	0
LAT = 35.23	TILT = 90:	1430.	1358.	1171.	845.	598.	511.	548.	733.	1037.	1375.	1434.	1412.	0
	KT:	.59	.60	.61	.63	.62	.64	.62	.63	.62	.63	.60	.58	0
	AMB TEMP:	36.0	39.7	45.6	56.5	65.6	74.6	78.7	77.6	69.8	59.5	46.3	38.5	57.4
	HTG DEG DAYS:	899	708	601	275	81	10	0	0	20	206	561	822	4183

STA NO:23048	HORIZ INSOL:	1009.	1297.	1712.	2098.	2314.	2484.	2349.	2164.	1829.	1443.	1073.	910.	1724.
TUCUMCARI	TILT = LAT:	1650.	1815.	2026.	2090.	2035.	2064.	2006.	2045.	2022.	1926.	1687.	1555.	0
NM	TILT = LAT+15:	1763.	1858.	1957.	1880.	1730.	1706.	1682.	1797.	1902.	1941.	1785.	1680.	0
LAT = 35.18	TILT = 90:	1530.	1434.	1240.	873.	609.	511.	551.	747.	1080.	1424.	1509.	1499.	0
	KT:	.62	.62	.64	.65	.64	.67	.64	.65	.64	.64	.62	.61	0
	AMB TEMP:	37.0	41.1	46.7	56.9	65.6	75.1	78.4	76.7	69.6	58.7	46.2	38.6	57.6
	HTG DEG DAYS:	868	669	567	260	57	7	0	0	20	217	564	818	4047

STA NO:23050	HORIZ INSOL:	1017.	1342.	1768.	2228.	2538.	2679.	2489.	2290.	1972.	1547.	1134.	928.	1827.
ALBUQUERQUE	TILT = LAT:	1659.	1887.	2098.	2227.	2227.	2214.	2120.	2167.	2198.	2091.	1804.	1587.	0
NM	TILT = LAT+15:	1773.	1935.	2030.	2003.	1886.	1819.	1772.	1903.	2071.	2114.	1913.	1715.	0
LAT = 35.05	TILT = 90:	1537.	1495.	1283.	914.	623.	502.	553.	772.	1168.	1556.	1623.	1530.	0
	KT:	.62	.64	.66	.69	.71	.72	.68	.68	.69	.69	.65	.62	0
	AMB TEMP:	35.2	40.0	45.8	55.8	65.3	74.6	78.7	76.6	70.1	58.2	44.5	36.2	56.8
	HTG DEG DAYS:	924	700	595	282	58	0	0	0	7	218	615	893	4292

STA NO:23051	HORIZ INSOL:	962.	1241.	1652.	2040.	2222.	2418.	2284.	2097.	1802.	1434.	1028.	861.	1670.
CLAYTON	TILT = LAT:	1623.	1770.	1974.	2038.	1953.	2006.	1948.	1984.	2010.	1960.	1664.	1520.	0
NM	TILT = LAT+15:	1734.	1811.	1906.	1832.	1662.	1659.	1634.	1743.	1890.	1977.	1760.	1641.	0
LAT = 36.45	TILT = 90:	1521.	1418.	1236.	887.	627.	539.	575.	762.	1105.	1479.	1506.	1478.	0
	KT:	.62	.61	.63	.64	.62	.65	.62	.63	.64	.66	.62	.60	0
	AMB TEMP:	33.1	36.1	40.4	50.8	60.0	69.2	73.6	72.4	65.0	54.8	42.3	35.1	52.7
	HTG DEG DAYS:	989	809	763	431	172	38	0	5	73	324	681	927	5212

TABLE A.4 Monthly Solar and Climatic Data for Locations in the Conterminous United States *(Continued)*

Station														
STA NO:23062	HORIZ INSOL:	840.	1127.	1530.	1879.	2135.	2351.	2273.	2044.	1727.	1301.	884.	732.	1568.
DENVER	TILT = LAT:	1550.	1709.	1890.	1894.	1873.	1938.	1929.	1946.	1980.	1871.	1539.	1420.	0
CO	TILT = LAT+15:	1655.	1748.	1821.	1698.	1591.	1601.	1613.	1705.	1859.	1884.	1625.	1531.	0
LAT = 39.75	TILT = 90:	1491.	1422.	1252.	908.	686.	610.	648.	830.	1167.	1471.	1429.	1411.	0
	KT:	.61	.61	.62	.60	.60	.63	.62	.62	.64	.64	.59	.59	0
	AMB TEMP:	29.9	32.8	37.0	47.5	57.0	66.0	73.0	71.6	62.8	52.0	39.4	32.6	50.1
	HTG DEG DAYS:	1088	902	868	525	253	80	0	0	120	408	768	1004	6016
STA NO:23063	HORIZ INSOL:	754.	1078.	1502.	1933.	2255.	2509.	2385.	2084.	1767.	1307.	869.	691.	1594.
EAGLE	TILT = LAT:	1338.	1612.	1845.	1951.	1979.	2063.	2021.	1985.	2030.	1878.	1500.	1308.	0
CO	TILT = LAT+15:	1421.	1644.	1777.	1750.	1677.	1697.	1686.	1739.	1907.	1892.	1582.	1407.	0
LAT = 39.65	TILT = 90:	1269.	1332.	1219.	931.	707.	621.	660.	841.	1195.	1476.	1387.	1290.	0
	KT:	.54	.58	.60	.62	.63	.67	.65	.63	.65	.64	.58	.55	0
	AMB TEMP:	18.0	23.3	31.1	41.9	51.3	58.9	65.9	63.7	55.6	44.8	30.9	20.3	42.2
	HTG DEG DAYS:	1457	1168	1051	693	425	190	43	79	285	626	1023	1386	8426
STA NO:23065	HORIZ INSOL:	789.	1056.	1424.	1829.	2062.	2357.	2319.	2046.	1643.	1268.	857.	695.	1529.
GOODLAND	TILT = LAT:	1405.	1558.	1726.	1836.	1810.	1945.	1968.	1945.	1862.	1796.	1457.	1301.	0
KS	TILT = LAT+15:	1494.	1588.	1660.	1647.	1539.	1607.	1645.	1705.	1746.	1806.	1535.	1398.	0
LAT = 39.37	TILT = 90:	1334.	1280.	1131.	875.	662.	602.	645.	821.	1088.	1400.	1340.	1278.	0
	KT:	.56	.56	.57	.58	.58	.63	.64	.62	.61	.62	.57	.55	0
	AMB TEMP:	27.6	31.5	36.3	48.7	58.9	69.1	75.8	74.1	64.3	52.8	38.5	30.1	50.6
	HTG DEG DAYS:	1159	938	890	489	216	55	0	0	108	387	795	1082	6119
STA NO:23066	HORIZ INSOL:	791.	1119.	1553.	1986.	2380.	2599.	2465.	2182.	1834.	1345.	918.	731.	1659.
GRAND JUNCTION	TILT = LAT:	1395.	1665.	1906.	2004.	2087.	2134.	2089.	2080.	2108.	1922.	1580.	1377.	0
CO	TILT = LAT+15:	1483.	1700.	1837.	1799.	1767.	1753.	1741.	1822.	1983.	1938.	1669.	1482.	0
LAT 39.12	TILT = 90:	1321.	1372.	1250.	940.	716.	612.	656.	859.	1227.	1502.	1461.	1357.	0
	KT:	.56	.59	.62	.63	.67	.69	.67	.66	.67	.65	.60	.57	0
	AMB TEMP:	26.6	33.6	41.2	51.7	62.2	71.3	78.7	75.4	67.2	54.9	39.8	29.5	52.7
	HTG DEG DAYS:	1190	879	738	404	133	20	0	0	60	324	756	1101	5605
STA NO:23090	HORIZ INSOL:	945.	1281.	1693.	2133.	2452.	2666.	2478.	2252.	1934.	1479.	1047.	837.	1766.
FARMINGTON	TILT = LAT:	1603.	1855.	2040.	2140.	2151.	2196.	2106.	2137.	2186.	2050.	1722.	1484.	0
NM	TILT = LAT+15:	1712.	1901.	1971.	1924.	1822.	1803.	1758.	1875.	2058.	2072.	1824.	1601.	0
LAT = 36.75	TILT = 90:	1505.	1499.	1287.	932.	663.	549.	596.	812.	1209.	1560.	1569.	1444.	0
	KT:	.61	.64	.65	.67	.68	.71	.68	.68	.69	.68	.64	.59	0
	AMB TEMP:	28.6	35.0	40.6	49.7	59.5	67.9	75.0	72.6	64.6	52.9	39.2	30.1	51.3
	HTG DEG DAYS:	1128	840	756	465	184	36	0	6	67	375	774	1082	5713

STA NO:23129	HORIZ INSOL:	928.	1215.	1610.	1938.	2064.	2140.	2300.	2100.	1701.	1326.	1004.	847.	1598.
LONG BEACH	TILT = LAT:	1420.	1629.	1659.	1913.	1823.	1800.	1971.	1979.	1845.	1699.	1491.	1350.	0
CA	TILT = LAT+15:	1507.	1660.	1792.	1723.	1560.	1504.	1657.	1743.	1734.	1704.	1568.	1449.	0
LAT = 33.82	TILT = 90:	1276.	1250.	1104.	780.	551.	474.	518.	698.	957.	1216.	1295.	1263.	0
	KT:	.54	.56	.59	.60	.57	.57	.63	.62	.59	.58	.56	.54	0
	AMB TEMP:	54.2	55.5	57.2	60.6	64.1	67.3	72.2	73.3	71.8	66.9	60.6	55.5	63.3
	HTG DEG DAYS:	339	273	247	148	71	23	0	0	7	48	155	295	1606

STA NO:23153	HORIZ INSOL:	918.	1274.	1777.	2251.	2577.	2788.	2703.	2438.	2043.	1521.	1031.	827.	1846.
TONOPAH	TILT = LAT:	1625.	1907.	2202.	2283.	2259.	2283.	2284.	2330.	2364.	2191.	1769.	1545.	0
NV	TILT = LAT+15:	1737.	1957.	2132.	2051.	1907.	1866.	1894.	2039.	2230.	2221.	1876.	1671.	0
LAT = 38.07	TILT = 90:	1546.	1571.	1429.	1023.	713.	582.	641.	901.	1347.	1711.	1638.	1527.	0
	KT:	.62	.65	.70	.71	.72	.74	.74	.74	.74	.72	.65	.62	0
	AMB TEMP:	30.2	34.6	39.6	48.1	56.9	65.3	73.0	70.7	63.5	52.1	39.8	31.9	50.5
	HTG DEG DAYS:	1079	851	787	512	269	92	0	13	108	407	756	1026	5900

STA NO:23154	HORIZ INSOL:	820.	1141.	1606.	2009.	2311.	2513.	2447.	2230.	1935.	1408.	926.	723.	1672.
ELY	TILT = LAT:	1470.	1714.	1987.	2030.	2027.	2067.	2073.	2129.	2249.	2045.	1609.	1365.	0
NV	TILT = LAT+15:	1567.	1752.	1918.	1822.	1717.	1701.	1728.	1865.	2118.	2067.	1701.	1470.	0
LAT = 39.28	TILT = 90:	1402.	1419.	1310.	955.	709.	612.	659.	879.	1316.	1611.	1493.	1347.	0
	KT:	.58	.61	.64	.64	.65	.67	.67	.68	.71	.69	.61	.57	0
	AMB TEMP:	23.8	27.9	32.8	41.3	50.0	57.7	67.2	65.5	56.7	46.0	34.0	26.2	44.1
	HTG DEG DAYS:	1283	1039	998	711	470	241	23	62	265	589	930	1203	7814

STA NO:23155	HORIZ INSOL:	766.	1102.	1595.	2095.	2509.	2749.	2684.	2421.	1992.	1458.	942.	677.	1749.
BAKERSFIELD	TILT = LAT:	1172.	1497.	1872.	2088.	2202.	2266.	2276.	2296.	2231.	1959.	1443.	1072.	0
CA	TILT = LAT+15:	1234.	1521.	1805.	1878.	1865.	1857.	1892.	2014.	2102.	1976.	1516.	1139.	0
LAT = 35.42	TILT = 90:	1048.	1163.	1147.	878.	631.	507.	562.	811.	1196.	1456.	1272.	991.	0
	KT:	.47	.53	.60	.65	.70	.74	.73	.72	.70	.65	.55	.46	0
	AMB TEMP:	47.5	52.4	56.6	62.7	69.8	76.9	83.9	81.6	76.6	66.9	56.0	47.9	64.9
	HTG DEG DAYS:	543	353	266	140	22	0	0	0	0	55	276	530	2185

STA NO:23159	HORIZ INSOL:	914.	1236.	1685.	2133.	2454.	2655.	2424.	2157.	1920.	1465.	1015.	818.	1740.
BRYCE CANYON	TILT = LAT:	1592.	1817.	2055.	2150.	2152.	2184.	2059.	2048.	2189.	2070.	1711.	1499.	0
UT	TILT = LAT+15:	1700.	1861.	1986.	1931.	1821.	1792.	1719.	1796.	2061.	2093.	1812.	1619.	0
LAT = 37.70	TILT = 90:	1507.	1483.	1319.	962.	689.	575.	616.	812.	1236.	1599.	1573.	1472.	0
	KT:	.61	.63	.66	.67	.68	.71	.66	.65	.69	.69	.64	.60	0
	AMB TEMP:	19.8	23.2	28.7	37.7	46.2	46.2	61.6	59.9	52.9	42.8	30.7	22.4	40.0
	HTG DEG DAYS:	1401	1170	1125	819	583	330	128	176	363	688	1029	1321	9133

TABLE A.4 Monthly Solar and Climatic Data for Locations in the Conterminous United States *(Continued)*

Station	Parameter													
STA NO:23160	HORIZ INSOL:	1099.	1432.	1864.	2363.	2671.	2730.	2341.	2183.	1979.	1602.	1208.	996.	1872.
TUCSON	TILT = LAT:	1669.	1918.	2156.	2343.	2345.	2269.	2012.	2055.	2152.	2064.	1792.	1567.	0
AZ	TILT = LAT+15:	1782.	1967.	2088.	2111.	1983.	1864.	1692.	1811.	2029.	2085.	1900.	1691.	0
LAT = 32.12	TILT = 90:	1500.	1462.	1243.	860.	538.	418.	482.	672.	1066.	1465.	1560.	1466.	0
	KT:	.61	.64	.67	.72	.74	.73	.64	.64	.67	.67	.64	.60	0
	AMB TEMP:	50.9	53.5	57.6	65.5	73.6	82.1	86.3	83.8	80.1	70.1	58.5	52.0	67.8
	HTG DEG DAYS:	442	333	243	61	0	0	0	0	0	29	221	403	1752
STA NO:23161	HORIZ INSOL:	958.	1281.	1772.	2274.	2591.	2766.	2603.	2383.	2008.	1516.	1085.	876.	1843.
DAGGETT	TILT = LAT:	1529.	1775.	2100.	2274.	2273.	2281.	2213.	2256.	2240.	2032.	1695.	1463.	0
CA	TILT = LAT+15:	1628.	1815.	2031.	2046.	1922.	1869.	1845.	1980.	2111.	2052.	1794.	1576.	0
LAT = 34.87	TILT = 90:	1401.	1393.	1279.	924.	620.	490.	549.	786.	1185.	1503.	1512.	1396.	0
	KT:	.58	.61	.66	.70	.72	.74	.71	.71	.70	.67	.62	.58	0
	AMB TEMP:	47.3	52.0	56.7	64.3	72.3	80.1	87.3	85.5	79.2	68.1	55.5	48.0	66.4
	HTG DEG DAYS:	549	371	271	118	14	0	0	0	0	57	296	527	2203
STA NO:23169	HORIZ INSOL:	978.	1340.	1823.	2319.	2646.	2778.	2588.	2355.	2037.	1540.	1086.	881.	1864.
LAS VEGAS	TILT = LAT:	1637.	1930.	2206.	2335.	2319.	2284.	2197.	2235.	2305.	2126.	1764.	1544.	0
NV	TILT = LAT+15:	1749.	1981.	2137.	2099.	1957.	1870.	1830.	1960.	2174.	2151.	1870.	1667.	0
LAT = 36.08	TILT = 90:	1530.	1553.	1380.	981.	658.	524.	582.	817.	1255.	1608.	1601.	1498.	0
	KT:	.62	.65	.69	.72	.74	.74	.71	.71	.72	.70	.64	.61	0
	AMB TEMP:	44.2	49.1	54.8	63.8	73.3	82.3	89.6	87.4	80.1	67.1	53.3	45.2	65.8
	HTG DEG DAYS:	645	451	324	126	10	0	0	0	0	74	357	614	2601
STA NO:23174	HORIZ INSOL:	926.	1214.	1619.	1951.	2060.	2119.	2307.	2080.	1681.	1317.	1004.	849.	1594.
LOS ANGELES	TILT = LAT:	1422.	1631.	1873.	1928.	1819.	1783.	1977.	1960.	1824.	1688.	1496.	1359.	0
CA	TILT = LAT+15:	1509.	1663.	1806.	1736.	1556.	1490.	1661.	1726.	1713.	1692.	1574.	1459.	0
LAT = 33.93	TILT = 90:	1280.	1253.	1115.	788.	552.	475.	521.	696.	948.	1210.	1302.	1274.	0
	KT:	.54	.56	.60	.60	.57	.57	.63	.62	.58	.57	.56	.54	0
	AMB TEMP:	54.5	55.6	56.5	58.8	61.9	64.5	68.5	69.5	68.7	65.2	60.5	56.9	61.7
	HTG DEG DAYS:	331	270	267	195	114	71	19	15	23	77	158	279	1819
STA NO:23179	HORIZ INSOL:	985.	1353.	1825.	2317.	2652.	2791.	2541.	2278.	2015.	1537.	1124.	913.	1861.
NEEDLES	TILT = LAT:	1578.	1895.	2170.	2319.	2324.	2301.	2163.	2154.	2246.	2064.	1767.	1538.	0
CA	TILT = LAT+15:	1683.	1943.	2102.	2086.	1964.	1883.	1806.	1892.	2118.	2085.	1873.	1660.	0
LAT = 34.77	TILT = 90:	1450.	1495.	1321.	935.	618.	484.	547.	762.	1185.	1527.	1582.	1474.	0
	KT:	.60	.64	.68	.72	.74	.75	.70	.68	.70	.68	.64	.60	0
	AMB TEMP:	51.6	56.5	61.6	70.4	79.6	88.3	95.4	93.3	86.9	74.3	60.7	52.7	72.6
	HTG DEG DAYS:	421	261	150	42	0	0	0	0	0	10	163	381	1428

STA NO:23183
PHOENIX
AZ
LAT = 33.43

HORIZ INSOL:	1021.	1374.	1814.	2355.	2677.	2739.	2486.	2293.	2015.	1576.	1150.	932.	1869.
TILT = LAT:	1584.	1875.	2121.	2346.	2347.	2269.	2124.	2164.	2220.	2073.	1749.	1505.	0
TILT = LAT+15:	1688.	1921.	2052.	2112.	1983.	1862.	1778.	1902.	2094.	2095.	1852.	1622.	0
TILT = 90:	1435.	1451.	1256.	902.	577.	453.	513.	727.	1135.	1504.	1541.	1421.	0
KT:	.59	.63	.67	.72	.74	.74	.68	.68	.70	.68	.63	.58	0
AMB TEMP:	51.2	55.1	59.7	67.7	76.3	84.6	91.2	89.1	83.8	72.2	59.8	52.5	70.3
HTG DEG DAYS:	428	292	185	60	0	0	0	0	0	17	182	388	1552

STA NO:23184
PRESCOTT
AZ
LAT = 34.65

HORIZ INSOL:	1016.	1335.	1777.	2275.	2629.	2762.	2309.	2092.	1955.	1543.	1140.	927.	1813.
TILT = LAT:	1636.	1858.	2101.	2273.	2305.	2279.	1976.	1973.	2168.	2068.	1794.	1561.	0
TILT = LAT+15:	1747.	1904.	2032.	2045.	1949.	1868.	1659.	1736.	2042.	2090.	1902.	1686.	0
TILT = 90:	1507.	1461.	1274.	917.	614.	484.	537.	716.	1142.	1528.	1606.	1497.	0
KT:	.61	.63	.66	.70	.73	.74	.63	.62	.68	.68	.65	.61	0
AMB TEMP:	37.1	40.5	44.3	52.0	60.4	69.1	75.5	73.0	68.1	57.2	45.8	38.6	55.1
HTG DEG DAYS:	865	686	642	394	165	33	0	0	23	254	576	818	4456

STA NO:23185
RENO
NV
LAT = 39.50

HORIZ INSOL:	800.	1150.	1649.	2159.	2523.	2701.	2692.	2406.	1998.	1431.	912.	706.	1761.
TILT = LAT:	1438.	1741.	2058.	2199.	2212.	2212.	2272.	2308.	2343.	2100.	1590.	1336.	0
TILT = LAT+15:	1531.	1781.	1989.	1974.	1867.	1811.	1883.	2019.	2209.	2125.	1680.	1437.	0
TILT = 90:	1371.	1447.	1365.	1032.	743.	625.	684.	940.	1378.	1665.	1477.	1318.	0
KT:	.57	.61	.66	.69	.71	.72	.74	.73	.74	.70	.61	.56	0
AMB TEMP:	31.9	37.1	40.3	46.8	54.6	61.5	69.3	66.9	60.2	50.3	40.1	33.0	49.4
HTG DEG DAYS:	1026	781	766	546	328	145	17	50	168	456	747	992	6022

STA NO:23188
SAN DIEGO
CA
LAT = 32.73

HORIZ INSOL:	976.	1266.	1632.	1937.	2003.	2062.	2186.	2057.	1717.	1373.	1063.	904.	1598.
TILT = LAT:	1465.	1676.	1866.	1907.	1773.	1744.	1882.	1936.	1851.	1738.	1554.	1415.	0
TILT = LAT+15:	1556.	1709.	1800.	1719.	1520.	1461.	1588.	1707.	1739.	1744.	1637.	1520.	0
TILT = 90:	1305.	1270.	1085.	753.	523.	451.	490.	663.	936.	1226.	1339.	1316.	0
KT:	.55	.57	.59	.59	.56	.55	.60	.61	.59	.58	.57	.55	0
AMB TEMP:	55.2	56.7	58.0	60.7	63.3	65.5	69.6	71.4	69.9	66.1	60.8	56.7	62.9
HTG DEG DAYS:	314	237	219	144	79	52	6	0	16	43	140	257	1507

STA NO:23194
WINSLOW
AZ
LAT = 35.02

HORIZ INSOL:	985.	1327.	1780.	2283.	2595.	2712.	2347.	2141.	1928.	1513.	1119.	894.	1802.
TILT = LAT:	1591.	1860.	2115.	2285.	2276.	2239.	2005.	2022.	2142.	2033.	1772.	1511.	0
TILT = LAT+15:	1697.	1906.	2046.	2056.	1924.	1838.	1681.	1778.	2017.	2053.	1879.	1630.	0
TILT = 90:	1466.	1470.	1292.	932.	624.	499.	547.	737.	1138.	1507.	1591.	1449.	0
KT:	.60	.63	.67	.71	.72	.73	.64	.64	.68	.67	.64	.59	0
AMB TEMP:	32.6	39.1	44.8	53.7	62.7	71.8	78.3	76.1	69.5	57.3	43.2	33.8	55.3
HTG DEG DAYS:	1004	725	626	348	124	14	0	0	19	252	654	967	4733

TABLE A.4 Monthly Solar and Climatic Data for Locations in the Conterminous United States *(Continued)*

STA NO:23195	HORIZ INSOL:	1096.	1443.	1919.	2413.	2728.	2814.	2453.	2329.	2051.	1623.	1215.	1000.	1924.
YUMA	TILT = LAT:	1691.	1958.	2243.	2401.	2392.	2329.	2101.	2196.	2250.	2117.	1832.	1604.	0
AZ	TILT = LAT+15:	1807.	2010.	2175.	2162.	2019.	1906.	1761.	1931.	2123.	2142.	1944.	1734.	0
LAT = 32.67	TILT = 90:	1532.	1507.	1310.	891.	551.	420.	495.	712.	1127.	1521.	1609.	1514.	0
	KT:	.62	.65	.70	.74	.76	.76	.67	.69	.70	.69	.65	.61	0
	AMB TEMP:	55.4	59.4	63.9	71.2	78.7	85.8	93.7	92.8	87.1	75.9	63.5	56.3	73.7
	HTG DEG DAYS:	308	192	97	24	0	0	0	0	0	5	108	276	1010
STA NO:23230	HORIZ INSOL:	708.	1018.	1456.	1922.	2211.	2350.	2322.	2053.	1701.	1212.	822.	647.	1535.
OAKLAND	TILT = LAT:	1144.	1429.	1735.	1923.	1942.	1947.	1976.	1945.	1907.	1640.	1305.	1102.	0
CA	TILT = LAT+15:	1205.	1451.	1668.	1727.	1651.	1611.	1654.	1708.	1790.	1644.	1367.	1175.	0
LAT = 37.73	TILT = 90:	1046.	1140.	1105.	874.	655.	564.	608.	782.	1077.	1240.	1167.	1049.	0
	KT:	.47	.52	.57	.61	.62	.63	.64	.62	.62	.57	.52	.48	0
	AMB TEMP:	48.6	51.9	53.7	56.1	58.9	61.9	63.1	63.5	64.5	61.1	55.3	49.9	57.4
	HTG DEG DAYS:	508	367	350	270	193	114	80	74	59	135	291	468	2909
STA NO:23232	HORIZ INSOL:	597.	939.	1458.	2004.	2435.	2684.	2688.	2368.	1907.	1315.	782.	538.	1643.
SACRAMENTO	TILT = LAT:	939.	1319.	1755.	2017.	2135.	2202.	2271.	2263.	2191.	1843.	1256.	890.	0
CA	TILT = LAT+15:	982.	1335.	1688.	1811.	1807.	1805.	1884.	1981.	2062.	1855.	1315.	940.	0
LAT = 38.52	TILT = 90:	846.	1054.	1134.	931.	708.	597.	654.	897.	1260.	1423.	1128.	832.	0
	KT:	.41	.49	.58	.64	.68	.72	.74	.72	.70	.63	.50	.41	0
	AMB TEMP:	45.1	49.8	53.0	58.3	64.3	70.5	75.2	74.1	71.5	63.3	53.0	45.8	60.3
	HTG DEG DAYS:	617	426	372	227	120	20	0	0	5	101	360	595	2843
STA NO:23234	HORIZ INSOL:	708.	1009.	1455.	1920.	2226.	2377.	2392.	2116.	1742.	1226.	821.	642.	1553.
SAN FRANCISCO	TILT = LAT:	1139.	1411.	1730.	1920.	1955.	1969.	2033.	2008.	1956.	1659.	1299.	1087.	0
CA	TILT = LAT+15:	1199.	1432.	1664.	1724.	1661.	1628.	1699.	1762.	1837.	1663.	1360.	1158.	0
LAT = 37.62	TILT = 90:	1039.	1123.	1099.	870.	654.	564.	612.	798.	1102.	1253.	1159.	1031.	0
	KT:	.47	.51	.57	.61	.62	.63	.65	.64	.63	.58	.51	.47	0
	AMB TEMP:	48.3	51.2	53.0	55.3	58.3	61.6	62.5	63.0	64.1	61.0	55.3	49.7	56.9
	HTG DEG DAYS:	518	386	372	291	210	120	93	84	66	137	291	474	3042
STA NO:23244	HORIZ INSOL:	738.	1037.	1485.	1944.	2277.	2453.	2441.	2167.	1760.	1248.	843.	660.	1588.
SUNNYVALE	TILT = LAT:	1193.	1453.	1767.	1944.	1999.	2029.	2074.	2057.	1975.	1689.	1334.	1118.	0
CA	TILT = LAT+15:	1259.	1476.	1701.	1746.	1698.	1675.	1732.	1804.	1855.	1694.	1399.	1192.	0
LAT = 37.42	TILT = 90:	1093.	1156.	1120.	875.	658.	563.	611.	807.	1108.	1274.	1192.	1062.	0
	KT:	.49	.52	.58	.61	.64	.65	.67	.65	.63	.58	.52	.48	0
	AMB TEMP:	0	0	0	0	0	0	0	0	0	0	0	0	0
	HTG DEG DAYS:	0	0	0	0	0	0	0	0	0	0	0	0	0

STA NO:23273	HORIZ INSOL:	854.	1141.	1582.	1921.	2141.	2349.	2341.	2106.	1730.	1353.	974.	804.	1608.
SANTA MARIA	TILT = LAT:	1323.	1544.	1844.	1902.	1886.	1960.	2000.	1988.	1896.	1773.	1481.	1312.	0
CA	TILT = LAT+15:	1400.	1571.	1777.	1712.	1609.	1626.	1678.	1748.	1781.	1781.	1557.	1407.	0
LAT = 34.90	TILT = 90:	1193.	1195.	1118.	801.	582.	503.	544.	726.	1007.	1295.	1301.	1237.	0
	KT:	.52	.54	.59	.60	.60	.63	.64	.63	.61	.60	.56	.53	0
	AMB TEMP:	50.5	52.0	52.8	54.9	57.1	59.6	62.1	62.3	62.6	60.4	56.1	51.8	56.9
	HTG DEG DAYS:	450	364	378	303	245	167	112	102	94	159	270	409	3053
STA NO:24011	HORIZ INSOL:	467.	776.	1168.	1459.	1848.	2060.	2184.	1877.	1354.	908.	507.	373.	1248.
BISMARCK	TILT = LAT:	998.	1326.	1538.	1489.	1616.	1681.	1840.	1817.	1626.	1438.	1005.	839.	0
ND	TILT = LAT+15:	1055.	1348.	1474.	1327.	1366.	1386.	1530.	1584.	1518.	1439.	1050.	894.	0
LAT = 46.77	TILT = 90:	984.	1168.	1126.	847.	750.	705.	789.	940.	1085.	1207.	960.	846.	0
	KT:	.47	.52	.54	.50	.53	.55	.60	.60	.56	.54	.46	.43	0
	AMB TEMP:	8.2	13.5	25.1	43.0	54.4	63.8	70.8	69.2	57.5	46.8	28.9	15.6	41.4
	HTG DEG DAYS:	1761	1442	1237	660	339	122	18	35	252	564	1083	1531	9044
STA NO:24013	HORIZ INSOL:	384.	656.	1044.	1461.	1846.	1975.	2098.	1600.	1277.	850.	439.	310.	1176.
MINOT	TILT = LAT:	830.	1123.	1380.	1505.	1614.	1607.	1765.	1748.	1549.	1386.	891.	715.	0
ND	TILT = LAT+15:	874.	1137.	1319.	1340.	1362.	1325.	1467.	1522.	1444.	1386.	928.	760.	0
LAT = 48.27	TILT = 90:	818.	993.	1026.	881.	778.	711.	797.	939.	1058.	1181.	855.	722.	0
	KT:	.42	.46	.50	.51	.53	.53	.58	.58	.54	.53	.43	.40	0
	AMB TEMP:	7.9	12.8	23.6	41.1	52.8	62.0	68.8	67.2	56.2	46.1	27.9	14.7	40.1
	HTG DEG DAYS:	1770	1462	1283	717	384	150	27	70	286	586	1113	1559	9407
STA NO:24018	HORIZ INSOL:	766.	1068.	1433.	1771.	1995.	2258.	2230.	1966.	1667.	1242.	823.	671.	1491.
CHEYENNE	TILT = LAT:	1458.	1662.	1786.	1787.	1749.	1860.	1890.	1874.	1931.	1828.	1482.	1355.	0
WY	TILT = LAT+15:	1555.	1698.	1718.	1601.	1486.	1537.	1579.	1641.	1812.	1840.	1564.	1460.	0
LAT = 41.15	TILT = 90:	1412.	1402.	1208.	891.	683.	629.	672.	837.	1170.	1461.	1390.	1357.	0
	KT:	.59	.60	.59	.57	.56	.60	.61	.60	.63	.63	.58	.57	0
	AMB TEMP:	26.6	29.0	31.6	42.7	52.4	61.3	69.1	67.6	58.2	47.9	35.5	29.2	45.9
	HTG DEG DAYS:	1190	1008	1035	669	394	156	22	31	225	530	885	1110	7255
STA NO:24023	HORIZ INSOL:	692.	958.	1333.	1724.	1988.	2267.	2277.	1990.	1565.	1177.	759.	605.	1445.
NORTH PLATTE	TILT = LAT:	1274.	1448.	1639.	1736.	1743.	1866.	1929.	1898.	1795.	1708.	1333.	1178.	0
NE	TILT = LAT+15:	1351.	1472.	1573.	1555.	1481.	1542.	1611.	1662.	1681.	1715.	1400.	1262.	0
LAT = 41.13	TILT = 90:	1219.	1207.	1103.	868.	681.	630.	680.	845.	1085.	1357.	1237.	1165.	0
	KT:	.53	.54	.55	.56	.56	.60	.62	.61	.59	.60	.54	.52	0
	AMB TEMP:	23.4	28.1	34.3	47.8	58.3	68.0	74.3	73.0	62.3	51.0	36.2	26.8	48.6
	HTG DEG DAYS:	1290	1033	952	522	238	65	7	8	141	439	864	1184	6743

TABLE A.4 Monthly Solar and Climatic Data for Locations in the Conterminous United States *(Continued)*

STA NO:24025	HORIZ INSOL:	530.	795.	1206.	1614.	1966.	2195.	2278.	1993.	1496.	1052.	623.	442.	1349.
PIERRE	TILT = LAT:	1039.	1258.	1532.	1643.	1722.	1797.	1923.	1920.	1770.	1616.	1179.	914.	0
SD	TILT = LAT+15:	1097.	1275.	1467.	1467.	1458.	1482.	1600.	1677.	1656.	1622.	1236.	974.	0
LAT = 44.38	TILT = 90:	1006.	1076.	1081.	886.	741.	687.	757.	932.	1137.	1331.	1117.	909.	0
	KT:	.47	.49	.53	.54	.56	.58	.63	.62	.59	.58	.51	.45	0
	AMB TEMP:	15.6	20.4	29.8	46.3	57.4	67.4	75.2	73.9	62.1	62.1	33.8	21.5	46.2
	HTG DEG DAYS:	1531	1249	1091	561	267	74	6	10	152	451	936	1349	7677
STA NO:24027	HORIZ INSOL:	735.	1089.	1530.	1944.	2344.	2574.	2547.	2240.	1833.	1306.	826.	651.	1635.
ROCK SPRINGS	TILT = LAT:	1409.	1729.	1945.	1983.	2056.	2106.	2151.	2155.	2170.	1973.	1519.	1330.	0
WY	TILT = LAT+15:	1501.	1769.	1876.	1777.	1737.	1727.	1785.	1684.	2041.	1993.	1604.	1433.	0
LAT = 41.60	TILT = 90:	1366.	1472.	1333.	992.	775.	677.	732.	952.	1331.	1597.	1433.	1335.	0
	KT:	.57	.62	.64	.63	.66	.69	.70	.69	.70	.67	.60	.57	C
	AMB TEMP:	19.2	23.4	28.9	40.1	50.4	58.9	68.2	66.1	56.4	44.7	30.7	22.6	42.5
	HTG DEG DAYS:	1420	1165	1119	747	453	198	18	49	269	629	1029	1314	8410
STA NO:24028	HORIZ INSOL:	676.	951.	1307.	1668.	1933.	2237.	2284.	2000.	1599.	1145.	723.	575.	1425.
SCOTTSBLUFF	TILT = LAT:	1275.	1464.	1619.	1681.	1694.	1840.	1932.	1912.	1855.	1681.	1288.	1141.	0
NE	TILT = LAT+15:	1353.	1489.	1554.	1505.	1440.	1520.	1612.	1673.	1738.	1687.	1352.	1222.	0
LAT = 41.87	TILT = 90:	1228.	1232.	1102.	856.	682.	641.	698.	868.	1138.	1346.	1200.	1132.	0
	KT:	.53	.54	.55	.54	.54	.60	.63	.62	.61	.59	.53	.51	0
	AMB TEMP:	24.9	29.5	34.3	46.2	56.5	65.9	73.7	71.6	61.2	50.2	36.2	27.6	48.2
	HTG DEG DAYS:	1243	994	952	564	280	91	0	8	160	459	864	1159	6774
STA NO:24029	HORIZ INSOL:	518.	788.	1205.	1537.	1883.	2156.	2329.	2006.	1502.	1005.	591.	441.	1330.
SHERIDAN	TILT = LAT:	1026.	1260.	1539.	1560.	1647.	1765.	1965.	1937.	1787.	1543.	1118.	934.	0
WY	TILT = LAT+15:	1083.	1277.	1475.	1393.	1396.	1456.	1633.	1691.	1672.	1546.	1171.	996.	0
LAT = 44.77	TILT = 90:	996.	1082.	1094.	850.	723.	687.	777.	948.	1156.	1272.	1058.	934.	0
	KT:	.47	.49	.53	.51	.53	.57	.64	.63	.60	.56	.49	.46	0
	AMB TEMP:	21.0	25.9	31.0	43.6	53.1	61.1	70.4	69.2	57.9	47.8	33.4	25.5	45.0
	HTG DEG DAYS:	1364	1095	1054	642	375	168	28	31	245	533	948	1225	7708
STA NO:24033	HORIZ INSOL:	486.	763.	1189.	1526.	1913.	2174.	2384.	2022.	1470.	987.	561.	421.	1325.
BILLINGS	TILT = LAT:	996.	1252.	1544.	1557.	1674.	1775.	2009.	1962.	1768.	1554.	1098.	935.	0
MT	TILT = LAT+15:	1052.	1269.	1480.	1389.	1416.	1462.	1667.	1712.	1653.	1558.	1150.	998.	0
LAT = 45.80	TILT = 90:	974.	1087.	1115.	867.	753.	712.	816.	985.	1164.	1298.	1048.	944.	0
	KT:	.46	.49	.54	.52	.54	.58	.66	.64	.59	.57	.49	.46	0
	AMB TEMP:	21.9	27.4	32.6	44.6	54.5	62.6	71.8	70.1	58.9	49.3	35.7	26.8	46.3
	HTG DEG DAYS:	1336	1053	1004	612	333	131	10	15	221	487	879	1184	7265

STA NO:24036 LEWISTOWN MT LAT = 47.05														
	HORIZ INSOL:	420.	692.	1128.	1444.	1807.	2059.	2288.	1901.	1372.	905.	502.	363.	1240.
	TILT = LAT:	873.	1151.	1482.	1475.	1579.	1679.	1927.	1845.	1658.	1445.	1007.	823.	0
	TILT = LAT+15:	919.	1165.	1419.	1314.	1335.	1384.	1599.	1609.	1548.	1446.	1052.	877.	0
	TILT = 90:	854.	1006.	1087.	843.	741.	710.	824.	960.	1113.	1218.	965.	832.	0
	KT:	.43	.47	.52	.49	.52	.55	.63	.61	.57	.54	.46	.43	0
	AMB TEMP:	19.1	23.8	27.5	40.1	49.6	56.6	65.5	64.4	54.0	45.5	32.2	24.5	41.9
	HTG DEG DAYS:	1423	1154	1163	747	477	265	70	94	348	605	984	1256	8586

STA NO:24037 MILES CITY MT LAT = 46.43														
	HORIZ INSOL:	457.	745.	1185.	1542.	1896.	2146.	2293.	1977.	1444.	961.	551.	399.	1300.
	TILT = LAT:	949.	1241.	1555.	1580.	1659.	1751.	1932.	1920.	1746.	1531.	1107.	905.	0
	TILT = LAT+15:	1000.	1259.	1491.	1410.	1402.	1442.	1604.	1674.	1632.	1535.	1160.	966.	0
	TILT = 90:	929.	1085.	1134.	891.	760.	719.	810.	981.	1162.	1287.	1063.	916.	0
	KT:	.45	.49	.54	.52	.54	.57	.63	.63	.59	.57	.49	.45	0
	AMB TEMP:	15.4	21.6	30.2	45.3	56.3	64.9	74.4	72.5	59.9	48.8	32.4	22.0	45.3
	HTG DEG DAYS:	1538	1215	1079	591	288	117	9	16	217	508	978	1333	7889

STA NO:24089 CASPER WY LAT = 42.92														
	HORIZ INSOL:	683.	1014.	1441.	1847.	2204.	2501.	2535.	2225.	1750.	1219.	765.	594.	1565.
	TILT = LAT:	1362.	1642.	1850.	1889.	1932.	2044.	2138.	2150.	2089.	1872.	1453.	1264.	0
	TILT = LAT+15:	1450.	1678.	1782.	1691.	1634.	1678.	1774.	1879.	1963.	1888.	1533.	1360.	0
	TILT = 90:	1331.	1412.	1291.	979.	775.	703.	767.	988.	1313.	1533.	1382.	1276.	0
	KT:	.57	.60	.61	.61	.62	.67	.70	.69	.68	.65	.58	.55	0
	AMB TEMP:	23.2	26.8	31.0	42.7	52.7	61.9	71.0	69.6	58.7	47.7	33.9	26.2	45.4
	HTG DEG DAYS:	1296	1070	1054	669	388	147	13	17	229	536	933	1203	7555

STA NO:24090 RAPID CITY SD LAT = 44.05														
	HORIZ INSOL:	542.	827.	1229.	1589.	1887.	2131.	2223.	1963.	1518.	1064.	647.	476.	1341.
	TILT = LAT:	1053.	1309.	1557.	1612.	1652.	1748.	1877.	1888.	1793.	1624.	1222.	995.	0
	TILT = LAT+15:	1112.	1328.	1492.	1440.	1401.	1444.	1564.	1649.	1678.	1630.	1282.	1063.	0
	TILT = 90:	1018.	1118.	1095.	864.	711.	668.	737.	909.	1145.	1333.	1157.	994.	0
	KT:	.47	.50	.53	.53	.53	.57	.61	.61	.60	.59	.52	.47	0
	AMB TEMP:	21.9	25.8	31.2	44.6	55.2	64.2	72.6	71.6	60.5	50.0	35.4	26.5	46.6
	HTG DEG DAYS:	1336	1098	1048	612	319	134	13	17	191	474	888	1194	7324

STA NO:24121 ELKO NV LAT = 40.83														
	HORIZ INSOL:	689.	1034.	1463.	1900.	2303.	2534.	2623.	2316.	1893.	1322.	812.	617.	1626.
	TILT = LAT:	1249.	1581.	1821.	1926.	2020.	2077.	2214.	2227.	2233.	1964.	1438.	1190.	0
	TILT = LAT+15:	1323.	1613.	1753.	1726.	1709.	1706.	1836.	1947.	2102.	1983.	1515.	1276.	0
	TILT = 90:	1189.	1324.	1227.	948.	746.	653.	719.	955.	1349.	1574.	1341.	1175.	0
	KT:	.52	.57	.60	.61	.65	.68	.72	.71	.71	.67	.57	.52	0
	AMB TEMP:	23.2	29.2	35.0	43.5	51.9	59.6	69.5	67.0	57.6	46.9	34.8	25.9	45.4
	HTG DEG DAYS:	1296	1002	930	645	406	190	27	60	248	561	906	1212	7483

TABLE A.4 Monthly Solar and Climatic Data for Locations in the Conterminous United States *(Continued)*

STA NO:24127	HORIZ INSOL:	639.	989.	1454.	1894.	2362.	2561.	2590.	2254.	1843.	1293.	788.	570.	1603.
SALT LAKE CITY	TILT = LAT:	1126.	1491.	1806.	1919.	2072.	2099.	2187.	2163.	2162.	1906.	1379.	1065.	0
UT	TILT = LAT+15:	1189.	1517.	1739.	1721.	1751.	1723.	1815.	1892.	2034.	1921.	1450.	1137.	0
LAT = 40.77	TILT = 90:	1061.	1240.	1215.	943.	756.	653.	713.	931.	1303.	1521.	1279.	1040.	0
	KT:	.48	.55	.60	.61	.66	.68	.71	.69	.69	.65	.55	.48	0
	AMB TEMP:	28.0	33.4	39.6	49.2	58.3	66.2	76.7	74.5	64.8	52.4	39.1	30.3	51.0
	HTG DEG DAYS:	1147	885	787	474	237	88	0	5	105	402	777	1076	5983
STA NO:24128	HORIZ INSOL:	691.	1028.	1472.	1967.	2362.	2569.	2678.	2348.	1907.	1322.	810.	618.	1648.
WINNEMUCCA	TILT = LAT:	1256.	1571.	1836.	2002.	2071.	2105.	2258.	2260.	2256.	1967.	1436.	1198.	0
NV	TILT = LAT+15:	1332.	1602.	1768.	1794.	1751.	1727.	1871.	1976.	2124.	1986.	1513.	1285.	0
LAT = 40.90	TILT = 90:	1198.	1315.	1239.	984.	760.	658.	725.	968.	1365.	1578.	1339.	1184.	0
	KT:	.52	.57	.61	.64	.66	.68	.73	.72	.72	.67	.57	.52	0
	AMB TEMP:	28.2	34.1	37.6	45.1	53.8	61.7	71.0	67.8	59.2	48.3	37.3	30.4	47.9
	HTG DEG DAYS:	1141	865	849	597	359	149	6	42	199	518	831	1073	6629
STA NO:24131	HORIZ INSOL:	485.	840.	1304.	1827.	2277.	2463.	2613.	2197.	1737.	1138.	628.	437.	1495.
BOISE	TILT = LAT:	880.	1315.	1659.	1874.	1997.	2012.	2202.	2126.	2091.	1746.	1149.	856.	0
ID	TILT = LAT+15:	922.	1334.	1593.	1676.	1666.	1652.	1823.	1857.	1964.	1756.	1203.	909.	0
LAT = 43.57	TILT = 90:	832.	1118.	1162.	986.	810.	714.	798.	996.	1330.	1433.	1078.	840.	0
	KT:	.41	.50	.56	.60	.64	.66	.72	.68	.68	.62	.49	.42	0
	AMB TEMP:	29.0	35.5	41.1	49.0	57.4	64.8	74.5	72.2	63.1	52.1	39.8	32.1	50.9
	HTG DEG DAYS:	1116	826	741	480	252	97	0	12	127	406	756	1020	5833
STA NO:24134	HORIZ INSOL:	490.	792.	1187.	1649.	2052.	2280.	2460.	2083.	1620.	1043.	594.	431.	1390.
BURNS	TILT = LAT:	893.	1220.	1482.	1674.	1798.	1867.	2076.	2008.	1924.	1563.	1064.	838.	0
OR	TILT = LAT+15:	936.	1234.	1419.	1497.	1522.	1539.	1723.	1754.	1804.	1566.	1111.	889.	0
LAT = 43.58	TILT = 90:	845.	1030.	1031.	886.	749.	685.	773.	949.	1221.	1271.	991.	821.	0
	KT:	.42	.48	.51	.55	.58	.61	.68	.65	.63	.57	.46	.42	0
	AMB TEMP:	25.2	31.0	36.1	44.2	52.2	59.0	68.4	66.1	58.2	47.3	35.8	27.9	46.0
	HTG DEG DAYS:	1234	952	896	624	402	205	30	68	226	549	876	1150	7212
STA NO:24137	HORIZ INSOL:	402.	688.	1128.	1485.	1883.	2045.	2287.	1897.	1352.	871.	480.	334.	1238.
CUT BANK	TILT = LAT:	911.	1215.	1530.	1536.	1647.	1663.	1925.	1853.	1667.	1449.	1036.	825.	0
MT	TILT = LAT+15:	962.	1233.	1466.	1368.	1389.	1368.	1595.	1614.	1556.	1452.	1085.	880.	0
LAT = 48.60	TILT = 90:	906.	1084.	1150.	905.	799.	736.	862.	1000.	1148.	1243.	1009.	844.	0
	KT:	.45	.49	.54	.52	.54	.55	.63	.61	.57	.55	.48	.44	0
	AMB TEMP:	16.2	22.4	26.8	39.5	49.6	56.5	64.4	62.6	53.2	44.1	29.7	21.4	40.5
	HTG DEG DAYS:	1513	1193	1184	765	477	267	82	125	368	648	1059	1352	9033

STA NO:24138 DILLON MT LAT = 45.25													
HORIZ INSOL:	527.	846.	1279.	1639.	1989.	2143.	2392.	2023.	1521.	1023.	602.	450.	1370.
TILT = LAT:	1080.	1406.	1670.	1679.	1742.	1753.	2017.	1959.	1827.	1601.	1175.	990.	0
TILT = LAT+15:	1142.	1431.	1604.	1499.	1473.	1446.	1674.	1710.	1710.	1607.	1233.	1059.	0
TILT = 90:	1057.	1225.	1201.	922.	766.	694.	804.	970.	1193.	1332.	1122.	1000.	0
KT:	.49	.54	.57	.55	.56	.57	.66	.64	.61	.58	.51	.48	0
AMB TEMP:	20.2	25.5	29.6	41.1	50.4	57.5	66.4	64.6	54.7	45.0	31.8	23.9	42.6
HTG DEG DAYS:	1389	1106	1097	717	453	238	54	85	325	620	996	1274	8354

STA NO:24143 GREAT FALLS MT LAT = 47.48													
HORIZ INSOL:	421.	720.	1170.	1489.	1848.	2101.	2329.	1933.	1379.	925.	498.	336.	1262.
TILT = LAT:	900.	1235.	1564.	1529.	1615.	1712.	1962.	1882.	1678.	1507.	1018.	759.	0
TILT = LAT+15:	948.	1253.	1500.	1363.	1365.	1409.	1626.	1640.	1567.	1511.	1065.	807.	0
TILT = 90:	885.	1091.	1158.	882.	764.	729.	846.	988.	1135.	1281.	981.	765.	0
KT:	.44	.50	.55	.51	.53	.56	.64	.62	.57	.56	.47	.41	0
AMB TEMP:	20.5	26.6	30.5	43.4	53.3	60.8	69.3	67.4	57.3	48.3	34.6	26.5	44.9
HTG DEG DAYS:	1380	1075	1070	648	367	162	18	42	260	524	912	1194	7652

STA NO:24144 HELENA MT LAT = 46.60													
HORIZ INSOL:	419.	709.	1146.	1487.	1850.	2040.	2334.	1930.	1412.	926.	521.	364.	1262.
TILT = LAT:	847.	1168.	1497.	1519.	1627.	1665.	1967.	1872.	1705.	1468.	1034.	801.	0
TILT = LAT+15:	890.	1182.	1433.	1354.	1376.	1374.	1631.	1632.	1593.	1470.	1081.	852.	0
TILT = 90:	823.	1018.	1091.	860.	751.	697.	824.	962.	1137.	1232.	989.	804.	0
KT:	.42	.47	.52	.51	.53	.54	.64	.61	.58	.55	.47	.42	0
AMB TEMP:	18.1	25.4	30.6	42.7	52.2	59.2	67.9	66.2	55.5	45.3	31.7	23.3	43.2
HTG DEG DAYS:	1454	1109	1066	669	401	194	33	57	304	611	999	1293	8190

STA NO:24149 LEWISTON ID LAT = 46.38													
HORIZ INSOL:	340.	609.	1020.	1435.	1842.	2015.	2336.	1931.	1435.	860.	413.	286.	1210.
TILT = LAT:	610.	944.	1292.	1458.	1611.	1646.	1968.	1871.	1731.	1324.	731.	545.	0
TILT = LAT+15:	632.	947.	1232.	1300.	1363.	1359.	1633.	1632.	1618.	1321.	754.	570.	0
TILT = 90:	572.	804.	930.	823.	741.	687.	819.	957.	1151.	1100.	676.	525.	0
KT:	.33	.40	.46	.49	.53	.54	.65	.61	.59	.51	.37	.32	0
AMB TEMP:	31.2	38.1	42.9	50.3	58.1	65.0	73.4	71.5	63.3	51.8	40.5	34.8	51.7
HTG DEG DAYS:	1048	753	685	441	232	84	0	17	124	409	735	936	5464

STA NO:24153 MISSOULA MT LAT = 46.92													
HORIZ INSOL:	312.	574.	982.	1382.	1783.	1933.	2327.	1881.	1358.	813.	410.	267.	1169.
TILT = LAT:	553.	888.	1245.	1403.	1557.	1579.	1961.	1823.	1634.	1251.	746.	508.	0
TILT = LAT+15:	570.	889.	1185.	1249.	1317.	1305.	1626.	1589.	1525.	1246.	771.	531.	0
TILT = 90:	515.	756.	901.	801.	730.	676.	831.	946.	1093.	1041.	696.	490.	0
KT:	.32	.39	.45	.47	.51	.52	.64	.60	.56	.49	.37	.31	0
AMB TEMP:	20.8	27.2	33.3	43.9	52.2	58.9	66.6	65.0	55.3	44.1	32.3	24.7	43.7
HTG DEG DAYS:	1370	1058	983	633	397	201	39	71	301	648	981	1249	7931

TABLE A.4 Monthly Solar and Climatic Data for Locations in the Conterminous United States *(Continued)*

STA NO:24155	HORIZ INSOL:	348.	614.	1044.	1503.	1926.	2144.	2396.	1994.	1502.	908.	438.	293.	1259.
PENDLETON	TILT = LAT:	607.	931.	1313.	1529.	1685.	1752.	2019.	1932.	1811.	1391.	770.	538.	0
OR	TILT = LAT+15:	627.	933.	1253.	1364.	1425.	1444.	1675.	1686.	1694.	1390.	795.	561.	0
LAT = 45.68	TILT = 90:	564.	785.	937.	850.	755.	703.	816.	968.	1191.	1150.	710.	514.	0
	KT:	.33	.40	.47	.51	.55	.57	.66	.63	.61	.52	.38	.32	0
	AMB TEMP:	32.0	38.9	43.8	50.9	58.5	65.6	73.5	71.5	64.0	52.6	41.4	35.7	52.4
	HTG DEG DAYS:	1023	731	657	423	220	70	6	13	97	384	708	908	5240
STA NO:24156	HORIZ INSOL:	539.	882.	1371.	1820.	2280.	2480.	2600.	2239.	1769.	1203.	689.	477.	1529.
POCATELLO	TILT = LAT:	987.	1373.	1743.	1859.	2000.	2027.	2192.	2165.	2118.	1841.	1262.	933.	0
ID	TILT = LAT+15:	1038.	1395.	1676.	1664.	1689.	1665.	1816.	1892.	1990.	1855.	1325.	993.	0
LAT = 42.92	TILT = 90:	938.	1163.	1212.	964.	794.	700.	777.	993.	1332.	1506.	1186.	918.	0
	KT:	.45	.52	.58	.60	.64	.66	.71	.69	.68	.64	.52	.45	0
	AMB TEMP:	23.2	29.4	35.4	45.3	54.4	61.8	71.5	69.5	59.4	48.4	35.7	26.9	46.7
	HTG DEG DAYS:	1296	997	918	591	336	138	0	20	192	515	879	1181	7063
STA NO:24157	HORIZ INSOL:	315.	606.	1041.	1495.	1918.	2083.	2357.	1942.	1435.	841.	398.	255.	1224.
SPOKANE	TILT = LAT:	589.	983.	1357.	1538.	1679.	1696.	1986.	1893.	1765.	1339.	743.	500.	0
WA	TILT = LAT+15:	610.	989.	1296.	1370.	1417.	1396.	1645.	1649.	1650.	1337.	768.	522.	0
LAT = 47.63	TILT = 90:	558.	853.	998.	889.	792.	727.	857.	997.	1199.	1129.	698.	485.	0
	KT:	.33	.42	.49	.51	.55	.56	.65	.62	.60	.51	.38	.31	0
	AMB TEMP:	25.4	32.2	37.5	46.1	54.7	61.5	69.7	68.0	59.6	47.8	35.5	29.0	47.3
	HTG DEG DAYS:	1228	918	853	567	327	144	21	47	196	533	885	1116	6835
STA NO:24172	HORIZ INSOL:	804.	1165.	1657.	2165.	2555.	2750.	2784.	2484.	2027.	1451.	929.	714.	1791.
LOVELOCK	TILT = LAT:	1482.	1800.	2087.	2213.	2240.	2246.	2346.	2393.	2403.	2172.	1666.	1396.	0
NV	TILT = LAT+15:	1580.	1844.	2018.	1985.	1888.	1836.	1939.	2092.	2267.	2201.	1765.	1505.	0
LAT = 40.07	TILT = 90:	1424.	1511.	1399.	1054.	769.	643.	705.	983.	1431.	1740.	1563.	1389.	0
	KT:	.59	.63	.67	.70	.72	.73	.76	.76	.75	.72	.63	.58	0
	AMB TEMP:	28.9	35.2	40.1	48.5	57.5	65.6	14.3	71.3	62.7	51.2	38.4	30.8	50.4
	HTG DEG DAYS:	1119	834	772	495	255	86	0	17	126	428	798	1060	5990
STA NO:24215	HORIZ INSOL:	561.	857.	1250.	1756.	2186.	2436.	2577.	2213.	1735.	1155.	659.	505.	1491.
MOUNT SHASTA	TILT = LAT:	967.	1264.	1523.	1773.	1917.	1999.	2176.	2125.	2027.	1676.	1114.	927.	0
CA	TILT = LAT+15:	1015.	1278.	1459.	1588.	1624.	1645.	1806.	1859.	1904.	1682.	1162.	985.	0
LAT = 41.32	TILT = 90:	902.	1042.	1024.	888.	732.	655.	728.	933.	1234.	1332.	1018.	897.	0
	KT:	.43	.48	.52	.57	.61	.65	.71	.68	.66	.59	.47	.44	0
	AMB TEMP:	33.6	37.8	40.4	46.3	53.3	60.0	67.8	66.0	61.2	51.4	41.7	35.5	49.6
	HTG DEG DAYS:	973	762	763	561	371	178	37	64	145	422	699	915	5890

Station														
STA NO:24216	HORIZ INSOL:	570.	893.	1354.	1910.	2375.	2600.	2672.	2311.	1845.	1228.	706.	511.	1581.
RED BLUFF	TILT = LAT:	941.	1289.	1646.	1931.	2083.	2131.	2255.	2216.	2148.	1757.	1167.	890.	0
CA	TILT = LAT+15:	985.	1304.	1580.	1731.	1761.	1749.	1869.	1939.	2021.	1766.	1219.	942.	0
LAT = 40.15	TILT = 90:	864.	1049.	1090.	934.	742.	640.	702.	931.	1279.	1381.	1058.	847.	0
	KT:	.42	.48	.55	.61	.67	.69	.73	.71	.69	.61	.48	.42	0
	AMB TEMP:	45.2	50.0	53.2	59.5	67.4	75.5	82.3	79.9	75.3	65.0	53.7	46.4	62.8
	HTG DEG DAYS:	614	420	366	218	64	8	0	0	0	82	339	577	2688
STA NO:24225	HORIZ INSOL:	407.	737.	1133.	1639.	2034.	2278.	2475.	2121.	1589.	982.	504.	337.	1353.
MEDFORD	TILT = LAT:	644.	1072.	1375.	1653.	1782.	1870.	2090.	2038.	1852.	1404.	806.	544.	0
OR	TILT = LAT+15:	664.	1078.	1313.	1479.	1511.	1543.	1737.	1782.	1736.	1401.	830.	563.	0
LAT = 42.37	TILT = 90:	578.	880.	934.	852.	718.	658.	743.	929.	1148.	1115.	719.	498.	0
	KT:	.33	.43	.48	.54	.57	.61	.68	.66	.61	.52	.38	.31	0
	AMB TEMP:	36.6	41.3	44.8	50.2	57.3	64.3	71.7	70.4	64.4	53.4	43.5	37.7	53.0
	HTG DEG DAYS:	880	664	626	444	250	94	11	21	89	360	645	846	4930
STA NO:24227	HORIZ INSOL:	269.	503.	845.	1255.	1632.	1693.	1913.	1549.	1157.	636.	339.	222.	1001.
OLYMPIA	TILT = LAT:	438.	738.	1033.	1260.	1422.	1386.	1611.	1478.	1352.	906.	562.	375.	0
WA	TILT = LAT+15:	446.	733.	978.	1121.	1206.	1151.	1345.	1289.	1256.	891.	573.	385.	0
LAT = 46.97	TILT = 90:	396.	617.	738.	721.	678.	615.	716.	779.	898.	733.	508.	347.	0
	KT:	.27	.34	.39	.43	.47	.45	.53	.49	.48	.38	.31	.26	0
	AMB TEMP:	37.2	41.0	43.2	48.2	54.0	58.9	63.6	62.8	58.6	50.6	43.3	39.5	50.1
	HTG DEG DAYS:	862	672	676	504	341	197	89	103	198	446	651	791	5530
STA NO:24229	HORIZ INSOL:	310.	554.	895.	1308.	1663.	1772.	2037.	1674.	1217.	724.	388.	260.	1067.
PORTLAND	TILT = LAT:	505.	806.	1085.	1310.	1452.	1455.	1719.	1600.	1411.	1034.	639.	441.	0
OR	TILT = LAT+15:	517.	803.	1026.	1167.	1232.	1209.	1434.	1397.	1313.	1021.	654.	455.	0
LAT = 45.60	TILT = 90:	458.	669.	762.	730.	667.	615.	724.	813.	918.	832.	577.	410.	0
	KT:	.29	.36	.40	.44	.47	.47	.56	.53	.49	.42	.33	.28	0
	AMB TEMP:	38.1	42.8	45.7	50.6	56.7	62.0	67.1	66.6	62.2	53.8	45.3	40.7	52.6
	HTG DEG DAYS:	834	622	598	432	264	128	48	56	119	347	591	753	4792
STA NO:24230	HORIZ INSOL:	491.	775.	1190.	1683.	2080.	2287.	2446.	2069.	1584.	999.	572.	425.	1383.
REDMOND	TILT = LAT:	928.	1212.	1504.	1719.	1822.	1871.	2063.	1999.	1891.	1510.	1044.	856.	0
OR	TILT = LAT+15:	975.	1226.	1440.	1536.	1541.	1540.	1712.	1745.	1772.	1511.	1090.	909.	0
LAT = 44.27	TILT = 90:	888.	1031.	1059.	923.	772.	702.	789.	962.	1214.	1235.	978.	846.	0
	KT:	.43	.48	.52	.56	.59	.61	.67	.65	.63	.55	.46	.43	0
	AMB TEMP:	30.2	35.8	38.6	38.6	51.3	58.2	65.7	63.8	57.7	48.4	39.0	33.4	47.2
	HTG DEG DAYS:	1079	818	818	618	425	220	55	102	233	515	780	980	6643

TABLE A.4 Monthly Solar and Climatic Data for Locations in the Conterminous United States *(Continued)*

STA NO:24232	HORIZ INSOL:	332.	588.	947.	1370.	1738.	1849.	2142.	1775.	1328.	769.	410.	277.	1127.
SALEM	TILT = LAT:	540.	856.	1151.	1375.	1519.	1519.	1808.	1700.	1551.	1099.	672.	467.	0
OR	TILT = LAT+15:	554.	854.	1093.	1227.	1289.	1261.	1507.	1485.	1446.	1088.	689.	483.	0
LAT = 44.92	TILT = 90:	490.	709.	804.	755.	680.	623.	737.	846.	1000.	882.	605.	434.	0
	KT:	.30	.37	.42	.46	.49	.49	.59	.56	.53	.43	.34	.29	0
	AMB TEMP:	38.8	42.9	45.2	49.8	55.7	61.2	66.6	66.1	61.9	53.2	45.2	40.9	52.3
	HTG DEG DAYS:	812	619	614	456	295	133	43	53	120	366	594	747	4852
STA NO:24233	HORIZ INSOL:	262.	495.	849.	1294.	1714.	1802.	2248.	1616.	1148.	656.	337.	211.	1053.
SEATTLE-TACOMA	TILT = LAT:	433.	735.	1048.	1306.	1495.	1472.	1893.	1550.	1348.	958.	573.	359.	0
WA	TILT = LAT+15:	441.	730.	993.	1162.	1265.	1218.	1572.	1351.	1252.	945.	585.	368.	0
LAT = 47.45	TILT = 90:	394.	618.	755.	754.	715.	651.	822.	823.	902.	784.	522.	333.	0
	KT:	.27	.34	.40	.44	.49	.48	.62	.52	.48	.40	.32	.26	0
	AMB TEMP:	38.2	42.3	44.1	48.7	54.9	59.8	64.5	63.8	59.6	52.2	44.6	40.5	51.1
	HTG DEG DAYS:	831	636	648	489	313	167	80	82	170	397	612	760	5165
STA NO:24243	HORIZ INSOL:	365.	666.	1122.	1598.	2009.	2169.	2358.	1975.	1483.	891.	444.	295.	1281.
YAKIMA	TILT = LAT:	689.	1074.	1459.	1645.	1759.	1769.	1987.	1919.	1806.	1394.	823.	579.	0
WA	TILT = LAT+15:	717.	1083.	1396.	1468.	1485.	1456.	1648.	1673.	1690.	1393.	853.	607.	0
LAT = 46.57	TILT = 90:	655.	928.	1061.	929.	801.	727.	830.	983.	1206.	1165.	771.	563.	0
	KT:	.36	.44	.51	.54	.57	.58	.65	.63	.61	.53	.40	.34	0
	AMB TEMP:	27.5	35.7	41.8	49.5	57.9	64.5	70.7	68.6	61.3	50.1	38.4	31.3	49.8
	HTG DEG DAYS:	1163	820	719	465	239	94	20	37	147	462	798	1045	6009
STA NO:24255	HORIZ INSOL:	283.	532.	918.	1345.	1760.	1820.	1981.	1593.	1173.	655.	357.	233.	1054.
WHIDBEY ISLAND	TILT = LAT:	523.	843.	1176.	1372.	1537.	1483.	1666.	1530.	1401.	984.	659.	457.	0
WA	TILT = LAT+15:	540.	844.	1118.	1220.	1298.	1226.	1387.	1333.	1303.	972.	679.	477.	0
LAT = 48.35	TILT = 90:	493.	727.	866.	806.	748.	670.	763.	830.	954.	816.	616.	443.	0
	KT:	.31	.38	.44	.47	.51	.49	.55	.51	.50	.41	.35	.30	0
	AMB TEMP:	0	0	0	0	0	0	0	0	0	0	0	0	0
	HTG DEG DAYS:	0	0	0	0	0	0	0	0	0	0	0	0	0
STA NO:24283	HORIZ INSOL:	529.	793.	1133.	1587.	1843.	1962.	1808.	1579.	1342.	936.	593.	470.	1214.
ARCATA	TILT = LAT:	879.	1133.	1349.	1586.	1616.	1625.	1539.	1489.	1503.	1280.	953.	824.	0
CA	TILT = LAT+15:	918.	1141.	1288.	1421.	1376.	1353.	1297.	1307.	1403.	1272.	988.	870.	0
LAT = 40.98	TILT = 90:	808.	920.	895.	795.	643.	581.	587.	685.	903.	989.	853.	784.	0
	KT:	.40	.44	.47	.51	.52	.52	.50	.48	.51	.47	.42	.40	0
	AMB TEMP:	0	0	0	0	0	0	0	0	0	0	0	0	0
	HTG DEG DAYS:	0	0	0	0	0	0	0	0	0	0	0	0	0

STA NO:24284	HORIZ INSOL:	439.	705.	1058.	1510.	1857.	1994.	2108.	1786.	1377.	893.	525.	381.	1219.
NORTH BEND	TILT = LAT:	755.	1041.	1287.	1519.	1626.	1641.	1782.	1704.	1589.	1279.	891.	694.	0
OR	TILT = LAT+15:	786.	1046.	1227.	1358.	1380.	1360.	1489.	1490.	1483.	1272.	923.	730.	0
LAT = 43.42	TILT = 90:	701.	864.	885.	805.	690.	629.	697.	817.	999.	1020.	814.	665.	0
	KT:	.37	.42	.45	.50	.52	.53	.58	.56	.54	.48	.41	.36	0
	AMB TEMP:	44.6	46.6	46.9	49.1	53.1	56.9	59.0	59.7	58.4	54.9	50.1	46.5	52.2
	HTG DEG DAYS:	632	515	561	477	369	243	188	168	201	313	447	574	4688

STA NO:25704	HORIZ INSOL:	231.	433.	716.	1033.	1180.	1182.	1120.	949.	759.	528.	308.	187.	719.
ADAK	TILT = LAT:	507.	734.	920.	1034.	1010.	955.	927.	870.	860.	830.	677.	454.	0
AK	TILT = LAT+15:	527.	734.	668.	914.	854.	797.	779.	754.	788.	818.	702.	477.	0
LAT = 51.88	TILT = 90:	496.	651.	699.	646.	557.	508.	504.	517.	603.	708.	657.	457.	0
	KT:	.32	.36	.37	.37	.34	.32	.31	.31	.34	.37	.37	.32	0
	AMB TEMP:	0	0	0	0	0	0	0	0	0	0	0	0	0
	HTG DEG DAYS:	0	0	0	0	0	0	0	0	0	0	0	0	0

STA NO:93037	HORIZ INSOL:	891.	1178.	1550.	1931.	2129.	2369.	2212.	2025.	1759.	1359.	944.	782.	1594.
COLORADO SPRINGS	TILT = LAT:	1610.	1762.	1892.	1942.	1869.	1957.	1881.	1923.	2002.	1933.	1621.	1484.	0
CO	TILT = LAT+15:	1721.	1803.	1824.	1743.	1589.	1617.	1576.	1687.	1881.	1949.	1714.	1603.	0
LAT = 38.82	TILT = 90:	1542.	1454.	1234.	907.	664.	590.	618.	800.	1157.	1505.	1498.	1470.	0
	KT:	.62	.62	.62	.61	.60	.63	.61	.61	.64	.65	.61	.60	0
	AMB TEMP:	28.6	31.3	35.3	46.2	55.5	64.6	70.7	69.1	60.9	50.5	37.5	31.0	48.4
	HTG DEG DAYS:	1128	944	921	564	301	103	9	13	155	456	825	1054	6473

STA NO:93044	HORIZ INSOL:	986.	1297.	1688.	2167.	2473.	2602.	2264.	2078.	1895.	1496.	1088.	893.	1744.
ZUNI	TILT = LAT:	1599.	1812.	1990.	2162.	2172.	2155.	1937.	1962.	2102.	2009.	1714.	1512.	0
NM	TILT = LAT+15:	1706.	1855.	1922.	1945.	1841.	1775.	1627.	1726.	1979.	2028.	1814.	1631.	0
LAT = 35.10	TILT = 90:	1476.	1430.	1215.	895.	620.	507.	544.	724.	1120.	1490.	1534.	1451.	0
	KT:	.60	.62	.63	.67	.69	.70	.62	.62	.67	.67	.63	.59	0
	AMB TEMP:	30.3	34.6	39.6	48.1	56.6	65.4	71.4	69.4	63.3	52.5	40.1	32.0	50.3
	HTG DEG DAYS:	1076	851	787	507	264	68	0	13	91	388	747	1023	5815

STA NO:93045	HORIZ INSOL:	1118.	1451.	1886.	2338.	2557.	2650.	2365.	2216.	1940.	1579.	1217.	1003.	1860.
TRUTH OR CONSEQUENCE	TILT = LAT:	1764.	1996.	2213.	2326.	2246.	2202.	2026.	2089.	2124.	2069.	1867.	1641.	0
NM	TILT = LAT+15:	1890.	2051.	2145.	2095.	1904.	1813.	1701.	1838.	2001.	2091.	1982.	1776.	0
LAT = 33.23	TILT = 90:	1616.	1552.	1307.	890.	573.	456.	507.	707.	1082.	1496.	1654.	1562.	0
	KT:	.64	.66	.69	.72	.71	.71	.65	.66	.67	.68	.66	.62	0
	AMB TEMP:	40.0	44.9	50.2	59.5	59.5	76.9	79.3	77.4	71.6	61.3	48.7	40.8	59.9
	HTG DEG DAYS:	775	563	459	188	19	0	0	0	5	144	489	750	3392

TABLE A.4 Monthly Solar and Climatic Data for Locations in the Conterminous United States *(Continued)*

Station														
STA NO:93058	HORIZ INSOL:	894.	1172.	1564.	1956.	2162.	2434.	2312.	2102.	1779.	1361.	954.	782.	1623.
PUEBLO	TILT = LAT:	1584.	1725.	1898.	1964.	1899.	2010.	1965.	1996.	2018.	1913.	1610.	1449.	0
CO	TILT = LAT+15:	1691.	1764.	1830.	1763.	1614.	1659.	1644.	1751.	1896.	1928.	1702.	1563.	0
LAT = 38.28	TILT = 90:	1506.	1412.	1226.	903.	659.	583.	619.	811.	1153.	1478.	1480.	1426.	0
	KT:	.61	.60	.62	.62	.60	.65	.63	.64	.65	.65	.61	.59	0
	AMB TEMP:	30.1	34.7	40.0	51.7	61.1	70.7	76.4	74.5	66.2	54.5	40.8	33.0	52.8
	HTG DEG DAYS:	1082	848	775	405	148	28	0	0	55	335	726	992	5394
STA NO:93101	HORIZ INSOL:	947.	1236.	1610.	1928.	2070.	2194.	2363.	2155.	1737.	1357.	1026.	869.	1625.
EL TORO	TILT = LAT:	1451.	1658.	1856.	1903.	1829.	1844.	2024.	2032.	1887.	1740.	1527.	1389.	0
CA	TILT = LAT+15:	1541.	1691.	1790.	1714.	1564.	1538.	1698.	1788.	1773.	1747.	1607.	1492.	0
LAT = 33.67	TILT = 90:	1304.	1271.	1099.	773.	548.	473.	517.	706.	974.	1246.	1327.	1301.	0
	KT:	.55	.57	.59	.59	.58	.59	.65	.64	.60	.59	.57	.55	0
	AMB TEMP:	0	0	0	0	0	0	0	0	0	0	0	0	0
	HTG DEG DAYS:	0	0	0	0	0	0	0	0	0	0	0	0	0
STA NO:93104	HORIZ INSOL:	909.	1229.	1735.	2233.	2549.	2747.	2612.	2616.	1980.	1473.	1034.	841.	1830.
CHINA LAKE	TILT = LAT:	1470.	1720.	2070.	2238.	2235.	2263.	2218.	2491.	2221.	1994.	1635.	1431.	0
CA	TILT = LAT+15:	1564.	1757.	2001.	2013.	1891.	1854.	1847.	2180.	2093.	2012.	1728.	1541.	0
LAT = 35.68	TILT = 90:	1353.	1360.	1280.	936.	641.	515.	571.	855.	1198.	1490.	1465.	1373.	0
	KT:	.57	.60	.66	.70	.71	.73	.71	.78	.70	.66	.61	.57	0
	AMB TEMP:	0	0	0	0	0	0	0	0	0	0	0	0	0
	HTG DEG DAYS:	0	0	0	0	0	0	0	0	0	0	0	0	0
STA NO:93111	HORIZ INSOL:	927.	1220.	1636.	1951.	2018.	2055.	2118.	1935.	1608.	1296.	1006.	856.	1552.
POINT MUGU	TILT = LAT:	1432.	1647.	1899.	1929.	1782.	1731.	1820.	1822.	1738.	1661.	1509.	1384.	0
CA	TILT = LAT+15:	1521.	1679.	1832.	1737.	1525.	1446.	1536.	1606.	1631.	1665.	1588.	1486.	0
LAT = 34.12	TILT = 90:	1292.	1270.	1135.	792.	551.	474.	513.	666.	909.	1192.	1317.	1302.	0
	KT:	.55	.57	.61	.60	.56	.55	.58	.58	.56	.57	.56	.55	0
	AMB TEMP:	0	0	0	0	0	0	0	0	0	0	0	0	0
	HTG DEG DAYS:	0	0	0	0	0	0	0	0	0	0	0	0	0
STA NO:93129	HORIZ INSOL:	682.	1180.	1636.	2092.	2467.	2706.	2503.	2241.	1968.	1460.	992.	786.	1743.
CEDAR CITY	TILT = LAT:	1521.	1715.	1984.	2106.	2164.	2222.	2123.	2131.	2253.	2061.	1660.	1421.	0
UT	TILT = LAT+15:	1621.	1752.	1916.	1892.	1831.	1822.	1770.	1869.	2123.	2083.	1757.	1531.	0
LAT = 37.70	TILT = 90:	1433.	1392.	1272.	945.	690.	574.	622.	836.	1273.	1591.	1522.	1388.	0
	KT:	.59	.60	.64	.66	.69	.72	.68	.68	.71	.69	.62	.58	0
	AMB TEMP:	28.7	33.1	38.4	47.1	56.2	65.0	73.2	71.3	63.2	51.5	38.8	30.8	49.8
	HTG DEG DAYS:	1125	893	825	537	281	86	0	6	114	424	786	1060	6137

STA NO:93193	HORIZ INSOL:	657.	1012.	1566.	2093.	2484.	2733.	2685.	2423.	1985.	1429.	889.	574.	1711.
FRESNO	TILT = LAT:	1003.	1390.	1863.	2097.	2179.	2247.	2273.	2306.	2253.	1966.	1397.	905.	0
CA	TILT = LAT+15:	1049.	1408.	1796.	1885.	1844.	1841.	1888.	2021.	2123.	1983.	1467.	955.	0
LAT = 36.77	TILT = 90:	892.	1091.	1170.	917.	667.	547.	602.	854.	1246.	1490.	1245.	831.	0
	KT:	.42	.50	.60	.66	.69	.73	.73	.73	.71	.66	.54	.41	0
	AMB TEMP:	45.3	49.9	53.9	60.3	67.4	73.9	80.6	78.3	73.8	64.2	53.5	45.8	62.3
	HTG DEG DAYS:	611	423	344	182	51	9	0	0	0	90	345	595	2650

STA NO:93721	HORIZ INSOL:	587.	840.	1162.	1488.	1714.	1879.	1823.	1600.	1330.	998.	660.	499.	1215.
BALTIMORE	TILT = LAT:	942.	1165.	1359.	1471.	1507.	1567.	1558.	1506.	1464.	1333.	1031.	825.	0
MD	TILT = LAT+15:	985.	1173.	1298.	1319.	1289.	1310.	1315.	1324.	1367.	1326.	1071.	870.	0
LAT = 39.18	TILT = 90:	855.	927.	876.	712.	582.	538.	562.	661.	849.	1009.	913.	770.	0
	KT:	.42	.44	.46	.47	.48	.50	.50	.49	.49	.49	.44	.39	0
	AMB TEMP:	33.4	34.8	42.8	53.8	63.7	72.4	76.6	74.9	68.5	57.4	46.1	35.3	55.0
	HTG DEG DAYS:	980	846	688	340	110	0	0	0	27	250	567	921	4729

STA NO:93729	HORIZ INSOL:	686.	952.	1326.	1774.	1962.	2036.	1921.	1705.	1470.	1137.	873.	659.	1375.
CAPE HATTERAS	TILT = LAT:	1011.	1250.	1516.	1751.	1731.	1710.	1652.	1603.	1589.	1450.	1305.	1028.	0
NC	TILT = LAT+15:	1057.	1260.	1453.	1576.	1481.	1430.	1399.	1414.	1488.	1446.	1366.	1090.	0
LAT = 35.27	TILT = 90:	885.	951.	918.	755.	564.	492.	513.	627.	853.	1046.	1135.	944.	0
	KT:	.42	.46	.50	.55	.55	.55	.53	.51	.52	.51	.51	.44	0
	AMB TEMP:	45.3	45.8	50.6	58.9	67.0	74.3	78.0	77.5	73.7	65.2	56.0	47.7	61.7
	HTG DEG DAYS:	611	538	458	188	47	0	0	0	0	76	277	536	2731

STA NO:93734	HORIZ INSOL:	572.	815.	1125.	1459.	1718.	1901.	1818.	1617.	1340.	1004.	651.	481.	1208.
WASHINGTON-STERLING	TILT = LAT:	901.	1116.	1305.	1439.	1511.	1585.	1554.	1523.	1474.	1337.	1003.	776.	0
DC	TILT = LAT+15:	940.	1122.	1245.	1291.	1293.	1325.	1312.	1339.	1375.	1330.	1041.	814.	0
LAT = 38.95	TILT = 90:	812.	881.	836.	695.	579.	538.	557.	664.	851.	1009.	883.	716.	0
	KT:	.40	.43	.45	.46	.48	.51	.50	.49	.49	.49	.43	.37	0
	AMB TEMP:	32.1	33.8	41.8	53.1	62.6	71.1	75.3	73.6	66.9	55.9	44.7	34.0	53.7
	HTG DEG DAYS:	1020	874	719	357	131	5	0	0	43	291	609	961	5010

STA NO:93805	HORIZ INSOL:	877.	1138.	1479.	1823.	1936.	1883.	1748.	1675.	1493.	1318.	1009.	813.	1433.
TALLAHASSEE	TILT = LAT:	1207.	1418.	1639.	1782.	1723.	1612.	1529.	1573.	1568.	1600.	1377.	1156.	0
FL	TILT = LAT+15:	1269.	1436.	1576.	1610.	1482.	1361.	1307.	1395.	1472.	1601.	1441.	1229.	0
LAT = 30.38	TILT = 90:	1020.	1020.	906.	667.	475.	409.	425.	538.	758.	1080.	1136.	1020.	0
	KT:	.46	.49	.52	.55	.54	.51	.48	.49	.50	.54	.51	.46	0
	AMB TEMP:	52.6	54.8	60.3	67.9	74.8	80.0	81.1	81.1	78.1	69.3	58.9	53.2	67.7
	HTG DEG DAYS:	408	323	187	34	0	0	0	0	0	31	204	376	1563

TABLE A.4 Monthly Solar and Climatic Data for Locations in the Conterminous United States *(Continued)*

STA NO:93814	HORIZ INSOL:	501.	738.	1027.	1399.	1672.	1837.	1771.	1634.	1312.	990.	589.	433.	1158.
CINCINNATI (COVINGTO	TILT = LAT:	754.	986.	1175.	1376.	1471.	1534.	1514.	1540.	1440.	1317.	881.	671.	0
OH	TILT = LAT+15:	779.	986.	1117.	1234.	1259.	1284.	1280.	1353.	1343.	1310.	908.	699.	0
LAT = 39.07	TILT = 90:	664.	769.	749.	669.	571.	530.	551.	672.	833.	994.	764.	607.	0
	KT:	.35	.39	.41	.45	.47	.49	.48	.50	.48	.48	.39	.34	0
	AMB TEMP:	31.1	33.3	41.7	53.9	63.2	72.1	75.6	74.4	67.8	56.8	43.8	33.7	54.0
	HTG DEG DAYS:	1051	888	722	341	138	9	0	0	44	271	636	970	5070
STA NO:93815	HORIZ INSOL:	489.	725.	1025.	1403.	1699.	1874.	1810.	1646.	1318.	969.	564.	408.	1161.
DAYTON	TILT = LAT:	754.	983.	1183.	1384.	1493.	1560.	1544.	1552.	1458.	1305.	856.	639.	0
OH	TILT = LAT+15:	780.	983.	1125.	1240.	1276.	1302.	1303.	1363.	1360.	1298.	882.	665.	0
LAT = 39.90	TILT = 90:	671.	775.	765.	685.	590.	549.	571.	691.	857.	996.	747.	580.	0
	KT:	.36	.39	.41	.45	.48	.50	.50	.50	.49	.48	.38	.33	0
	AMB TEMP:	28.1	30.4	39.0	51.4	61.6	71.3	74.6	73.0	66.3	55.5	41.8	30.9	52.0
	HTG DEG DAYS:	1144	969	806	413	166	13	0	7	63	307	696	1057	5641
STA NO:93817	HORIZ INSOL:	574.	823.	1151.	1501.	1783.	1983.	1920.	1735.	1403.	1087.	683.	499.	1262.
EVANSVILLE	TILT = LAT:	876.	1106.	1327.	1480.	1569.	1655.	1642.	1636.	1540.	1447.	1037.	785.	0
IN	TILT = LAT+15:	912.	1111.	1266.	1328.	1342.	1381.	1386.	1439.	1440.	1443.	1076.	824.	0
LAT = 38.05	TILT = 90:	778.	862.	838.	698.	579.	534.	560.	688.	875.	1086.	906.	718.	0
	KT:	.39	.42	.45	.47	.50	.53	.53	.52	.51	.52	.43	.37	0
	AMB TEMP:	32.6	35.9	44.3	56.7	65.7	74.7	77.8	76.2	69.1	58.2	44.9	35.3	56.0
	HTG DEG DAYS:	1004	815	653	263	95	5	0	0	34	236	603	921	4629
STA NO:93819	HORIZ INSOL:	496.	747.	1037.	1399.	1688.	1868.	1806.	1643.	1324.	977.	579.	417.	1165.
INDIANAPOLIS	TILT = LAT:	763.	1017.	1197.	1373.	1433.	1556.	1542.	1550.	1463.	1314.	882.	654.	0
IN	TILT = LAT+15:	790.	1018.	1139.	1235.	1268.	1300.	1301.	1361.	1365.	1307.	903.	682.	0
LAT = 39.73	TILT = 90:	678.	802.	773.	680.	584.	545.	567.	687.	858.	1001.	771.	595.	0
	KT:	.36	.40	.42	.45	.47	.50	.49	.50	.49	.48	.39	.33	0
	AMB TEMP:	27.9	30.7	39.7	52.3	62.2	71.7	75.0	73.2	66.3	55.7	41.7	30.9	52.3
	HTG DEG DAYS:	1150	960	784	387	159	11	0	5	63	302	699	1057	5577
STA NO:93820	HORIZ INSOL:	546.	780.	1100.	1479.	1747.	1897.	1850.	1685.	1362.	1044.	657.	486.	1219.
LEXINGTON	TILT = LAT:	818.	1032.	1258.	1457.	1538.	1586.	1584.	1587.	1490.	1377.	986.	756.	0
KY	TILT = LAT+15:	848.	1034.	1198.	1308.	1317.	1327.	1339.	1396.	1391.	1372.	1021.	792.	0
LAT = 38.03	TILT = 90:	719.	798.	792.	688.	572.	523.	548.	671.	845.	1029.	857.	688.	0
	KT:	.37	.40	.43	.47	.49	.51	.51	.51	.49	.49	.42	.36	0
	AMB TEMP:	32.9	35.3	43.6	55.3	64.7	73.0	76.2	75.0	68.6	57.8	44.6	35.5	55.2
	HTG DEG DAYS:	995	832	673	302	106	8	0	0	40	246	612	915	4729

STA NO:93821	HORIZ INSOL:	546.	789.	1102.	1467.	1720.	1904.	1838.	1680.	1361.	1042.	653.	488.	1216.
LOUISVILLE	TILT = LAT:	822.	1052.	1263.	1444.	1514.	1591.	1573.	1583.	1490.	1378.	982.	766.	0
KY	TILT = LAT+15:	852.	1055.	1204.	1296.	1296.	1331.	1329.	1392.	1392.	1373.	1017.	803.	0
LAT = 38.18	TILT = 90:	724.	817.	797.	685.	568.	526.	549.	672.	848.	1032.	854.	699.	0
	KT:	.37	.41	.43	.46	.48	.51	.50	.51	.49	.50	.42	.37	0
	AMB TEMP:	33.3	35.8	44.0	55.9	64.8	73.3	76.9	75.9	69.1	58.1	45.0	35.6	55.6
	HTG DEG DAYS:	983	818	661	286	105	5	0	0	35	241	600	911	4645
STA NO:93822	HORIZ INSOL:	585.	861.	1143.	1515.	1865.	2097.	2058.	1806.	1454.	1068.	677.	490.	1302.
SPRINGFIELD	TILT = LAT:	961.	1221.	1344.	1503.	1638.	1737.	1751.	1710.	1629.	1470.	1089.	828.	0
IL	TILT = LAT+15:	1007.	1233.	1283.	1347.	1396.	1443.	1470.	1500.	1523.	1468.	1135.	873.	0
LAT = 39.83	TILT = 90:	882.	985.	875.	737.	628.	581.	615.	748.	958.	1133.	978.	779.	0
	KT:	.42	.46	.46	.49	.52	.56	.56	.55	.54	.53	.46	.39	0
	AMB TEMP:	26.7	30.4	39.4	53.1	63.4	72.9	76.1	74.4	67.2	56.6	41.9	30.5	52.7
	HTG DEG DAYS:	1187	959	794	363	132	12	0	8	48	282	693	1070	5558
STA NO:93987	HORIZ INSOL:	794.	1069.	1376.	1624.	1867.	2055.	2006.	1864.	1531.	1349.	963.	768.	1439.
LUFKIN	TILT = LAT:	1090.	1337.	1523.	1584.	1660.	1747.	1740.	1751.	1618.	1663.	1328.	1102.	0
TX	TILT = LAT+15:	1140.	1350.	1462.	1431.	1429.	1466.	1476.	1549.	1519.	1666.	1388.	1169.	0
LAT = 31.23	TILT = 90:	917.	967.	855.	624.	483.	427.	454.	591.	795.	1142.	1103.	975.	0
	KT:	.43	.47	.49	.49	.52	.55	.55	.55	.52	.56	.50	.45	0
	AMB TEMP:	48.8	52.2	58.0	67.3	74.1	80.3	83.0	83.1	77.5	68.2	57.2	50.8	66.7
	HTG DEG DAYS:	509	371	256	56	0	0	0	0	0	52	256	440	1940
STA NO:94008	HORIZ INSOL:	388.	671.	1105.	1488.	1828.	2047.	2193.	1863.	1340.	877.	479.	334.	1218.
GLASGOW	TILT = LAT:	841.	1157.	1479.	1536.	1597.	1666.	1846.	1814.	1640.	1443.	1007.	799.	0
MT	TILT = LAT+15:	885.	1172.	1416.	1368.	1349.	1371.	1532.	1580.	1531.	1445.	1054.	851.	0
LAT = 48.22	TILT = 90:	829.	1024.	1103.	898.	770.	729.	825.	971.	1122.	1232.	975.	813.	0
	KT:	.42	.47	.52	.52	.52	.55	.61	.60	.56	.55	.47	.43	0
	AMB TEMP:	9.2	15.2	25.2	42.8	54.2	62.0	70.5	69.0	57.2	46.4	29.0	17.1	41.5
	HTG DEG DAYS:	1730	1394	1234	666	344	151	15	30	263	577	1080	1485	8969
STA NO:94224	HORIZ INSOL:	315.	545.	866.	1253.	1608.	1626.	1746.	1499.	1183.	713.	387.	261.	1000.
ASTORIA	TILT = LAT:	534.	804.	1050.	1253.	1402.	1335.	1473.	1424.	1374.	1030.	657.	461.	0
OR	TILT = LAT+15:	549.	801.	995.	1115.	1190.	1112.	1234.	1242.	1278.	1018.	674.	478.	0
LAT = 46.15	TILT = 90:	491.	671.	743.	706.	657.	586.	653.	740.	901.	835.	599.	434.	0
	KT:	.31	.36	.39	.42	.46	.43	.48	.47	.48	.42	.34	.29	0
	AMB TEMP:	40.6	43.6	44.4	47.8	52.3	56.5	60.0	60.3	58.4	52.8	46.5	42.8	50.5
	HTG DEG DAYS:	756	599	639	516	394	255	163	151	201	378	555	688	5295

TABLE A.4 Monthly Solar and Climatic Data for Locations in the Conterminous United States *(Continued)*

STA NO:94701	HORIZ INSOL:	476.	710.	1016.	1326.	1620.	1817.	1749.	1486.	1260.	890.	503.	403.	1105.
BOSTON	TILT = LAT:	806.	1019.	1208.	1313.	1419.	1504.	1486.	1401.	1417.	1240.	803.	711.	0
MA	TILT = LAT+15:	840.	1023.	1149.	1173.	1210.	1254.	1251.	1227.	1320.	1232.	827.	747.	0
LAT = 42.37	TILT = 90:	745.	833.	814.	686.	605.	577.	596.	670.	871.	974.	716.	675.	0
	KT:	.38	.41	.43	.43	.46	.48	.48	.46	.48	.47	.37	.37	0
	AMB TEMP:	29.2	30.4	38.1	48.6	58.6	68.0	73.3	71.3	64.5	55.4	45.2	33.0	51.3
	HTG DEG DAYS:	1110	969	834	492	218	27	0	8	76	301	594	992	5621
STA NO:94725	HORIZ INSOL:	391.	620.	978.	1343.	1613.	1779.	1751.	1484.	1124.	736.	388.	294.	1042.
MASSENA	TILT = LAT:	693.	921.	1197.	1345.	1408.	1463.	1480.	1405.	1275.	1038.	619.	514.	0
NY	TILT = LAT+15:	720.	922.	1138.	1199.	1197.	1216.	1242.	1227.	1183.	1026.	632.	535.	0
LAT = 44.93	TILT = 90:	648.	769.	839.	739.	641.	607.	636.	711.	816.	829.	552.	484.	0
	KT:	.36	.39	.43	.45	.46	.47	.48	.47	.45	.42	.32	.31	0
	AMB TEMP:	14.5	16.7	27.6	42.2	54.1	64.3	64.3	66.7	59.2	48.5	35.9	20.1	43.2
	HTG DEG DAYS:	1566	1352	1159	684	350	78	22	57	192	512	873	1392	8237
STA NO:94728	HORIZ INSOL:	500.	721.	1037.	1364.	1636.	1710.	1688.	1483.	1214.	895.	533.	404.	1099.
NEW YORK CITY (CENTR	TILT = LAT:	807.	997.	1212.	1346.	1435.	1425.	1439.	1394.	1338.	1206.	818.	657.	0
NY	TILT = LAT+15:	839.	998.	1153.	1205.	1227.	1194.	1216.	1224.	1245.	1196.	841.	686.	0
LAT = 40.78	TILT = 90:	732.	796.	796.	680.	587.	534.	559.	644.	798.	924.	717.	607.	0
	KT:	.38	.40	.43	.44	.46	.46	.46	.45	.46	.45	.37	.34	0
	AMB TEMP:	32.2	33.4	41.1	52.1	62.3	71.6	76.6	74.9	68.4	58.7	47.4	35.5	54.5
	HTG DEG DAYS:	1017	885	741	387	137	0	0	0	29	209	528	915	4848
STA NO:94823	HORIZ INSOL:	424.	625.	943.	1317.	1602.	1762.	1689.	1510.	1209.	895.	505.	347.	1069.
PITTSBURGH	TILT = LAT:	631.	821.	1078.	1295.	1406.	1467.	1442.	1421.	1328.	1198.	750.	518.	0
PA	TILT = LAT+15:	648.	815.	1022.	1159.	1203.	1228.	1219.	1248.	1236.	1188.	768.	533.	0
LAT = 40.50	TILT = 90:	552.	638.	699.	652.	574.	539.	555.	650.	788.	914.	648.	458.	0
	KT:	.32	.34	.38	.42	.45	.47	.46	.46	.45	.45	.35	.29	0
	AMB TEMP:	28.1	29.3	38.1	50.2	59.8	68.6	71.9	70.2	63.8	53.2	41.3	30.5	50.4
	HTG DEG DAYS:	1144	1000	834	444	208	26	7	16	98	372	711	1070	5930
STA NO:94830	HORIZ INSOL:	435.	680.	997.	1384.	1717.	1878.	1849.	1616.	1276.	911.	498.	355.	1133.
TOLEDO	TILT = LAT:	685.	944.	1168.	1372.	1505.	1556.	1572.	1527.	1427.	1255.	767.	567.	0
OH	TILT = LAT+15:	707.	944.	1110.	1227.	1283.	1296.	1322.	1339.	1330.	1247.	788.	588.	0
LAT = 41.60	TILT = 90:	614.	757.	776.	704.	620.	576.	606.	710.	865.	977.	674.	517.	0
	KT:	.34	.39	.41	.45	.48	.50	.51	.50	.48	.47	.36	.31	0
	AMB TEMP:	24.8	27.1	35.8	48.4	58.8	68.9	72.3	70.8	63.8	53.0	39.6	28.0	49.3
	HTG DEG DAYS:	1246	1061	905	498	229	32	5	18	99	379	762	1147	6381

STA NO:94849	HORIZ INSOL:	362.	617.	1028.	1407.	1720.	1879.	1885.	1583.	1156.	743.	382.	270.	1086.
ALPENA	TILT = LAT:	622.	917.	1276.	1418.	1503.	1542.	1592.	1506.	1320.	1055.	609.	454.	0
MI	TILT = LAT+15:	643.	919.	1216.	1264.	1276.	1279.	1332.	1315.	1226.	1043.	621.	468.	0
LAT = 45.07	TILT = 90:	575.	767.	900.	780.	677.	632.	674.	760.	848.	845.	543.	420.	0
	KT:	.33	.39	.46	.47	.49	.50	.52	.50	.46	.42	.32	.28	0
	AMB TEMP:	17.8	18.3	26.2	40.1	50.5	60.9	65.5	64.2	56.3	47.3	34.9	23.4	42.1
	HTG DEG DAYS:	1463	1308	1203	747	455	150	75	110	265	549	903	1290	6518
STA NO:94860	HORIZ INSOL:	370.	648.	1014.	1412.	1755.	1956.	1914.	1676.	1262.	858.	446.	311.	1135.
GRAND RAPIDS	TILT = LAT:	572.	918.	1214.	1409.	1537.	1613.	1622.	1591.	1428.	1199.	691.	494.	0
MI	TILT = LAT+15:	586.	917.	1155.	1259.	1307.	1339.	1361.	1393.	1330.	1190.	707.	510.	0
LAT = 42.88	TILT = 90:	509.	746.	825.	741.	652.	612.	643.	759.	886.	945.	609.	450.	0
	KT:	.31	.38	.43	.46	.50	.52	.53	.52	.49	.46	.34	.29	0
	AMB TEMP:	23.2	24.5	33.1	46.5	57.1	67.4	71.5	70.0	62.4	52.0	38.7	27.4	47.8
	HTG DEG DAYS:	1296	1134	989	555	270	44	8	27	114	409	789	1166	6801
STA NO:94918	HORIZ INSOL:	634.	892.	1223.	1558.	1873.	2123.	2107.	1858.	1373.	1050.	644.	511.	1321.
NORTH OMAHA	TILT = LAT:	1144.	1331.	1484.	1558.	1642.	1751.	1787.	1768.	1549.	1490.	1082.	946.	0
NE	TILT = LAT+15:	1209.	1349.	1421.	1395.	1397.	1452.	1496.	1548.	1446.	1489.	1128.	1005.	0
LAT = 41.37	TILT = 90:	1086.	1104.	997.	789.	657.	613.	654.	801.	937.	1173.	986.	918.	0
	KT:	.49	.50	.51	.51	.53	.57	.58	.57	.52	.54	.46	.44	0
	AMB TEMP:	20.2	25.5	34.6	50.0	60.9	70.2	75.1	73.7	64.4	54.4	37.9	25.7	49.4
	HTG DEG DAYS:	1389	1106	942	456	186	33	7	10	99	342	813	1218	6601

*NOTES: 1. Climatic data and horizontal solar flux taken from V. Cinquemani, et al., *Input Data for Solar Systems*, National Climatic Center, U.S. Department of Commerce, Asheville, NC, 1978. Blank table entries indicate data unavailable from this source.

2. Solar flux on tilted surfaces (tilts equal to latitude, latitude +15°, and vertical are used) calculated using the method described in "Solar Radiation Calculations" in Chap. 2.

3. Tabulated solar flux values on tilted surfaces *do not include radiation reflected from the foreground*. To include this effect add the term $\overline{H}_h \rho \sin^2(\beta/2)$ where $\overline{H}_h$ is the horizontal flux, ρ is the diffuse foreground reflectance, and β is the surface tilt angle.

4. Units used in the table are Btu/(day • ft²) for solar flux, °F for temperatures, and °F-days for degree days. $\overline{K}_T$ has no units. The degree-day basis is 65°F. For other units see Table A.1.

TABLE A.5 Insulation Data for Mineral Wool, Industrial Felt, and Fiberglas[*]

Material[a]	Nominal density, lb/ft³	Temperature limitation, °F	Mean temperature thermal conductivity		Federal specification compliance	Producer	Standard sizes (variable)	Cost,[c] dollars per board foot
			°F	Btu·in/(hr·ft²·°F)[b]				
				Mineral wool				
No. 10 insulation	10.0	1200	200 350 500	0.26 0.32 0.375	HH-I-558 B form A, class 4	Forty-Eight Insulations	2′ × 4′ (board) Thk: 1–3″ (½″ inc.)	0.125–0.14 (dist)[d] Carload: 30,000 bd ft
LTR insulation	8.0	1000	200 350 500	0.27 0.32 0.385	HH-I-558 B form A, class 4	Forty-Eight Insulations	2′ × 4′ (board) Thk: 1–4″ (½″ inc.)	0.105–0115 (dist) Carload: 30,000 bd ft
I-T insulation	6.0	850	200 350 500	0.27 0.34 0.45	HH-I-558 B form A, class 3	Forty-Eight Insulations	2′ × 4′ (board) Thk: 1–4″ (½″ inc.)	0.095–0.10 (dist) Carload: 30,000 bd ft
MT-board (MT-10)	10.0	1050	200 350 500	0.25 0.333 0.445	HH-I-558 B form A, class 1, 2, 3	Eagle-Picher	2′ × 4′ (board) Thk: 1–3″ (½″ inc.)	0.13–0.14 (dist) Carload: 36,000 bd ft
MT-board (MT-8)	8.0	1050	200 350 500	0.255 0.350 0.470	HH-I-558 B form A, class 1, 2, 3	Eagle-Picher	2′ × 4′ (board) Thk: 1–4″ (½″ inc.)	0.107–0.12 (dist) Carload: 36,000 bd ft
MT-board (MT-6)	6.0	1050	200 350 500	0.270 0.373 0.495	HH-I-558 B form A, class 1, 2, 3	Eagle-Picher	2′ × 4′ (board) Thk: 1–4″ (½″ inc.)	0.085–0.10 (dist) Carload: 36,000 bd ft
				Industrial felt[e]				
Thermafiber (SF-234)	8.0	1000	200 350 500	0.27 0.36 0.48	HH-I-558 B form A, class 1, 2, 3	United States Gypsum	Thk: 1–2″ (N.J.) Length: 60″ Thk: 1–2½″ (Ind.) Length: 48″	0.131 (dist) 7,000–38,000 bd ft
Thermafiber (SF-240)	6.0	1000	200 350 500	0.27 0.37 0.50	HH-I-558 B form A, class 1, 2, 3	United States Gypsum	Thk: 1–2½″ (Tex.) Length: 90″ Thk: 1–3½″ (Ind.) Length: 48″	0.095–0.113 (dist) 7,000–38,000 bd ft

Thermafiber (SF-250)	4.5	800	200 350 500	0.29 0.415 0.55	HH-I-558 B form A, class 1, 2	United States Gypsum	Thk: 1–4″ (Tex.) Length: 90″ Thk: 1–5″ (Ind.) Length: 48″	0.081–0.10 (dist) 7,000–38,000 bd ft
Thermafiber (SF-252)	4.0	800	200 350 500	0.30 0.435 0.59	HH-I-558 B form A, class 1, 2	United States Gypsum	Thk: 1–4″ (Tex.) Length: 90″ Thk: 1–5″ (Ind.) Length: 48″	0.07–0.087 (dist) 7,000–38,000 bd ft
Thermafiber (SF-256)	3.5	600	200 350 500	0.33 0.47 0.62	HH-I-558 B form A, class 1, 2	United States Gypsum	Thk: 1–4″ (Tex.) Length: 90″ Thk: 1–6″ (Ind.) Length: 48″	0.066–0.084 (dist) 7,000–38,000 bd ft
Thermafiber (SF-260)	3.0	500	200 350 500	0.35 0.50 0.65	HH-I-558 B form A, class 1, 2	United States Gypsum	Thk: 1–4″ (Tex.) Length: 90″ Thk: 1–6″ (Ind.) Length: 48″	0.064–0.82 (dist) 7,000–38,000 bd ft
Thermafiber (SF-270)	2.5	400	200 350 500	0.39 0.56 —	No data provided	United States Gypsum	Thk: 1–4″ (Tex.) Length: 90″ Thk: 1–5″ (Ind.) Length: 48″	0.06–0.078 (dist) 7,000–38,000 bd ft
				Foamglas and Fiberglas insulation[f]				
Foamglas[g]	8.5	600	200 350 500	0.46 0.58 0.74	HH-I-551 D (Fed) ASTMC 552-73	Pittsburgh Corning	1′ × 1.5′ (board) 1½′ × 2′ (board) Thk: 1½–4″ (½″ inc.)	0.22–0.24 (Corning)[h] Carload: 36,000 bd ft
701 Fiberglas[i]	1.6	450	200 350	0.33 0.51	HH-I-558 B form A, class I HH-I-558 B, type I form B, class 7	Owens-Corning Fiberglas	2′ × 4′ (board) Thk: 1½–4″ (½″ inc.)	0.07–0.08 (dist) Carload: 30,000– 35,000 bd ft
703 Fiberglas[i]	3.0	450	200 350	0.30 0.41	HH-I-558 B form A, class 1, 2	Owens-Corning Fiberglas	2′ × 4′ (board) Thk: 1–2″ (½″ inc.)	0.14–0.15 (dist) Carload: 30,000– 35,000 bd ft
705 Fiberglas[i]	6.0	450	200 350	0.27 0.38	HH-I-558 B form A, class 1, 2	Owens-Corning Fiberglas	2′ × 4′ (board) Thk: 1–2″ (½″ inc.)	0.25–0.27 (dist) Carload: 30,000– 35,000 bd ft

TABLE A.5 Insulation Data for Mineral Wool, Industrial Felt, and Fiberglas* (Continued)

Material[g]	*Nominal density, lb/ft³*	*Temperature limitation, °F*	*Mean temperature thermal conductivity*		*Federal specification compliance*	*Producer*	*Standard sizes (variable)*	*Cost,[c] dollars per board foot*
			°F	*Btu·in/(hr·ft²·°F)*[b]				
Thermal insulating wool type I[i]	1.25	1000	200 350 500	0.41 0.65 0.85	HH-I-558 B form B, type 1, class 6	Owens-Corning Fiberglas	Rolls Width: 2′ or 3′ Thk: 2″, 3″, 4″ Length: 76′, 52′, 38′	0.04–0.06 (Corning) ½ Carload: 35,000 bd ft
Thermal insulating wool type II[i]	2.4	1000	200 350 500	0.30 0.44 0.60	HH-I-558 B form B, type 1, class 7, 8	Owens-Corning Fiberglas	2′ × 8′ 2′ × 4′ (board) Thk: 1–3″ (½″ inc.)	0.08–0.09 (Corning) 0.14–0.15 (dist) Carload: 35,000 bd ft
IS board[i]	4.0	800	200 350 500	0.30 0.44 0.61	HH-I-558 B form A, class 3	Owens-Corning Fiberglas	2′ × 4′, 3′ × 4′ 4′ × 8′ (board) Thk: 1–6″ (½″ inc.)	0.10–0.13 (Corning) 0.18–0.20 (dist) Carload: 35,000 bd ft

*Abstracted from Ratzel, R. C., and R. B. Bannerot, Optimal Material Selection for Flat-Plate Solar Energy Collectors Utilizing Commercially Available Materials, presented at ASME–AIChE Natl. Heat Transfer Conf., 1976.

[a]All insulations listed will not cause or aggravate corrosion and will absorb less than 1% moisture. All insulations listed appear as semirigid board composed of silica-base refractory fibers bonded with special binders for service in indicated temperature ranges.

[b]Units are consistently employed within the insulation industry. Conductivity measurements consider a test specimen 1 in thick and 1 ft² normal area.

[c]Cost data current through Oct. 30, 1975. Costs are based on carload purchases and include freight where necessary to move insulation to Houston.

[d]Cost from Houston distributors noted by (dist).

[e]Industrial felt is preformed mineral fiber felt, which will not cause or sustain corrosion. It absorbs less than 1% moisture by weight and is rated noncombustible. Insulation to be ordered in varying thicknesses and lengths; standard width of 24 in employed (see column 8).

[f]All codes in specification compliance column are federal specifications unless otherwise noted.

[g]Foamglas is an impermeable, incombustible, rigid insulation composed of completely sealed glass cells with no binder material. Its rigid form may allow for Foamglas being implemented as the collector box.

[h]Cost from Corning Houston warehouse noted by (Corning); cost from Houston distributor noted by (dist).

[i]Insulations are made of inorganic glass fibers preformed into semirigid to rigid rectangular boards (T1WI in blankets). Insulations will not accelerate nor cause corrosion and will absorb less than 1% moisture (by volume).

TABLE A.6 G-Chart Coefficients for Solar Heating Systems with Air Working Fluid*

		A	B	C	D	E	F	Lo
ABILENE	TX	.4466E-01	-.1840E-02	-.1183E-04	-.2183E-03	.8963E-05	.9167E-06	63.07
ALBANY	NY	.1290E-01	-.8575E-03	-.8409E-05	-.2767E-04	.1840E-05	.5586E-06	129.60
ALBUQUERQUE	NM	.3686E-01	-.1093E-02	.7717E-05	-.5499E-04	-.6823E-05	.3106E-06	89.23
AMARILLO	TX	.3374E-01	-.1107E-02	.3930E-06	-.5661E-04	-.5443E-05	.3866E-06	87.53
AMES	IA	.1365E-01	-.8160E-03	-.1568E-04	-.1860E-04	.5386E-06	.6068E-06	128.61
AMHERST	MA	.9909E-02	-.9562E-03	.8698E-05	-.1366E-04	.2913E-05	.2389E-06	124.75
ANNAPOLIS	MD	.1796E-01	-.1143E-02	.3146E-05	-.3508E-04	.2576E-05	.4951E-06	93.21
APALACHICOLA	FL	.7609E-01	-.3074E-02	-.7428E-05	-.5807E-03	.2555E-04	.1649E-05	42.82
ASHEVILLE	NC	.2581E-01	-.1252E-02	-.2569E-04	-.6739E-04	.1655E-05	.1151E-05	88.37
ASTORIA	OR	.1676E-01	-.1461E-02	-.4538E-05	-.4747E-04	.7418E-05	.6108E-06	104.83
ATLANTA	GA	.3081E-01	-.1597E-02	.2226E-05	-.1143E-03	.6885E-05	.5391E-06	70.62
ATLANTIC CITY	NJ	.2042E-01	-.1060E-02	.2441E-05	-.3639E-04	-.8035E-06	.4916E-06	95.47
BIG SPRING	TX	.4541E-01	-.1889E-02	-.6195E-05	-.2482E-03	.1253E-04	.5796E-06	62.78
BILLINGS	MT	.1653E-01	-.7642E-03	.1416E-05	-.1810E-04	-.1888E-05	.3481E-06	135.46
BINGHAMPTON	NY	.1042E-01	-.9159E-03	-.2741E-05	-.1420E-04	.2602E-05	.2757E-06	135.78
BIRMINGHAM	AL	.3151E-01	-.1769E-02	.1430E-05	-.1353E-03	.1054E-04	.5150E-06	66.71
BISMARCK	ND	.1353E-01	-.7344E-03	-.2355E-04	-.1884E-04	.4151E-06	.5460E-06	163.13
BLUE HILL	MA	.1274E-01	-.8961E-03	-.1320E-04	-.1548E-04	.1219E-05	.6075E-06	121.50
BOISE	ID	.1899E-01	-.1143E-02	-.1559E-04	-.5291E-04	.4178E-05	.4383E-06	112.82
BOSTON	MA	.1211E-01	-.9887E-03	-.1134E-04	-.1790E-04	.2343E-05	.6620E-06	110.10
BOULDER	CO	.2115E-01	-.8627E-03	-.4025E-06	-.1087E-04	-.6477E-05	.7299E-06	108.64
BROWNSVILLE	TX	.9212E-01	-.2870E-02	-.3029E-04	-.5324E-03	-.9016E-04	.9504E-05	32.59
CAPE HATTERAS	NC	.4393E-01	-.1798E-02	-.1209E-04	-.2173E-03	.8177E-05	.9872E-06	64.95
CARIBOU	ME	.1068E-01	-.7104E-03	-.1758E-04	-.1325E-04	.5208E-06	.4306E-06	174.38
CHARLESTON	SC	.4831E-01	-.2478E-02	-.4187E-04	-.2980E-03	.2372E-04	.1421E-05	54.10
CHARLOTTE	NC	.3185E-01	-.1549E-02	-.2533E-06	-.1192E-03	.6374E-05	.5613E-06	72.52
CHATTANOOGA	TN	.2447E-01	-.1486E-02	-.4523E-05	-.7670E-04	.6511E-05	.5472E-06	76.99
CHICAGO	IL	.1337E-01	-.9110E-03	-.1105E-04	-.2023E-04	.1533E-05	.5493E-06	117.77
CLEVELAND	OH	.1063E-01	-.1091E-02	-.5355E-05	-.1765E-04	.4133E-05	.2239E-06	118.19
COLUMBIA	MO	.1833E-01	-.1084E-02	-.1486E-05	-.3739E-04	.2360E-05	.4012E-06	100.96
COLUMBUS	OH	.1245E-01	-.1113E-02	-.1323E-04	-.2266E-04	.3895E-05	.3901E-06	110.51
CORPUS CHRISTI	TX	.8408E-01	-.3511E-02	-.1428E-04	-.9637E-03	.4528E-04	.3040E-05	36.94
CORVALLIS	OR	.1575E-01	-.1550E-02	-.3636E-05	-.4408E-04	.8908E-05	.1260E-06	97.97
DALLAS	TX	.4056E-01	-.2037E-02	.1724E-04	-.2117E-03	.1439E-04	.1665E-06	58.09
DAVIS	CA	.3843E-01	-.2289E-02	-.2977E-04	-.2501E-03	.2443E-04	.7077E-06	61.39
DAYTON	OH	.1444E-01	-.1043E-02	-.1895E-06	-.2579E-04	.2832E-05	.3508E-06	110.21
DENVER	CO	.2468E-01	-.7770E-03	-.8975E-05	.1612E-04	-.1010E-04	.8112E-06	116.04
DES MOINES	IA	.1311E-01	-.8989E-03	-.8340E-05	-.1948E-04	.1833E-05	.3495E-06	126.83
DETROIT	MI	.1106E-01	-.1025E-02	-.1295E-04	-.1741E-04	.3296E-05	.3489E-06	122.31
DODGE CITY	KA	.2709E-01	-.9953E-03	-.6475E-05	-.4465E-04	-.3231E-05	.6065E-06	100.96
DULUTH	MN	.1027E-01	-.7616E-03	-.1318E-04	-.1425E-04	.1458E-05	.3508E-06	174.20
EAST LANSING	MI	.9811E-02	-.9447E-03	.1145E-04	-.1329E-04	.2903E-05	.1804E-06	129.85
EL PASO	TX	.5801E-01	-.1734E-02	-.8989E-06	-.3385E-03	.7476E-05	.6266E-06	64.47
ELY	NV	.2227E-01	-.6523E-03	.3829E-05	.1064E-04	-.9039E-05	.4511E-06	142.74
FARGO	ND	.9665E-02	-.7276E-03	.1874E-05	-.1076E-04	.1337E-05	.1943E-06	166.66
FORT SMITH	AR	.2512E-01	-.1470E-02	-.9007E-05	-.7833E-04	.5318E-05	.6873E-06	74.36
FORT WAYNE	IN	.1309E-01	-.9979E-03	-.3275E-05	-.2193E-04	.2795E-05	.3278E-06	119.04
FORT WORTH	TX	.4531E-01	-.1958E-02	-.2241E-04	-.2528E-03	.1139E-04	.1561E-05	59.88
FRESNO	CA	.4164E-01	-.2102E-02	-.1202E-04	-.2856E-03	.2178E-04	.4311E-06	63.09
GAINESVILLE	FL	.7786E-01	-.2888E-02	.6966E-05	-.3927E-03	-.1859E-04	.2434E-05	39.29
GLASGOW	MT	.1512E-01	-.6820E-03	-.7041E-05	-.1633E-04	-.1202E-05	.2754E-06	162.39
GRAND JUNCTION	CO	.2431E-01	-.9295E-03	-.2149E-04	-.4009E-04	-.1973E-05	.8810E-06	110.18
GRAND LAKE	CO	.1623E-01	-.6358E-03	-.5608E-06	-.1668E-05	-.3843E-05	.3592E-06	190.47
GREAT FALLS	MT	.1613E-01	-.8038E-03	-.1066E-04	-.2175E-04	-.7685E-06	.4739E-06	143.01
GREEN BAY	WI	.1021E-01	-.8001E-03	-.3193E-05	-.1284E-04	.1655E-05	.2857E-06	148.42
GREENSBORO	NC	.2464E-01	-.1338E-02	-.4384E-05	-.6843E-04	.3631E-05	.6974E-06	81.97
GRNVLE-SPTNBRG	NC	.3135E-01	-.1596E-02	-.1850E-05	-.1216E-03	.7361E-05	.5675E-06	71.67
GRIFFIN	GA	.3709E-01	-.1804E-02	.3921E-05	-.1749E-03	.1089E-04	.4482E-06	66.03
HARTFORD	CT	.1439E-01	-.8335E-03	-.2677E-04	-.1676E-04	-.7109E-06	.8649E-06	121.23
HOUSTON	TX	.6308E-01	-.3357E-02	-.3746E-04	-.6251E-03	.5670E-04	.9140E-06	44.78
INDIANAPOLIS	IN	.1271E-01	-.1098E-02	-.1625E-04	-.2291E-04	.3661E-05	.4132E-06	109.21
INYOKERN	CA	.7724E-01	-.1951E-02	-.3133E-04	-.7153E-03	.1489E-04	.2547E-05	59.34
ITHACA	NY	.8764E-02	-.9623E-03	.1258E-04	-.1025E-04	.3078E-05	.1827E-06	132.15
JACKSON	MS	.3794E-01	-.2096E-02	.1049E-04	-.2083E-03	.1700E-04	.3492E-06	58.25
JACKSONVILLE	FL	.6445E-01	-.3125E-02	-.2419E-04	-.4292E-03	.2711E-04	.2454E-05	43.12
KANSAS CITY	MO	.1760E-01	-.1066E-02	-.6036E-05	-.3418E-04	.2328E-05	.4190E-06	102.74
KEY WEST	FL	.1139E-00	-.6171E-03	-.6758E-05	.5536E-02	-.9855E-03	.3210E-04	23.48
LAKE CHARLES	LA	.6319E-01	-.3320E-02	-.2870E-04	-.5799E-03	.5137E-04	.9146E-06	45.17
LANDER	WY	.2095E-01	-.6305E-03	.2441E-05	.8686E-05	-.7579E-05	.2571E-06	144.88
LANSING	MI	.1048E-01	-.9049E-03	.1361E-05	-.1359E-04	.2272E-05	.3034E-06	129.85
LARAMIE	WY	.1702E-01	-.5775E-03	-.8726E-05	.1201E-04	-.7029E-05	.6612E-06	159.95
LAS VEGAS	NV	.5580E-01	-.1736E-02	-.4749E-05	-.3357E-03	.8223E-05	.8465E-06	64.61
LEMONT	IL	.1310E-01	-.9135E-03	-.7720E-06	-.1742E-04	.1600E-05	.3725E-06	117.77
LEXINGTON	KY	.2059E-01	-.1176E-02	.8363E-07	-.5421E-04	.3505E-05	.3626E-06	96.03
LINCOLN	NE	.1748E-01	-.1008E-02	-.2425E-04	-.2859E-04	.1413E-05	.6812E-06	113.72
LITTLE ROCK	AR	.2563E-01	-.1553E-02	-.1259E-04	-.8581E-04	.6717E-05	.6800E-06	72.54
LOS ANGELES	CA	.6201E-01	-.1885E-02	-.1011E-04	.1730E-03	-.8229E-04	.5314E-05	50.77
LOUISVILLE	KY	.1648E-01	-.1238E-02	-.1778E-04	-.3650E-04	.4452E-05	.5495E-06	94.64
LYNN	MA	.1117E-01	-.1062E-02	.1019E-04	-.1792E-04	.3710E-05	.2698E-06	110.10
MACON	GA	.4441E-01	-.2144E-02	-.1041E-04	-.2612E-03	.1739E-04	.7882E-06	57.32
MADISON	WI	.1191E-01	-.8035E-03	-.1378E-04	-.1744E-04	.1271E-05	.4761E-06	142.70
MANHATTAN	KA	.1824E-01	-.9951E-03	-.4470E-06	-.3213E-04	.8497E-06	.4629E-06	103.07
MEDFORD	OR	.1831E-01	-.1411E-02	-.7568E-06	-.5986E-04	.8448E-05	.1298E-06	99.15
MEMPHIS	TN	.2655E-01	-.1561E-02	-.9039E-05	-.9404E-04	.7142E-05	.6427E-06	72.66
MIAMI	FL	.1063E-00	-.6337E-03	-.6621E-04	.4396E-02	-.8540E-03	.3815E-04	25.81
MIDLAND	TX	.4735E-01	-.1819E-02	.3879E-05	-.2442E-03	.8670E-05	.4500E-06	62.78
MILWAUKEE	WI	.1162E-01	-.8812E-03	-.5096E-05	-.1701E-04	.2122E-05	.3079E-06	138.25
MINN-ST. PAUL	MN	.1000E-01	-.8019E-03	.3437E-05	-.1143E-04	.169E-05	.2153E-06	149.36
MT. WEATHER	VA	.1646E-01	-.9149E-03	-.2144E-04	-.2572E-04	-.1390E-06	.9895E-06	110.63
NASHVILLE	TN	.2005E-01	-.1489E-02	-.2204E-04	-.5739E-04	.6705E-05	.6480E-06	79.96

TABLE A.6 G-Chart Coefficients for Solar Heating Systems with Air Working Fluid* (*Continued*)

NATICK	MA	.1369E-01	-.9031E-03	-.1813E-04	-.2141E-04	.1298E-05	.6628E-06	118.03
NEW ORLEANS	LA	.4925E-01	-.3253E-02	.4851E-05	-.3375E-03	.3818E-04	.5379E-06	44.02
NEWPORT	RI	.1577E-01	-.9508E-03	-.9771E-05	-.1936E-04	-.1164E-06	.6830E-06	112.74
NEW YORK	NY	.1187E-01	-.118E-02	-.4983E-07	-.1869E-04	.3847E-05	.4577E-06	97.30
NORFOLK	VA	.2673E-01	-.1412E-02	-.5594E-05	-.7837E-04	.3832E-05	.7583E-06	76.72
NORTH OMAHA	NE	.1683E-01	-.8824E-03	-.1854E-04	-.2273E-04	.3432E-06	.6574E-06	125.31
OAK RIDGE	TN	.1960E-01	-.1455E-02	-.9783E-05	-.5484E-04	.6848E-05	.4731E-06	83.82
OKLAHOMA CITY	OK	.3150E-01	-.1301E-02	-.1707E-04	-.9207E-04	.9306E-06	.1154E-05	79.94
PAGE	AZ	.3114E-01	-.9061E-03	-.7686E-05	-.1121E-04	-.9528E-05	.7016E-06	106.15
PARKERSBURG	WV	.1352E-01	-.1242E-02	-.1619E-04	-.2626E-04	.4875E-05	.4592E-06	97.39
PASADENA	CA	.6335E-01	-.2183E-02	.9446E-05	-.8437E-04	-.4753E-04	.1860E-05	48.82
PENSACOLA	FL	.6056E-01	-.3232E-02	-.2603E-04	-.5394E-03	.5018E-04	.5761E-06	47.02
PEORIA	IL	.1346E-01	-.9927E-03	-.3445E-05	-.2236E-04	.2620E-05	.3273E-06	117.31
PHOENIX	AZ	.0164E-01	-.2835E-02	-.1515E-04	-.1054E-02	.3509E-04	.1544E-05	46.62
PHILADELPHIA	PA	.1793E-01	-.1084E-02	-.1802E-04	-.3914E-04	.1993E-05	.8790E-06	98.14
PITTSBURGH	PA	.1535E-01	-.1082E-02	-.7393E-05	-.2941E-04	.3001E-05	.4528E-06	104.56
POCATELLO	ID	.1877E-01	-.9066E-03	-.2662E-05	-.3913E-04	.1195E-05	.3686E-06	132.33
PORT ARTHUR	TX	.5613E-01	-.3261E-02	-.3324E-04	-.4921E-03	.4931E-04	.7548E-06	46.09
PORTLAND	ME	.1454E-01	-.7919E-03	-.1293E-04	-.1613E-04	-.8572E-06	.6340E-06	136.18
PORTLAND	OR	.1304E-01	-.1427E-02	-.2526E-05	-.2653E-04	.6965E-05	.3795E-06	97.00
PROSSER	WA	.1951E-01	-.1251E-02	-.8884E-05	-.6534E-04	.7015E-05	.2254E-06	109.69
PUEBLO	CO	.2786E-01	-.8481E-03	-.2233E-05	-.6087E-05	-.8690E-05	.4999E-06	106.36
PULLMAN	WA	.1649E-01	-.1116E-02	-.1332E-04	-.4260E-04	.4293E-05	.4608E-06	125.50
PUT-IN-BAY	OH	.1027E-01	-.9663E-03	.3950E-05	-.1478E-04	.2870E-05	.3299E-06	114.52
RALEIGH	NC	.2876E-01	-.1366E-02	-.1487E-04	-.6752E-04	.6995E-06	.1045E-05	77.13
RALEIGH-DURHAM	NC	.2760E-01	-.1377E-02	-.2577E-04	-.8777E-04	.2988E-05	.1370E-05	77.13
RAPID CITY	SD	.1898E-01	-.7420E-03	-.1418E-04	-.1577E-04	-.3614E-05	.6544E-06	136.71
RENO	NV	.2868E-01	.8408E-03	.1878E-05	-.2479E-04	-.9975E-05	.4873E-06	116.13
RICHLAND	WA	.1570E-01	-.1399E-02	-.2367E-05	-.4369E-04	.7697E-05	.1681E-07	98.56
RICHMOND	VA	.2233E-01	-.1306E-02	.4345E-06	-.5848E-04	.3931E-05	.5014E-06	83.74
RIVERSIDE	CA	.6842E-01	-.2348E-02	-.1011E-04	-.2411E-03	-.1451E-04	.9672E-06	52.33
ROCHESTER	NY	.1145E-01	-.9415E-03	-.4870E-05	-.1689E-04	.2603E-05	.3305E-06	126.97
SACRAMENTO	CA	.3831E-01	-.1990E-02	.3004E-06	-.2394E-03	.1824E-04	.4194E-06	66.69
ST. CLOUD	MN	.1211E-01	-.6769E-03	-.1717E-04	-.1220E-04	-.3299E-06	.5187E-06	160.40
ST. LOUIS	MO	.1794E-01	-.1169E-02	-.1001E-04	-.4071E-04	.3451E-05	.4819E-06	96.35
SALT LAKE CITY	UT	.1825E-01	-.1039E-02	-.2080E-04	-.4391E-04	.2429E-05	.6508E-06	115.53
SAN ANTONIO	TX	.6537E-01	-.3001E-02	-.2100E-04	-.5690E-03	.4095E-04	.9328E-06	46.52
SAN DIEGO	CA	.5838E-01	-.1994E-02	-.9571E-05	.2662E-03	-.9598E-04	.6829E-05	45.91
SAN FRANCISCO	CA	.3631E-01	-.5952E-03	-.2176E-04	.2744E-03	-.9152E-04	.5872E-05	70.38
SAN JOSE	CA	.4341E-01	-.2223E-02	-.3513E-04	-.2545E-03	.1152E-04	.2601E-05	58.75
SANTA MARIA	CA	.4982E-01	-.1060E-02	.4438E-04	.3572E-03	-.8923E-04	.8805E-06	68.62
SAVANNAH	GA	.4986E-01	-.2568E-02	-.2125E-04	-.3383E-03	.2708E-04	.8887E-06	52.84
SAULT ST. MARIE	MI	.1154E-01	-.8043E-03	-.2355E-04	-.2085E-04	.1553E-05	.5560E-06	163.20
SCHENECTADY	NY	.8697E-02	-.8778E-03	-.1984E-05	-.8643E-05	.2339E-05	.3224E-06	128.50
SEATTLE	WA	.1405E-01	-.1564E-02	.3860E-04	-.3490E-04	.9232E-05	-.2197E-06	91.29
SHREVEPORT	LA	.3886E-01	-.2192E-02	.1245E-04	-.2120E-03	.1840E-04	.2694E-06	56.45
SILVER HILL	MD	.1995E-01	-.1212E-02	-.1058E-04	-.4563E-04	.3111E-05	.6201E-06	87.97
SPOKANE	WA	.1448E-01	-.1023E-02	-.1898E-04	-.3324E-04	.3057E-05	.4496E-06	128.78
STATE COLLEGE	PA	.1112E-01	-.1030E-02	-.8899E-05	-.1694E-04	.3288E-05	.3594E-06	117.85
STILLWATER	OK	.2739E-01	-.1346E-02	-.7877E-05	-.8297E-04	.2972E-05	.7149E-06	78.95
SUMMIT	MT	.1014E-01	-.8722E-03	-.1578E-04	-.1566E-04	.2480E-05	.2888E-06	187.77
SYRACUSE	NY	.9797E-02	-.9759E-03	-.4789E-05	-.1307E-04	.3090E-05	.2905E-06	126.33
TALLAHASSEE	FL	.6411E-01	-.2808E-02	-.1310E-04	-.3232E-03	.1563E-04	.1890E-05	46.79
TAMPA	FL	.8726E-01	-.1007E-02	-.2516E-04	.1457E-02	-.4250E-03	.1993E-04	33.65
TRENTON	NJ	.1806E-01	-.1065E-02	-.1095E-04	-.3488E-04	.1555E-05	.7469E-06	99.42
TUCSON	AZ	.8202E-01	-.2683E-02	.1085E-04	-.7821E-03	.3423E-04	-.2742E-06	50.48
TULSA	OK	.2449E-01	-.1322E-02	-.6363E-05	-.6403E-04	.3052E-05	.6694E-06	79.71
TWIN FALLS	ID	.1678E-01	-.1014E-02	-.3673E-04	-.4280E-04	.2346E-05	.9722E-06	120.83
WASHINGTON	DC	.1804E-01	-.1217E-02	.3038E-06	-.4018E-04	.3354E-05	.5473E-06	88.17
WICHITA	KS	.2305E-01	-.1120E-02	-.2002E-04	-.5336E-04	.1384E-05	.8987E-06	95.37
YUMA	AZ	.1080E-00	-.2809E-02	-.3479E-04	-.1033E-02	-.3144E-04	.6555E-05	38.13

TABLE A.7 G-Chart Coefficients for Solar Heating Systems with Liquid Working Fluid*

		A	B	C	D	E	F
ABILENE	TX	.6434E-01	-.2886E-02	-.1067E-04	-.5819E-03	.2848E-04	.1802E-05
ALBANY	NY	.1688E-01	-.1492E-02	.1677E-05	-.4632E-04	.6591E-05	.6190E-06
ALBUQUERQUE	NM	.5337E-01	-.1762E-02	-.9273E-05	-.2563E-03	-.8658E-05	.1811E-05
AMARILLO	TX	.4881E-01	-.1827E-02	-.2434E-04	-.2370E-03	-.4112E-05	.2238E-05
AMES	IA	.1873E-01	-.1346E-02	-.9207E-05	-.4797E-04	.3736E-05	.9341E-06
AMHERST	MA	.1309E-01	-.1506E-02	.2097E-04	-.2753E-04	.7467E-05	.3757E-06
ANNAPOLIS	MD	.2509E-01	-.1899E-02	-.1855E-04	-.9530E-04	.9920E-05	.1441E-05
APALACHICOLA	FL	.1142E-00	-.4805E-02	-.2252E-04	-.1765E-02	.8529E-04	.3735E-05
ASHEVILLE	NC	.3641E-01	-.1959E-02	-.1561E-04	-.1798E-03	.7663E-05	.1760E-05
ASTORIA	OR	.2190E-01	-.2369E-02	.1366E-04	-.8222E-04	.1840E-04	.6129E-06
ATLANTA	GA	.4365E-01	-.2570E-02	-.1586E-04	-.2900E-03	.2145E-04	.1812E-05
ATLANTIC CITY	NJ	.2882E-01	-.1853E-02	-.2738E-04	-.1147E-03	.6069E-05	.2110E-05
BIG SPRING	TX	.6539E-01	-.3018E-02	-.2521E-04	-.6278E-03	.3471E-04	.2182E-05
BILLINGS	MT	.2337E-01	-.1349E-02	-.3504E-04	-.6828E-04	.6992E-06	.1750E-05
BINGHAMPTON	NY	.1380E-01	-.1480E-02	.8170E-05	-.3035E-04	.7057E-05	.4277E-06
BIRMINGHAM	AL	.4460E-01	-.2880E-02	-.1468E-04	-.3237E-03	.3089E-04	.1527E-05
BISMARCK	ND	.1867E-01	-.1162E-02	-.1103E-04	-.4728E-04	.2469E-05	.7172E-06
BLUE HILL	MA	.1737E-01	-.1504E-02	-.7208E-05	-.4165E-04	.5454E-05	.1064E-05
BOISE	ID	.2602E-01	-.1840E-02	-.1921E-04	-.1144E-03	.1200E-04	.8673E-06
BOSTON	MA	.1641E-01	-.1628E-02	-.9357E-05	-.4022E-04	.7730E-05	.9990E-06
BOULDER	CO	.3039E-01	-.1465E-02	-.3371E-04	-.8644E-04	-.7214E-05	.2856E-05
BROWNSVILLE	TX	.1408E-00	-.4746E-02	-.1044E-03	-.2082E-02	-.1055E-03	.3016E-04
CAPE HATTERAS	NC	.6306E-01	-.2893E-02	-.2277E-04	-.5616E-03	.2853E-04	.2118E-05
CARIBOU	ME	.1426E-01	-.1189E-02	-.1851E-04	-.2993E-04	.3104E-05	.9159E-06
CHARLESTON	SC	.7034E-01	-.3768E-02	-.5013E-04	-.7686E-03	.5921E-04	.2767E-05
CHARLOTTE	NC	.4513E-01	-.2511E-02	-.1266E-04	-.3014E-03	.2053E-04	.1537E-05
CHATTANOOGA	TN	.3425E-01	-.2416E-02	-.7688E-05	-.1867E-03	.1961E-04	.1238E-05
CHICAGO	IL	.1830E-01	-.1491E-02	-.9007E-05	-.4964E-04	.5894E-05	.9241E-06
CLEVELAND	OH	.1356E-01	-.1665E-02	.3060E-04	-.2943E-04	.9385E-05	.1451E-06
COLUMBIA	MO	.2561E-01	-.1796E-02	-.1503E-04	-.9998E-04	.9150E-05	.1156E-05
COLUMBUS	OH	.1648E-01	-.1715E-02	.1428E-04	-.4560E-04	.9797E-05	.4305E-06
CORPUS CHRISTI	TX	.1267E-00	-.5557E-02	-.7684E-04	-.2551E-02	.1412E-03	.9736E-05
CORVALLIS	OR	.2024E-01	-.2392E-02	.4256E-04	-.7259E-04	.1982E-04	-.3130E-06
DALLAS	TX	.5834E-01	-.3311E-02	.1115E-0v	-.5351E-03	.4402E-04	.1032E-05
DAVIS	CA	.5362E-01	-.3536E-02	-.9929E-05	-.5140E-03	.5811E-04	.3262E-06
DAYTON	OH	.2002E-01	-.1667E-02	.2533E-05	-.6474E-04	.8789E-05	.7161E-06
DENVER	CO	.3584E-01	-.1263E-02	-.1264E-04	-.6683E-04	-.1598E-04	.2089E-05
DES MOINES	IA	.1813E-01	-.1443E-02	-.2342E-05	-.5127E-04	.6178E-05	.6655E-06
DETROIT	MI	.1456E-01	-.1584E-02	.1403E-04	-.3485E-04	.8292E-05	.3859E-06
DODGE CITY	KA	.3843E-01	-.1639E-02	-.4109E-05	-.1504E-03	-.1186E-05	.1334E-05
DULUTH	MN	.1348E-01	-.1211E-02	.1158E-04	-.2791E-04	.4314E-05	.3775E-06
EAST LANSING	MI	.1306E-01	-.1478E-02	.1884E-04	-.2770E-04	.7216E-05	.3448E-06
EL PASO	TX	.8390E-01	-.2676E-02	.1637E-05	-.9013E-03	.1995E-04	.1364E-05
ELY	NV	.3249E-01	-.1115E-02	-.2904E-04	-.6266E-04	-.1431E-04	.2039E-05
FARGO	ND	.1320E-01	-.1199E-02	-.8666E-05	-.2729E-04	.4271E-05	.4892E-06
FORT SMITH	AR	.3549E-01	-.2463E-02	.7699E-06	-.1997E-03	.2043E-04	.1147E-05
FORT WAYNE	IN	.1790E-01	-.1618E-02	-.1002E-04	-.5199E-04	.8216E-05	.7147E-06
FORT WORTH	TX	.6511E-01	-.3142E-02	-.1178E-04	-.6300E-03	.3722E-04	.1998E-05
FRESNO	CA	.5837E-01	-.3455E-02	-.2235E-04	-.5944E-03	.5846E-04	.8361E-06
GAINESVILLE	FL	.1184E-00	-.4673E-02	-.3846E-04	-.1585E-02	.1683E-04	.9750E-05
GLASGOW	MT	.2138E-01	-.1157E-02	-.1764E-04	-.5703E-04	.9862E-06	.9567E-06
GRAND JUNCTION	CO	.3445E-01	-.1495E-02	-.1922E-04	-.1312E-03	-.5707E-06	.1438E-05
GRAND LAKE	CO	.2300E-01	-.1071E-02	-.3201E-04	-.4162E-04	-.5681E-05	.1709E-05
GREAT FALLS	MT	.2279E-01	-.1336E-02	-.3647E-04	-.7082E-04	.1750E-05	.1630E-05
GREEN BAY	WI	.1366E-01	-.1333E-02	-.5262E-06	-.2931E-04	.5365E-05	.5232E-06
GREENSBORO	NC	.3461E-01	-.2189E-02	-.1321E-04	-.1771E-03	.1359E-04	.1576E-05
GRNVLE-SPTNBRG	NC	.4438E-01	-.2580E-02	-.1488E-04	-.3026E-03	.2277E-04	.1648E-05
GRIFFIN	GA	.5279E-01	-.2888E-02	-.1163E-04	-.4301E-03	.3128E-04	.1598E-05
HARTFORD	CT	.1993E-01	-.1380E-02	-.2164E-04	-.5041E-04	.2150E-05	.1557E-05
HOUSTON	TX	.9176E-01	-.5259E-02	-.3156E-04	-.1446E-02	.1421E-03	.7918E-06
INDIANAPOLIS	IN	.1701E-01	-.1711E-02	.1313E-04	-.4820E-04	.9662E-05	.4642E-06
INYOKERN	CA	.1111E-00	-.3077E-02	.3345E-05	-.1659E-02	.3974E-04	.3659E-06
ITHACA	NY	.1157E-01	-.1485E-02	.2907E-04	-.2090E-04	.7443E-05	.2279E-06
JACKSON	MS	.5390E-01	-.3449E-02	-.2382E-05	-.4824E-03	.4891E-04	.1056E-05
JACKSONVILLE	FL	.9650E-01	-.4854E-02	-.3612E-04	-.1305E-02	.8187E-04	.4995E-05
KANSAS CITY	MO	.2470E-01	-.1756E-02	-.1213E-04	-.9346E-04	.9012E-05	.1022E-05
KEY WEST	FL	.1727E-00	-.9641E-03	-.1211E-03	.1165E-01	-.2193E-02	.1014E-03
LAKE CHARLES	LA	.9298E-01	-.5247E-02	-.3442E-04	-.1436E-02	.1341E-03	.1488E-05
LANDER	WY	.3026E-01	-.1024E-02	-.2032E-04	-.4779E-04	-.1323E-04	.1622E-05
LANSING	MI	.1413E-01	-.1469E-02	.1695E-05	-.3167E-04	.6690E-05	.5723E-06
LARAMIE	WY	.2468E-01	-.1038E-02	-.3049E-04	-.3437E-04	-.9412E-05	.1946E-05
LAS VEGAS	NV	.8061E-01	-.2745E-02	-.6506E-05	-.8711E-03	.2470E-04	.1730E-05
LEMONT	IL	.1815E-01	-.1494E-02	-.9824E-05	-.4920E-04	.6010E-05	.9197E-06
LEXINGTON	KY	.2898E-01	-.1990E-02	-.2-52E-04	-.1357E-03	.1315E-04	.1087E-05
LINCOLN	NE	.2441E-01	-.1558E-02	-.6636E-05	-.8058E-04	.5251E-05	.1018E-05
LITTLE ROCK	AR	.3609E-01	-.2611E-02	-.2994E-05	-.2107E-03	.2418E-04	.1267E-05
LOS ANGELES	CA	.9270E-01	-.3031E-02	-.1366E-04	-.2531E-03	-.1508E-03	.1218E-04
LOUISVILLE	KY	.2237E-01	-.1913E-02	.1516E-04	-.8193E-04	.1209E-04	.6736E-06
LYNN	MA	.1481E-01	-.1697E-02	.2417E-04	-.3601E-04	.9647E-05	.3909E-06
MACON	GA	.6380E-01	-.3450E-02	-.1195E-04	-.6398E-03	.5053E-04	.1463E-05
MADISON	WI	.1605E-01	-.1339E-02	-.6051E-05	-.3848E-04	.4875E-05	.6774E-06
MANHATTAN	KA	.2554E-01	-.1668E-02	-.1623E-04	-.9248E-04	.6164E-05	.1304E-05
MEDFORD	OR	.2412E-01	-.2315E-02	.1713E-05	-.1048E-03	.1986E-04	-.1316E-06
MEMPHIS	TN	.3744E-01	-.2621E-02	.2799E-05	-.2284E-03	.2484E-04	.9286E-06
MIAMI	FL	.1611E-00	-.9134E-03	-.9214E-04	.8851E-02	-.1897E-02	.8552E-04
MIDLAND	TX	.6817E-01	-.2922E-02	-.7205E-05	-.6407E-03	.2812E-04	.1615E-05
MILWAUKEE	WI	.1570E-01	-.1446E-02	-.4001E-05	-.3940E-04	.6439E-05	.5793E-06
MINN-ST. PAUL	MN	.1369E-01	-.1294E-02	-.4412E-05	-.2944E-04	.5043E-05	.5209E-06
MT WEATHER	VA	.2288E-01	-.1538E-02	-.2237E-04	-.6964E-04	.3689E-05	.1587E-05
NASHVILLE	TN	.2783E-01	-.2292E-02	.3438E-04	-.1316E-03	.1917E-04	.4377E-06

TABLE A.7 G-Chart Coefficients for Solar Heating Systems with Liquid Working Fluid* ***(Continued)***

NATICK	MA	.1862E-01	-.1513E-02	-.1475E-04	-.5080E-04	.5852E-05	.1061E-05
NEW ORLEANS	LA	.7253E-01	-.5185E-02	-.1234E-04	-.8705E-03	.1050E-03	.1393E-05
NEWPORT	RI	.2215E-01	-.1622E-02	-.3569E-04	-.6768E-04	.4724E-05	.1847E-05
NEW YORK	NY	.1572E-01	-.1845E-02	.1438E-04	-.3897E-04	.1107E-04	.6466E-06
NORFOLK	VA	.3764E-01	-.2305E-02	-.7795E-05	-.2064E-03	.1489E-04	.1707E-05
NORTH OMAHA	NE	.2351E-01	-.1368E-02	-.2321E-05	-.6621E-04	.2510E-05	.8479E-06
OAK RIDGE	TN	.2668E-01	-.2280E-02	.6036E-05	-.1192E-03	.1773E-04	.7303E-06
OKLAHOMA CITY	OK	.4483E-01	-.2124E-02	-.7578E-05	.2554E-03	.8826E-05	.1499E-05
PAGE	AZ	.4499E-01	-.1432E-02	.6852E-05	-.1362E-03	-.1676E-04	.8911E-06
PARKERSBURG	WV	.1797E-01	-.1897E-02	.2440E-04	-.5360E-04	.1204E-04	.4313E-06
PASADENA	CA	.9371E-01	-.3622E-02	-.1810E-04	-.6393E-03	-.5623E-04	.7618E-05
PENSACOLA	FL	.8882E-01	-.5027E-02	-.4162E-04	-.1317E-02	.1258E-03	.1208E-05
PEORIA	IL	.1857E-01	-.1577E-02	.5389E-06	-.5561E-04	.7852E-05	.6442E-06
PHOENIX	AZ	.1364E-00	-.4556E-02	-.3700E-04	-.2786E-02	.1207E-03	.3880E-05
PHILADELPHIA	PA	.2465E-01	-.1789E-02	-.1805E-04	-.9071E-04	.8349E-05	.1380E-05
PITTSBURGH	PA	.2100E-01	-.1768E-02	-.8224E-05	-.7026E-04	.9540E-05	.9243E-06
POCATELLO	ID	.2604E-01	-.1504E-02	-.1999E-04	-.9769E-04	.5303E-05	.1087E-05
PORT ARTHUR	TX	.8110E-01	-.5010E-02	-.5533E-04	-.1126E-02	.1179E-03	.1647E-05
PORTLAND	ME	.2028E-01	-.1379E-02	-.3394E-04	-.5367E-04	.2458E-05	.1597E-05
PORTLAND	OR	.1675E-01	-.2167E-02	.5456E-04	-.4615E-04	.1594E-04	.1139E-06
PROSSER	WA	.2605E-01	-.2053E-02	-.8086E-05	-.1210E-03	.1696E-04	.1588E-06
PUEBLO	CO	.4010E-01	-.1434E-02	-.4517E-x5	-.1047E-03	-.1211E-04	.1291E-05
PULLMAN	WA	.2203E-01	-.1836E-02	-.1255E-04	-.8249E-04	.1153E-04	.6351E-06
PUT-IN-BAY	OH	.1379E-01	-.1624E-02	-.6055E-07	-.3124E-04	.8583E-05	.5069E-06
RALEIGH	NC	.4121E-01	-.2130E-02	-.1597E-04	-.2154E-03	.6481E-05	.2234E-05
RALEIGH-DURHAM	NC	.3849E-01	-.2243E-02	-.5223E-05	-.2102E-03	.1302E-04	.1659E-05
RAPID CITY	SD	.2684E-01	-.1203E-02	-.2196E-04	-.6703E-04	-.4484E-05	.1562E-05
RENO	NV	.4129E-01	-.1439E-02	-.3094E-04	-.1397E-03	-.1461E-04	.2332E-05
RICHLAND	WA	.2055E-01	-.2164E-02	.3932E-04	-.7612E-04	.1718E-04	-.5540E-06
RICHMOND	VA	.3122E-01	-.2131E-02	-.1170E-04	-.1491E-03	.1344E-04	.1449E-05
RIVERSIDE	CA	.1005E-00	-.3570E-02	-.3800E-04	-.9571E-03	-.1430E-04	.6730E-05
ROCHESTER	NY	.1543E-01	-.1543E-02	-.3620E-05	-.3837E-04	.7561E-05	.5941E-06
SACRAMENTO	CA	.5397E-01	-.3353E-02	-.3239E-04	-.5086E-03	.5091E-04	.1482E-05
ST. CLOUD	MN	.1672E-01	-.1107E-02	-.7041E-05	-.3489E-04	.1467E-05	.7639E-06
ST. LOUIS	MO	.2476E-01	-.1896E-02	-.9113E-05	-.9819E-04	.1125E-04	.1001E-05
SALT LAKE CITY	UT	.2519E-01	-.1669E-02	-.2287E-04	-.1005E-03	.8214E-05	.1140E-05
SAN ANTONIO	TX	.9561E-01	-.4641E-02	-.3060E-04	-.1426E-02	.1023E-03	.2609E-05
SAN DIEGO	CA	.8889E-01	-.3333E-02	-.6475E-04	-.1561E-03	-.1572E-03	.2115E-04
SAN FRANCISCO	CA	.5538E-01	-.1432E-02	-.1200E-03	.1030E-03	-.1486E-03	.1873E-04
SAN JOSE	CA	.6154E-01	-.3665E-02	-.5180E-04	-.5832E-03	.4362E-04	.5070E-05
SANTA MARIA	CA	.7396E-01	-.1834E-02	-.4302E-06	.2540E-03	-.1641E-03	.8281E-05
SAVANNAH	GA	.7216E-01	-.3963E-02	-.4932E-04	-.8316E-03	.6800E-04	.2732E-05
SAULT ST. MARIE	MI	.1498E-01	-.1352E-02	.5634E-05	-.3536E-04	.5131E-05	.5376E-06
SCHENECTADY	NY	.1134E-01	-.1400E-02	.1934E-04	-.1838E-04	.6377E-05	.4378E-06
SEATTLE	WA	.1778E-01	-.2485E-02	.8812E-04	-.5538E-04	.2091E-04	-.6951E-06
SHREVEPORT	LA	.5555E-01	-.3563E-02	.8560E-05	-.5062E-03	.5176E-04	.8107E-06
SILVER HILL	MD	.2767E-01	-.1985E-02	-.1413E-04	-.1159E-03	.1109E-04	.1410E-05
SPOKANE	WA	.1942E-01	-.1665E-02	-.2136E-04	-.6525E-04	.8881E-05	.8182E-06
STATE COLLEGE	PA	.1465E-01	-.1595E-02	.1692E-04	-.3466E-04	.8351E-05	.4378E-06
STILLWATER	OK	.3867E-01	-.2237E-02	-.4697E-06	-.2195E-03	.1446E-04	.1367E-05
SUMMIT	MT	.1295E-01	-.1307E-02	.1811E-04	-.2713E-04	.5500E-05	.2597E-06
SYRACUSE	NY	.1276E-01	-.1519E-02	.2045E-04	-.2565E-04	.7699E-05	.3332E-06
TALLAHASSEE	FL	.9602E-01	-.4361E-02	-.1218E-04	-.1124E-02	.5357E-04	.2812E-05
TAMPA	FL	.1331E-00	-.1951E-02	.8175E-04	.2129E-02	-.8096E-03	.1678E-04
TRENTON	NJ	.2504E-01	-.1754E-02	-.1406E-04	-.9014E-04	.7348E-05	.1396E-05
TUCSON	AZ	.1212E-00	-.4146E-02	-.2523E-04	-.2095E-02	.9134E-04	.1884E-05
TULSA	OK	.3462E-01	-.2179E-02	-.5700E-05	-.1757E-03	.1341E-04	.1420E-05
TWIN FALLS	ID	.2255E-01	-.1688E-02	-.8680E-05	-.8239E-04	.8698E-05	.8826E-06
WASHINGTON	DC	.2495E-01	-.2018E-02	-.1338E-04	-.9951E-04	.1208E-04	.1366E-05
WICHITA	KS	.3230E-01	-.1797E-02	-.1194E-04	-.1402E-03	.6735E-05	.1341E-05
YUMA	AZ	.1635E-00	-.4248E-02	-.6694E-04	-.3288E-02	-.3523E-04	.1675E-04

TABLE A.8 Interest Factors for Various Discount Rates: (*a*) Present-worth Factor $P = F\frac{1}{(1+i)^N}$*

	Discount rates						
N	*6%*	*8%*	*10%*	*12%*	*15%*	*20%*	*25%*
1	.943	.926	.909	.893	.870	.833	.800
2	.890	.857	.826	.797	.756	.694	.640
3	.840	.794	.751	.712	.658	.579	.512
4	.792	.735	.683	.636	.572	.482	.410
5	.747	.681	.621	.567	.497	.402	.328
6	.705	.630	.565	.507	.432	.335	.262
7	.665	.584	.513	.452	.376	.279	.210
8	.627	.540	.467	.404	.327	.233	.168
9	.592	.500	.424	.361	.284	.194	.134
10	.558	.463	.386	.322	.247	.162	.107
11	.527	.429	.351	.288	.215	.135	.086
12	.497	.397	.319	.257	.187	.112	.069
13	.469	.368	.290	.229	.163	.094	.055
14	.442	.341	.263	.205	.141	.078	.044
15	.417	.315	.239	.183	.123	.065	.035
16	.394	.292	.218	.163	.107	.054	.028
17	.371	.270	.198	.146	.093	.045	.022
18	.350	.250	.180	.130	.081	.038	.018
19	.331	.232	.164	.116	.070	.031	.014
20	.312	.215	.149	.104	.061	.026	.012
21	.294	.199	.135	.093	.053	.022	.009
22	.278	.184	.123	.083	.046	.018	.007
23	.262	.170	.112	.074	.040	.015	.006
24	.247	.158	.102	.066	.035	.013	.005
25	.233	.146	.092	.059	.030	.010	.004
26	.220	.135	.084	.053	.026	.009	.003
27	.207	.125	.076	.047	.023	.007	.002
28	.196	.116	.069	.042	.020	.006	.002
29	.185	.107	.063	.037	.017	.005	.002
30	.174	.099	.057	.033	.015	.004	.001

*A future sum of money *F* times PWF for appropriate discount rate *i* gives present value *P* of *F*.

TABLE A.8 Interest Factors for Various Discount Rates: (*b*) Capital Recovery Factor $A = P\frac{i(1+i)^N}{(1+i)^N - 1}$*

	Discount rates						
N	*6%*	*8%*	*10%*	*12%*	*15%*	*20%*	*25%*
1	1.060	1.080	1.100	1.120	1.150	1.200	1.250
2	.545	.561	.576	.592	.615	.655	.694
3	.374	.388	.402	.416	.438	.475	.512
4	.289	.302	.315	.329	.350	.386	.423
5	.237	.250	.264	.277	.298	.334	.372
6	.203	.216	.230	.243	.264	.301	.339
7	.179	.192	.205	.219	.240	.277	.316
8	.161	.174	.187	.201	.223	.261	.300
9	.147	.160	.174	.188	.210	.248	.289
10	.136	.149	.163	.177	.199	.239	.280
11	.127	.140	.154	.168	.191	.231	.273
12	.119	.133	.147	.161	.184	.225	.268
13	.113	.127	.141	.156	.179	.221	.265
14	.108	.121	.136	.151	.175	.217	.262
15	.103	.117	.131	.147	.171	.214	.259
16	.099	.113	.128	.143	.168	.211	.257
17	.095	.110	.125	.140	.165	.209	.256
18	.092	.107	.122	.138	.163	.208	.255
19	.090	.104	.120	.136	.161	.206	.254
20	.087	.102	.117	.134	.160	.205	.253
21	.085	.100	.116	.132	.158	.204	.252
22	.083	.098	.114	.131	.157	.204	.252
23	.081	.096	.113	.130	.156	.203	.251
24	.080	.095	.111	.128	.155	.203	.251
25	.078	.094	.110	.128	.155	.202	.251
26	.077	.013	.109	.127	.154	.202	.251
27	.076	.091	.108	.126	.154	.201	.251
28	.075	.090	.107	.125	.153	.201	.250
29	.074	.090	.107	.125	.153	.201	.250
30	.073	.089	.106	.124	.152	.201	.250

*A present sum of money *P* times CRF for appropriate discount rate *i* gives an end-of-period payment *A* equivalent over *N* discount periods to *P*.

TABLE A.9 P-Chart Coefficients *A* and *B**

Station name		With R9 night insulation						Without night insulation						Degree days
		I Direct gain		*II* Masonry wall		*III* Water wall		*IV* Direct gain		*V* Masonry wall		*VI* Water wall		
		A	B	A	B	A	B	A	B	A	B	A	B	
Washington-Sterling	DC	.459	.242	.534	.172	.725	.120	.261	.204	.370	.125	.414	.106	5010
Birmingham	AL	.549	.180	.584	.141	.813	.097	.455	.142	.534	.091	.586	.085	2844
Prescott	AZ	.597	.269	.611	.224	.977	.127	.580	.175	.645	.119	.900	.084	4456
Sacramento	CA	.465	.299	.488	.232	.601	.184	.453	.206	.493	.136	.577	.121	2843
San Francisco	CA	.543	.343	.528	.299	.705	.213	1.316	.094	.717	.127	.933	.103	3042
Denver	CO	.555	.305	.596	.239	.924	.141	.481	.197	.538	.142	.643	.118	6016
Hartford	CT	.424	.255	.499	.186	.659	.132	.094	.544	.249	.174	.237	.167	6350
Wilmington	DE	.474	.234	.535	.175	.738	.119	.288	.186	.394	.117	.440	.102	4940
Tallahassee	FL	.618	.147	.626	.120	.959	.073	.468	.153	.622	.078	.688	.076	1563
Atlanta	GA	.537	.190	.567	.151	.833	.095	.420	.155	.511	.096	.567	.088	3095
Boise	ID	.392	.439	.439	.322	.509	.278	.279	.336	.340	.221	.376	.199	5833
Chicago	IL	.403	.292	.488	.199	.622	.149	.121	.478	.260	.186	.252	.180	6127
Indianapolis	IN	.412	.258	.500	.175	.647	.129	.124	.424	.262	.168	.253	.165	5577
Des Moines	IA	.421	.306	.500	.216	.664	.152	.130	.515	.260	.214	.255	.207	6710
Wichita	KS	.505	.261	.566	.192	.811	.125	.332	.237	.448	.136	.475	.130	4687
Louisville	KY	.454	.230	.525	.164	.720	.112	.227	.240	.347	.130	.354	.126	4645
Shreveport	LA	.562	.170	.590	.133	.884	.083	.483	.140	.584	.082	.599	.086	2167
Caribou	ME	.378	.378	.441	.289	.498	.257	N-R	N-R	N-R	N-R	N-R	N-R	9632
Baltimore	MD	.481	.228	.547	.168	.745	.117	.312	.175	.411	.113	.462	.098	4729
Boston	MA	.434	.259	.509	.185	.681	.130	.168	.323	.293	.162	.312	.141	5621
Detroit	MI	.401	.265	.469	.196	.590	.148	.075	.798	.223	.198	.195	.214	6228
Minneapolis-St. Paul	MN	.392	.314	.445	.250	.539	.198	N-R	N-R	N-R	N-R	N-R	N-R	8159
Kansas City	MO	.469	.257	.528	.193	.721	.132	.232	.282	.350	.154	.368	.143	5357
Billings	MT	.406	.404	.477	.282	.577	.229	.206	.384	.309	.220	.313	.212	7265
Helena	MT	.358	.496	.430	.336	.496	.291	.136	.588	.253	.273	.250	.264	8190
North Omaha	NE	.435	.306	.515	.215	.713	.143	.160	.431	.285	.203	.288	.191	6601
Las Vegas	NV	.583	.239	.562	.212	.792	.141	.552	.184	.640	.111	.717	.105	2601
Concord	NH	.419	.274	.482	.210	.612	.158	N-R	N-R	.213	.224	.185	.234	7360
Newark	NJ	.464	.237	.529	.175	.724	.120	.258	.205	.370	.124	.402	.110	5034
Albuquerque	NM	.580	.257	.621	.198	.968	.117	.615	.147	.644	.107	.793	.088	4292

TABLE A.9 P-Chart Coefficients *A* and *B** (*Continued*)

Station name		With R9 night insulation						Without night insulation						Degree days
		I Direct gain		II Masonry wall		III Water wall		IV Direct gain		V Masonry wall		VI Water wall		
		A	B	A	B	A	B	A	B	A	B	A	B	
Farmington	NM	.543	.307	.586	.239	.866	.150	.414	.238	.505	.153	.671	.111	5713
Buffalo	NY	.391	.243	.453	.191	.521	.164	N-R	N-R	N-R	N-R	N-R	N-R	7285
New York City (La Guardia)	NY	.477	.227	.539	.170	.727	.119	.273	.196	.380	.121	.414	.108	4909
Rochester	NY	.390	.246	.499	.193	.523	.163	N-R	N-R	N-R	N-R	N-R	N-R	6719
Charlotte	NC	.538	.202	.569	.160	.810	.105	.434	.160	.512	.102	.574	.093	3218
Bismarck	ND	.379	.392	.436	.299	.515	.247	N-R	N-R	.190	.337	N-R	N-R	9044
Columbus	OH	.411	.247	.487	.177	.625	.130	.095	.551	.243	.172	.222	.178	5702
Burns	OR	.402	.436	.472	.302	.559	.254	.226	.385	.319	.230	.353	.199	7212
Salem	OR	.385	.373	.437	.267	.494	.241	.224	.381	.308	.213	.347	.185	4852
Harrisburg	PA	.447	.240	.522	.172	.701	.121	.207	.243	.328	.135	.352	.119	5224
Providence	RI	.443	.263	.521	.188	.705	.130	.173	.312	.297	.162	.334	.131	5972
Columbia	SC	.558	.186	.600	.141	.855	.093	.503	.138	.584	.087	.611	.088	2598
Pierre	SD	.389	.385	.465	.268	.579	.205	.104	.792	.236	.273	.218	.285	7677
Nashville	TN	.483	.197	.543	.145	.749	.099	.279	.199	.393	.111	.403	.110	3696
Lubbock	TX	.574	.243	.622	.183	.996	.104	.594	.150	.640	.103	.804	.084	3545
Salt Lake City	UT	.444	.381	.500	.277	.623	.217	.437	.228	.389	.193	.443	.167	5983
Richmond	VA	.498	.219	.541	.170	.739	.117	.368	.168	.454	.108	.514	.095	3939
Charleston	WV	.439	.227	.511	.162	.674	.116	.203	.251	.320	.134	.330	.126	4590
La Crosse	WI	.402	.294	.461	.227	.579	.170	N-R	N-R	.198	.256	N-R	N-R	7417
Cheyenne	WY	.529	.337	.584	.259	.876	.159	.395	.235	.488	.158	.586	.128	7255
Sheridan	WY	.412	.405	.484	.285	.610	.217	.194	.406	.306	.224	.314	.210	7708

TABLE A.10 Average Monthly Temperature at Source for Cold-Water Supply, °F*

State	City	Jan.	Feb.	Mar.	Apr.	May	June	July	Aug.	Sep.	Oct.	Nov.	Dec.	Deep well temperature
Alabama	Montgomery	50	52	58	63	73	78	82	81	79	69	56	50	67
Alaska	Anchorage	—	—	—	—	—	—	—	—	—	—	—	—	—
Arizona	Phoenix	48	48	50	52	57	59	63	75	79	69	59	54	67
Arkansas	Little Rock	48	47	57	68	77	85	88	86	82	77	64	58	65
California	Fresno	72	72	72	72	72	72	72	72	72	72	72	72	72
	Los Angeles	50	50	54	63	68	73	74	76	75	69	61	55	65
	San Diego	50	50	54	63	68	73	74	76	75	69	61	55	66
	San Francisco	60	60	60	60	60	60	60	60	60	60	60	60	60
Colorado	Denver	39	40	43	49	55	60	63	64	63	56	45	37	50
Connecticut	Hartford	40	40	40	44	54	59	66	70	69	59	50	40	51
Delaware	Wilmington	50	48	50	52	68	73	78	79	73	69	60	55	57
D.C.	Washington	42	42	52	56	63	67	67	78	79	68	55	46	57
Florida	Jacksonville	71	71	75	79	81	85	86	87	82	77	75	72	71
	Miami	70	70	70	70	70	70	70	70	70	70	70	70	78
	Tampa	75	75	76	80	85	87	85	85	83	80	77	75	75
Georgia	Atlanta	43	48	53	59	72	78	84	80	78	70	60	48	65
Hawaii	Honolulu	80	80	80	80	80	80	80	80	80	80	80	80	80
Idaho	Boise	52	52	52	52	52	52	52	52	52	52	52	52	52

TABLE A.10 Average Monthly Temperature at Source for Cold-Water Supply, °F* (*Continued*)

State	*City*	*Jan.*	*Feb.*	*Mar.*	*Apr.*	*May*	*June*	*July*	*Aug.*	*Sep.*	*Oct.*	*Nov.*	*Dec.*	*Deep well temperature*
Illinois	Chicago	32	32	34	42	51	57	65	67	62	57	45	35	52
	Peoria	50	50	50	54	54	56	56	56	54	50	50	50	53
Indiana	Indianapolis	50	50	50	53	68	73	80	82	77	70	60	50	55
Iowa	Des Moines	40	40	40	44	49	58	66	73	71	62	50	45	52
Kansas	Topeka	50	52	57	64	70	76	84	80	74	64	58	50	56
Kentucky	Louisville	40	40	45	49	69	77	82	82	77	70	60	50	58
Louisiana	New Orleans	55	55	60	66	77	86	89	90	90	80	70	60	70
	Shreveport	55	55	60	67	78	88	91	89	84	78	69	60	68
Maine	Portland	35	35	40	45	50	56	64	66	64	58	47	40	47
Maryland	Baltimore	40	40	43	47	53	61	66	70	64	58	50	46	57
Massachusetts	Boston	32	36	39	52	58	71	74	67	60	56	48	45	50
Michigan	Detroit	35	35	38	41	56	64	75	74	68	63	55	45	47
Minnesota	Minneapolis	32	32	35	41	61	69	80	73	68	60	50	40	44
Mississippi	Jackson	60	60	65	70	75	80	82	74	74	69	65	61	68
Missouri	St. Louis	47	47	50	53	69	77	85	83	75	70	62	55	57
Montana	Great Falls	42	42	42	42	42	42	42	42	42	42	42	42	42
Nebraska	Lincoln	40	40	45	51	56	68	81	79	69	60	50	45	53
Nevada	Las Vegas	73	73	73	73	73	73	73	73	73	73	73	73	73
	Reno	52	52	52	52	52	52	52	52	52	52	52	52	52

New Jersey	Newark	40	40	45	45	56	64	69	71	71	65	60	50	55
	Trenton	35	35	38	40	68	71	79	77	72	65	55	45	55
New Hampshire	Manchester	35	35	38	40	45	52	58	67	65	55	45	40	47
New Mexico	Albuquerque	72	72	72	72	72	72	72	72	72	72	72	72	62
New York	Albany	32	32	35	40	52	60	56	66	65	55	45	40	49
	Buffalo	32	32	32	32	37	62	71	73	66	55	45	35	47
	New York City	36	35	36	39	47	54	58	60	61	57	48	45	52
	Syracuse	35	35	39	44	50	60	66	70	68	60	50	40	47
North Carolina	Greensboro	55	55	55	60	67	77	83	82	79	72	67	60	62
North Dakota	Bismarck	42	42	42	42	42	42	42	42	42	42	42	42	42
Ohio	Cincinnati	40	40	45	49	66	76	82	81	77	70	60	50	55
	Columbus	38	38	40	46	64	72	76	76	74	65	55	45	54
	Toledo	35	35	40	48	66	72	87	85	72	60	50	40	52
Oklahoma	Oklahoma City	45	45	50	55	68	73	77	77	72	65	55	50	62
Oregon	Portland	35	35	38	44	50	56	62	55	52	45	40	35	50
Pennsylvania	Philadelphia	35	35	38	40	68	71	79	77	72	60	50	40	53
	Pittsburgh	38	38	40	46	66	75	81	81	75	68	55	40	52
Rhode Island	Providence	40	40	45	48	56	62	64	65	63	50	45	40	52
South Carolina	Charleston	55	58	60	65	75	81	83	88	80	75	70	65	67
South Dakota	Rapid City	55	55	55	55	55	55	55	55	55	55	55	55	47
Tennessee	Knoxville	45	45	50	59	76	82	84	84	80	70	60	50	55

TABLE A.10 Average Monthly Temperature at Source for Cold-Water Supply, °F* (*Continued*)

State	*City*	*Jan.*	*Feb.*	*Mar.*	*Apr.*	*May*	*June*	*July*	*Aug.*	*Sep.*	*Oct.*	*Nov.*	*Dec.*	*Deep well temperature*
	Memphis	55	55	58	60	60	68	68	70	70	65	60	60	62
	Nashville	46	46	53	66	63	69	71	75	75	71	58	53	55
Texas	Brownsville	70	70	70	75	80	84	86	87	89	85	80	75	75
	Dallas/Fort Worth	56	49	57	70	75	81	79	83	81	72	56	46	67
	El Paso	70	70	75	80	82	84	85	85	84	80	75	70	65
	San Antonio	65	65	65	68	70	75	78	80	77	75	70	65	67
Utah	Salt Lake City	35	37	38	41	43	47	53	52	48	43	38	37	52
Vermont	Burlington	32	32	38	45	50	58	63	66	68	60	50	40	47
Virginia	Norfolk	50	50	55	62	70	78	83	83	80	70	60	55	62
Washington	Seattle	39	37	43	45	48	57	60	68	66	57	48	43	52
	Spokane	47	47	47	47	47	47	47	47	47	47	47	47	52
West Virginia	Charleston	50	50	55	60	70	79	83	82	77	70	60	55	55
Wisconsin	Madison	35	35	40	45	50	53	52	53	52	50	48	40	47
Wyoming	Lander	45	45	45	45	45	45	45	45	45	45	45	45	45
Puerto Rico	San Juan	80	80	80	80	80	80	80	80	80	80	80	80	80

*Based on *Handbook of Air Conditioning System Design.*

TABLE A.11 Liquid Transfer Fluid Properties*†

Property	*Water*	*50% ethylene glycol/water*	*50% propylene glycol/water*	*Silicone fluid*	*Aromatics*	*Paraffinic oil*
Freezing point °F (°C)	32 (0)	−33 (−36)	−28 (−33)	−58 (−50)	−100 to −25 (−73 to −32)	
Boiling point °F (°C) at atmospheric pressure	212 (100)	230 (110)		None	300–400 (149–204)	700 (371)
Fluid stability	Requires pH or inhibitor monitoring	Requires pH or inhibitor monitoring	Requires pH or inhibitor monitoring	Good	Good	Good
Flash point °F (°C)‡	None	None	600 (315)	600 (315)	145–300 (63–149)	455 (235)
Specific heat (73°F), Btu/(lb·°F)	1.0	0.80	0.85	0.34–0.48	0.36–0.42	0.46
Viscosity, (centistokes at 77°F)	0.9	21	5	50–50,000	1–100	
Toxicity	Depends on inhibitor used	Depends on inhibitor used	Depends on inhibitor used	Low	Moderate	

*From Ref. 33.

†These data are extracted from manufacturers' literature to illustrate the properties of a few types of liquid that have been used as transfer fluids.

‡It is important to identify the conditions of tests for measuring flash point. Since the manufacturers' literature does not always specify the test, these values may not be directly comparable.

TABLE A.12 Solar Absorptance of Various Passive Wall Materials

Material	Absorptance
Optical flat black paint	0.98
Flat black paint	0.95
Black lacquer	0.92
Dark-gray paint	0.91
Black concrete	0.91
Dark-blue lacquer	0.91
Black oil paint	0.90
Stafford blue bricks	0.89
Dark-olive-drab paint	0.89
Dark-brown paint	0.88
Dark-blue-gray paint	0.88
Azure blue or dark-green lacquer	0.88
Brown concrete	0.85
Medium-brown paint	0.84
Medium-light-brown paint	0.80
Brown or green lacquer	0.79
Medium-rust paint	0.78
Light-gray oil paint	0.75
Red oil paint	0.74
Red bricks	0.70
Uncolored concrete	0.65
Moderately light buff bricks	0.60
Medium-dull-green paint	0.59
Medium-orange paint	0.58
Medium-yellow paint	0.57
Medium-blue paint	0.51
Medium-Kelly-green paint	0.51
Light-green paint	0.47
White semigloss paint	0.30

NOTES: 1. Materials with absorptance values below 0.50 are of little use for passive wall outer surfaces.

2. This table is meant to serve as a guide only. Variations in texture, tone, overcoats, pigments, binders, etc., can vary these values.

3. Data derived from G. G. Gubareff et al., *Thermal Radiation Properties Survey*, 2d ed., Honeywell Research Center, Minneapolis-Honeywell Regulator Company, Minneapolis, Minnesota, 1960; from S. Moore, Los Alamos Scientific Laboratory, Solar Energy Group, unpublished data; and from Ref. 57 with permission.

The altitude and azimuth of the sun are given by

$$\sin a = \sin \phi \sin \delta + \cos \phi \cos \delta \cos h \tag{1}$$

and

$$\sin \alpha = -\cos \delta \sin h / \cos a \tag{2}$$

where a = altitude of the sun (angular elevation above the horizon)
ϕ = latitude of the observer
δ = declination of the sun
h = hour angle of sun (angular distance from the meridian of the observer)
α = azimuth of the sun (measured eastward from north)

From Eqs. (1) and (2) it can be seen that the altitude and azimuth of the sun are functions of the latitude of the observer, the time of day (hour angle), and the date (declination).

Figure A1 (*b-g*) provides a series of charts, one for each 5° of latitude (except 5°, 15°, 75°, and 85°) giving the altitude and azimuth of the sun as a function of the true solar time and the declination of the sun in a form originally suggested by Hand. Linear interpolation for intermediate latitudes will give results within the accuracy to which the charts can be read.

On these charts, a point corresponding to the projected position of the sun is determined from the heavy lines corresponding to declination and solar time.

To find the solar altitude and azimuth:

1. Select the chart or charts appropriate to the latitude.
2. Find the solar declination δ corresponding to the date.
3. Determine the *true solar time* as follows:
   (a) To the *local standard time* (zone time) add 4′ for each degree of longitude the station is east of the standard meridian or subtract 4′ for each degree west of the standard meridian to get the *local mean solar time.*
   (b) To the *local mean solar time* add algebraically the equation of time; the sum is the required *true solar time.*
4. Read the required altitude and azimuth at the point determined by the declination and the true solar time. Interpolate linearly between two charts for intermediate latitudes.

It should be emphasized that the solar altitude determined from these charts is the true geometric position of the center of the sun. At low solar elevations terrestrial refraction may considerably alter the apparent position of sun. Under average atmospheric refraction the sun will appear on the horizon when it actually is about 34′ below the horizon; the effect of refraction decreases rapidly with increasing solar elevation. Since sunset or sunrise is defined as the time when the upper limb of the sun appears on the horizon, and the semidiameter of the sun is 16′, sunset or sunrise occurs under average atmospheric refraction when the sun is 50′ below the horizon. In polar regions especially, unusual atmospheric refraction can make considerable variation in the time of sunset or sunrise.

Altitude and azimuth in southern latitudes. To compute solar altitude and azimuth for southern latitudes, change the sign of the solar declination and proceed as above. The resulting azimuths will indicate angular distance from *south* (measured eastward) rather than from north.

(*a*)

FIG. A.1 Sun-path diagrams for various latitudes.

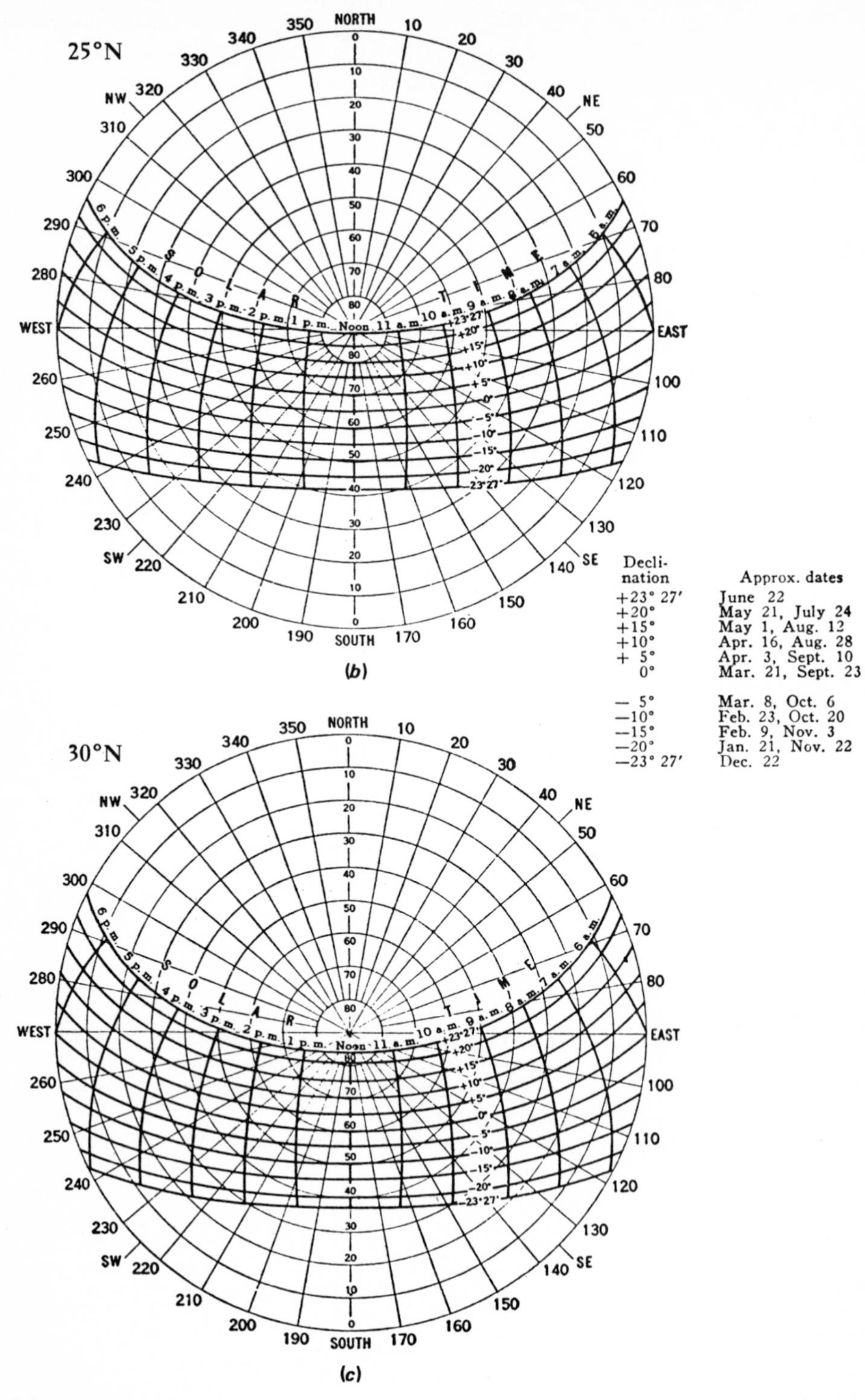

FIG. A.1 *(Continued)*

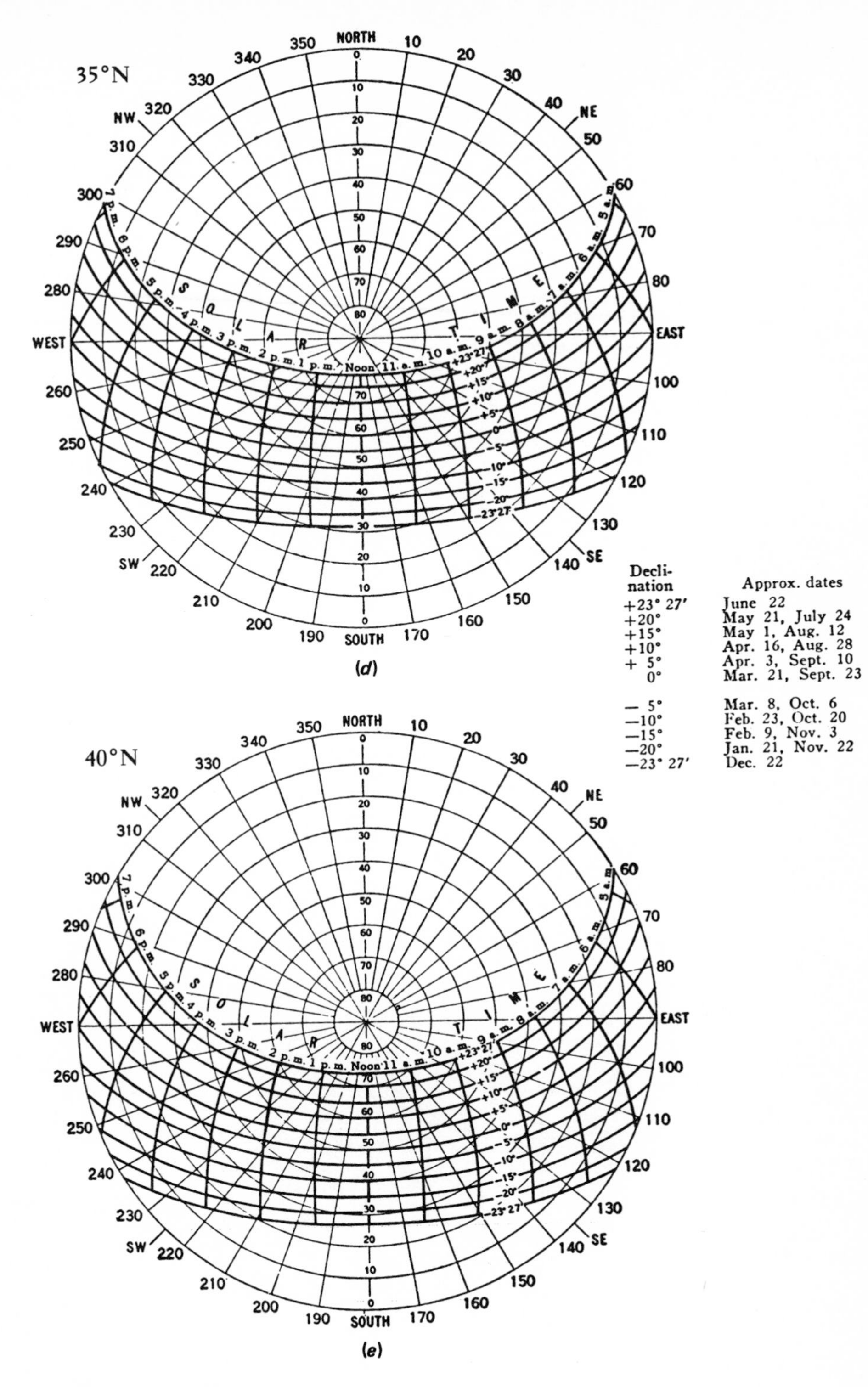

Declination	Approx. dates
+23° 27′	June 22
+20°	May 21, July 24
+15°	May 1, Aug. 12
+10°	Apr. 16, Aug. 28
+ 5°	Apr. 3, Sept. 10
0°	Mar. 21, Sept. 23
− 5°	Mar. 8, Oct. 6
−10°	Feb. 23, Oct. 20
−15°	Feb. 9, Nov. 3
−20°	Jan. 21, Nov. 22
−23° 27′	Dec. 22

FIG. A.1 *(Continued)*

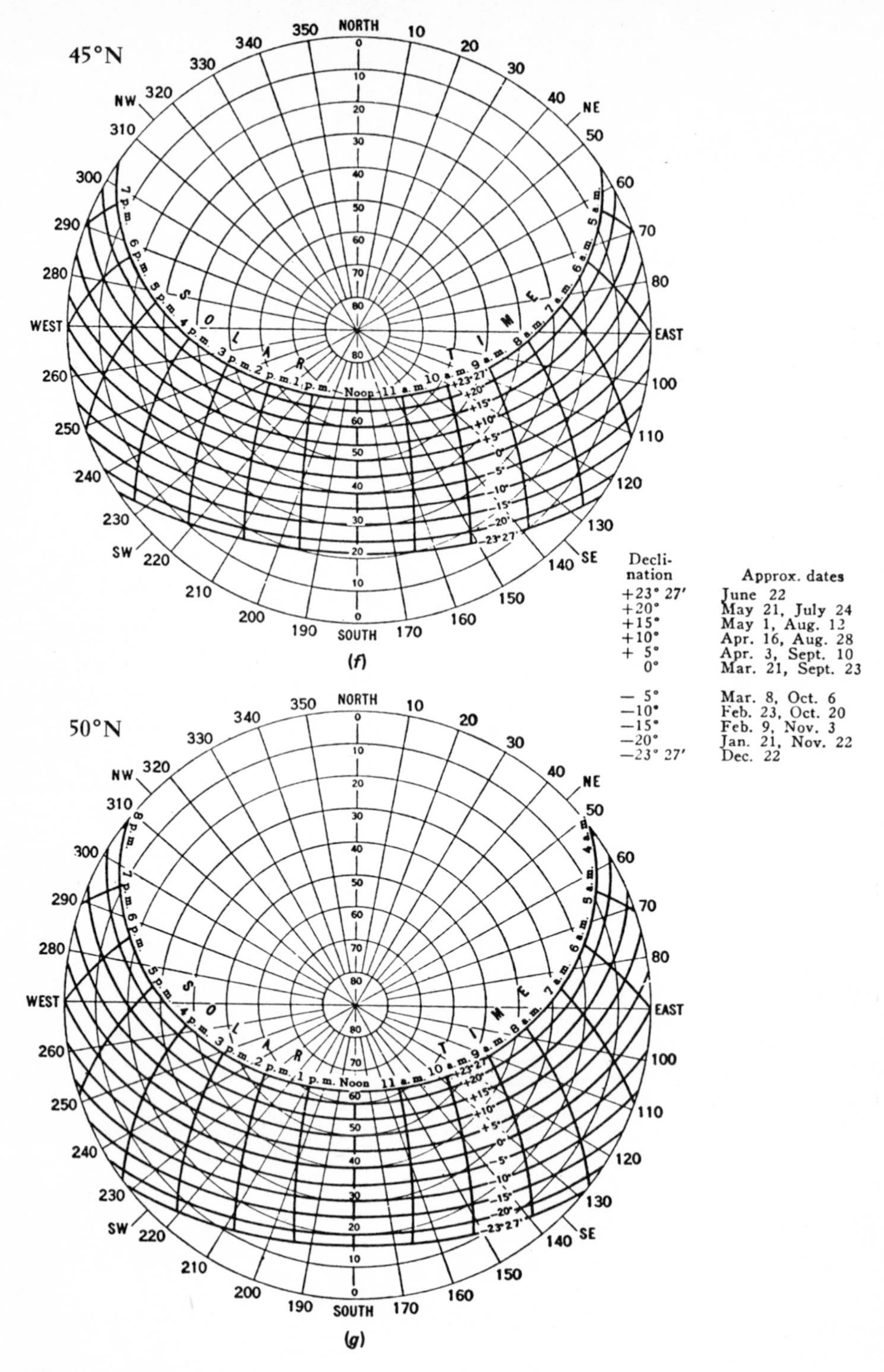

FIG. A.1 *(Continued)*

REFERENCES

1. F. Kreith and J. F. Kreider, *Principles of Solar Engineering* (New York: McGraw-Hill, 1978).
2. Commonwealth Scientific and Industrial Research Organization, *Solar Water Heating* (Melbourne: CSIRO, 1964).
3. U.S. Department of Energy, *Conference Proceedings—Solar Heating and Cooling Systems Operational Results*, DOE No. Solar/0500-79-00, SERI/TP-49-063 (Springfield, VA 22161: National Technical Information Service, 5285 Port Royal Road, 1979).
4. S. A. Klein, *Solar Energy*, vol. 17, 1975, p. 79.
5. K. A. Reed, *Concentrating Collectors* (Atlanta, GA: Georgia Institute of Technology, 1977), p. 5.59.
6. American Society of Heating, Refrigerating and Air Conditioning Engineers, *Handbook of Fundamentals* (New York: ASHRAE, 1977).
7. W. M. Kays and A. L. London, *Compact Heat Exchangers*, 2d ed. (New York: McGraw-Hill, 1964).
8. F. deWinter, *Solar Energy*, vol. 17, 1975, p. 335.
9. J. F. Kreider (ed. in chief), *Solar Energy Handbook* (New York: McGraw-Hill, 1981).
10. ASHRAE, *Systems* (New York: ASHRAE, 1976).
11. W. E. Buckles, S. A. Klein, and J. A. Duffie, *Proc. ISES Silver Jubilee Congress*, American Section of ISES, vol. 2, 1979, p. 959.
12. M. Villecco et al., *AIA Journal*, American Institute of Architecture, September 1979.
13. J. F. Kreider and F. Kreith, *Solar Heating and Cooling*, 2d ed. (New York: McGraw-Hill, 1981).
14. F. S. Dubin and C. G. Long, Jr., *Energy Conservation Standards* (New York: McGraw-Hill, 1978).
15. C. D. Barley, *Solar Energy*, vol. 23, 1979, p. 149.

16. B. Y. H. Liu and R. C. Jordan, *Solar Energy*, vol. 4, 1960, p. 1; see also other articles by the same authors.

17. G. B. Hayes, *Solar Access Law* (Cambridge, MA: Ballinger, 1979).

18. R. H. Montgomery, *The Solar Decision Book* (Midlands, MI: Dow Corning Corp., 1978).

19. E. Mazria, *The Passive Solar Energy Book* (Emmaus, PA: Rodale Press, 1978).

20. J. D. Balcomb and R. D. McFarland, *Passive Solar State of the Art*, D. Prowler, ed. (Philadelphia: Mid-Atlantic Solar Energy Association, 1978), p. 377.

21. G. F. Lameiro and P. Bendt, *The GFL Method*, SERI-30 (Golden, CO: Solar Energy Research Institute, 1978).

22. J. F. Kreider, *Medium and High-Temperature Solar Processes* (New York: Academic Press, 1979).

23. U.S. Department of Energy, *Project Experience Handbook*, DOE No. DOE/CS-0045/D, NTIS, 1978.

24. U.S. Department of Energy, *DOE Facilities Solar Design Handbook*, DOE/AD-0006/1, NTIS, 1978.

25. W. O. Wray et al., *Proc. Third National Passive Solar Conf.*, American Section of ISES, 1979, p. 395.

26. J. D. Chase, *Chemical Engineering Reprints* (New York: McGraw-Hill, 1970).

27. K. M. Guthrie, *Chemical Engineering Reprints* (New York: McGraw-Hill, 1969).

28. R. V. Dunkle, *Mech. Chem. Engrg. Trans. Inst. Engrs. (Australia), MC-1*, 1975, p. 73.

29. Sheet Metal and Air Conditioning Contractors National Association, *Manual for Balancing and Adjustment* (Arlington, VA: SMACNA, 1967); see also other SMACNA manuals on duct construction and design.

30. ASHRAE, *Equipment Handbook* (New York: ASHRAE, 1979).

31. Ref. 3, p. 305.

32. Union Carbide, *Glycols*, Dept. F-41515A 7/71-12M (Charleston, WV: Union Carbide, 1971).

33. U.S. Department of Housing and Urban Development, *Intermediate Minimum Property Standards for Solar Heating*, no. 4930.2, 1977.

34. American Section of ISES, *Proc. Third National Passive Solar Conf.*, AS-ISES, 1979.

35. American Section of ISES, *Proc. Fourth National Passive Solar Conf.*, AS-ISES, 1980.

36. W. Lam, *Perception and Lighting as Formgivers for Architecture* (New York: McGraw-Hill, 1977).

37. J. D. Balcomb, *Proc. Third National Passive Solar Conf.*, American Section of ISES, 1979, p. 56.

38. D. A. Guerin, *Proc. Third National Passive Solar Conf.*, American Section of ISES, 1979, p. 735.

39. S. W. Churchill, *Chem. Eng.*, vol. 99, 1977, p. 91.

40. Ref. 24, p. 61.

41. P. L. Versteegen and D. E. Cassel, *Proc. ISES Silver Jubilee Congress,* American Section of ISES, vol. 1, 1979, p. 869.

42. W. A. Beckman et al., *Solar Heating Design* (New York: John Wiley, 1978).

43. J. D. Balcomb et al.: Various LASL reports including LAUR 78-1159; LAUR 78-2570; also Ref. 34, p. 257 (a good summary of the SLR method).

44. R. S. Means Co., *Mechanical and Electrical Cost Data* (Kingston, MA: R. S. Means, issued annually).

45. A. M. Khasab, *HVAC Systems Estimating Manual* (New York: McGraw-Hill, 1977).

46. U.S. Department of Housing and Urban Development, *Installation Guidelines for Solar DHW Systems,* HUD PDR-407, 1979.

47. U.S. Department of Energy, *Fundamentals of Solar Heating,* HCP/M4038-01 (Springfield, VA: NTIS, 1978).

48. P. J. Potter, *Power Plant Theory and Design* (New York: Ronald Press, 1959).

49. D. M. Utzinger and S. A. Klein, *Solar Energy*, vol. 23, 1979, p. 369.

50. J. V. Anderson, J. W. Mitchell, and W. A. Beckman, *Proc. ISES Silver Jubilee Congress,* American Section of ISES, vol. 1, 1979, p. 763.

51. P. J. Hughes et al., *Proceedings of the ISES Silver Jubilee Congress,* American Section of ISES, vol. 1, 1979, p. 772; see also L. M. Murphy, *SAT Program, Annual Progress Report,* SERI/PR-351-419 (Golden, CO: Solar Energy Research Institute, 1979), p. 23.

52. W. S. Harris et al., *ASHRAE Journal,* vol. 7, 1965, p. 50.

53. W. A. Beckman, *Solar Energy*, vol. 21, 1978, p. 531.

54. A. V. Sebald et al., *Solar Energy*, vol. 23, 1979, p. 479.

55. R. L. Oonk et al., *Solar Energy*, vol. 23, 1979, p. 535.

56. M. Arney, P. Seward, and J. F. Kreider, *Proc. AEE Congress* (Atlanta, GA: AEE, 1980).

57. J. D. Balcomb, *Passive Solar Design Handbook*, vol. 2 (Washington, D.C.: U.S. Department of Energy, 1980).

58. J. D. Balcomb, *Proc. ISES Congress,* American Section of ISES, vol. 4, 1976, p. 281.

59. W. A. Shurcliff, *Superinsulated Houses and Double-Envelope Houses,* published by the author, 1980.

60. National Bureau of Standards, *Interim Performance Criteria for Solar Heating and Cooling Systems in Commercial Buildings,* NBSIR 76-1187, 1976.

61. U.S. Departments of the Air Force, Army, and Navy, *Engineering Weather Data,* AFM 88-29, 1978.

62. Colorado State University, *Design of Systems,* no. 003-011-00084-4 (U.S. Government Printing Office, 1977).

63. W. Shurcliff, *Thermal Shutters and Shades* (Andover, MA: Brick House Publishers, 1980).

64. W. Fuller, *Solar Age*, vol. 5, 1980, pp. 26–27.

65. U.S. Department of Energy, *Recommended Requirements to Code Officials for Solar Heating, Cooling and Hot Water Systems*, DOE No. CS/34281-01, 1980.

66. A. Eden, *Analysis of Solar Collector Array Systems Using Thermography*, SERI-TR-351, no. 494 (Golden, CO: Solar Energy Research Institute, 1980).

67. W. A. Monsen, S. A. Klein, et al., *Proc. Fourth National Passive Solar Conf.*, American Section of ISES, 1979, p. 119.

68. Underwriter's Laboratories, *Proposed First Edition of the Standard for Safety for Solar Collectors*, UL 1279, Northbrook, IL, 1979.

69. Versar, Inc., *Survey and Evaluation of Available Thermal Insulation Materials for Use on Solar Heating and Cooling Systems*, 6621 Electronic Drive, Springfield, VA, 1980.

70. W. H. Mink, *Chemical Engineering*, vol. 103, 1981, p. 93.

71. National Concrete Masonry Assoc., *Passive Solar Construction Book* (Herndon, VA: 1981).

INDEX